The
GEOMETRY
of MUSICAL
RHYTHM

What Makes a "Good" Rhythm Good?

The GEOMETRY of MUSICAL RHYTHM

What Makes a "Good" Rhythm Good?

Second Edition

Godfried T. Toussaint

CRC Press
Taylor & Francis Group
Boca Raton London New York

CRC Press is an imprint of the
Taylor & Francis Group, an **informa** business

A CHAPMAN & HALL BOOK

CRC Press
Taylor & Francis Group
6000 Broken Sound Parkway NW, Suite 300
Boca Raton, FL 33487-2742

© 2020 by Taylor & Francis Group, LLC
CRC Press is an imprint of Taylor & Francis Group, an Informa business

No claim to original U.S. Government works

Printed and bound by CPI Group (UK) Ltd, Croydon, CR0 4YY

International Standard Book Number-13: 978-0-8153-7097-0 (Paperback)
International Standard Book Number-13: 978-0-8153-5038-5 (Hardback)

Visit the Taylor & Francis Web site at
http://www.taylorandfrancis.com

and the CRC Press Web site at
http://www.crcpress.com

For the love of my life –
Eva Rosalie Toussaint

Contents

Preface to the First Edition

IT WOULD NOT BE INCORRECT TO SAY THAT this book offers a view of musical rhythm through the eyes of mathematics and computer science. However, mathematics and computer science are extremely broad subjects, and it is easy even within these disciplines to give accounts of musical rhythm that bear little resemblance to each other. The mathematics employed may be continuous or discrete, deterministic or probabilistic, algebraic, combinatorial, or geometric. A computer science approach could focus on software packages, programming principles, electronic production, information retrieval, design and analysis of algorithms, or artificial intelligence. To be more accurate, this book offers a description of my personal mathematical and computational predilections for analyzing musical rhythm. As such, it is in part a record of my own recent investigations into questions about rhythm that have inspired me, in particular, questions such as: How does one measure rhythm similarity? How do rhythms evolve over time? And what is it that makes a "good" rhythm good? However, this book is more than that. Before describing my general approach to the study of musical rhythm as well as the material that is included and left out, it is appropriate to give a brief outline of where I am coming from, what my background is, what my main goals are, who my target audience is, and how I ended up writing this book.

Three of my passions when I was a teenager attending a boarding school in England were geometry, music, and designing, building, and flying model airplanes. The first two are directly related to this book. At St. Joseph's College in Blackpool, I took three years of geometry classes in which we proved many theorems taken from Euclid's book *The Elements*. The beauty of the problems tackled in this book using the straight edge and compass, and the puzzle-like nature of creating new proofs for the theorems therein, made a deep and lasting impression on me. At St. Joseph's, I also sang in the school choir, collected 45-rpm records by Buddy Holly, and longed for classical guitar lessons. Unfortunately, I was not permitted to study classical guitar because, as the headmaster explained to my father, the guitar was considered a vulgar instrument forbidden on the school premises, and therefore, I had to wait until I graduated from high school before taking up this instrument.

In university, I focused my studies on information theory, pattern recognition, machine learning, and artificial intelligence, eventually obtaining a PhD in electrical engineering at the University of British Columbia. Then I switched fields and joined the School of Computer Science at McGill University, where I rediscovered my passion for Euclidean geometry—this time, wearing a computational hat—and where my interest in the guitar was supplanted by a new passion for African drumming and percussion. In the year 2000, I began to create an interdisciplinary academic bridge between my professional interests in discrete mathematics and computational geometry, and my leisure time enthusiasm for rhythm and percussion. I started to read the literature on the mathematics of music and learned that not much research had been done on rhythm, when compared with other aspects of music such as pitch, chords, melody, harmony, scales, and tuning. This discovery propelled me to investigate musical rhythm more ardently, using the conceptual and computational tools at my disposal. My computer science mindset spurred me to view rhythm, in its simplest reductionist terms, as a binary sequence of ones and zeros denoting sounds and silences, respectively. My colleagues, Michael Hallett and David Bryant at McGill University, graciously introduced me to phylogenetic analysis and molecular biology. Where I heard rhythms I saw DNA sequences, and I was immediately inspired by the possibilities of applying bioinformatics tools to the analysis of musical rhythm. I promptly initiated a research project on the phylogenetic analysis of the musical rhythms of the world. Such a project offered a variety of challenging problems, such as the development of measures of rhythm similarity, and rhythm complexity, to which I applied my knowledge of pattern recognition and information theory.

While at McGill University, I was invited by Stephen McAdams, the director of the Centre for Interdisciplinary Research in Music Media and Technology (CIRMMT), to become one of its members in the group on Music Information Retrieval headed by Ichiro Fujinaga. At that time, I had already been organizing workshops on computational geometry at McGill University's Bellairs Research Institute in Barbados since 1986, and I invited Dmitri Tymoczko, a music theorist and composer at Princeton University, to join us there to explore connections between computational geometry and music. Since then, Dmitri and I have jointly organized annual workshops on mathematics and music at Bellairs.

From McGill, I moved to Harvard University, when in 2009 I was awarded a Radcliffe Fellowship to extend my project on the phylogenetics of rhythm at the Radcliffe Institute for Advanced Study. It was there, after giving a public lecture, that I met Luke Mathews of the Department of Human Evolutionary Biology. Our mutual interest in cultural evolution spawned a fruitful ongoing collaboration on the comparison and application of phylogenetics methods. Luke introduced me to Charlie Nunn, the director of the Comparative Primatology Research Group (CPRG) at Harvard, who invited me to join their weekly seminar meetings. At these meetings, I learned about Bayesian phylogenetics techniques and obtained useful feedback on my ideas for their application to musical rhythm. It was also during my residence at the Radcliffe Institute, in the fall of 2009, that my wife, Eva, encouraged me to write a book on the geometry of musical rhythm, a project on which I immediately embarked with zest. The following year, I joined the Music Department at Harvard University as a visiting scholar. There, I benefitted from a great music library and stimulating colleagues. Christopher Hasty invited me to attend his graduate seminar course on rhythm and made time available from his busy schedule to discuss my research ideas and offer his opinions and suggestions. I also began an ongoing collaboration with Olaf Post to model the cognitive aspects of rhythm similarity.

Then, in August of 2011, I was offered a position as research professor of computer science at New York University in Abu Dhabi, United Arab Emirates, where a teaching-free fall semester gave me time to finish writing the book, and a spring course on Computers and Music allowed me to test much of the material on a group of undergraduate students with diverse backgrounds.

My goals in writing this book were multidimensional. First of all, I wanted to illustrate how the study of the mathematical properties of musical rhythm generate common mathematical problems that arise in a variety of seemingly disparate fields other than music theory, such as number theory in mathematics, combinatorics of words and automatic sequence generation in theoretical computer science, molecule reconstruction in crystallography, the restriction scaffold assignment problem in computational biology, timing in spallation neutron source accelerators in nuclear engineering, spatial arrangement of telescopes in radio astronomy, auditory illusions in the psychology of perception, facility location problems in operations research, leap year calculations in astronomical calendar design, drawing straight lines in computer graphics, and the Euclidean algorithm for computing the greatest common divisor of two integers in the design and analysis of algorithms.

Second, I wanted the book to be accessible to a wide audience of academicians and musicians, classical violinists or drummers, with diverse backgrounds and musical activities, and with a minimal set of prerequisites. I hope that mathematicians and computer scientists will find interest in the connections I make with other fields and will obtain motivation for working on open questions inspired by the problems considered here. I also hope the book is useful to composers, producers of electronic dance music, music technologists, and teachers. Although the book does not include a set of exercises and problems, it should be suitable as a text for an undergraduate interdisciplinary course on music technology, music and computers, or music and mathematics.

Third, I wanted to introduce to the music community the distance approach to *phylogenetic analysis*, and illustrate its application to the study of musical rhythm. In spite of the fact that phylogenetic analysis has been applied not only to cultural anthropology for some 40 years—most notably to linguistics, but also to other cultural objects such as stone tools, carpet designs, and musical instruments—its application to musical rhythm is just beginning. While I have applied phylogenetic analysis to several different corpora of rhythms from different parts of the world, here I give just one example of my approachins, using the flamenco meters of southern Spain.

Having described the essential features of what I tried to do with this book, I should add a few words about what I did not cover and the reasons for leaving such material out. First of all, I am not concerned with the vast domain of music theory that has traditionally used discrete mathematics (especially combinatorics) to analyze chords and scales. This topic, which may be described as "musical set theory," generally involves a number of issues such as various symmetries, relations between complementary sets, and measuring chord similarity. There already exist many books dealing with this musical set theory, and therefore, no effort is made to review this area here. It is well known that there exists an isomorphic relation between pitch and rhythm, which several authors have pointed out from time to time. Also, this book is not a general introduction to what might be called "rhythmic set theory," even though connections (and pointers to the literature) between the rhythmic concepts covered here and the corresponding pitch set theory ideas are made at points scattered throughout the book. Instead, this book offers a more geometric and computational approach to the study of musical rhythm. However, I do not go as far as to propose a "theory of rhythm." Nevertheless, I hope that the tools provided here will help in the eventual development of such a theory.

The study of musical rhythm may also be divided into two general strategies according to how the input rhythms are represented: acoustic and symbolic. In the first approach, rhythms are given as acoustic waveforms recorded or produced with electronic equipment. In this method, rhythm analysis belongs to the domains of acoustics and signal processing, involves problems that include beat induction and automatic transcription, and uses tools including trigonometry and more advanced mathematics such as Fourier analysis. Many books have already been published that deal with these topics, and I have therefore also left all that material out. In the second approach, rhythms are represented as symbolic sequences, much like the text written on this page. In this approach, rhythm analysis falls in the domain of discrete mathematics. My emphasis in writing this book has been on new symbolic geometric approaches, while making connections to existing methods where appropriate.

By means of copious use of illustrative figures, I have tried to emphasize a visual geometric treatment of musical rhythm and its underlying structures to make it more accessible to a wide audience of musicians, computer scientists, mathematicians, cultural anthropologists, composers, ethnomusicologists, psychologists, researchers, academics, tradespeople, professionals, and teachers. I have emphasized a new methodology, namely distance geometry and phylogenetic analysis, for research in comparative musicology, ethnomusicology, and the new field of evolutionary musicology, and I have tried to strengthen the bridge between these disciplines and mathematical music theory.

There are many concepts in the book that, mainly for pedagogical reasons, I have illustrated with examples using a select group of six distinguished rhythms that feature prominently in world music, and one in particular, which is known around the world mainly by its Cuban name: the *clave son*. Thus, this book also includes an ethnological investigation into the prominence of this rhythm. Part of my reasoning for this approach is my belief that if we can understand what makes the *clave son* such a good rhythm, we will gain insight into what makes rhythms good in general.

I should also mention that this is not a cookbook of only the best methods available for measuring rhythmic similarity, rhythm complexity, syncopation, irregularity, goodness, or what have you. First of all, the field is too new, and the problems are too difficult to afford such categorical descriptions. Second, I have written this book wearing a scientific hat, and thus am interested in considering methods that are bad as well as good, with the hope that comparison of these methods will yield further insight into the nature of the problems considered.

One of the main themes of the book is the exploration of mathematical properties of good rhythms. Therefore, the reader will notice upon reading this book that I often introduce a mathematical property of a sequence, then point out one or more musical rhythms used in practice that have this property, and subsequently speculate on how the mathematical property may encapsulate psychological characteristics that contribute to their attractiveness or salience as rhythm timelines. This form of argumentation is not meant to imply that the rhythm exists or is successful merely because of its mathematical properties. For a mathematician, mathematical properties of rhythms are easy to find. What is more rare and interesting from both the musicological and mathematical points of view is to obtain a *characterization* of a given rhythm in the mathematical sense, or in other words, to discover a collection of properties possessed *only* by the rhythm in question. In the book, I illustrate this characterization approach with two examples: the *clave son* timeline and the rhythmic pattern used by Steve Reich in his piece *Clapping Music*. While characterizing a rhythm in this way helps to understand the structure of the rhythm, it does not imply that the structure so obtained is the whole story. Indeed, other mathematical characterizations may exist that are even more useful to the musicologist. However, such characterizations suggest avenues for musicological discourse and the design of psychological experiments to determine the perceptual validity of the properties in question.

The book has a relatively large number of short chapters, rather than a few long ones. This feature serves to highlight the importance and variety of the individual topics covered. The organization of the chapters is partly determined by the logical progression of the topics. The first eight chapters should be read in the order given. However, after Chapter 8, the chapters are fairly self-contained and could be read in almost any order with some judicious flashbacks to fill in a few gaps. The reader will not be handfed the larger narrative of the book at every chapter, and I hope she or he will make a creative effort to connect the dots between the narrative and the individual chapter topics. Furthermore, a long list of books and articles that I found useful and relevant to the topics covered is provided at the end, where the reader may find much additional material at a variety of different levels of exposition.

I would like to thank several ethnomusicologists, music theorists, and drummers who, in the past decade, have generously lent me their ears and were open to the rather strange mathematical meanderings of an outsider wandering into their backyards, learning the ropes along the way. Kofi Agawu invited me to participate in his 2012 Workshop on African Music at Princeton University and made me feel at home among a group of scholars of African Music who offered candid and useful suggestions for my future research. In his book, *Representing African Music*, Kofi Agawu asks the question "How not to analyze African Music?" On page 196, he provides the answer: "Any and all ways are acceptable." These words have always been an inspiration to me. Through email correspondence, Jeff Pressing shared his insightful views on the cognitive complexity of rhythm. I met Willie Anku during one of my visits to Simon Fraser University in Burnaby (Vancouver, Canada), where we presented a joint performance and geometric analysis of African timelines, and brainstormed on the benefits of a mathematical analysis of African rhythm. Enrique Pla, the drummer for the group *IRAKERE*, was a model host and drum set teacher in Havana, Cuba, in 2004 before, during, and after our joint presentation of a geometric analysis of Afro-Cuban rhythms at *PERCUBA: The 15th International Percussion Festival*. Simha Arom patiently taught me about African rhythm and ethnomusicology by sharing his wonderful stories of the listening experiments he performed with the Aka Pygmies of Central Africa. With an open mind, David Locke invited me to lecture in his African Music class at Tufts University. Jay Rahn at York University sent me all his wonderful papers, where I found detailed documentation of many of the rhythms that appear in this book. Richard Cohn welcomed me to Yale University at the 2009 conference that he organized there on Mathematics and Computation in Music/The John Clough Memorial Conference. Rolando Pérez-Fernandez explained his theory of the binarization of ternary rhythms and shared the latest data he had collected. At several music–math conferences, I had illuminating discussions with Jack Douthett and Richard Krantz.

I would like to thank Dmitri Tymoczko for reading and evaluating a preliminary version of the manuscript and through back-and-forth email discussions providing me with a great deal of feedback and numerous germane constructive suggestions for improving the book. I incorporated almost all of his recommendations. However, any errors, omissions, and shortcomings present are entirely my responsibility.

I would also like to thank the team at CRC Press of Taylor and Francis for their help with the logistics of the preparation of the book, especially Sunil Nair, Rachel Holt, and Amber Donley.

My deepest gratitude goes to my wife, Eva Rosalie Toussaint, for suggesting that I write this book in the first place, and for her continuous support and encouragement, in spite of the fact that the three years of writing stole too many hours of my time that we could otherwise have spent together.

Godfried Toussaint
Abu Dhabi
May 26, 2012

Preface to the Second Edition

Since the publication of the first edition of *The Geometry of Musical Rhythm* in January 2013, the geometric approach to the analysis of musical rhythm has experienced an ever-growing flurry of activity. I have summarized a sample of the new results by expanding some chapters and inserting several new ones. This edition contains some 100 additional pages, 93 more figures, and more than 200 new references. Even so, the added material does not do justice to the amount of noteworthy relevant new research that has been carried out in the past 6 years. Therefore, in numerous places, I offer only pointers to the latest literature, for lack of space. During the process of reviewing new material for inclusion in this second edition, I made new discoveries which I inserted in the appropriate new and old chapters. The new chapters are titled: Meter and Metric Complexity, Rhythmic Grouping, Phase Rhythms: The "Good," the "Bad," and the "Ugly," and Grouping and Meter as Features of Rhythm Similarity.

This edition benefited from a number of reviews of the first edition that appeared in a variety of media. Several reviews brought to my attention descriptions of concepts that were not understood by the reader in the manner intended. These reviews provided me with an opportunity to clarify these notions. One reviewer suggested that I should have used Western music notation in the first edition, rather than box, circular, and polygon notations. The reviewer wrote: "Many of these alternative notations are difficult to read, or mysterious to the uninitiated" (Stover, 2009, pp. 17–28; Gerstin, p. 6). Of course, Western music notation is equally (I would argue more) difficult to read and mysterious to the uninitiated in Western music notation. After careful consideration of whether I should switch to Western notation, since I consider my target audience to be unfamiliar with Western notation, I decided to keep the notation the same as in the first edition. What clinched my decision on this issue was a quotation of Steve Reich, who used box (graph) notation, in spite of his familiarity with Western notation, in an interview of Reich by Russell Hartenberger. "I knew that [Morton] Feldman and others were using graph notation, and I used it because I could jot rhythms down succinctly. Using graph-paper, I could space things out and look at them in a way that is different from looking at them in regular Western notation" (Hartenberger, 2016, p. 154).

I would like to thank the team at CRC Press of Taylor and Francis for their help with the logistics of the preparation of the second edition of this book, in particular, the acquiring editor, Callum Fraser.

My deepest gratitude goes to my wife, Eva Rosalie Toussaint, for her continual support, patience, and encouragement during the preparation of this expanded second edition, which not only occupied time that we could have spent together but required the postponement of work on a jointly authored book project.

Godfried T. Toussaint
Abu Dhabi
February 3, 2019

Author

Godfried T. Toussaint was a Canadian computer scientist born in Belgium. He was Professor and Head of the Computer Science Program at the University of New York, Abu Dhabi, United Arab Emirates, and a researcher in the Center for Inter-disciplinary Research in Music Media and Technology (CIRMMT) in the Schulich School of Music at McGill University in Montreal, Canada. After receiving a PhD in electrical engineering from the University of British Columbia in Vancouver, Canada, he taught and did research at the School of Computer Science at McGill University, in the areas of information theory, image processing, pattern recognition, pattern analysis and design, computational geometry, instance-based machine learning, music information retrieval, and computational music theory. In 1978, he received the Pattern Recognition Society's Best Paper of the Year Award, and in 1985, he was awarded a Senior Killam Research Fellowship by the Canada Council for the Arts. In May 2001, he was awarded the David Thomson Award for excellence in graduate supervision and teaching at McGill University. Dr. Toussaint was a founder and cofounder of several international conferences and workshops on computational geometry, and was an editor of several journals. He appeared on television programs to explain his research on the mathematical analysis of flamenco rhythms, and published several books and more than 400 papers. In 2009, he was awarded a Radcliffe Fellowship by the Radcliffe Institute for Advanced Study at Harvard University, for the 2009–2010 academic year, to carry out a research project on the phylogenetic analysis of the musical rhythms of the world. After spending an additional year at Harvard University, in the music department, he moved in August 2011 to New York University in Abu Dhabi. In 2017, he was honored with a Lifetime Achievement Award by the Canadian Association for Computer Science.

This book was published posthumously, with permission from Godfried's wife and daughter, Eva and Stephanie.

CHAPTER 1

What is Rhythm?

RHYTHM IS A FUNDAMENTAL FEATURE of all aspects of life.[1] Creating music, listening to music, and dancing to the rhythms of music are practices cherished in cultures all over the world. Although the function of music as a survival strategy in the evolution of human species is a hotly debated topic, there is little doubt that music satisfies a deep human need.[2] To the ancient philosopher Confucius, good music symbolized the harmony between heaven and earth.[3] The nineteenth-century philosopher Friedrich Nietzsche puts it this way: "without music life would be a mistake."[4] And the Blackfoot people roaming the North American prairies "traditionally believed that they could not live without their songs."[5]

Of the many components that make up music, two stand tall above all others: rhythm and melody. Rhythm is associated with time and the horizontal direction in a typical Western music score. Melody, on the other hand, is associated with pitch and the vertical direction. Rhythm can do very well without melody, but melody cannot exist without rhythm. Although rhythm and melody may be studied independently, in music, they generally interact together and influence each other in complex ways.[6] Experimental results have shown that melody and rhythm (pitch and time) can be encoded in the human brain, either independently or in a combined manner, which depends on the structure of the melody as well as the experience of the listener.[7] Of these two properties, rhythm is considered by many scholars to be the most fundamental of the two, and it has been argued that the development of rhythm predates that of melody in evolutionary terms.[8] "Rhythm is music's central organizing structure."[9] The ancient Greeks maintained that without rhythm, melody lacked strength and form. Martin L. West writes: "rhythm is the vital soul of music,"[10] the philosopher Andy Hamilton notes that "rhythm is the one indispensable element of all music,"[11] and Ton de Leeuw considers that "rhythm is the highest and most autonomous expression of time-conciousness."[12] Joseph Schillinger writes: "The temporal flow of music is primarily a matter of rhythm."[13] Christopher Hasty offers a concise universal definition of music as the "rhythmization of sound."[14] From the scientific perspective, psychological experiments designed to assess the dimensional features of the music space, based on similarity judgments of pairs of melodic fragments, suggest that the major dimensions are rhythmic rather than melodic.[15] The American composer George Gershwin believed that the public loved his music because of its rhythm, and in analyzing his rhythms, Isabel Morse Jones writes: "Gershwin has found definite laws of rhythm as mathematical and precise as any science."[16]

Curt Sachs asks the question: "What *is* rhythm?" and replies: "The answer, I am afraid is, so far just—a word: a word without a generally accepted meaning. Everybody believes himself entitled to usurp it for an arbitrary definition of his own. The confusion is terrifying indeed."[17] In other words, there is no simple answer to this question. Christopher Hasty cautions that "rhythm is often regarded as one of the most problematic and least understood aspects of music."[18] James Beament echoes this sentiment when he writes: "Rhythm is often considered the most difficult feature of music to understand."[19] For Robert Kauffman "The difficulties of dealing with rhythm are immense."[20] Wallace Berry writes: "The awesome complexity of problems of rhythmic structure and analysis can be seen when one appreciates that rhythm is a generic factor."[21] Berry goes on to note that another consideration that makes studying rhythm difficult is the fact that meanings ascribed to terms such as "rhythm," "meter," "accent," "duration," and "syncopation" are vague and used inconsistently. Elsewhere he writes more concisely: "Rhythm is: everything."[22] In spite of some of these difficulties, or perhaps because of them, many definitions of rhythm have been offered throughout the centuries. Already in 1973, Kolinski wrote that more than 50 definitions of rhythm could be found in the music literature.[23] Before diving into the geometric intricacies of rhythm that are explored in this book, it is instructive to review a few examples of definitions and characterizations of rhythm, both ancient and modern.

Plato: "An order of movement."[24]

Baccheios the Elder: "A measuring of time by means of some kind of movement."[25]

Phaedrus: "Some measured thesis of syllables, placed together in certain ways."[26]

Aristoxenus: "Time, divided by any of those things that are capable of being rhythmed."[27]

Nichomacus: "Well marked movement of 'times'."[28]

Leophantus: "Putting together of 'times' in due proportion, considered with regard to symmetry amongst them."[29]

Didymus: "A schematic arrangement of sounds."[30]

Aristides Quintilianus: "Rhythm is a scale of chronoi compounded according to some order, and the conditions of these we call arsis and thesis, noise and quietude."[31]

Vincent d'Indy: "Rhythm is the primordial element. One must consider it as anterior to all other elements of music."[32]

S. Hollos and J. R. Hollos: "In its most general form rhythm is simply a recurring sequence of events."[33]

S. K. Langer: Rhythm is "The setting-up of new tensions by the resolution of former ones."[34]

H. W. Percival: "The character and meaning of thought expressed through the measure or movement in sound or form, or by written signs or words."[35]

D. Wright: "*Rhythm* is the way in which time is organized within measures."[36]

A. C. Lewis: "Rhythm is the language of time."[37]

J. Martineau: "Rhythm is the component of music that punctuates time, carrying us from one beat to the next, and it subdivides into simple ratios."[38]

A. C. Hall: "Rhythm is made by durations of sound and silence and by accent."[39]

T. H. Garland and C. V. Kahn: "Rhythm is created whenever the time continuum is split up into pieces by some sound or movement."[40]

J. Bamberger: "The many different ways in which *time* is organized in music."[41]

J. Clough, J. Conley, and C. Boge: "Patterns of duration and accent of musical sounds moving through time."[42]

G. Cooper and L. B. Meyer: "Rhythm may be defined as the way in which one or more unaccented beats are grouped in relation to an accented one."[43]

D. J. Levitin: "Rhythm refers to the durations of a series of notes, and to the way that they group together into units."[44]

P. Vuust and M. A. G. Witek: "Rhythm is a pattern of discrete durations and is largely thought to depend on the underlying perceptual mechanisms of grouping."[45]

A. D. Patel: "The systematic patterning of sound in terms of timing, accent, and grouping."[46]

R. Parncutt: "A musical rhythm is an acoustic sequence evoking a sensation of pulse."[47]

C. B. Monahan, and E. C. Carterette: "Rhythm is the perception of both regular and irregular accent patterns and their interaction."[48]

M. Clayton: "Rhythm, then, may be interpreted either as an alternation of stresses or as a succession of durations."[49]

B. C. Wade: "A rhythm is a specific succession of durations."[50]

J. London: "A sequential pattern of durations, relatively independent of metre or phrase structure."[51]

S. Chashchina: "Rhythm is a sequence of durations of sounds, disregarding their pitch."[52]

A. J. Milne, and R. T. Dean: "A sequence of sonic events arranged in time, and thus primarily characterized by their inter-onset intervals."[53]

S. Arom: "For there to be rhythm, sequences of audible events must be characterized by contrasting features."[54] Arom goes on to specify that there are three types of contrasting features that may operate in combination: *duration*, *accent*, and *tone colour* (*timbre*). Contrast in each of these may be present or absent, and when accentuation or tone contrasts are present they may be regular or irregular. With these marking parameters Arom generates a combinatorial classification of rhythms.[55]

C. Egerton Lowe writes: "There is, I think, no other term used in music over which more ambiguity is shown." Then he provides a discussion of a dozen definitions found in the literature.[56]

The reader must have surely noticed that the various definitions enumerated earlier emphasize different properties of the term rhythm. Perhaps, the most distinct of these definitions (and certainly the shortest) is the earliest one by Plato, in terms of movement, which we can interpret as dance. W. T. Fitch and Rosenfeld, A. J. (2016) argue that the important aspects of musical rhythm "cannot be properly understood without reference to movement and dance, and that the persistent tendency of 'art music' to divorce itself from motion and dance is a regrettable phenomenon to be resisted by both audiences and theorists."[57]

Some of these definitions imply that a rhythm must be "good" to qualify as being a rhythm. Leophantus, for example, insists that the durations that make up a rhythm must exhibit due *proportions* and *symmetry*. While the question of what makes a "good" rhythm good is a central concern in this book, the definition adopted here is neutral on this issue. A more relevant property of the definitions listed earlier discriminates between rhythms that are either general or specific. Perhaps the most general definition is that of Hollos and Hollos as "a recurring sequence of events." Here it is not specified if the events are visual, aural, or time dependent. Most definitions do involve the notion of time. The *Harvard Dictionary of Music* distinguishes explicitly between general and specific rhythms. Its general definition of rhythm echoes that of Plato with the notion of time thrown in: "The pattern of movement in time." Its specific definition of "*a rhythm*" is: "A patterned configuration of attacks."[58] When we listen to a piece of music such as "Hey, Bo Diddley," we hear several instruments, each playing a different rhythm. Some instruments are playing solos with rhythmic patterns that vary, while other instruments repeat rhythms throughout. The singer adds yet another layer of rhythm. What is the rhythm of such a piece? The answer to this question corresponds to the *general* definition of rhythm given by the *Harvard*

Dictionary of Music, and is exceedingly difficult to ascertain. This book is not concerned with either the objective rhythmic signal, or its subjective perception, that result when a *group* of rhythms is played and heard simultaneously, but rather with the *specific* definition of a rhythm given by the *Harvard Dictionary of Music*: a "patterned configuration of attacks." Furthermore, the emphasis is on rhythm considered as a sequence of durations, disregarding not only meter and pitch, but everything else as well. In addition, the focus is on a particular class of distinguished rhythms: those that are repeated throughout most or all of a piece of music.

Apart from the many conceptual definitions of rhythm listed earlier, we all know experientially what rhythm is, because it is a natural phenomenon, an inherent aspect of nature. Even before we come into this world, while we are still in the womb, we are already bathed in the steady comforting rhythm of our mother's thumping heartbeat and her smooth breathing.[59] Figure 1.1 (top) shows a greatly simplified schematic diagram of the waveform that shows up in an electrocardiogram of a beating heart. The horizontal axis measures time, and the evenly spaced spikes indicate instants of time at which a healthy heartbeat is heard. Since we are only interested in the points in time at which the spikes occur (and not their height), the waveform may be represented as a string of elements (Figure 1.1—middle) in which a mark is made wherever a spike occurs. However, the white rectangles representing the spaces between the spikes are now longer than the black squares representing the spikes. It is most convenient for the analysis of rhythmic patterns to divide these long interspike intervals (silences) into smaller silent units that have the same duration as the sounded units (Figure 1.1—bottom). In this way, the heartbeat has been reduced to a *pulsation*, a binary string (sequence) of evenly spaced pulses, some of which are sounded (attacks) while others remain silent (rests). The term pulse is used in this book to denote the location at which a sound or attack may be realized. This representation of rhythm is also called box notation.[60] A convenient way to write box-notation in running text is to use the symbol [x] for a black square (attack) and the period symbol [.] for a white square (rest). Thus, the rhythm at the bottom of Figure 1.1 may be written in box notation as [x . . . x . . . x . . . x . . . x . . . x . . .].

It is often said that rhythm is in the mind and not in the acoustic signal. Grosvenor Cooper and Leonard B. Meyer write: "Rhythmic grouping is a mental fact, not

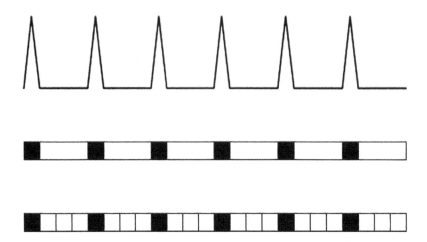

FIGURE 1.1   The idealized heartbeat represented as a sequence of binary elements.

a physical one."[61] This statement should of course not be taken literally. Besides the fact that at present we still do not know that the mind is not physical, and that there exist machines such as functional magnetic resonance imaging and magnetoencephalography that record physical manifestations of rhythm in the brain, the fact is that whether or not grouping is physical, what we perceive as rhythm emanates from an acoustic physical signal. Although William Sethares is right to point out that "many of the most important rhythmic structures are present only in the mind's ear,"[62] the converse is also true: many of the most important rhythmic structures are present only in the physical signal or the symbolic score. Indeed, music psychologists have found compelling evidence that everyone, whether trained musicians or not, can discriminate among the styles of western classical music based solely on the variability of the durations of the notes, as measured by their standard deviation and normalized pairwise variability index.[63] Therefore, it is more accurate to characterize rhythm as a manifestation of a *process* that emerges from the amalgamation of a physical signal with perceptual and cognitive structures of the mind. Such a broad definition naturally leaves open the door to consideration, for analysis, any of the multitude of complex features that make up rhythmic patterns. As such, rhythm may be studied at any level in between these two extremes, ranging from the purely objective mathematical and scientific[64] approaches to the experiential "mythopoetic explanations"[65] as well as its spiritual roots.[66] Knowledge gained from such studies helps us to understand the totality of rhythm.

In this book, musical rhythm is studied predominantly at one extreme of the earlier panorama: rhythm is considered purely in durational terms as a symbolic binary sequence of isochronous elements representing sounds and silences. Simha Arom writes: "In the absence of accentuation or differences in tone color, contrasting durations are the only criterion for the determination of rhythm."[67] This is the simplest definition of rhythm possible. Since rhythm is considered to be such a difficult topic, it behooves us to understand it at this level well, especially when exploring and evaluating new tools, before moving on to higher ground. We should first understand precisely what we lose by confining ourselves to such a skeletal definition of rhythm as well as how much we can gain from it. In this book, I attempt to demonstrate what we can gain by combining geometric methods with such a simple and unambiguous objective definition of rhythm. Some researchers have argued that rhythm must be studied in a cultural and social context.[68] This is a perfectly valid endeavor, particularly so if the main interest is sociology or anthropology, just as it is of interest to study Einstein's theory of general relativity or his equation $E = mc^2$, to gain insight into the structure of scientific revolutions, in a sociocultural context. However, the physical laws of the universe are independent of culture, although their description may be culturally determined, and arguably so are the physical laws of rhythmic patterns. John McLaughlin puts it this way: "The mathematics of rhythm is universal. They don't belong to any particular culture."[69] Here we take the position of Kofi Agawu with respect to the application of analytic methods to ethnomusicology, which may be extended to musicology in general: "Given the relative paucity of analyses, erecting barriers against one or another approach seems premature."[70] Furthermore, although thinking about musical rhythm in a mathematical way, using mathematical terminology,[71] may be quite inharmonious with certain cultural traditions,[72] it facilitates another goal of this book, the exploration of geometric rhythmic universals.[73] However, it should be emphasized that employing mathematical terminology in no way implies gratuitous use of the language of numbers and abstruse symbols. Rigorous mathematical discourse is possible with simple and clear English language, and this is the method favored in this book, to make the concepts within reach of a wide audience.[74]

Just as music is made up of many components, rhythm being one of them, so is rhythm. Three of rhythm's principal elements are meter, beat, and tempo.[75] It is possible that, in the past, rhythm has been difficult to dissect, because not enough attention has been given to its components and their interaction, or its myriad definitions have been too vague, or too general, or because it has not received enough attention from a purely objective mathematical point of view. To quote Curt Sach's, "Rhythm weakens the more we widen its concept and scope."[76] It is hoped that the analysis presented in this book, of "rhythms" as ideal, narrow, purely mathematical, culturally independent, and binary symbolic sequences, will stimulate future progress in the systematic and comparative study of rhythm's subconstituent elements, and rhythm as a whole, in the context of perception, cognition, culture, and *world music theory*.[77] This is not to suggest that the psychological and cultural aspects of rhythm should be ignored in future research on rhythm. Indeed, "for a true understanding, the power of mathematics should be applied to the process of musical behavior, not merely to its product."[78] The stance taken here is that such a humanistic endeavor is best left to the behavioral psychologists. In this book, on the other hand, the main focus is on the mathematical properties of the *product* and the mathematical models of the *processes* that generate it.

NOTES

1　Glass, L. & Mackey, M. C., (1988).
2　Huron, D., (2009), Bispham, J., (2006), and Cross, I., (1999, 2001).
3　Lau, F., (2008), p. 119.
4　Nietzsche, F., (1889). This quotation comes from the 33rd Maxims and Arrows in the book *Die Götzen-Dämmerung* (*Twilight of the Gods*).
5　Nettl, B., (2005), p. 23.
6　Monahan, C. B. & Carterette, E. C., (1985). However, in many parts of the world, such as India, Iran, and the Arab world "musical rhythm is a highly artistic element, self-contained in its rich and most intricate composition, and conceived quite independently of the melodic line." See Gerson-Kiwi, E., (1952), p. 18.
7　Boltz, M. G., (1999), p. 67.
8　Benzon, W. L., (1993).
9　Thaut, M. H., Trimarchi, P. D., & Parsons, L. M., (2014), p. 429.
10　West, M. L., (1992), p. 129.
11　Hamilton, A., (2007), p. 122.
12　De Leew, T., (2005), p. 38.
13　Schillinger, J., (2004), p. vi.
14　Hasty, C. F., (1997), p. 3. See Gow, G. C., (1915) for arguments that rhythm is the life of music.
15　Monahan, C. B. & Carterette, E. C., (1985), p. 1. It has also been shown experimentally that rhythmic structures serve a principal function in the perception of melodic similarity (Casey, M., Veltkamp, R., Goto, M., Leman, M., Rhodes, C., & Slaney, M., (2008), p. 687). This view is apparently not held in some legal circles in the context of music-copyright infringement resolution. According to Cronin, C., (1997–1998), p. 188, "For federal courts at least, originality—the *sine qua non* of copyright—in music lies in melody." He quotes from the case of *Northern Music Corporation v. King Record Distribution Co.* that "Rhythm is simply the tempo in which a composition is written."
16　Jones, I. M., (1937), p. 245.
17　Sachs, C., (1952), p. 384.
18　Hasty, C. F., (1997), p. 3.
19　Beament, J., (2005), p. 139.
20　Kauffman, R., (1980), p. 393.
21　Berry, W., (1985), p. 303.
22　Berry, W., (1985), p. 33.
23　Kolinski, M., (1973), p. 494.
24　*Ibid.*
25　Abdy Williams, C. F., (2009), p. 24.
26　*Ibid.*
27　*Ibid.* It is clear that Aristoxenus considers rhythm as a general phenomenon that is not restricted to music, but also includes speech and dance, among other "things".
28　*Ibid.*
29　*Ibid.*
30　*Ibid.*
31　Mathiesen. T. J., (1985), p. 161. The word "chronoi" in this definition refers to a duration of time.
32　D'Indy, V., (1902), p. 20. Also quoted in Mocquereau, A., (1932), p. 44.
33　Hollos, S. & Hollos, J. R., (2014), p. 7.
34　Langer, S. K., (1957), p. 51.
35　Percival, H. W., (1946), p. 1006.
36　Wright, D., (2009), p. 23. See also Hughes, J. R., (2000), for a review of this book in the context of the interdisciplinarity of mathematics and music.
37　Lewis, A. C., (2005), p. 1.2.
38　Martineau, J., (2008), p. 12.
39　Hall, A. C., (1998), p. 6.
40　Garland, T. H. & Kahn, C. V., (1995), p. 6.

41 Bamberger, J., (2000), p. 59.

42 Clough, J., Conley, J., & Boge, C., (1999), p. 470.

43 Cooper, G. W. & Meyer, L. B., (1960), p. 6.

44 Levitin, D. J., (2006), p. 15.

45 Vuust, P. & Witek, M. A. G., (2014).

46 Patel, A. D., (2008), p. 96.

47 Parncutt, R., (1994), p. 453.

48 Monahan, C. B. & Carterette, E. C., (1985), p. 4.

49 Clayton, M., (2000), p. 38.

50 Wade, B. C., (2004), p. 57.

51 London, J., (2003), p. 277.

52 Chashchina, S., (2016), p. 146, ascribes this definition of rhythm to the Russian musicology literature.

53 Milne, A. J. & and Dean, R. T., (2016), p. 36.

54 Arom, S., (1991), p. 202.

55 Rivière, H., (1993). Rivière proposes an alternative classification of rhythms in terms of the parameters: *intensity, timbre,* and *duration.* See also the commentary by Arom, S., (1994).

56 Lowe, C. E., (1942), p. 202.

57 Fitch, W. T. & Rosenfeld, A. J. (2016).

58 Randel, D. M., Ed., (2003), p. 723.

59 Ayres, B., (1972), describes research that uncovers a significant correlation between preferences for regular rhythms and infant-carrying practices that involve bodily contact with the mother. Wang, H.-M., Lin, S.-H., Huang, Y.-C., Chen, I.-C., Chou, L.-C., Lai, Y.-L., Chen, Y.-F., Huang, S.-C., and Jan, M.-Y., (2009), showed that listening to certain rhythms can also change the interbeat time intervals of the heart of the listener.

60 Kaufman Shelemay, K., (2000), p. 35. Koetting, J., (1970), p. 117, is greatly responsible for popularizing box notation among ethnomusicologists, which he called Time Unit Box System (TUBS). Although, Koetting credits Philip Harland, the assistant head of the UCLA drum ensemble at the time, as the originator of TUBS, this notation has been in use in Korea for hundreds of years; see the paper by Lee, H.-K., (1981). The TUBS system notates only the time or duration information of rhythms. This is not a problem for the timelines considered here, where the attacks are almost always isotonic. However, in African drumming the timbre of the drums is also important. Therefore, the TUBS system has been extended by Serwadda, M. & Pantaleoni, H., (1968), and Ngumu, P.-C. & A. T., (1980), to take into account information other than interattack durations, that may be contained in drum attacks.

61 Cooper, G. W. & Meyer, L. B., (1960), p. 9.

62 Sethares, W. A., (2007), p. 75.

63 Dalla Bella, S. & Peretz, I., (2005), p. B66. The nPVI will be treated in more depth in Chapter 17: Rhythm Complexity.

64 Cross, I., (1998).

65 Cook, N., (1990).

66 Redmond, L., (1997).

67 Arom (1991), *op. cit.*

68 Avorgbedor, D., (1987), p. 4. If there is such a thing as "Western" mathematics, some have argued that its application to non-Western material is a form of cultural imperialism. See Bishop, A. J., (1990) for such a view. I believe there is no such thing as "Western" mathematics. Mathematics is the discovery of *patterns,* and no matter who discovers the patterns, or how they are discovered, they compose the fabric of the universe.

69 Prasad, A., (1999). This quotation is part of the answer of John McLaughlin to Anil Prasad's interview question: "How did you go about balancing the mathematic equations of Indian rhythmic development with the less-studied, more chaos-laden leanings of jazz?"

70 Agawu (1987), *op. cit.,* p. 196.

71 Rahn, J., (1983), p. 33, discusses the problems inherent in three approaches that deal with terminology when analyzing world music: the use of Western terms, the use of non-Western terms, and the avoidance of both, necessitating the introduction of new terminology. Needless to say, all three approaches have their drawbacks. Nevertheless, one may argue that the third approach makes more sense and that the mathematical language is the most objective. Once such terminology is agreed upon, and both Western and non-Western terms may be translated to the mathematical terms on equal footing.

72 Agawu, K., (1987), p. 403.

73 Honingh, A. K. & Bod, R., (2011, 2005).

74 Marsden, A., (2012).

75 Wang, H.-M. & Huang, S.-C., (2014) consider the most significant features of rhythm to be "tempo, complexity (regularity), and energy (intensity, strength, dynamic loudness, and volume)."

76 Sachs, C., (1953), p. 17.

77 Tenzer, M., (2006), p. 33 and Hijleh, M., (2008), p. 88.

78 Wiggins, G. A., (2012), p. 111.

CHAPTER 2

Isochrony, Tempo, and Performance

ISOCHRONOUS RHYTHMS

IMAGINE A STEADY HEARTBEAT going boom, boom, boom, boom, … or a grandfather's clock ticking *tick, tick, tick, tick,* … without end. It is natural for most people to consider these sequences to be examples of the simplest kinds of rhythms. They certainly satisfy several of the definitions of rhythm given in Chapter 1, such as those of Baccheios the Elder, Nichomacus, Didymus, Parncutt, Wade, and Wright. However, some music theorists and musicologists[1] would say that this isochronous sequence is not a rhythm at all because it contains no discernible audible *pattern*, by which they mean that in such a sequence there are no contrasting features. H. Riviere asserts that, "A succession of sounds of equal duration, with invariable intensity and identical timbre, do not constitute a rhythmic event."[2] A more appropriate term for such a sequence is arguably an "isochronous pattern"[3] or an "undifferentiated pulsation."[4] Nevertheless, it has been suggested that, in the context of human behavior, musical pulsation (periodic production) is a uniquely human trait that appears to have evolved specifically for music.[5] Furthermore, other musicologists express a contrary view; the folklorist Alan Lomax refers to pulsations as "one beat" rhythms.[6] From the mathematical, psychological, and biological points of view (and the position adopted in this book), it makes perfect sense to consider pulsations as *bona fide* members of the universe of rhythms.

The human brain would quickly tire of a monotonous sequence of isochronous sounds. Indeed, music psychologists discovered more than hundred years ago that if a human subject in a laboratory setting is presented with a sequence of identically sounding evenly spaced ticks, such as *tick, tick, tick, tick, tick, tick*…, the mind, being so thirsty for patterns, often perceives the sequence as *tick, tock, tick, tock, tick, tock*… instead.[7] In other words, the mind converts the repetition of single-sound *ticks* into the repetition of two-tone *tick-tock* patterns.[8] The same phenomenon, called the *perceptual center*, has been observed with speech rhythm, and is hypothesized to be a *rhythm universal*: "a sequence of spoken digits with evenly spaced acoustic onsets was judged to be uneven by listeners."[9] These psychological phenomena underscore, in the simplest possible manner, the fact that the perceived rhythms in the human mind are not veridical representations of the written score or its realization by the human voice or a musical instrument. Rhythm perception emerges from the interplay between the bottom-up, data-driven, external stimuli emanating from the world, and the top-down, conceptually driven, inner response mechanisms of the mind's ear. In spite of this, it is useful to focus exclusively on the written score, which is more objective than human perception of the production of the score.[10] In the geometric analysis of musical rhythms developed in this book, the psychological aspects of music perception, while not completely ignored, play second fiddle, as do the acoustic aspects of music production.[11] The focus instead is on purely durational symbolically notated rhythms. Therefore, in this setting, to create a more interesting rhythm out of the pulsation *tick, tick, tick, tick, tick, tick*…, we should use two different tones to create *tick, tock, tick, tock, tick, tock*…. This could also be accomplished by accenting every other *tick* in some way, such as making it louder or changing its timbre. However, the sequence becomes an even more interesting rhythm if the durations between the inceptions of two adjacent notes are not all the same. These durations are called *interonset intervals* or IOIs, for short.

In the discipline of rhythmology, rhythms composed of equidistant onsets (accents) have been called "qualitative" rhythms, in contrast to "quantitative" rhythms composed of unequal IOIs. Curt Sachs considered these terms to be vague and confusing.[12] Here, the former will be referred to as *isochronous* (also *regular*) rhythms, and the latter as *nonisochronous* (also *irregular*) rhythms. "The intervals between the onsets of successive notes (IOIs) are the main component of a rhythm. Hence, two rhythms sound similar even if one is a sequence of staccato notes each followed by a rest, and the other is a series of legato notes that each last until the onset of the next note."[13] In staccato notes, each sound is a sharp impulse clearly separated from the others. This terminology comes from a more general setting in which melodies are represented by sustained notes that start and end at fixed positions in time, and the notes do not necessarily end at the positions where other notes begin, as pictured in Figure 2.1. The starting and ending times of the notes are the *onsets* and *offsets*, respectively. In the case of rhythms consisting of sharp attacks, it is assumed that there are no sustained notes, and thus we dispense with the offsets altogether. Furthermore, even in the case of sustained notes, psychological experiments have shown that the duration between the onset and offset of a note has a negligible effect on the perception of rhythmic organization.[14] In this setting, the interonset-duration intervals are simply the durations between two consecutive attacks (onsets). In the physical world where the notes are acoustic signals, the notes would of course not look like the isothetic[15] rectangles depicted in Figure 2.1, with perfectly vertical lines denoting the onsets. Instead, the acoustic signal would be a much more complicated waveform.[16] Furthermore, with acoustic input in the real world, the exact placement of the attacks exhibits deviations caused by factors such as interpretation,[17]

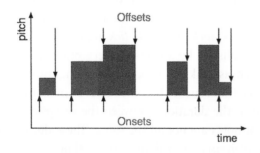

FIGURE 2.1 *Idealized onsets* and *offsets* of sustained notes with differing pitches.

time warping due to expressive timing on the part of the performer,[18] the physical distance between the drummer and some of the drums,[19] or by purposeful design, to test theories of perception.[20] Here, however, we are dealing with idealized symbolic notated rhythms, and hence the model adopted is justified.

TEMPO

In musical terminology, *tempo* refers to the speed at which a rhythm (or music) is intended to be performed. Although the relative durations between notes or beats remain constant with small changes in tempo, the perception of the rhythm changes. Furthermore, with large changes in tempo, even the performance of rhythms changes. Thus, tempo is a salient feature of rhythm in some contexts. Whereas certain animals, such as *zebra finches* (vocal learning songbirds), are able to distinguish between regular and irregular rhythms, they fail to recognize the similarity of two rhythms when the tempo of one rhythm is changed. In effect, the finches do not solve this task using features that characterize the notion of rhythm irregularity. Humans, on the other hand, can generalize the recognition of rhythm similarity even when the tempo is changed.[21] Nevertheless, even in humans, the effects of changes in tempo vary between trained musicians and novice listeners. The perception by novice listeners is more easily affected by changes in tempo, when compared with trained musicians. One explanation for this variation is that trained musicians perceive the deeper underlying structure of a rhythm with respect to an isochronous beat, whereas the novice listeners tend to focus more on the surface grouping structure of the rhythm.[22] In this book, it is assumed that rhythms exist at tempos that are not too fast or too slow to cause misperceptions. Furthermore, since the effects of tempo are a behavioral manifestation, they will be mostly disregarded in the mathematical analysis of rhythms represented as binary sequences.

PERFORMANCE

Performance and tempo influence each other in complex ways, causing a variety of illusions observed in laboratory experiments.[23] For instance, tapping along with music can alter its perceived tempo, so that "the faster one taps, the faster the music will seem to move."[24] Furthermore, just as the perception of rhythms by listeners deviates from the objective exact written score even at a medium tempo, so does the production of rhythms by performers strays from the perfect subdivisions of the beats in the score under a variety of circumstances. Even trained musicians will systematically alter the production of rhythms by distorting IOIs when the tempo is changed. For instance, in one recent experiment, it was found that, as the tempo of a rhythm increased, the performers systematically shortened the longest IOIs and lengthened the shortest IOI, thereby reducing the standard deviation of the IOIs of the rhythm.[25] Small deviations from a score are called *microtiming* (also *expressive timing*[26]). They can occur either intentionally or unintentionally as a result of the complexity of certain rhythms. They can be caused by factors such as melodic grouping,[27] mere "motor noise," or can be produced systematically with the intention of creating effects such as adding tension in a solo performance. They can also be caused by environmental factors such as the distance that a hand must travel to reach a more distant conga drum. While the analysis of the performance of rhythms is not a priority in this book, one excursion in this direction is taken in Chapter 29, which provides a mathematical analysis of the expressive timing of several performances of Steve Reich's *Clapping Music*, as an example of how the tools developed in this book may be applied. Sometimes, microtiming can create a "groove" and thus make a rhythm sound "better," but in other circumstances, it can make a rhythm sound "worse."[28] The concept of "groove" is not well defined, but generally it is described as the feeling of wanting to move or dance when hearing music.[29] How to induce "groove" is an ongoing research area, but as yet there are no firm conclusions regarding its necessary and sufficient physical correlates. It is hoped that the geometric properties of rhythm explored in this book, which help to make a "good" rhythm good, will provide at least some necessary conditions for rhythms to induce "groove."

NOTES

1 I use the term *musicologist* not in the narrow sense, restricted to Western music history, but rather in the broadest sense possible, to denote a scholar of music in any musically relevant discipline in either the sciences, arts, or humanities.

2 Rivière, H., (1993), p. 243. See also Sachs, C., (1953), p. 16, for further references on this view as well as the opposing view.

3 Ravignani, A. & Madison, G., (2017).

4 Sachs, C., (1953), p. 387.

5 Bispham, J., (2006), p. 131. See Swindle, P. F., (1913), for early arguments about whether the human skill of producing pulsations is inherited or acquired.

6 Lomax, A. & Grauer, V., (1964).

7 Bolton, T. L., (1894). Some participants also perceive a sequence such as *tick, tick, tick, tick, tick, tick, tick, tick, tick, tick, tick, tick,* as either *tick, tick, **tok**, tick, tick, **tok**, tick, tick, **tok**, tick, tick, **tok**, tick, tick, **tok**,* or as *tick, tick, tick, **tok**, tick, tick, tick, **tok**, tick, tick, tick, **tok**.*

8 Parncutt, R., (1994), p. 418, refers to this perceptual grouping of isochronous sound events as *subjective rhythmization*. Since a two-tone pulsation qualifies as the simplest of melodies, in a practical sense rhythm cannot exist without melody. Bååth, R., (2012) reports results on two effects of subjective rhythmization: (1) perceived grouping tends to increase as the intervals between stimuli decrease and (2) even groupings are more common than odd groupings.

9 Hoequist, C. J., (1983), p. 368.

10 In spite of the fact that certain aspects of rhythm are psychological in nature, Dahlig-Turek, E., (2009), obtained very useful results on the evolution of the morphology of Polish rhythms using mathematical features of the pure durational patterns, thereby ignoring all other information of the rhythms. On p. 131 he writes: "Thanks to the applied method, it was possible to back up the discussion on "Polish rhythms" using solid ("objective") arguments rather than emotional ("subjective") statements, typical of many previous studies."

11 Dannenberg, R. B. & Hu, N., (2002). Recognition of musical structure from audio recordings is a central problem in music technology that uses tools that are often quite different from the tools used here. Indeed, one of the main goals in analyzing audio signals is the transcription of music to the types of symbolic representations covered here.

12 Sachs, C., (1953), p. 393.

13 Cao, E., Lotstein, M., & Johnson-Laird, P. N., (2014), p. 446.

14 Handel, S., (2006).

15 A geometric object such as a rectangle or polygonal chain is isothetic, provided all its sides are either vertical or horizontal.

16 For a tutorial on techniques for detecting onsets in acoustic signals, see Bello, J. P., Daudet, L., Abdallah, S., Duxbury, C., Davies, M., & Sandler, M. B., (2005), and the references therein. Acoustic onset detection differs from perceptual onset detection. Indeed, Wright, M. J., (2008), argues that musicians do not learn to make their physical attacks have a certain rhythm, but rather to make their perceptual attacks exhibit that rhythm. He proposes maximum likelihood estimation methods to estimate the perceptual attack times.

17 During, J., (1997), p. 26. See also Repp, B. H. & Marcus, R. J., (2010), for a discussion of perceptual illusions concerning onsets and offsets, such as the *sustained sound illusion* (SSI): a continuous sound that seems longer than the silence of equal duration.

18 Benadon, F., (2009b), p. 2.

19 Alén, O., (1995), p. 69.

20 Gjerdingen, R. O., (1993), p. 503, explore "smooth" rhythms as probes of rhythmic entrainment. "Smooth" rhythms have no well-defined time points that determine the onsets and offsets.

21 Van Der Aa, J., Honing, H., & Ten Cate, C., (2015), p. 37.

22 Duke, R. A., (1994), p. 28.

23 Boltz, M. G., (2011).

24 London, J. & Cogsdill, E., (2012).

25 Barton, S., Getz, L., & Kubovy, M., (2017), p. 303.

26 Sethares, W. A. & Toussaint, G. T., (2014).

27 Goldberg, D., (2015). See also Polak, R., (2015).

28 Davies, M., Madison, G., Silva, P., & Gouyon, F., (2012), p. 497.

29 Madison, G. & Sioros, G., (2014). See Butterfield, M. W., (2006), for concepts related to "groove" including "engendered feeling," "swing," and "vital drive."

CHAPTER 3

Timelines, Ostinatos, and Meter

IN MUCH TRADITIONAL, CLASSICAL, and contemporary music around the world, one hears a distinctive, repetitive, and characteristic rhythm that appears to be an essential feature of the music, which stands out above other rhythms, and which repeats throughout most if not the entire piece. Sometimes, this essential feature will be merely an isochronous pulsation without any recognizable repetitive pattern. At other times, the music will be characterized by one or more distinctive such patterns. These special rhythms are customarily called *timelines*.[1] Timelines should be distinguished from the more general term *rhythmic ostinatos*. A rhythmic ostinato (from the word *obstinate*) refers to a rhythm or phrase that is continually repeated during a musical piece. An example of a type of ostinato is the *isorhythmic motet*. An isorhythm (also *talea*) is: "The repetition of a rhythmic pattern throughout a voice part."[2] Timelines on the other hand are more characteristic ostinatos that are easily recognized and remembered, play a distinguished role in the music, and serve the functions of conductor and regulator by signaling to other musicians the fundamental cyclic structure of the piece. Thus, timelines act as an orienting device that facilitates musicians to stay together and helps soloists navigate the rhythmic landscape offered by the other instruments. Indeed, Royal Hartigan considers timelines akin to the "*heartbeat*" of music.[3] Wendell Logan refers to the timeline as a *lifeline*.[4] In the context of *takada* dance, drumming the timeline played on a bell has been called the "*keystone*" of the rhythm.[5] Since timelines repeat over and over in a cyclical manner, they are periodic within the entire piece of music, and thus it is natural to represent them on a circle.[6]

The simplest (shortest) timelines that create repeating cycles consist of two onsets in a cycle of three pulses.[7] In box notation, they are [x x .], [. x x], and [x . x], shown in circle notation in Figure 3.1. The onsets (or sounded pulses) are shown as black-filled circles and the silent pulses as white-filled circles. It is assumed throughout the book that a rhythm starts at the pulse labeled zero and that time flows in a clockwise direction. There are only three such rhythms, and they can be considered as circular rotations of each other 120° apart. Note that timelines can start on a silent pulse, such as the rhythm in the middle diagram of Figure 3.1. Borrowing the terminology from prosody rather than music, I will call the property of a silent first beat, *anacrusis*, since mathematically the rhythm starts at the silent first beat, and we dispense with the so-called "bar lines" denoting measures. In music, anacrusis refers to unstressed (or pickup notes) before the first strong beat, and it plays a powerful role in the creation of "groove."[8] To appreciate the differences in the perception of these three rhythms, it suffices to tap your foot as a regular isochronous and isotonic beat at position 0, in each of the three cycles, and then play each of the three 2-onset rhythms by clapping your hands. Such isochronous beats constitute an example of a concept often referred to as the *meter* of the rhythm. The three rhythms will sound (and feel) very different from each other. Tapping regularly with your foot, as in this example, imposes an isochronous

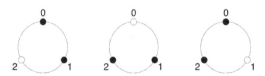

FIGURE 3.1 The three possible two-onset, three-pulse timelines with beats (accents) at position 0.

meter on the rhythm and "has a striking effect on the perception of rhythms."[9] The evasive concept of meter will be illustrated further and analyzed in more detail in Chapters 18 and 37.

Timelines can have much larger numbers of pulses than three. Figure 3.2 illustrates two examples of well-known timelines: one is relatively short, consisting of three onsets in a cycle of eight pulses, and the other is relatively long, made up of 24 onsets in a cycle of 36 pulses. Furthermore, Indian *talas* can have cycles as long as 128 pulses.

The timeline on the left in Figure 3.2 is made up of three adjacent interonset intervals (IOIs or durations). The first duration occurs from the first onset at pulse zero to the second onset at pulse three and thus has duration equal to three units of time. Similarly, the second and third IOIs have durations, respectively, of three units (from pulse three to pulse six) and two units (from pulse six to pulse zero). Another useful notation for rhythms, which complements the circle notation just described, consists of expressing the sequence of adjacent IOIs present in the rhythm as a list. This rhythm may thus be notated as [3-3-2]. Breslauer refers to this representation of rhythms as "durational patterns."[10] In this book, the terms "durational pattern," "IOI structure," and IOI pattern, will be used interchangeably.

The durational pattern [3-3-2], popular in Central Africa, is most famously known as the Cuban *tresillo* (*tres* in Spanish means *three*), is widely used in the circum-Caribbean,[11] and plays a prominent role in the Fandango de Huelva genre of flamenco music.[12] However, it forms part of almost every music tradition throughout the world,[13] and dates back historically to at least thirteenth-century Bagdad, where it was called *al-thakil al-thani*.[14] It is played with the Chinese *naobo*

(cymbals) in Peking Opera.[15] It is a traditional rhythm played on the banjo in bluegrass music.[16] It is a bass-drum pattern often used in the "second line" drumming of New Orleans funeral marches.[17] It was also used extensively in the American rockabilly music of the 1950s for instruments such as bass, saxophone, or piano. It is sometimes referred to as the *habanera* rhythm or *tumbao* rhythm, although the term habanera usually refers to the four-onset rhythm [3-1-2-2], which is less syncopated than the tresillo, since it inserts an additional attack (onset) in the middle of the cycle.[18] The habanera rhythm is also known as the *tumba francesa*.[19] If to the habanera a fifth onset is inserted at pulse five, then the resultant rhythm [3-1-1-1-2] is the *bomba* from Puerto Rico.[20] On the other hand, if the third onset of the tresillo pattern is deleted, one obtains the prototypical lively Charleston rhythm: [x . . x . . . .].[21] An eightfold repetition of this two-onset timeline played at a slower tempo on the electric guitar comprises the entire solo section in the middle of the song "*Don't Worry Baby*" first recorded by the rock band The Beach Boys and released in 1964 by Capitol records. The surprisingly effective guitar solo must surely hold the world record as the simplest of all guitar solos in rock music. A couple of typical examples of rockabilly songs that use the tresillo pattern are "*Shake, Rattle, and Roll*" by Bill Haley, and "*Hound Dog*" by Elvis Presley. In 2005, the Greek singer Helena Paparizou won the *Eurovision* contest with a song titled "*My Number One*" written by Christos Dantis and Natalia Germanou,[22] which used the tresillo timeline on the snare drum. We will revisit this quintessential rhythm frequently with further details of its unique geometric properties.

The timeline on the right in Figure 3.2, with IOI structure [3-1-1-1-3-1-1-1-3-3-3-1-1-1-3-1-1-1-1-1-1-1-1] is

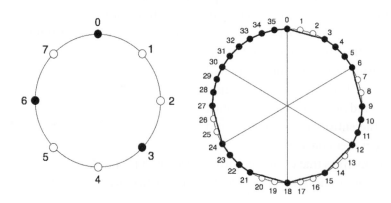

FIGURE 3.2 The Cuban *tresillo* timeline (left) and the ostinato in Ravel's "*Bolero*" (right).

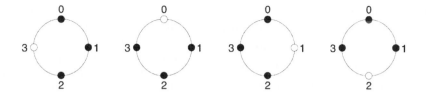

FIGURE 3.3   The four possible three-onset, four-pulse timelines with beats (accents) at position 0.

the ostinato percussion pattern from Ravel's *Bolero*, usually played on one or more snare drums.[23] Also shown in Figure 3.2 are the three lines connecting the six main beats in the cycle at pulses 0, 6, 12, 18, 24, and 30.

In the circular representation of rhythms, such as those illustrated in Figure 3.2, we shall not be concerned with the *absolute* length of time between two consecutive numbered pulses or the entire repeating cycle. For example, in the tresillo pattern [3-3-2], the sum of the three durations, 3 + 3 + 2 equals 8. The circle could have been divided into 16 pulses instead of eight, yielding the IOI structure [6-6-4], implying a slower tempo at which the rhythm is performed. However, since we are interested mainly in the *relative* durations of IOIs, and not concerned with the *tempo* at which a rhythm is played, we will in general use the smallest number of pulses that will accommodate the required integer IOIs. Such pulses are also called *elementary pulses*. The number of elementary pulses used in the cycle is sometimes called the *form number* or *cycle number* of the rhythm.[24] It should be noted that some books define the elementary pulses as the *smallest* time unit (IOI) present in a rhythm. For example, the smallest time unit in the tresillo timeline [3-3-2] is two. However, here the elementary pulse for this rhythm is *one*, and is defined in general as the *largest* time unit that evenly divides into all IOIs contained in the rhythm. Thus, the largest number that evenly divides both two and three is one. In the case of irregular rhythms, we shall also be concerned almost exclusively with aperiodic (nonperiodic) timelines. Therefore, rhythms such as [3-3-2–3-3-2], which is a repetition (concatenation) of the tresillo [3-3-2] + [3-3-2], will be mostly disregarded.

The importance of meter for the perception of rhythm was hinted at with the simple example of the three-pulse, two-onset rhythms of Figure 3.1. To further elaborate on isochronous meters, let us consider one more example with three onsets, in a cycle of four pulses with the metric beat once per cycle at pulse zero.

These constraints yield the four possible timelines pictured in Figure 3.3.

The three rhythm timelines that start with an onset are common not just in music, but as rhythmic stressed (accented) words in the English (and other natural languages) as well. The rhythm with IOI structure [x x x .] is realized by words such as algebra and logical, which can be written as **al**-ge-bra and **lo**-gi-cal, separating the syllables with a dash, and indicating the stressed syllables in bold. Similarly, words corresponding to the musical rhythm [x . x x] include en-gi-**neer** and kan-ga-**roo**. Finally, the rhythm [x x . x] corresponds to words such as spa-**ghe**-tti and po-**ta**-to. Alternately, rather than holding the metric beat fixed at pulse zero, one can view the rhythm as being fixed as [x x x .], and vary the meter by moving the accented downbeat to different positions in the cycle, as illustrated in Figure 3.4, where the metric beat is highlighted with gray-filled boxes. "The perceptions of rhythm and of meter are therefore mutually interdependent: the intuitive system infers meter from a rhythm, but the nature of the rhythm itself depends on the meter."[25] This "bootstrapping" phenomenon is akin to the visual perception of a figure in a background, in which both participate to yield the resultant perception, the figure and ground playing the roles of rhythm and meter, respectively.[26]

In the ethnomusicology literature, the use of the word *timeline* is generally limited to asymmetric durational patterns of sub-Saharan origin, such as the tresillo in Figure 3.2 (left). In this book, however, the term is expanded to cover similar notions used in other cultures such as *compás* in the flamenco music of southern Spain,[27] *tala* in India,[28] *loop* in electronic dance music,[29]

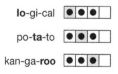

FIGURE 3.4   The four possible three-onset, four-pulse timelines with metric beats (accents) at pulses 0, 1, and 2, respectively.

and memorable rhythmic ostinatos in any type of music. Some of these timelines are likened to several definitions of the concept of *meter* found in the literature. Although there is less vagueness and variance present in the published definitions of meter, as are the definitions of rhythm listed in Chapter 1, from a mathematical point of view, definitions of meter are still lacking.[30] There is also much discussion about the differences between meter and rhythm.[31] Meter is often defined in terms of a hierarchy of accent patterns, and considered to be more (if not perfectly) regular than rhythm.[32] On the other hand, Michael Keith has provided an online Dictionary of Exotic Rhythms, in which he lists a score of irregular (nonisochronous) meters, for which the number of pulses varies between 2 and 33, and all the IOIs have just two durations consisting of 2's and 3's, with examples of pieces from popular and classical music in which they appear.[33] The meters in this list certainly qualify as rhythms. Some music, such as sub-Saharan African music, is sometimes claimed to have only pulsation as a temporal reference, and no meter in the strict sense of the word.[34] Christopher Hasty's book, titled *Meter as Rhythm*[35], considers meter to be a special case of rhythm. In this book, the use of the word timeline is expanded to include all those meters used in music around the world, that function as timekeepers, or ostinatos, and determine the predominant underlying rhythmic structure of a piece. Furthermore, the meter in a piece is viewed as another rhythm that may be sounded or merely felt by the performer or listener. Therefore, how the presence of meter, sounded or imagined, affects rhythm, generally functions in a similar manner as how the presence of one rhythm affects another rhythm being played simultaneously ("double rhythms"[36]). While consideration of a *metric* context is indispensable for a complete understanding of rhythm, as is a *rhythmic* context in which any given rhythm is played, the underlying presupposition in this book is that it is profitable to focus on the interonset durational issues of rhythms heard and analyzed in isolation, before tackling the more complex case of two rhythms being played simultaneously. Nevertheless, Chapter 29 offers an excursion in that direction by analyzing Steve Reich's *Clapping Music*.

NOTES

1 Agawu, K., (2006), p. 1, also refers to a *timeline* as a "bell pattern, bell rhythm, guideline, timekeeper, *topos*, and phrasing referent," and characterizes it as a "rhythmic figure of modest duration that is played as an ostinato throughout a given dance composition." According to Kofi Agawu, the term *timeline* was introduced in 1963 by Kwabena Nketia. *ibid*, p. 3. Nketia, J. H. K., (1962), p. 78, characterizes timeline as a "short but persistent" rhythm that acts as "constant point of reference." He considers the simplest types of timelines to be isochronous and isotonic sequences performed with a gong or handclapping. Flatischler, R., (1992), p. 119, uses the term "guideline" for the same concept. Agawu's 2006 paper discusses timelines at length and provides many useful references on the topic. A similar concept to timeline, called a hook, is present in Mande drumming. See Polak, R. & London, J., (2014). "Mande hook lines mostly involve two-tone melodies and also allow for significant variation."

2 Isorhythm: *Harvard Dictionary of Music*, p. 423. In the context of fifteenth century motets, a rhythmic pattern is also called a *talea*. p. 530.

3 Hartigan, R., Adzenyah, A., & F. Donkor, F., (1995), p. 63.

4 Logan, W., (1984), p. 193.

5 Ladzepko, S. K. & Pantaleoni, H., (1970), p. 10.

6 Anku, W., (2000), Benadon, F., (2007). London, J., (2000), also finds it useful to represent meters on a circle. Collins, J., (2004), p. 58, uses a similar circular notation to analyze polyrhythms, which he calls "circular graphical TUBS figures." The acronym TUBS stands for Time Unit Box System.

7 Cohn, R., (2016b), provides a detailed and exhaustive combinatorial analysis of short timelines with less than 12 pulses.

8 Butterfield, M. W., (2006).

9 Cao, E., Lotstein, M., & Johnson-Laird, P. N., (2014), p. 447. A meter can also emerge from a rhythm, as when a listener spontaneously taps his or her feet while listening to music.

10 Breslauer, P., (1988), p. 2. In Breslauer's notation, this rhythm would actually be written as [3 3 2] rather than [3-3-2]. The latter modification is used here to avoid confusion, because some timelines may have IOIs of duration greater than ten units, and the *dash* symbol provides greater iconic value than the *space*. Hook, J. L., (1998), develops an algebra of durational patterns and applies it to the analysis of the music of Messiaen.

11 Johnson, H. S. F. & Chernoff, J. M., (1991), p. 67. Gerard, G., (1998), p. 69, and Floyd, Jr., S. A., (1999), p. 9. Uribe, E., (1993), p. 126. It is the foundation pattern played on the bass drum for the Baião music of Brazil. It is also used in the Jamaican *mento* song "Sly Mongoose." See Logan, W., (1984), p. 194. Manuel, P., (1985), p. 250, analyzes its influence in salsa music as the typical Afro-Cuban bass line. Sandroni, C., (2000), p. 61, analyzes the influence that the tresillo rhythm had in Latin America.

12 De Cisneros Puig, B. J., (2017), p. 19.

13 Apel, W., (1960), Toussaint, G. T., (2005c), and Leake, J., (2009). Acquista, A., (2009) documents the impact that the tresillo has had on a variety of different rhythmic styles.

14 Wright, O., (1978). p. 216. The cycle of *al-thakil al-thani* actually has 16 pulses and thus the complete rhythm is [3-3-2-3-3-2].

15 Srinivasamurthy, A., Repetto, R. C., Sundar, H., and Serra, X., (2014). http://compmusic.upf.edu/bo-perc-patterns Accessed December 30, 2017.

16 Keith, M., (1991), p. 126.

17 Stewart, J., (2018), p. 170.

18 Brewer, R., (1999), p. 303, refers to the tresillo [3-3-2] as the habanera rhythm. However, according to Orovio, H., (1992), p. 237, and Rey, M., (2006), p. 192, [3-1-2-2] is the habanera, and the tresillo is a habanera-derived rhythm. Note that the habanera rhythm is the same as the drumstick timekeeping rhythm played with a stick on the side of a drum in the *sabar* drumming of Senegal. See Tang, P., (2007), p. 98. Sandroni, C., (2000), p. 61, hypothesizes that the *habanera* [x . . x x . x .] was born independently in different parts of Latin America from a marriage of the African *tresillo* [x . . x . . x .] with the Spanish-Portuguese pattern [x . . . x . x .], since it is the *resultant* (union) of both rhythms. On the other hand, Manuel, P., (1985), p. 250, writes: "Its European and predominantly bourgeois origin is obvious." It is interesting to note that if the *habanera* is rotated by a half-cycle so that if it becomes [x . x . x . . x], it is then the *whaî* rhythm of the *kanak* people of New Caledonia in the South Pacific. See Ammann, R., (1997), p. 242.

19 Fernandez, R. A., (2006), p. 9.

20 *Ibid*, p. 7.

21 Kleppinger, S. V., (2003), p. 82.

22 Released in Greece by Sony BMG Music Entertainment on March 24, 2005.

23 Tanguiane, A. S., (1994), p. 478, Tanguiane, A. S., (1993), p. 149, and Blades, J., (1992), p. 374. Tanguiane tested a machine perception model with a set of experiments aimed at recognizing the rhythm timeline of Ravel's *Bolero*. Asada, M. & Ohgushi, K., (1991), on the other hand, tested and analyzed the human perceptual impressions of the 18 pieces in Ravel's *Bolero*.

24 Kubik, G., (2010a), p. 42.

25 See Cao, E., Lotstein, M., & Johnson-Laird, P. N., (2014). p. 448, and the references therein: Essens, P., (1995), Palmer, C. & Krumhansl, C. L., (1990), Shmulevich, I. & Povel, D.-J., (2000a, 2000b).

26 Vecera, S. P. & O'Reilly, R. C., (1998), p. 449.

27 Parra, J. M., (1999).

28 Morris, R., (1998).

29 Butler, M. J., (2006), p. 90.

30 See for example Arom, S., (1991), p. 184 and Temperley, D., (2002), p. 77, for contrasting views.

31 London, J., (2000, 2004), Arom, S., (1989), Kenyon, M., (1947), Ku, L. H., (1981), Palmer, C., & Krumhansl, C. L., (1990), and Lehmann, B., (2002).

32 Chatman, S., (1965).

33 Keith, M., Dictionary of Exotic Rhythms: www.cadaeic.net/meters.htm.

34 Arom, (1991), p. 206.

35 Hasty, C. F., (1997).

36 Gow, G. C., (1915), p. 646.

CHAPTER 4

The Wooden Claves

IMAGINE A SCORE OF DRUMMERS playing loud dance music at a festival in a village somewhere in sub-Saharan Africa. If the musician playing the timeline is to fulfill the role of a conductor and timekeeper, then all the drummers, including those playing far from each other and separated by other drummers, and especially the soloists who will improvise during their flights of fancy, must be able to hear the timeline, so that when they depart on their rhythmic improvisational adventures, they can find their way back to home base. For this reason, the instruments that play the timeline are designed so as to produce a sound that cuts through the intense booming of all those drums. Traditionally, these instruments consist of either two sticks, 20–30 cm in length, made of a hard wood such as ebony, that are clicked together, or a metallic object usually made of iron, such as a bell that is struck with another piece of metal or stick of wood. Sometimes, a pair of metal axe blades is chinked together.[1] The slaves in the Caribbean improvised with the tools they had available, and for some of their music used equipment for farming sugar cane or construction of their Chattel houses. In Cuba, "Iron percussion instruments are not used for all categories of Lucumí music. They appear now and again, but seem to be mainly identified with songs and dances for the deity Changó. The favored object is a hoe blade, called agogó or agogoró, which is struck with a heavy nail or other iron object."[2] In Afro-Cuban music, the wooden sticks are called *claves*. Clave is the Spanish word for *key* or *code*. The charcoal drawing in Figure 4.1 illustrates a typical pair of wooden claves.

FIGURE 4.1   A pair of wooden *claves*. (Courtesy of Yang Liu.)

The quintessential timeline rhythm played with the claves, described as "the essence of periodicity in Cuban music,"[3] is also called *clave*, suggesting that the *key* to the piece of music lies in the rhythm itself. This term has been extended to apply to other similar rhythms from Brazil and elsewhere.[4] The name *claves* was also accorded to groups of singers who in nineteenth-century Havana would play in the streets during carnival festivities. Eventually, the songs the musicians played during these occasions were also called *claves*, or sometimes "*los palitos son*," (the little sticks).[5] Today, the word *clave* has taken even broader significance. Chris Washburn considers the term to refer to the rules that govern the rhythms played with the claves.[6] Bertram Lehman regards the clave as a concept with wide-ranging theoretical syntactic implications for African music in general,[7] and for David Peñalosa, the *clave matrix* is a comprehensive system for organizing music.[8]

As simple as this instrument looks, there is a certain amount of technique required to bring out its magical sound that will frustrate the novice. It will not do to just hold a stick in each hand and strike the ends together. Such an approach will produce a short dull dry sound. First one clave must be laid on one tightly cupped hand balanced between the wrist and the tips of the fingers. In this configuration, this clave called the "female" in Cuba is not only free to resonate, but the cupped hand provides a chamber much like a miniature kettle drum that acts as an amplifier. This resting clave is struck near its center by the first clave called the "male."[9] If the result is a sound that appears to be produced by a material that resonates somewhat like glass crystal and a little like metal, then success has been achieved.

There exist similar instruments in other cultures around the world. For example, the Australian aboriginal people strike together a similar pair of wooden sticks called *clap sticks*. Clap sticks are larger and heavier than claves, tapered to a rounded point at their ends, and often decorated with an aboriginal dot art. Instruments such as claves and clap sticks belong to the family of instruments called *clappers* which have a long history dating back more than 5,500 years ago to ancient China, where they were originally made of bone.[10] As the names of these instruments suggests, claves, clap sticks, or clappers, are in fact an evolutionary extension of perhaps the oldest rhythmic instruments dating back millions of years: the human hands. Kofi Agawu writes: "The energy displayed in hand-clapping, the metrical and rhythmic reinforcement that they bring, and the variety of patterns that they engender all justify recognizing colliding palms as musical instruments."[11]

NOTES

1 Jones, A. M., (1954b), p. 58.
2 Courlander, H., (1942), p. 230.
3 Wright, M., Schloss, W. A., & Tzanetakis, G., (2008), p. 647.
4 Toussaint, G. T., (2002).
5 Ortiz, F., (1995), pp. 5–8. Mauleón, R., (1997) traces the origins and development of the clave in world music.
6 Washburne, C., (1995, 1998).
7 Lehmann, B., (2002).
8 Peñalosa, D., (2009).
9 Orovio, H., (1992), p. 110.
10 Logan, O., (1879), p. 690.
11 Agawu, K., (2016), p. 89.

CHAPTER 5

The Iron Bells

IN ADDITION TO HARD wood, a variety of metal bells and gongs are used for playing rhythm timelines in traditional African and Afro-Cuban music.[1] Perhaps the most noteworthy metal bell is the *gankogui*, a hand-held forged iron instrument composed of two bells of different pitches, attached together as shown in Figure 5.1. Such double bells have more "talking ability" than single bells, "because they possess minimal tonal differentiation that allows players to generate melorhythmic

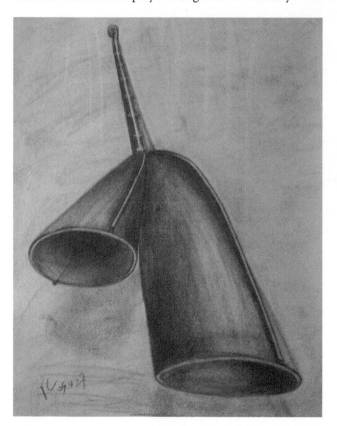

FIGURE 5.1 The iron *gankogui* double-bell. (Courtesy of Yang Liu.)

patterns."[2] These bells come in a variety of sizes that produce a wide range of tones and textures. The instrument is held with one hand without muting the sound, and the bells are struck with either a wooden stick or a metal bar. The first onset of the timeline is often played on the low-pitched bell and the remaining onsets on the high-pitched bell. However, in the *takada* dance drumming of Ghana "the smaller of the two bells is used almost exclusively."[3]

In the drum ensemble music of the Ewe people of Ghana, there is a popular dance music called *gahu*, which makes use of the following 16-pulse timeline played with the first onset on the large bell and the other four on the small bell [x . . x . . x . . . x . . . x .].[4]

The *dawuro* (also called *atoke*,[5] banana, or boat bell) is shaped somewhat like a canoe or taco shell, as pictured in Figure 5.2. To play it, the bell is balanced delicately in the palm of one hand, and the edges of the bell are struck with a metal rod. The sound is a piercing reverberation that resembles a whistle that can cut through the sound of a score of drums. Furthermore, by muting the sides of the bell with the thumb after striking it, a variety of interesting sustained sound effects may be produced. Two traditional 12-pulse timelines performed with this bell in the *adowa* drum music of the Ashanti ethnic group of southern Ghana are [x . x . x . . x . x x .] and [x . . . x . . x . x . .].[6]

The *frikyiwa* illustrated in Figure 5.3 is a type of metal castanet used to play the timeline in the *sikyi* rhythms of the Ashanti people.[7] The walnut-shaped object is held with the middle finger inserted under the bridge that connects the two halves of the object. The ring-shaped part is worn on the thumb, which is used to strike the object. Although made of metal, the sound

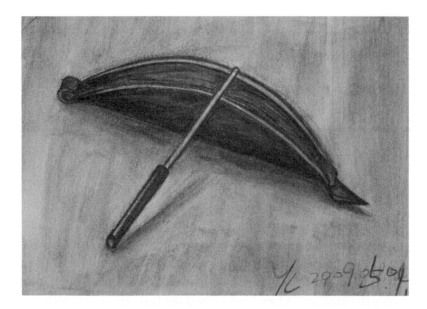

FIGURE 5.2 The iron dawuro bell. (Courtesy of Yang Liu.)

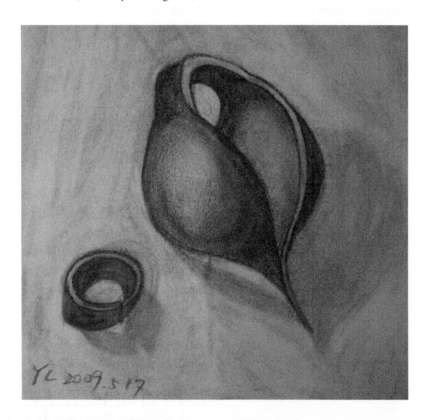

FIGURE 5.3 The *frikyiwa*, a metallic castanet-like bell. (Courtesy of Yang Liu.)

produced by the frikyiwa is similar to that made by the hard, wooden claves.

A typical timeline played on the frikyiwa castanet in the *sikyi* rhythms of the Ashanti people is given by [. . . x . x . x]. Note that in contrast to the other timelines introduced so far, this one starts not on a sounded pulse, but on a silent pulse instead. Figure 5.4 shows this rhythm in polygon notation on a circular lattice of eight pulses, superimposed on the regular four-beat meter. The metric beats (gray-filled circles) fall on pulses 0, 2, 4, and 6 (connected by dashed lines to form a square). The highlife timeline has onsets at

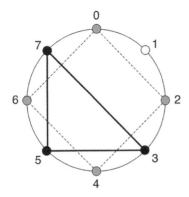

FIGURE 5.4 The *highlife* timeline pattern (bold solid lines) superimposed on the regular four-beat metric pattern (thin dashed lines).

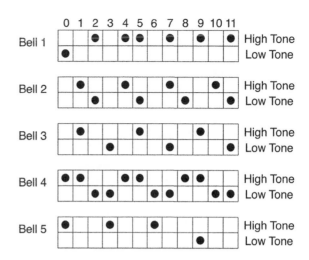

FIGURE 5.5 The *Gamamla* bell ensemble patterns.

pulses 3, 5, and 7 (connected by solid lines to form a triangle). The most important aspect of this timeline is that it is situated completely off the beat, exactly in the middle offbeat positions of the metric beats. When played in the context of the four-beat music, this property gives the timeline a certain "magical" feel.[8]

Although the metal bell is used in African music predominantly to play timelines to mark the time for drumming ensembles, it is sometimes used as a "master instrument"[9] in its own right, without any other musical instruments. An example in Ghana is the *gamamla*, played with five gankogui bells of varying sizes. A different rhythmic 12-pulse ostinato is played on each bell, which uses both high and low tones, resulting in a unique and rich overall composition.[10] Figure 5.5 shows the five bell patterns in box notation. Bell 5 plays the *regulative beat* that divides the cycle into four intervals of equal duration.[11]

NOTES

1 Bells are sometimes made of nonmetallic materials. Söderberg, B., (1953), p. 49, documents bells made from fruit shells in Central Africa, and Fagg, B., (1956), p. 6, describes rock gongs discovered in the rocky hills of Nigeria, that were used as percussion instruments.
2 Agawu, K., (2016), p. 97. See also Chernoff, J. M., (1991), p. 1096. Agawu, K., (2006).
3 Ladzepko, S. K. & Pantalleoni, H., (1970), p. 10.
4 Locke, D., (1998), p. 17.
5 *Ibid*, p. 13.
6 Hartigan, R., Adzenyah, A., & Donkor, F., (1995), p. 35.
7 *Ibid*, p. 16.
8 Agawu, K., (2016), p. 176, provides a brief discussion of the history and meaning of the highlife pattern. For a detailed discussion of the highlife pattern in a cultural context, and its various metrical interpretations, see Agawu, K., (2006), pp. 24–28.
9 Nzewi, O., (2000), p. 25.
10 Klőwer, T., (1997), p. 175.
11 Anku, W., (2007) and Kwabena Nketia, J. H., (1974).

CHAPTER 6

The *Clave Son*

A Ubiquitous Rhythm

As we have seen earlier, the Cuban tresillo has duration pattern [3-3-2] consisting of three attacks in a cycle of eight pulses. At an abstract level, all rhythms can be classified into families described by these two numbers: the *onset number* and the *pulse number*.[1] Let the integers k and n denote, respectively, the onset number and pulse number of any rhythm. Among the timelines used in traditional and contemporary music all over the world, the values of k and n vary across cultures. In Western music, n is usually less than or equal to 24. The fourth century BC Greek statesman Aristides wrote that it is not possible to perceive rhythm when n is greater than 18. However, the Aka Pygmies of central Africa use timelines with values of n as high as 24. Furthermore, some of the largest values of n are found in Indian classical music, where the timelines are called *talas* and the value of n can be as high as 128.[2] In western and sub-Saharan African music, the value of n is usually an even number such as 4, 6, 8, 12, or 16. There are of course exceptions. The San Bushmen of Namibia use a most unusual 20-pulse timeline with four onsets and duration interval pattern [4-6-4-6].[3] In Eastern Europe, North Africa, and the Middle East, n is often an odd number, and sometimes a *prime* number such as 5, 7, or 11 (a number that can be divided without remainder only by n and 1). In the Black Atlantic region, on the other hand, the pulse number is almost never a prime number.[4]

For reasons that will be explored in this book, the value of k is usually slightly higher or slightly lower than *half* the value of n. In sub-Saharan Africa, a value of $n = 12$ is preferred. The most popular value of n in the world appears to

be 16. Furthermore, for rhythms with 16 pulses, a value of k equal to five is preferred. Specifying the values of k and n allows the calculation of the number of theoretically possible rhythms in the family. For example, the number of rhythms with five onsets and 16 pulses corresponds to the number of ways of selecting five items from among 16. This is equal to the number of different two-symbol sequences that contain k ones and $(n–k)$ zeros. There is a well-known combinatorial formula for this type of calculation that yields the solution.[5] The general formula is given by $(n!)/[(k!)(n - k)!]$, which for $n = 16$ and $k = 5$ becomes $(16!)/[(5!)(11!)]$, where the symbol "!" is a pronounced *factorial*, and $k!$ denotes the product of k terms $(k)(k - 1)(k - 2)(k - 3)...(1)$. Evaluating this formula yields the number 4,368. Note that the number 11 in this formula is the difference between the number of pulses (16) and the number of onsets (5). Figure 6.1 shows, in box notation, a dozen arbitrary members of this very large family of rhythms.

Of course, not every rhythm in this family is considered to be a "good" rhythm, in the sense that it has been adopted as a timeline pattern in traditional music somewhere on the planet. Indeed, the ancient Greek music scholar Aristoxenus of Tarentum wrote in his *Elements of Rhythm* in the sixth century BC that not every division of a time span yields a rhythm that is rhythmical, by which he meant "good." Interestingly, Aristoxenus reserved the term *eurhythmical* for those rhythms that were *beautifully* rhythmical. However, he did not provide any algorithms for generating eurhythmic rhythms, and was content to emphasize that for examples of eurhythmic rhythms, one should turn to the compositions of

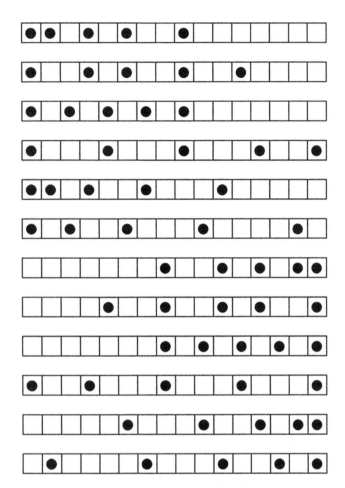

FIGURE 6.1 A dozen examples of the 4,368 rhythms with five onsets and 16 pulses.

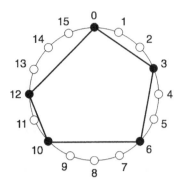

FIGURE 6.2 The *clave son* in polygon (circular) notation.

great masters.[6] In contrast, one of the main goals of this book is precisely the identification of mathematical properties that can be applied to the design of algorithms for generating good rhythms. Nevertheless, the suggestion by Aristoxenus to turn to the compositions of the "great masters" provides inspiration to explore the properties of the most successful timelines that have been adopted by cultures all over the world, under the assumption that such timelines evolved at the hands of an unknown and perhaps unconscious and collective "great master."

Figure 6.2 displays an illustrious member of the family of rhythms with five onsets and 16 pulses.[7] In Cuba, it goes by the name *clave son*.[8] In USA, the *clave son* became known as the Bo Diddley beat, since the late 1950s, but "black shoeshine boys of the nineteenth century called it the hambone."[9] In Ghana, it is called the *kpanlogó*, and in thirteenth century Bagdad, it was called *al-thaquīl al-awwal*. It has interonset-interval sequence [3-3-4-2-4]. In addition to marking the onsets at positions 0, 3, 6,

10, and 12 with black-filled circles, Figure 6.2 shows how a rhythm may be represented as a convex polygon by connecting each onset to its adjacent onsets in time with a straight-line segment, in this case, yielding a pentagon (five-sided polygon). Such a geometric representation is useful for a multitude of purposes, as will become clear in subsequent chapters.

To begin with, a natural question that comes to mind, besides the reason for the choices of $n = 16$ and $k = 5$, is how, out of the 4,368 possible rhythms with five onsets from among 16 pulses, this particular configuration of interonset intervals [3-3-4-2-4] managed to become such a catchy and widely used rhythm.[10] What is so seductive about this rhythm that has been described as "the elegantly insinuating syncopated rhythm that defines Cuban *son* and *salsa*,"[11] that it should win the hearts and minds of people all over the world? Attempting to answer this question using a wide variety of tools ranging from geometry, information theory, and complexity theory, to musicology, psychology, and phylogenetic analysis, is one of the main themes of this book.

Studying the evolution of rhythms is not an easy task. Until recently, many musical traditions in the world were oral, and so the rhythms used by musicians were handed down from teacher to apprentice without leaving written records. This is particularly so in the sub-Saharan African tradition, which gave less importance to who was responsible for creating rhythms, or when they were created, than how well and on what occasions they were performed.[12] In addition, sailors, soldiers, and explorers throughout history traveled constantly back and forth from one place to another, carrying with them their songs, musical instruments, and rhythms that were either borrowed intact or perhaps transformed by intercultural exchanges.[13] A few historical markers about the *clave son* are known. According to Peter Manuel, the rhythm was common in Afro-Cuban music since 1850s.[14] In the early twentieth century, it was played in eastern Cuba in the area of Santiago in a style of music called *son*. This music made its way to Havana in the early part of the twentieth century. Between 1930 and 1960, many Latin musicians such as Tito Puente incorporated the *clave son* and other Afro-Cuban rhythms into popular music based on European harmonies.[15] During the 1950s, the son traveled northward from Cuba to the ports of New Orleans and New York. In New Orleans, it influenced rockabilly musicians such as Bo Diddley, and in New York, with the Puerto Rican influence, it was transformed to what is called *salsa*. Finally, from New York the rhythm infiltrated virtually all parts of the world.[16]

In Ghana, the *clave son* (*kpanlogó*) timeline is played on an iron bell.[17] The BaYaka Pygmies of central Africa also use this rhythm as a timeline.[18] There is evidence that the rhythm travelled with *gome* music from central Africa to Ghana in the early part of the twentieth century.[19] The earliest historical documentation of this rhythm appears in a book about rhythm written in Bagdad by the Persian scholar Safi al Din in the middle of the thirteenth century.[20] It is also noteworthy that in his writings about rhythm, Safi al Din used a circular notation similar to that used here. Apart from these snapshots it is difficult to determine where the rhythm originated from, and how it traveled between Persia, Central Africa, and Cuba. We will return to this topic later in the book after exploring its structure, its relationship to other rhythms, and the phylogenetic tools useful for the exploration of the evolution and migration of musical rhythms.

NOTES

1 Cohn, R., (1992a), p. 195, refers to the pulse number of a rhythm as the *span*.
2 Šimundža, M., (1987, 1988) and Clayton, M., (2000).
3 Poole, A., (2018). This is an example of a *well-formed* rhythm, a topic covered in Chapter 21.
4 Pressing, J., (2002), p. 289.
5 Keith, M., (1991), p. 17.
6 Abdy Williams, C. F., (2009), pp. 34–35.
7 Vurkaç, M., (2011), p. 30, Toussaint, G. T., (2011), Waterman, R. A., (1948), p. 36. Kauffman, R., (1980), Table 1, p. 409. Gerard, G., (1998), p. 33.
8 Chernoff, J. M., (1979), p. 145. Johnson, H. S. F. & Chernoff, J. M., (1991), p. 67, and Fernandez, R. A., (2006), p. 15. Robbins, J., (1990), p. 189, explores the social and historical contexts in which the son evolved, and offers some possible reasons for its great success. My daughter Stephanie refers to the *clave son* as the *knock-of-death*, because while she was growing up I used it on her bedroom door to wake her up for school at 6:00 am.
9 Palmer, P., (2009), p. 116.
10 Vurkaç, M., (2012) and Toussaint, G. T., (2011).
11 Zigel, L. J., (1994), p. 131.
12 Arom, S., (1988), p. 1.
13 For example, traditional Turkish rhythms with five and seven pulses have been imported into popular Western music: Dave Brubeck's *Take Five* has a five-pulse meter, and Pink Floyd's *Money* has a seven-pulse meter. See Keith, M., (1991), p. 126.
14 Manuel, P., with Bilby, K. & Largey, M., (2006), p. 50. Mauleón, R., (1997), p. 9, whose thesis examines the evolution of the *clave son* rhythm, and expounds its worldwide influence on a variety of musical genres, also traces the emergence of this pattern in Cuba to the nineteenth century.
15 Goines, L. & Ameen, R., (1990), p. 6.
16 Washburne, C., (1997), p. 66, documents the influence of the *clave son* timeline on jazz music. Rey, M., (2006), illustrates with many examples the incorporation of the *clave son* rhythm on the art music of Cuba. See also Quintero-Rivera, A. G. & Márquez, R., (2003).
17 Rentink, S., (2003), p. 45.
18 Poole, A., (2018).
19 *Ibid*, p. 35.
20 Safi al-Din al-Urmawî, (1252).

CHAPTER 7

Six Distinguished Rhythm Timelines

Given that the number of possible rhythms with five onsets and 16 pulses is 4,368, a natural question that arises is: how many of these rhythms are used as timelines in traditional music practice around the world? In other words, what is the competition? Naturally, the vast majority of the rhythms in this family, such as [x x x x x . . . .] and [. x x x x x . . . .], for example does not appear to be very interesting as timelines. Therefore, most people might venture a guess of perhaps a number between 50 and 100. However, even this number is too large. In ethnomusicology literature, it is difficult to find more than a dozen traditional five-onset, 16-pulse timelines. Of these, six have made a significant mark as timelines in the music of the world. These distinguished rhythms are shown in box notation in Figure 7.1, and for pedagogical reasons, I have chosen these to illustrate many of the concepts explored in this book. This list is in no way suggestive that the other 4,362 rhythms are not good for some function in music. For example, by permuting the durations in these six rhythms, one may obtain other rhythms that could serve quite well as timelines in some musical contexts. Many more could be used as rhythmic solos, and variations in a large variety of music would sound attractive to the modern ear. However, these six are the rhythms that have been adopted over vast expanses of historical time to serve as *timelines* in the traditional music of cultures scattered around the world. Because they are distinguished in this sense, it is safe to assume that Aristoxenus would agree that they are worthy to be studied in depth. When a rhythm is described as "good" in this book, the word is intended to denote that it is effective as a timeline, as judged by cultural traditions in some parts of the world and the test of time.

A word is in order concerning the names attached to these rhythms. As noted in Chapter 6, what is called the *clave son* in Cuba is called the *kpanlogó bell pattern* in Ghana.[1] All these rhythms have different names in disparate parts of the world where they are used. The names adopted here reflect the terminology perhaps most well known in the western popular media, and I use them here purely for convenience rather than the establishment of any historical priority or cultural entitlement.

THE SHIKO TIMELINE

The timeline at the top in Figure 7.1, shown in polygon notation in Figure 7.2, is a common rhythm in West Africa and the Caribbean. The dashed lines in this (and the remaining figures in this chapter) connect pulses 0 to 8, and 4 to 12, indicating the four isochronous beats of the quadruple meter of these 16-pulse rhythms. In Nigeria, this rhythm goes by the name *shiko*. It has a durational pattern [4-2-4-2-4], and has three onsets that coincide with the metric beats, and two onsets that fall halfway between these beats. Note that these intervals are divisible by two, and thus the rhythm could be

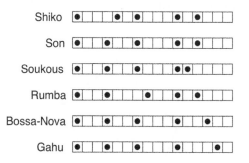

FIGURE 7.1 The six distinguished timelines with five onsets and 16 pulses.

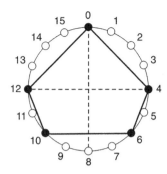

FIGURE 7.2   The *shiko* timeline in polygon notation.

notated as [2-1-2-1-2], or in box notation as [x . x x . x x .].[2] In this form it is commonly called by its Cuban name, the *cinquillo*.[3] It is played on the *xiaoluo* (small gong) in Peking Opera.[4] As a timeline, it is played in the *moribayasa* rhythm among the *Malinke* people of Guinea, and in the *Banda* rhythm used in *voudoo* ceremonies in Haiti. In Martinique, it is the *tibwa* timeline.[5] In Cuba, it is played on a wooden block in the *makuta* rhythm, and in Romania it is a folk-dance rhythm.[6] This rhythm is also started on the second, third, and fourth onsets. For example, the *timini* rhythm in Senegal is [x x . x x . x .], which is equivalent to starting the shiko on the second onset. This pattern is played on a bell for the *adzogbo* dance of the *Fon* people of Benin, and is also frequently encountered in the Persian Gulf region.[7] When started on the third onset, it becomes the Rumanian folk-dance rhythm [x . x x . x . x].[8] On the other hand, the *kromanti* rhythm of Surinam is [x x . x . x x .], which is equivalent to starting the shiko on the fourth onset, and is also a Rumanian folk-dance rhythm.[9] Starting on the fourth onset yields the durational pattern [2-4-4-2-4], which is a timeline played by Mbuti Pygmies in the Democratic Republic of Congo and the San Bushmen of Botswana.[10]

The shiko timeline is also found as the first part of longer rhythms. A well-known Arabic rhythm, the *wahda kebira* given by [**X** . x x . x x . **X** . **X** . x . x x], contains the shiko as its first part. Since this is played on a drum, rather than bell, the boldface uppercase **X**'s denote low-pitched sounds (*dum*) and the small x's high-pitched slaps (*tek* or *tak*). Cyclic shifts of the shiko timeline also appear in other longer rhythms. For example, the *kassa* from Guinea given by [x x . x . x x . x . x . x . x .] has the kromanti as its first bar. Finally, we remark that starting the shiko on the second onset is also a popular pattern found in Arabic rhythms played on a drum. Here again, some notes are low sounds,

whereas others are high pitched. The *maqsum* is given by [**X** x . x **X** . x .], and the *baladi* by [**X X** . x **X** . x .]. The *masmudi* is a slow baladi, and the *sáidi* has duration intervals [**X X** . **X X** . x .].[11] The durational pattern in these last three Arabic rhythms is the same, and it is only the **pitch** (or timbre) of the drum notes that varies from one rhythm to another.

THE CLAVE RUMBA

The timeline with interonset intervals (IOIs) [3-4-3-2-4] shown in Figure 7.3 in polygon notation, has also traveled through much of the world and goes mainly by its Cuban name: *clave rumba*.[12] It has two onsets that coincide with metric beats and one onset that falls halfway between beats. The rumba is one of the most well-known Afro-Cuban folkloric song-and-dance styles popular at large feasts. There are three styles of rumba music: the 12-pulse *fume-fume*[13] and the two 16-pulse rhythms: the fast *guaguancó* and the slower *yambú*. It is these two last styles that use the clave rumba timeline, which is played on the wooden claves.[14] This pattern is also used in several other Cuban rhythms such as the *conga de comparsa* and the *mozambique*, both employed mainly for carnivals. The same timeline pattern is played on a bell in a processional music of the *Ibo* and *Yoruba* peoples of Nigeria.

THE SOUKOUS TIMELINE

Rhythm did not travel with the slaves along a one-way street from Africa to America. After the Second World War, Cuban music became popular in Central Africa. Rumba in particular hits a sympathetic chord with dancers, and was speeded up to create what was first called *congolese* and later became *soukous*, with IOIs [3-3-4-1-5] shown in Figure 7.4 in polygon notation. It has only one onset that coincides with a metric beat and

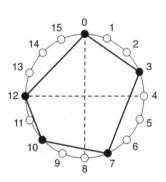

FIGURE 7.3   The *rumba* clave in polygon notation.

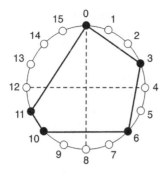

FIGURE 7.4   The *soukous* timeline in polygon notation.

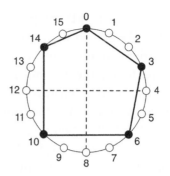

FIGURE 7.5   The *gahu* timeline in polygon notation.

two onsets that fall halfway between beats. The word soukous comes from the French word *secouer* meaning to "shake," and what may account for the strong heartbeat pulse, is the timeline pattern, with its last two onsets close together. Gary Stewart writes about soukous that "Its intricate, invigorating rhythms set feet to tapping."[15] This pattern is often played on either a wood block or a snare drum.

It is not surprising that Africans would resonate well with a new music that had African roots. Neither is it surprising that they would speed it up to suit their more energetic dances. What may seem surprising at first glance is that they would change the rhythm itself to such an extent that the resulting timeline ended up closer to the son than the rumba. To transform the son to soukous, one merely plays the last onset of the son one pulse earlier. On the other hand, to transform the rumba to soukous one must play both the third and fifth onsets of the rumba one pulse earlier. Even when the tempo of the rumba is increased, the relatively late third onset has the subjective effect of slowing it down because the second IOI is longer than the first. Advancement of this third onset by one pulse as in the son and soukous gives the resulting rhythms a more rolling drive.

THE GAHU TIMELINE

Gahu is a polyrhythmic drumming music of the Ewe people of Ghana.[16] The word gahu means either "money dance" or "airplane." It appears to have been created by *Yoruba* speakers of Benim and Nigeria as a form of satirical commentary on the modernization in Africa and was first taken to Ghana in the early 1950s. The Gahu timeline is played on a *gankogui* double bell. It has IOIs [3-3-4-4-2], and is shown in polygon notation in Figure 7.5. It has only one onset that coincides with a metric beat, and three onsets that fall halfway between beats at pulses

6, 10, and 14. Note that these three onsets are precisely the onsets of the three-onset *highlife* timeline shown in Figure 5.4 in Chapter 5. Thus, there is a very close connection between the gahu and *highlife* timelines.

THE BOSSA-NOVA TIMELINE

The timeline with IOIs [3-3-4-3-3] shown in Figure 7.6 in polygon notation is sometimes called the *bossa-nova* rhythm[17] and is played in the slower bossa-nova music, the faster *samba* music, and in Afro-Brazilian folk music from Bahia.[18] It has only one onset that coincides with a metric beat, and two onsets that fall halfway between beats. The bossa-nova, an offspring of samba, is a style of music that was developed in the late 1950s in Rio de Janeiro by musicians Joao Gilberto and Stan Getz. The bossa-nova clave was originally played either with the claves or wood block. However, in more contemporary music, it is played on cymbals or a snare drum, such as in Dave Brubeck's *Bossa Nova U.S.A.*[19]

Several rotations of this pattern are sometimes used in traditional African and Brazilian music, including the rotations [3-3-3-4-3] and [3-3-3-3-4]. The [3-3-3-3-4] version is used in the form of a synthesized rhythmic ostinato sound by the Manchester-based British

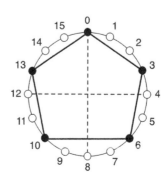

FIGURE 7.6   The *bossa-nova* timeline in polygon notation.

electronic dance music (EDM) group *808 State*, in their piece titled *Cubik*. Their rather strange name was inspired by one of the earliest commercially available programmable drum machines, the Roland TR-808 *Rhythm Composer*.[20] It is one of the signature rhythms of EDM. It may be viewed as a dilation of the tresillo: [3-3-2] by doubling the [3-3] and [2] sections to [3-3-3-3] and [2-2], respectively, to obtain the rhythm [3-3-3-3-2-2] and then merging the last two onsets to obtain [3-3-3-3-4]. For this reason, it is also referred to as the *double tresillo*.[21] In ragtime music, the rhythm is known as the *secondary rag*. Richard Cohn traces back the written record of this term to 1926.[22] There also exists a timeline consisting of 16 IOIs of three durations, yielding the 48-pulse rhythm: [3-3-3-3-3-3-3-3-3-3-3-3-3-3-3-3].[23]

One property that these distinguished rhythms have in common is that, excluding the soukous, they use adjacent interonset durations with values equal to 2, 3, or 4. Interestingly, ancient Chinese philosophers at the time of Confucius regarded these three numbers, including the number 1, as "the source of all perfection."[24] The soukous has additional intervals of durations 1 and 5. Of course there are other rhythms with five onsets among 16 pulses that use only these five durations. Although no *perfect universals* of rhythm have been found yet, several *statistical universals* have been recently discovered. One of them is that rhythms are based on fewer than five duration intervals.[25] Among the six distinguished timelines, no rhythm has more than four distinct duration intervals. In the remaining chapter of this book, we will expound on additional features that contribute to making these six rhythms so distinguished.

NOTES

1 Rentink, S., (2003) and Unruh, A. J., (2000).
2 As mentioned earlier, the emphasis in this book is on the *proportions* (or ratios) of the interonset durations of the rhythms rather than their absolute values, as done by Pearsall, E., (1997).
3 Fernandez, R. A., (2006), p. 7. Gerard, G., (1998), p. 69. Sandroni, C., (2000) and Peñalosa, D., (2009) analyze the influence that the *cinquillo* and *tresillo* had on Latin American music.
4 Srinivasamurthy, A., Repetto, R. C., Sundar, H., & Serra, X., (2014). http://compmusic.upf.edu/bo-perc-patterns Accessed December 30, 2017.
5 Gerstin, J., (2017), p. 55.
6 Proca-Ciortea, V., (1969), rhythm No. 32 in group VII, p. 186.
7 Olsen, P. R., (1967), p. 32.
8 Proca-Ciortea, V., (1969), rhythm No. 33 in group VII, p. 186.
9 Proca-Ciortea, V., (1969), rhythm No. 31 in group VII, p. 186.
10 Poole, A., (2018).
11 Wade, B. C., (2004), p. 70.
12 Johnson, H. S. F., Chernoff, J. M., (1991), p. 68, and Crook, L., (1982). This timeline is sometimes called the clave *guaguancó*. See Gerard, G., (1998), p. 83. It is also played in the Afro-Cuban religious *batá* drumming, Moore, R. & Sayre, E., (2006), p. 128.
13 Klőwer, T., (1997), p. 176.
14 In the guaguancó style of rumba, the clave pattern is played starting on the second half of the measure; thus, [. . x . x . . . x . . x . . x .], Manuel, P. with Bilby, K. & Largey, M., (2006), p. 29.
15 Stewart, G., (1989), p. 19.
16 Locke, D., (1998), Agawu, K., (2003), p. 81, and Reich, S., (2002), p. 60.
17 Kernfeld, B., (1995), p. 23.
18 Van der Lee, P., (1998) and Morales, E., (2003). According to Gerstin, J., (2017), p. 20, the [3-3-4-3-3] timeline is not the main timeline of samba.
19 Columbia Records—CS 8798, Vinyl, LP, U.S. 1963.
20 Butler, M. J., (2001).
21 Biamonte, N., (2014).
22 Cohn, R., (2016a).
23 Béhague, G., (1973), p. 221.
24 Jeans, J., (1968), p. 155.
25 Savage, P. E., Brown, S., Sakai, E., & Currie, T. E., (2015), p. 8989.

CHAPTER 8

The Distance Geometry of Rhythm

IN *THE BIRTH OF TRAGEDY*, Friedrich Nietzsche wrote: "Music is like geometric figures and numbers, which are the universal forms of all possible objects of experience."[1] Traditionally, musicologists have analyzed music from the numerical, historical, sociological, psychological, anthropological, neurological as well as music theory and praxis points of view. The geometric approach used here permits a new kind of analysis of rhythms that yields novel insights and thus augments the traditional tools utilized by musicologists.[2] This is not to imply that geometry has not been employed in the past as a music-theoretic tool. Indeed, geometric images have served multiple purposes for illustrating a variety of musical concepts since antiquity.[3] The circular notation for cyclic rhythms goes back at least to thirteenth-century Baghdad.[4] In modern times, geometric structures in two and higher dimensions are applied to a variety of different aspects of music analysis with increasing frequency.[5] Furthermore, the visualization of rhythms as cyclic polygons allows instant recognition of many structural features of the rhythms that are more difficult to perceive with standard Western music notation or even box notation. For example, suppose we want to know whether the *clave son* has the *palindrome* property: it contains an onset from which one can start playing the rhythm either forwards or backwards so that it sounds the same. With Western music notation, the novice requires some reflection to come up with the answer. On the other hand, with polygon notation, the answer is instantly revealed. Consider the six distinguished timelines described previously and pictured in polygon notation in Figure 8.1.

The son polygon has a solid line connecting the pulse at position 3 with the pulse at position 11. This is an axis of mirror (reflection) symmetry for the polygon. The polygon looks the same on both sides of this line. Therefore, the son rhythm has the palindrome property when started on the second onset (at pulse position 3). Note that among the six timelines two other rhythms have the palindrome property, the shiko and the bossa-nova, both of which have mirror symmetry about the vertical line connecting pulses 0 and 8. This example highlights the ease with which humans perceive visual reflection symmetry, especially with the polygon notation, as compared to Western music notation. By contrast, Stephen Handel claims that "it is extremely difficult to perceive temporal symmetry."[6] Such a statement is true for temporal symmetries, in general, and depends on the particular representations of the stimuli tested.[7] Recent experiments have shown that recognizing acoustic reflection symmetry of melodies about a vertical axis (palindromic symmetry) is easier than detecting other types of temporal symmetries. Experiments have also been done in matching visual and acoustic patterns (called mirror forms) to test for cross-modal correspondence, and it was found that the perception of symmetry about a vertical axis was even more marked for acoustic stimuli than for visual stimuli, and for nonisochronous (rhythmic) melodies than for isochronous ones. With regard to the melodies tested, it was found that performing the acoustic task first increased the sensitivity to the visual equivalent stimulus, whereas performing the visual task first did not increase the sensitivity to its acoustic equivalent.[8] In their study, researchers did not use the circular polygonal representations of rhythms. Herein lies an opportunity for training the human ear to perceive temporal symmetry by comparing the polygonal representations

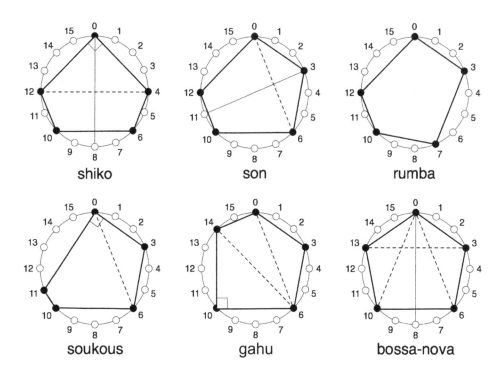

FIGURE 8.1   Some geometric properties of the six distinguished timelines.

of rhythms while simultaneously listening to their acoustic counterparts. Although research has been done on how well listeners can perceive melodic symmetries, it would be interesting to determine how easily listeners can learn to discern acoustic reflection symmetry of pure duration patterns using such a visualization tool.[9] The study of such a skill would contribute to a broadening of the theory of rhythm perception. We shall cover symmetric rhythms in more detail in Chapter 33.

The rhythm polygons in Figure 8.1 contain two other markers of noteworthy geometric and musical properties. All but the rumba have a dashed line connecting some pairs of onsets: the shiko, son, and soukous each has one such line, the gahu has two, and the bossa-nova has three. Geometrically, these lines determine isosceles triangles, together with the two adjacent edges on the polygons. Musically, such a triangle indicates that there are two adjacent interonset intervals of the same duration. Three of the rhythm polygons contain vertices that make an interior right angle (90°), indicated by small squares. The shiko and soukous have a right angle at their first onset at pulse zero. The gahu has a right angle at the fourth onset at pulse ten. Geometrically, a right angle indicates that there exists a pair of onsets diametrically opposite each other.[10] Musically, this means that there are two onsets that break the cycle into two equal half cycles, introducing a certain degree of *regularity*, an

important musical property that we shall return to in more detail in Chapter 38.

These geometric properties can provide new explanations that complement certain conclusions that musicologists have made about rhythms, on the basis of musicological properties alone. For instance, consider the clave rumba and the *clave son*, which differ only in the position of the third onset. Musicologists agree that the clave rumba rhythm is more complex than the *clave son* rhythm, because the rumba has its third onset on a weak beat (pulse seven) and a silence on a strong beat (pulse six), whereas the opposite is true for the son.[11] These two properties of rhythms (the presence of onsets on weak beats, and the absence of onsets on strong beats) are features of the musical notion called *syncopation*.[12] Comparing the two rhythm polygons in Figure 8.1, we observe that, unlike the son, the rumba has no isosceles triangles and no axes of mirror symmetry. Therefore, from a purely geometric point of view the rumba is less structured and thus more complex than the son. In other words, the geometrical properties correlate with the musical principles and psychological perceptual observations.

When the mind is presented with a rhythm, such as the *clave son*, that is repeated continuously throughout a piece of music, and that has a cycle that lasts only a few seconds, it is natural to ask whether it perceives durations other than those that occur between adjacent onsets. There exists evidence and consensus that the "conscious present" (also called "specious present"[13]) lasts for about 3 s. This phenomenon is known as the "three-second window of temporal integration."[14] Therefore, it is most likely that the mind also perceives (perhaps unconsciously) the durations between *all* the other pairs of onsets, in rhythms that last less than 3 s.[15]

A list of all the interonset intervals is called the full *interval content* of the rhythm. Figure 8.2 shows each of the five onsets of the *clave son* connected to all the others with straight lines labeled with numbers. The line connecting the first onset at pulse zero with the third onset at pulse six has the label "6" attached to it, indicating that the time duration between these two onsets is six units. This number is the shortest distance along the circle that connects pulse zero to pulse six. Note that the *clockwise* distance along the circle starting from pulse six and ending at pulse zero is ten. However, we use the shorter

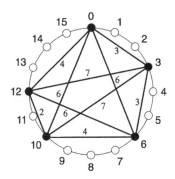

FIGURE 8.2 The full *interval content* of the *clave son* rhythm using geodesic distances.

of the two distances (be it clockwise or counterclockwise) as the distance between the pair of onsets, since that is the likely perceptual distance between two onsets in a cyclic rhythm that is repeated.[16] In geometry, such a distance is called the *geodesic* distance.[17]

Therefore, the full interval content of the *clave son* contains ten distances in total, some of which occur more than once. One numerical way to represent the interval content of a rhythm is by listing how many times each possible distance occurs. In the case of 16-pulse rhythms, the possible distances range from one to eight, and therefore, the interval content may be written as (0, 1, 2, 2, 0, 3, 2, 0). This is sometimes called the *interval vector* of the rhythm.[18] A more visually compelling representation of the interval vector is as a histogram. Figure 8.3 shows the histograms of the six distinguished timelines. Useful information about the rhythms may be gleaned from the properties of their interval-content histograms. For example, shiko uses only four different distances: two, four, six, and eight. On the other hand, gahu uses seven different distances ranging from two to eight. In Chapter 17, we shall explore the concept of rhythm complexity and its application to the creation of music and to the comparison of a variety of measures of mathematical, perceptual, and performance complexities. In the present context, one measure of complexity of a rhythm is the total number of different distances that it generates. Therefore, one would expect the gahu to be more complex than the shiko and perhaps more challenging to learn as well. The difference between son and rumba is not as pronounced; son contains five different intervals and rumba six. Nevertheless, it is

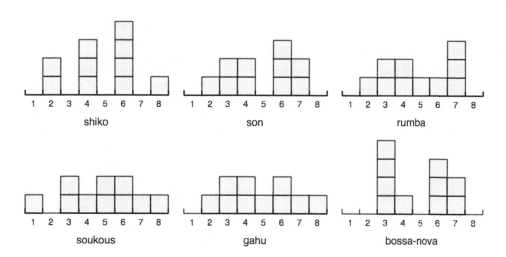

FIGURE 8.3 The *full* interval-content histograms of the six distinguished timelines.

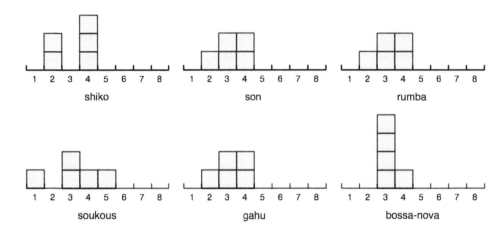

FIGURE 8.4 The *adjacent* interval-content histograms of the six distinguished timelines.

observed again that this property of histograms (their density of occupied cells) suggests, like the previous geometric features, as well as syncopation, that the rumba is more complex than the son. Furthermore, the higher the number of different distances that a rhythm contains, the flatter the histogram will tend to be, since the histogram bins have to spread themselves out. Therefore, the shape of the histogram is relevant to a variety of musical properties of rhythms. We will return to consider shape properties of interval-content histograms other than flatness in Chapter 27.

It is instructive to compare the histograms that contain *all* interonset intervals shown in Figure 8.3 with the histograms that contain only *adjacent* intervals, shown in Figure 8.4. The latter histograms are equivalent to Pearsall's *duration sets*.[19] Although these histograms have their strengths, here we see one of their weaknesses. The son, rumba, and gahu all have identical histograms, and thus cannot be distinguished from each other based only on this information. Indeed, these three rhythms are permutations of their adjacent interonset intervals. It will be seen, however, that even the full interval histograms have some serious drawbacks for characterizing rhythms, and thus, for one of the most important problems in musicology, the measurement of rhythm similarity.

NOTES

1 Johnson, I., (2009), p. 16.
2 See Toussaint, G. T., (2003a, 2004a, 2005a, 2005b), for a more detailed and deeper discussion of computational geometric tools that may be exploited by musicologists.
3 Christensen, T., (2002), p. 280.
4 Liu, Y. & Toussaint, G. T., (2010a).
5 Tymoczko, D., (2011). Bhattacharya, C. & Hall, R. W., (2010). Hall, R. W., (2008). Hook, J., (2006). Rappaport, D., (2005) and Yust, J., (2009) explore the geometry of harmony. McCartin, B. J., (1998). Don, G. W., Muir, K. K., Volk, G. B., & Walker, J. S., (2010). Hodges, W., (2006). Cohn, R., (2000, 2001, 2003). Andreatta, M., Noll, T., Agon, C., & Assayag, G., (2001) explore rhythmic canons. See also Mazzola, G., (2002, 2003), Honingh, A. K. & Bod, R., (2005), and Wild, J., (2009).
6 Handel, S., (2006), p. 188.
7 Kanaya, S., Kariya, K., & Fijisaki, W., (2016), p. 1099.
8 Bianchi, I., Burro, R., Pezzola, R., & Savardi, U., (2017), p. 14.
9 Mongoven, C. & Carbon, C.-C., (2016), p. 1. These authors found that listeners could recognize temporal symmetry better for longer melodies with fewer tones and that older participants were more accurate in their judgments.
10 Patsopoulos, D. & Patronis, T., (2006), p. 59. The theorem asserting that if a triangle inscribed in a circle has one of its sides as the diameter of the circle, then the angle opposite that side is a right angle, is attributed to Thales, the pre-Socratic Greek philosopher from Miletus, who is considered by many to be the "Father of the Scientific Method" for introducing the notion of a mathematical proof by means of deductive reasoning. If a rhythm does not contain two onsets that are diametrically opposite each other, then the rhythm is said to have the "rhythmic oddity property," a topic to be considered in Chapter 15.
11 Velasco, M. J. & Large, E. W., (2011), p. 185.
12 Fitch, W. T. & Rosenfeld, A. J., (2007) and Longuet-Higgins, H. C. & Lee, C. S., (1984).
13 Phillips, I., (2008), p. 182.
14 Pöppel, E., (1989), p. 86.

15 There is as yet no experimental evidence that durations between nonadjacent attacks play a role in the perception of rhythm similarity. The analog question in the pitch domain has been investigated experimentally using chords, throwing doubt on the perceptual validity of some music theoretical assumptions such as octave equivalence: see Gibson, D., (1993). Nevertheless, experiments performed by Quinn, I., (1999), show that the relations between nonadjacent pitch tones (the *combinatorial* model) do affect the judgments of perceptual similarity, but to a lesser degree, than the relations between adjacent tones (the *note-to-note* model).

16 This assumption is speculative. I am not familiar with any studies that try to determine whether the perceived distance between two onsets in a repeating rhythm matches the geodesic distance better than the clockwise distance.

17 Points that lie on a circle are called *cyclotomic sets* in crystallography, where they serve as models of one-dimensional periodic molecules of crystals. The actual models are straight line segments of one period, but the ends are tied together into a circle to facilitate the visualization of all distances between the pairs of points (atoms), Buerger, M. J., (1978). See also Senechal, M., (2008) for related material. Tymoczko, D., (2009) analyses the relationship between three different musical distances and the musicogeometrical spaces they inhabit.

18 Lewin, D., (2007), p. 98. The term *interval vector* is normally used to describe the pitch intervals in chords and scales. The terms *interval-class content, interval function*, and *pitch class content* are also used: see Isaacson, E. J., (1990), Lewin, D., (1959, 1977), Rogers, D. W., (1999), and Block, S. & Douthett, J., (1994). Much additional work has been done exploring interval vector relations in the pitch domain for chords and scales.

19 Pearsall, E., (1997).

CHAPTER 9

Classification of Rhythms

THE CLASSIFICATION OF OBJECTS into categories is a universal preoccupation of human beings all over the world. Besides seeming to provide untold pleasures in creating order around us, classification assists us in an uncountable number of more specific ways. For one, it improves our ability to remember large amounts of information. It aids librarians to catalog musical material for archiving and efficiently retrieving information.[1] It helps doctors prescribe the right medication if they classify a patient's disease correctly. In the domain of music, musicologists classify almost everything they can: musical instruments,[2] drum sounds,[3] music notes on score sheets,[4] music patterns,[5] folk tunes,[6] scales,[7] chords,[8] keys,[9] meters,[10] spans,[11] complexity classes of meters,[12] rhythms,[13] Indian *talas*,[14] melodies,[15] contours,[16] genres,[17] styles,[18] dance music,[19] and other types[20] of music. Clearly, classification is a primary concern in almost all aspects of music. For each of these applications, there exist suitable features and a variety of tools available for classification. Simha Arom classifies the family of *aksak* rhythms from the Balkan region into three classes depending on the properties of the number of pulses contained in the rhythm's cycle.[21] All aksak rhythms are composed of a string of interonset intervals (durations) of lengths two and three, which Arom calls *binary* and *ternary* cells. He calls an aksak rhythm *authentic*, provided that its pulse number is a *prime* number. Some authentic aksak rhythms include: [2-3], [2-2-3], and [2-2-2-3-2-2], with pulse numbers 5, 7, and 13, respectively. A rhythm is *quasi-aksak*, provided that its pulse number is *odd, not prime*, and *divisible by three*. Some instances of quasi-aksak rhythms are: [2-2-2-3] and [2-2-2-2-3-2-2], with pulse numbers 9 and 15, respectively. Finally, a rhythm is called *pseudo-aksak* if its pulse number is even. Thus, the rhythms [2-3-3], [2-2-3-3], and [2-2-2-3-3], with pulse numbers 8, 10, and 12, respectively, are pseudo-aksak. Arom lists 33 confirmed and documented aksak rhythms that are played in practice, which have pulse numbers ranging from 5 to 44, excluding the numbers 6, 20, 31, 36, 38, 40, and 43. In this chapter, I illustrate several general approaches to solving the problem of classification, by using the six distinguished timelines as a pedagogical toy exemplar.

Geometric properties such as those described in the preceding chapters can be used to design methods for the categorization and automatic classification of rhythms. Starting from the acoustic signal produced by an instrument, there are several stages in any musical rhythm recognition system. A fundamental and difficult first step is the analysis of the acoustic waveform, to detect and estimate the locations of onsets. Once these onsets are established, a matching is sought between the query rhythm to be classified and the stored templates. This matching problem is made easier if the underlying *fundamental beat* is also known. Fundamentally, beat is meant to be a series of perceived salient pulses marking equal durations of time. Intuitively, the fundamental beat is what most people do unconsciously when they tap their feet to music that is playing in the background. However, this problem of automatically determining by computer, at what points in time people tap their feet, also known as *beat induction*, is not an easy problem.[22] One approach is to look for a match over all cyclic shifts between the unknown pattern and the stored templates by means of a *decision tree*.[23] This idea is illustrated in Figure 9.1 for the six distinguished timelines, without knowledge of where the "start" of the rhythm is. However, here the input is not acoustic, but assumed to be known and symbolically

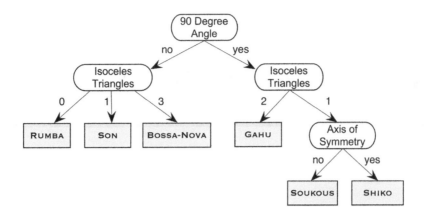

FIGURE 9.1 A decision tree classification using geometric features of the rhythm polygons.

notated.[24] From this input, it is straightforward to compute the polygonal representations of rhythms.

A decision tree classifier may be designed in a variety of ways using different features. The first tree described in the following uses geometric features of the rhythm polygons (refer to Figure 9.1). A node in the tree corresponds to the measurement of a feature. The leaves of the tree (with no offspring) correspond to the rhythms to be classified. First determine whether the polygon has a vertex (corresponding to an onset) with a 90° interior angle. If the answer is NO, then we know that the rhythm must be the rumba, son, or bossa-nova. These three cases can be differentiated by computing the number of isoceles triangles contained at vertices of the polygon: zero for rumba, one for son, and three for bossa-nova. If the polygon does contain a 90°angle, then we also compute the number of isoceles triangles. If there are two isoceles triangles, we have isolated the gahu rhythm. Otherwise, if there is only one isoceles triangle, then determine whether the polygon has an axis of symmetry. If the answer is YES, we have identified the shiko; otherwise, we have found the soukous. Note that no measurement depends on knowing which is the starting note of the rhythm, because these properties are invariant to rotations of the polygons.

Another approach measures *global* shape features of the full interonset-interval histograms of Figure 8.3 to obtain a decision tree such as the one pictured in Figure 9.2. Here, two measurements are made: the height of the histogram and the number of connected components that make up the histogram. A component is considered connected if it consists of a set of cells of height at least one, not separated by an empty cell. The *clave son* has a histogram of height three (see Figure 8.3) and is made up of two connected components, one consisting of cells with distances two, three, and four, and another of cells with distances six and seven.

A third approach uses the presence and absence of certain distances in the interval histogram, as illustrated in Figure 9.3. In this case the *clave son* is identified by the fact that it contains a distance of two, but no distances equal to five or eight.

A fourth illustration of decision trees uses the *number* of distinct interonset distances that occur in the histogram as well as the *ranges* that these values take.

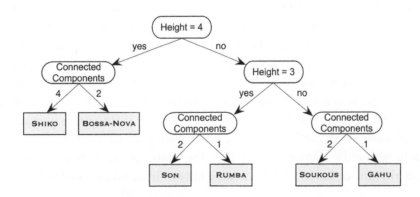

FIGURE 9.2 A decision tree classification using global shape features of the interval histogram.

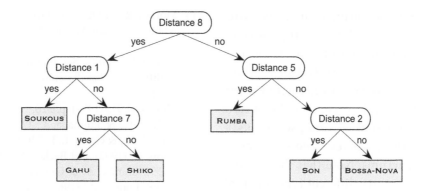

FIGURE 9.3   A decision tree classification using interonset-interval distances.

One possible decision tree that uses these features is shown in Figure 9.4. Here, the range is calculated as the difference between the highest and lowest values of distances present in the histogram.

Other types of geometric features may be computed from the polygons that represent the rhythms, including features based on symmetry properties as well as statistical moments,[25] and moments of inertia.[26] Furthermore, once a set of features has been chosen, there exists a plethora of metric equations that can be used to obtain classifications.[27] Such features may also be used to generate rhythms for use in performances.[28]

An alternate approach to classification constructs proximity trees using a suitable measure of distance or dissimilarity.[29] This method is illustrated in Figure 9.5 for the *aksak* rhythms listed earlier, and the six Afro-Cuban timelines, using a proximity tree called *BioNJ*, frequently employed in bioinformatics.[30] In this approach, the distance between every pair of rhythms is first calculated (here with a distance measure called the *edit distance* that we shall revisit in more detail in Chapter 36). The tree is then computed using the resulting distance matrix. In Figure 9.5 the rhythms are designated by the labeled nodes in the tree, and the edit distance between each pair of rhythms, is portrayed by the shortest path in the tree, between the corresponding nodes. From this tree, it can be clearly observed that the aksak and Afro-Cuban rhythms form two

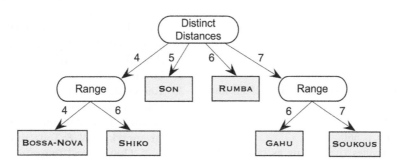

FIGURE 9.4   A decision tree classification using histogram distinct distances and their ranges.

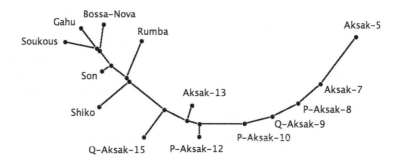

FIGURE 9.5   The *BioNJ* proximity tree of the Afro-Cuban and *aksak* rhythms.

separate groups. This is no surprise since the Afro-Cuban rhythms have 16 pulses, the aksak rhythms have fewer than 16 pulses, and the edit distance is sensitive to this parameter. Within the Afro-Cuban rhythms, the bossa-nova, gahu, and soukous form their own separate subgroup. The "Q" and "P" prefixes in the labels of aksak rhythms denote quasi-aksak and pseudo-aksak, respectively, whereas the postfixes denote the numbers of pulses of the rhythms. Note that the edit distance does not group the three different types of aksak rhythms into separate branches of the tree.

There also exist approaches to classification that use probability, statistics, and Bayesian decision theory. This topic, however, is beyond the scope of this book, and the reader is referred to the references listed.[31]

NOTES

1 Casey, M., Veltkamp, R., Goto, M., Leman, M., Rhodes, C., & Slaney, M., (2008).
2 Kartomi, M. J., (1990), tackles the classification methods used to classify instruments all over the world in the past and present, whereas Lo-Bamijoko, J. N., (1987), focuses on the classification of Igbo musical instruments in Nigeria, based on two considerations: how the instruments are played and the function that the instruments serve in a cultural context. See also Lawergren, B., (1988) and Hornbostel, E. M. von & Sachs, C., (1992).
3 Herrera, P., Yeterian, A., & Gouyon, F., (2002).
4 Bainbridge, D., (2001).
5 Coyle, E. J. & Shmulevich, I., (1998).
6 Elschekova, A., (1966), Kolata, G. B., (1978), Lomax, A., (1959), Lomax, A. & Grauer, V., (1964), and Lomax, A., (1972).
7 Clough, J., Engebretsen, N., & Kochavi, J., (1999).
8 Callender, C., Quinn, I., & Tymoczko, D., (2008) and Rowe, R., (2001), p. 19.
9 Temperley, D., (2002).
10 Christensen, T., (2002), p. 660.
11 Cohn, R., (2001), classifies time spans as pure if the pulse numbers are a power of a single prime number. Thus, the number 16 is pure because it may be written as 2^4, where 2 is a prime number. Similarly, the number 9 is pure because it may be written only as 3^2, where the number 3 is a prime number. The former is a "pure duple" span, whereas the latter is a "pure triple" span. On the other hand, the number 12 is not pure because it does not afford such an expression.
12 London, J., (1995), p. 70.
13 Arom, S., (1989), p. 92 and Toussaint, G. T., (2003a). Arom, S., (2004), p. 41, classifies 44 *aksak* rhythms. Butler, M. J., (2006), p. 81, classifies the rhythms of electronic dance music into three categories: even, diatonic, and syncopated.
14 Morris, R., (1998), Akhtaruzzaman, Md., (2008), and Akhtaruzzaman, Md., Rashid, M. M., & Ashrafuzzaman, Md., (2009).
15 Vetterl, K., (1965). See Bhattacharya, C. & Hall, R. W., (2010), for a classification of North Indian *thaats* and *raags*.
16 Morris, R. D., (1993), p. 220.
17 McKay, C. & Fujinaga, I., (2006) and Correa, D. C., Saito, J. H., & Costa, L. F., (2010).
18 Blum, S., (1992) and Backer, E., (2005). See Lomax, A., (1959), Lomax, A. & Grauer, V., (1964), Grauer, V. A., (1965), Lomax, A., (1968), and Lomax, A., (1972), for an ambitious program to classify all the folk song styles of the world. Cilibrasi, R., Vitányi, P., & de Wolf, R., (2004) are able, surprisingly, to classify types of music disregarding all music information by merely using data compression algorithms from information theory.
19 Chew, E., Volk, A., & Lee, C.-Y., (2005).
20 Levitin, D. J., (2008). Dan Levitin classifies the music of the world into six types (friendship, joy, comfort, knowledge, religion, and love) and explains how music contributed to the evolution of society, science, and art.
21 Arom, S., (2004), p. 45. See also Brăiloiu, C., (1951).
22 Sethares, W. A., (2007). See also Desain, P. & Honing, H., (1999). Dixon, S., (1997).
23 Breiman, L., Friedman, J. H., Olshen, R. A., & Stone, C. J., (1984). See Duda, R. O., & Hart, P. E., (1973) for additional classification methods.
24 Wright, M., Schloss, W. A., & Tzanetakis, G., (2008), present a variety of tools for analyzing rhythm in audio recordings. In particular, they propose an original method for beat tracking in Afro-Cuban music by using knowledge specific to this type of music, namely, the clave pattern itself. Their technique highlights the benefits that may be accrued by incorporating domain-specific knowledge about the musical style and culture under study.
25 Boveiri, H. R., (2010), p. 17.
26 Toussaint, G. T., (1974).
27 Polansky, L., (1996).
28 Sampaio, P. A., Ramalho, G., & Tedesco, P., (2008).
29 Jaromczyk, J. W. & Toussaint, G. T., (1992), Toussaint, G. T., (2005d), and Toussaint, G. T., (1980, 1988).
30 Huson, D. H. & Bryant, D., (2006), Dress, A., Huson, D., & Moulton, V., (1996), Gascuel, O., (1997), and Huson, D. H., (1998).
31 Lin, X., Li, C., Wang, H., & Zhang, Q., (2009). Temperley, D., (2007). Gómez, E. & Herrera, P., (2008), apply probabilistic methods to compare music audio recordings from western and nonwestern traditions by automatically extracting tonal features. Toussaint, G. T., (2005d), provides a survey of nonparametric methods for machine learning and data mining, including nearest neighbor methods that approximate Bayesian decisions by incorporating proximity graphs. For the application of proximity graph methods used in discriminating western from non-Western music, see Toussaint, G. T. & Berzan, C., (2012).

CHAPTER 10

Binary and Ternary Rhythms

OST OF THE RHYTHMS CONSIDERED IN the previous chapters were determined by cycles that had either 8 or 16 pulses. Such rhythms are referred to here as *binary* rhythms, as are those with cycles of 2, 4, or 32 pulses. Note that all these numbers can be evenly divided by two, but not by three. Rhythms with cycles of 16 pulses are popular all over the world. In addition to 16, there is another pulse that also figures prominently in the music of many parts of the world, most notably in sub-Saharan Africa and Southern Spain, and this is the number 12. Such rhythms are called *ternary* rhythms. Rhythms with 3, 6, 9, and 24 pulses also belong to the family of ternary rhythms. The smallest binary and ternary rhythms with two and three pulses (also called *duple* and *triple* rhythms), and their combinations, form the building blocks of most rhythms of the world, leading some scholars to label them as music universals. The Reverend A. M. Jones writes: "When Europeans sing or play, their music will consist of rhythms which are essentially duple or triple or a combination of both. So it is with Africans. It is a fundamental natural law of rhythm and is therefore universal."[1]

The most illustrious ternary rhythm timelines in sub-Saharan Africa and the Caribbean have either five or seven onsets, with durational patterns [2-2-3-2-3] and [2-2-1-2-2-2-1], respectively, in a cycle of 12 pulses. They are shown as polygons in Figure 10.1. Both rhythms have been called the *standard* pattern in the literature.[2] E. D. Novotney reviews the evolution of the terminology used for these rhythms and proposes his own: the "*five-stroke key pattern*" and the "*seven-stroke key pattern*," respectively, on the basis that they play in sub-Saharan African music, the same function that the "clave" rhythms play in Cuban music, and the word "clave" means "key."[3] The rhythm on the left is sometimes called the *fume–fume* timeline[4] and the one on the right the *bembé* pattern.[5] The fume–fume is often used as a clap pattern in West African music, and A. M. Jones writes that: "This little clap-pattern is quite charming."[6]

The reader has surely noticed that the ternary fume–fume clave has a geometric structure similar to that of the binary *clave son*. The two rhythms are shown together in Figure 10.2 for easy comparison.[7] In a subsequent chapter, both rhythms will be mapped to a

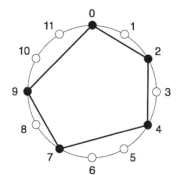
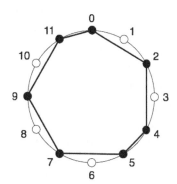

FIGURE 10.1   The *fume–fume* and *bembé* ternary timelines.

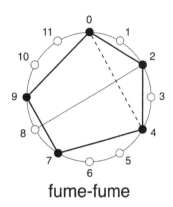
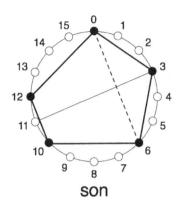

FIGURE 10.2 The similar geometry of the *fume–fume* and *son* clave rhythms.

48-pulse clock to quantify the absolute values of the differences between their corresponding attack points. For now, it suffices to remark that visually the two pentagons are almost identical in shape and orientation. To be more precise, both rhythms contain an isosceles triangle rooted at the second onset, both exhibit mirror symmetry about an axis that bisects the isosceles triangle at its apex, and therefore both are *balanced* in the sense that they are partitioned into two identical groups by a diameter of the circle, i.e., the line through the center of the circle connecting pulses 2 and 8 in the fume–fume, and 3 and 11 in the son.[8] In addition, both rhythms contain *obtuse* isosceles triangles, which means that the triplets (onsets at pulses 0, 2, and 4 in the fume–fume and at pulses 0, 3, and 6 in the son) are well separated from the twins (7 and 8 in the fume–fume, and 10 and 12 in the son) by the diagonals (11, 5) and (15, 7), respectively.

There are other musicological structural similarities between the two rhythms. For example, both numbers 12 and 16 may be evenly divided into four equal durations (quarter measures) without requiring additional pulses, by selecting the "north," "south," "east," and "west" pulses numbered 0, 3, 6, and 9 in the ternary case and 0, 4, 8, and 12 in the binary case. These are the four most salient locations for regular metric beats in families of rhythms with 12 and 16 pulses. The fume–fume and son rhythms both have their first and last onsets on their "north" and "west" metric pulses, respectively. Since both regular meters, [3-3-3-3] and [4-4-4-4], can be easily aligned with each other, and the two rhythms are so similar, they can easily be interchanged during the performance of a piece, as is done in the *Highlife* music of West Africa. Jeff Pressing calls such timelines with unequal values of pulses in their cycles, but with similar interonset-interval structures, *transformational*

analogs,[9] and Fernando Benadon explores their use as compositional and analytical expressive transformations of each other.[10]

In addition to the fact that the two rhythms are quite similar to each other with respect to the exact locations of their attacks, they are in fact identical to each other if they are represented by their *rhythmic contours*. The rhythmic contour of a rhythm is obtained by coding the change in the durations of two adjacent interonset intervals using zero, +1, and −1 to stand for equal, greater, and smaller, respectively. The durational patterns of the fume–fume and son timelines are, respectively, [2-2-3-2-3] and [3-3-4-2-4]. Therefore, both rhythms have the same rhythmic contour: [0, +1, −1, +1, −1]. Rhythmic contours are relevant from the perceptual point of view because humans have an easier time-perceiving qualitative relations such as "less than" or "greater than" or "equal to" than quantitative relations such as the second interval is four-thirds the duration of the first interval. It has also been found that often the reduced information contained in the contour is sufficient to effectively describe certain types of music.[11] On the other hand, two rhythms with the same contour may also sound different, as is the case for the 16-pulse and 11-pulse rhythms with interonset intervals [4-3-2-3-4] and [3-2-1-2-3], respectively.[12] Therefore, used in isolation or in a context where the intervals can vary widely, the rhythmic contour suffers from severe drawbacks as a representation from which to extract meaningful rhythmic similarity features.[13]

John Chernoff has suggested that for all practical purposes there is not much difference between the binary and ternary versions of the "standard" African bell pattern (fume–fume) when perceived relative to their underlying quadruple metric beats, [4-4-4-4] and

[3-3-3-3], and that these regular beats play a perceptually important role. However, while there is little doubt that these metric beats influence perception, it is not at all clear that this influence propels the listeners' judgments of the two versions towards greater similarity. It may be argued to the contrary that the quadruple underlying beats flesh out rather than camouflage their differences. Figure 10.3 shows the binary and ternary versions of the five-onset standard pattern superimposed on their quadruple metric structures, to more accurately examine their perceptual role. For either rhythm, imagine playing the metric beats (each highlighted with a ring) with the left hand on a bass drum, and the rhythm with the right hand on a woodblock. Let us denote with the letters **R**, **L**, and **U** the events consisting of striking the instruments with the right hand, left hand, and both hands in unison, respectively. While it is true that the sequence of onsets that describes the union of the metric beats and rhythm onsets yields the same alternating pattern for both the *clave son* and the fume–fume, namely [**U-R-L-R-L-R-U**], and although both rhythms start and end on the first and last beats of the cycle, these properties by themselves are not sufficient to engender greater perceived similarity. On the contrary, feeling the quadruple meter makes the listener more keenly aware of the differences in the placements of the third and fourth onsets of the rhythms, which in the *clave son* fall squarely in the middle of the interbeat intervals, whereas in the standard pattern fall closer to the beats, creating the perception of greater syncopation. In the fume–fume pattern, the third onset is twice as close to the second beat than to the third beat, and the fourth onset is twice as close to the third beat than to the fourth beat.

Furthermore, examples may be constructed of quite dissimilar rhythms that satisfy both of these properties. To this end, consider the two rhythms in Figure 10.4. The left rhythm is a rotation of the bossa-nova clave, and the right is a mutation of the standard pattern, in which the second onset is moved from pulse two to pulse one, and the fourth onset is moved from pulse seven to pulse eight. Both rhythms have the same onset-to-meter placement pattern [**U-R-L-R-L-R-U**], and both have their first and last onsets on the first and last beats of the cycle, but they sound quite different from each other.

In closing this chapter, it should be noted that the terms *binary* and *ternary* are also used in music to

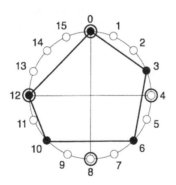
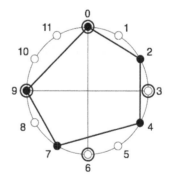

FIGURE 10.3   The *clave son* (left) and *fume–fume* (right) embedded in a duple (regular) meter.

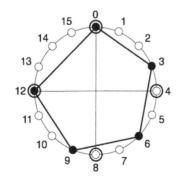
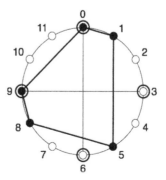

FIGURE 10.4   A rotation of the bossa-nova rhythm (left) and a mutation of the fume–fume (right) embedded in a quadruple meter.

describe the form of compositions as a whole. In this context, form refers to the manner in which sections of the piece are structured. In binary form, two main sections of the work are repeated to create patterns such as AABB, whereas in ternary form, the sections are organized into patterns such as ABA.

NOTES

1 Jones, A. M., (1949), p. 293.
2 Agawu, K., (2006), King, A., (1960), p. 51, and Kubik, G., (1999), p. 53.
3 Novotney, E. D., (1998), p. 165.
4 Klöwer, T., (1997), p. 176. This timeline pattern is also a clapping rhythm used by Ewe children in Ghana, who sometimes dance four evenly spaced steps marking out the rhythm [x . . x . . x . . x . .]. See also Kubik, G., (2010b), p. 55, Kubik, G., (2010a), p. 45, Jones, A. M., (1959), p. 3, Collins, J., (2004), p. 29, Akpabot, S., (1972), p. 62, and Logan, W., (1984), p. 194. Bettermann, H., Amponsah, D., Cysarz, D., & Van Leeuwen, P., (1999), p. 1736, call this rhythm by the name of *inyimbo*.
5 King, A., (1960), p. 51, Malabe, F. & Weiner, B., (1990), p. 8. Logan, W., (1984), p. 194. Waterman, R. A., (1948), p. 28. Kauffman, R., (1980), p. 397. Stone, R. M., (2005), p. 82. Kubik, G., (2010a), p. 44. Johnson, H. S. F., & Chernoff, J. M., (1991), p. 67. Chernoff, J. M., (1979), p. 145.
6 Jones, A. M., (1954a), p. 33.
7 Both of these rhythms are played in the Highlife popular dance rhythm of West Africa. See Chernoff, J. M., (1979), p. 145.
8 Other definitions of balanced rhythms will be covered in Chapter 20.
9 Pressing, J., (1983), p. 43.
10 Benadon, F., (2010).
11 Hutchinson, W. & Knopoff, L., (1987), p. 281, hypothesize that music style may be effectively described and discriminated on the basis of syntactic structures of three symbols used to code rhythmic contours, namely R (repetition), S (shortening), and L (lengthening), in effect, a three-letter alphabet for temporal groupings.
12 See Marvin, E. W., (1991), for the application of rhythmic contours to composition analysis. Contours have also been explored in the pitch domain, where they are called *pitch contours* or *melodic contours*. See Schultz, R., (2008), for the application of melodic contours to the analysis of the *nonretrogradable* structure in the birdsong music of Olivier Messiaen. A structure is nonretrogradable if it is palindromic, i.e., has the same structure when played forwards or backwards. Freedman, E. G., (1999), p. 365, writes that "musically experienced listeners can recognize both the contour and interval information, whereas musically inexperienced listeners rely predominantly on the contour information." See also Callender, C., Quinn, I., & Tymoczko, D., (2008), for a more recent discussion on contour in the pitch domain.
13 Whether or not contours will play a significant role in music theory, they have already spawned interesting problems in computer science. Demaine, E. D., Erickson, J., Krizanc, D., Meijer, H., Morin, P., Overmars, M., & Whitesides, S., (2008), consider the problem of reconstructing rhythms from the full and partial contour information.

CHAPTER 11

The Isomorphism Between Rhythms and Scales

Twelve is a special number. There are 12 months in a year. The Chinese use a 12-year cycle in their calendar. Near the equator, there are 12 h of daylight and 12 h of darkness. Jesus had 12 apostles. There are 12 days of Christmas, 12 in. in a foot, and we buy a dozen eggs. To a mathematician, 12 is the sum of three smallest integers $a = 3$, $b = 4$, and $c = 5$, that satisfy the famous Theorem of Pythagoras ($a^2 + b^2 = c^2$). Here $3 + 4 + 5 = 12$, and $3^2 + 4^2 = 5^2$ or $9 + 16 = 25$. From the musical point of view, 12 has an important property that it is a small number that contains many (four) divisors other than 1 and 12, in particular, 2, 3, 4, and 6. For comparison, the larger number 16 has only three divisors: 2, 4, and 8, other than 1 and 16. The fact that 12 can be divided without remainder by both an even number (such as 2) and an odd number (such as 3) gives it the powerful distinction that it can be easily manipulated to feel like either a binary or ternary rhythm. Twelve is also the number of different pitches in the chromatic scale or octave and of the modern piano keyboard that consists of 12 pitch intervals called semitones (see Figure 11.1). "At least since ancient Greece, thinkers about music have intuited a deep analogy between pitch and time."[1] Because a note that is transposed by an octave may be considered to be the same note, we may wrap the piano octave onto a circle consisting of 12 intervals, as in Figures 11.2 and 11.3 using the labels C, C#, D, Eb, E, F, F#, G, G#, A, Bb, and B.[2] The white keys on the modern piano keyboard correspond to the seven pitches without the sharps and flats, i.e., C, D, E, F, G, A, B, as shown in polygon notation[3] in Figure 11.3 (left), and have been immortalized in song with the words *Do-Re-Mi-Fa-Sol-La-Ti*, which will bring us back to *Do*.

FIGURE 11.1 A modern piano keyboard. (Courtesy of Yang Liu.)

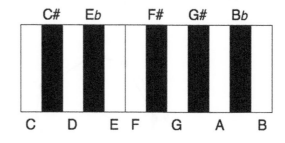

FIGURE 11.2 The diatonic and chromatic scales on the piano keyboard.

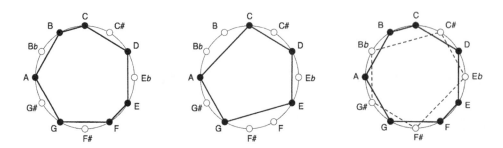

FIGURE 11.3 The *diatonic* scale (left), a *pentatonic* scale (center), and a superposition of a diatonic and a pentatonic scale (right).

Note that this polygon is identical to the *bembé* ternary rhythm timeline polygon of Figure 10.1 (right) with interonset-interval duration pattern [2-2-1-2-2-2-1]. Thus, the *bembé* rhythm and the diatonic scale (or heptachord) are interval patterns that are mathematically *isomorphic* to each other, that is, they form the same pattern of long and short intervals, one expressed in time intervals and the other in pitch intervals.[4]

The diatonic scale is a *heptatonic* (seven-note) scale.[5] For this reason, the interval pattern [2-2-1-2-2-2-1] is sometimes called the "diatonic pattern," even when referring to rhythm. The pentatonic (five-note) scale is found throughout the world, prompting Simha Arom to suggest that it is a musical universal.[6] One of the favorite pentatonic scales contains no semitones and consists of tones C, D, E, G, and A, as shown in polygon notation[7] in Figure 11.3 (center). In ancient classical Chinese music, this was the only scale used until the Chou dynasty more than 3,000 years ago.[8] Observe that this scale is isomorphic to the fume–fume timeline of Figure 10.1 (left). Also, worthy of note is that if this scale is rotated in a half circle, one obtains the complement of the diatonic scale or the black keys of the modern piano keyboard, corresponding to the white circles in

Figure 11.3 (left). The diagram on the right shows both scales superimposed on each other.

Another noteworthy pair of isomorphic rhythm-scale structures consists of the eight onsets in a cycle of 12 pulses shown in Figure 11.4. The rhythm on the left with durational pattern [2-1-2-1-2-1-2-1] is a common rhythmic ostinato used in many parts of the world. It is played on the *kenkeni* drum for the traditional circumcision song *kéné foli* in Guinea.[9] The *kanak* people of New Caledonia call it the *pilou* rhythm.[10] It is also the "ancestral" pattern obtained from a phylogenetic analysis of all the rotations of the rhythmic ostinato pattern [x x x . x x . x . x x .] used in Steve Reich's *Clapping Music*.[11] Its isomorphic pitch counterpart shown on the right in Figure 11.4 is the *octatonic* scale.[12] It is one of the four most important scales used in jazz, where it is called the *diminished* scale.[13] This pattern possesses many rotational and mirror symmetries, indeed, as many as those possessed by a square.[14] It has four axes of mirror symmetry about the lines through pulse pairs (1, 7) and (10, 4), and the orthogonal bisectors of the four short edges of the octagon, and it can be rotated by four different angles to correspond with itself.[15]

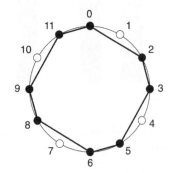
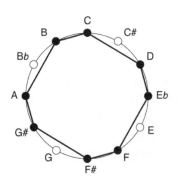

FIGURE 11.4 The *kéné foli* rhythm (left) and the *octatonic* scale (right).

The octatonic durational pattern also has two noteworthy geometric extremal properties. The first of these involves the magnitude of the areas of polygons that represent the scales and rhythms. To introduce the terminology, consider all possible heptagons (seven-sided polygons) inscribed in the circular lattice consisting of 12 elements. In other words, these heptagons are the polygonal models of seven-note scales or seven-onset rhythms in a cycle of 12 notes or pulses, respectively. Recall from Chapter 6 that the general formula to determine how many possible seven-note scales there are in a 12-note universe is given by $(n!)/[(k!)(n-k)!]$, which for $n = 12$ and $k = 7$ becomes $(12!)/[(5!)(7!)] = 792$. Six of these 792 are shown in Figure 11.5 in circular polygon notation, where in addition, an edge is drawn between every vertex of the polygon and the center of the circle, thus partitioning the polygon into triangles of differing sizes. These triangles facilitate the comparison of the polygons in terms of their areas. The smallest-area triangle is determined by two adjacent vertices. Of all the 792 polygons, the one with *minimum* area, in the upper-left diagram, is made up of six smallest triangles, with interval structure [1-1-1-1-1-1-6] modulo rotation. The polygons in the upper-center diagram with intervals [1-1-1-1-2-4-2] and upper-right diagram with intervals [1-1-1-2-2-3-2] have larger areas. Of all the 792 polygons, the ones with *maximum* area are the three in the

bottom row. For this reason, we refer to them as the *maximum-area* heptagons. To see that all three have the same area note that each is composed of two identical smallest-area triangles and five identical equilateral triangles. These three are the only three (modulo rotation) that realize the maximum area. Therefore, area is not a distinguishing feature of these three polygons. What does discriminate between the three is the position of the smallest triangles with respect to each other. On the left, the two smallest triangles are next to each other, and in the center, they are separated by at least one equilateral triangle, and on the right, they are separated by at least two such triangles. The rightmost polygon is of course the diatonic scale (Figure 11.3) as well as the *bembé* rhythm timeline (Figure 10.1). In the following chapters, we shall see that these three polygons may also be differentiated by means of several other pertinent mathematical properties.

We can now turn to the first geometric extremal property of the octatonic scale shown in solid lines and black-filled vertices in Figure 11.6 (left). Also shown in dashed lines and gray-filled vertices is the maximum-area four-note scale (a diminished seventh chord). Among all the *maximum-area* eight-note scales, the octatonic scale is the only one whose complementary four-note scale is a *maximum-area* four-note scale.[16] By contrast, the eight-note scale in the diagram on the right, which is also a maximum-area octagon, has

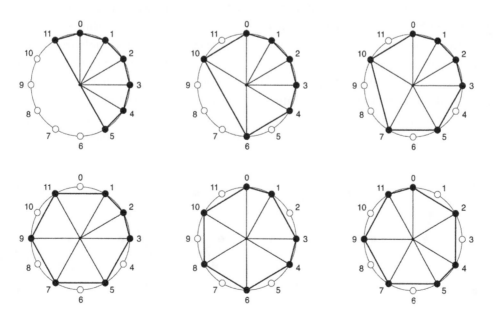

FIGURE 11.5 Illustrating the areas of 6 of the 792 possible heptagons in the 12-element circular lattice. The maximum area is realized by the three heptagons in the bottom row.

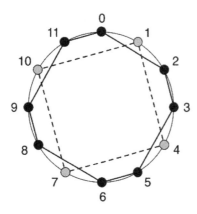
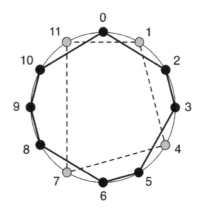

FIGURE 11.6 The maximum-area eight-note scale (solid lines) and its complementary maximum-area square four-note scale (dashed lines) superimposed (left), and another maximum-area eight-note scale with a nonsquare four-note scale (right).

a complementary polygon that is a nonsquare quadrilateral with vertices at positions 1, 4, 7, and 11. Since the maximum-area quadrilateral inscribed in a circle is realized only by a square, it follows that this complementary four-note scale is not a maximum-area four-note scale.[17]

The second extremal property concerns the partitioning of the onset points into unions of disjoint pairs of points separated by distance d. Such pairs are called *dyads* in the context of scales and chords. In this case, the possible a priori values that d can take are the values one, two, three, four, five, and six. Figure 11.7 shows the partitions of the octatonic pattern into unions of *disjoint* dyads of durations one, two, four, and five. The reader is invited to generate the remaining partitions. The extremal combinatorial-geometric property of the octatonic pattern concerns dyads, and is known as Cohn's Theorem in music theory.[18] It states that the octatonic pattern is the only one among 12-point cycles that can be partitioned into dyads of all six durations. Brian McCartin obtained a simple geometric proof of Cohn's theorem.[19]

Jeff Pressing has expounded on several fascinating parallels between pitch and time. However, Justin London maintains that the two are not isomorphic concepts.[20] Milton Babbitt, while recognizing the limitations of the pitch-time analogy, nevertheless developed methods for transferring pitch-class operations to the rhythmic domain.[21] Indeed, exploring the extent to which the comparative analysis of pitch and rhythm provides insight that can be transferred from one modality to the other is a fruitful endeavor. On the one hand, it is nice to be able to apply the tools developed for one domain to the other, and if some concepts do not transfer successfully, these provide us with insight about their differences. Furthermore, there exists a variety of tools that are equally applicable to both domains.[22]

We close this chapter with a simple example in which a concept is not transferable from the pitch to the time domain. Consider the question of consonance and dissonance in pitch and rhythm. The three chords shown in Figure 11.8 sound quite dissonant as chords in the pitch domain. On the other hand, as rhythms in the time domain they are stable.

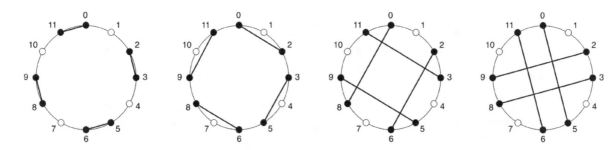

FIGURE 11.7 Dyad partitions of the octatonic pattern with $d = 1, 2, 4,$ and 5.

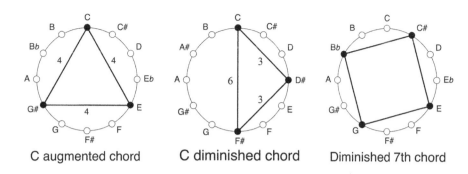

C augmented chord   C diminished chord   Diminished 7th chord

FIGURE 11.8   These three dissonant chords determine stable rhythms.

NOTES

1 Cohn, R., (2015), p. 6.
2 Note that C# = D♭, D# = E♭, F# = G♭, G# = A♭, and A# = B♭.
3 McCartin, B. J., (2007), p. 2420, refers to such polygons as pitch-class polygons.
4 Pressing, J., (1983). Rahn, J., (1983) devotes Chapter 5 to the issue of isomorphism of pitch and time. See also Fitch, W. T., (2012), Rahn, J., (1987), and Carey, N. & Clampitt, D., (1996). No distinction is made in this chapter between the interval patterns of *scales* and *chords*. Chords are subsets of scales that are usually played at the same time. The issue of pitch-time isomorphism can be analyzed at different levels of mathematical, psychological, and neurological models. London, J., (2002) showed that graph theory models of pitch and time are not mathematically isomorphic. Whether they are perceptually isomorphic is another matter. Wen, O. X. & Krumhansl, C. L., (1916), found experimentally using the bembé rhythm and its isomorphic diatonic scale that pitch and time interact perceptually.
5 Loy, G., (2006), p. 17. The diatonic scale is considered to be the prototype of all scale systems in the West. See Kappraff, J., (2010) for a mathematical treatment of ancient scales.
6 Arom, S., (2001), p. 28. For contrary views concerning the pentatonic scale, see Bradby, B., (1987). There exists a fair amount of variability in the number of tones in scales used around the world. A more constant universal of scales appears to be the preference for unequal step intervals. See Trehub, S., (2001), p. 435. See Meyer, L. B., (1998), for biologically constrained music universals.
7 Representing musical scales with a "clock-face" is quite common in the literature; see Jeans, J., (1968), p. 163. Krenek, E., (1937), was one of the first writers to represent chords and scales as polygons as done here, and thus such polygons are also referred to as *Krenek* diagrams. See also McCartin, B. J., (1998), Rappaport, D., (2005), Ashton, A., (2007), p. 43, and Martineau, J., (2008), p. 17.
8 Certainly, by the year 433 BC, the 12-note chromatic scale was well established in China, as evidenced by the bronze bells of Hubei discovered in 1977 (see Benson, D. J., (2007), p. 139).
9 Konaté, F. & Ott, T., (2000), p. 89.
10 Ammann, R., (1997), p. 241.
11 Colannino, J., Gómez, F., & Toussaint, G. T., (2009).
12 McCartin, B. J., (2007), p. 2424.
13 Levine, M., (1995).
14 Cohn, R., (1991).
15 In the pitch domain, mirror symmetry (or reflection) is called *inversion* and rotation is termed *transposition*. See Tymoczko, D., (2011), p. 35. These symmetries are important for music. See also Coxeter, H. S. M., (1968).
16 Rappaport, D., (2005), p. 67.
17 Niven, I., (1981), p. 113, gives a simple elegant proof that the maximum-area quadrilateral inscribed in a circle must be a square, without the need to prove a priori that a solution exists.
18 Cohn, R., (1991).
19 McCartin, B. J., (2007), p. 2431.
20 London, J., (2002).
21 Christensen, T., (2002), p. 720.
22 Amiot, E. & Sethares, W. A., (2011).

CHAPTER 12

Binarization, Ternarization, and Quantization of Rhythms

IMAGINE A SEVENTEENTH-CENTURY Spanish sailor crossing the Atlantic Ocean on a galleon full of Aztec and Inca gold on a regular run from the port of Havana in Cuba to the port of Sevilla in Spain. Let us assume that this sailor was brought up in a village where the popular music always incorporated rhythms that used a 16-pulse cycle. One day, some freed slaves from sub-Saharan West Africa show up on the ship, with drums and an iron bell, and they play the fume–fume rhythm [x . x . x . . x . x . .]. Since this rhythm is so similar to the *clave son* [x . . x . . x . . . x . x . . .], it is quite reasonable that our sailor would perceive it as being the *clave son*.[1] To more accurately compare these two rhythms, it helps to put them together on the same clock diagram so that both complete cycles take the same amount of real time. For this, it is convenient to use a clock with a number of pulses that

is divided evenly (without remainder) by both 12 and 16. The smallest such number is 48: it is equal to 4×12 and 3×16. Figure 12.1 (left) shows the son and fume–fume rhythms embedded in such a 48-pulse clock. The son is indicated with larger white circles on pulses 0, 9, 18, 30, and 36, whereas the fume–fume is made up of the smaller black circles on pulses 0, 8, 16, 28, and 36. As pointed out in a previous chapter, the first and last onsets of both rhythms are in unison. The second onsets differ by one forty-eighth of a cycle, and the other two onsets differ by one twenty-fourth of a cycle. If we assume that the entire cycle lasts for about 2 s, it means the second onsets differ by one twenty-fourth of a second, and the two others by one twelfth of a second. Thus, even in perceptual terms, the rhythms may be considered to be quite similar with respect to their corresponding onset alignments.

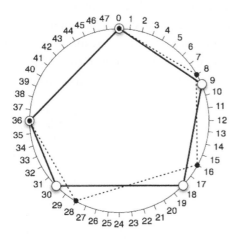 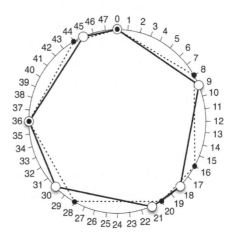

FIGURE 12.1   Left: The binary son (solid lines) and ternary fume–fume (dotted lines) on a 48-pulse clock. Right: The binary *bembé* (solid lines) and its ternary version (dotted lines).

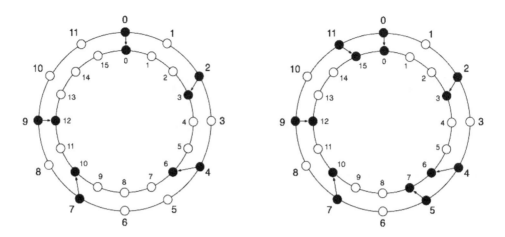

FIGURE 12.2 Binarization of the *fume–fume* timeline to the *clave son* (left) and the ternary *bembé* to its binary counterpart (right), obtained by rounding up.

The ternary *bembé* timeline and its binarized version[2] are shown in the right diagram of Figure 12.1. For the ternary rhythm, the onsets consist of small black circles on pulses 0, 8, 16, 20, 28, and 36, which are connected by dotted lines. The binarized rhythm (solid lines) has onsets indicated with larger white circles on pulses 0, 9, 18, 21, 30, and 36. For these two rhythms also, the onsets at the first and last of the four main beats coincide, and the second, fourth, and seventh ternary onsets precede their binary counterparts by one forty-eighth of a cycle, and the third and fifth onsets by one twenty-fourth of a cycle.

The Cuban ethnomusicologist Rolando Pérez Fernández hypothesized that the African ternary rhythms were binarized by means of cultural blending caused by human migrations, and that the ternary fume–fume rhythm was thus converted to the binary *clave son* rhythm.[3] One might wonder if such a hypothetical perceptually- and culturally based transformation may be translated into distance geometric terms. At first glance, one might even venture an intuitive guess that the onsets in the ternary 12-pulse clock should snap to the nearest pulses of the binary 16-pulse clock.[4] To obtain insight into the nature of such a geometric process, refer to Figure 12.2, where the two clocks are drawn one inside the other. From these diagrams, it may be observed that snapping the ternary onsets to their nearest binary pulses neither convert the fume–fume timeline to the *clave son*, nor the ternary *bembé* to its binary counterpart. Instead, one possible snapping rule that yields the desired result is that: (1) if the onset of the ternary rhythm coincides with a pulse of the binary rhythm, then the onset stays where it is and (2) if the onset of the ternary rhythm falls anywhere in between two binary pulses, then it snaps to the binary pulse that follows it (rounding "up" or rounding in a clockwise direction). These operations are indicated in Figure 12.2 with arrows pointing from the onsets of the binary rhythms to the corresponding pulses of the ternary cycles.[5] This example appears to challenge the efficacy of the nearest pulse hypothesis as a geometric model of the perceptual process underlying the binarization of the fume–fume and rhythm quantization in general. But before speculating further on this issue, it should be noted that there exists an equivalence relation between the nearest integer of a number and rounding that number down to the next integer, as explained later.

Figure 12.3 shows a circle with 16 unit pulses marked on it. Each unit is also divided into three smaller intervals for convenience. Consider the two points on the circle corresponding to the arrows *a* and *b*, occurring at locations 0.67 and 4.33. Rounding these two numbers to their nearest integers takes *a* to 1.0

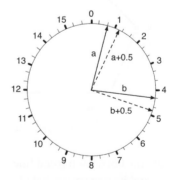

FIGURE 12.3 Rounding a number *x* to the *nearest* integer is the same as rounding *down* the number $x + 0.5$.

and *b* to 4.0. Now add 0.5 to both numbers to obtain $a + 0.5$ and $b + 0.5$, occurring at locations 1.17 and 4.83, respectively. Rounding down these two numbers to the next lowest integer also yields the numbers 1.0 and 4.0, respectively. This will always happen as long as the notion of nearest is well defined, that is, as long as an attack does not occur exactly halfway between two pulses.[6]

The earlier observations about the problems incurred when an attack of the ternary rhythm lies either exactly on a pulse or halfway in between two pulses of the 16-point lattice, makes one wonder what happens with all the remaining infinitude of positions at which a ternary rhythm may lie. How many different nearest-pulse binarizations of the fume–fume exist when none of its onsets lies either on a pulse or halfway between two adjacent pulses in the 16-point lattice? And more importantly, is the *clave son* one of these binarizations? How good are the others? And how are they related to the *clave son*? The answers to these questions may be obtained by examining Figure 12.4. For convenience, the circle is partitioned into 48 integers, making up the 16 binary pulses required for binarization. This provides two fiducial points (indicated by thin marks) in between every pair of adjacent pulses in the 16-pulse cycle. The 16 pulses are shown in bold lines, at 0, 3, 6, 9, etc. In the top-left diagram, the fume–fume (in dashed lines) is positioned in its standard mode at pulses 0, 8, 16, 28, and 36 to give the duration pattern [2-2-3-2-3]. The binarization of this mode has attacks at positions 0, 9, 15, 27, and 36, yielding the duration pattern [3-2-4-3-4]. Proceeding from left to right and top to bottom, the remaining four diagrams each show the fume–fume rotating clockwise by one forty-eighth of a cycle. In all the diagrams, the rhythms obtained by snapping the fume–fume rhythm using the nearest-pulse rule are indicated with white circles connected by solid lines. When the fume–fume is rotated by one forty-eighth of a cycle to positions 1, 9, 17, 29, and 37, as in the upper right diagram, the binarization obtained is the *clave son* at positions 0, 9, 18, 30, and 36 with duration pattern [3-3-4-2-4].[7] Rotating the fume–fume another one forty-eighth of a cycle to positions 2, 10, 18, 30, and 38 yields, in the middle left diagram, the binarization at positions 3, 9, 18, 30, and 39, with duration pattern [2-3-4-3-4]. Rotating the fume–fume again another one forty-eighth of a cycle to positions 3, 11, 19, 31, and 39 yields, in the middle right diagram, the binarization at positions 3, 12, 18, 30, and

39 having duration pattern [3-2-4-3-4]. Finally, rotating the fume–fume one forty-eighth of a cycle to positions 4, 12, 20, 32, and 40 yields for a second time, at the bottom diagram, the binarization at positions 3, 12, 21, 33, and 39 with duration pattern [3-3-4-2-4]. Note that it is not necessary to rotate the fume–fume one complete revolution around the circle. By the rotational symmetry of the 16-pulse cycle, once a particular binarization is obtained for a second time, the set of binarizations thus far observed will be repeated. Indeed, the fourth pattern in the series is already a rotation of the first, with duration pattern [3-2-4-3-4]. Even so, since each rotation always takes the fume–fume to a position in which at least one of its onsets coincides with a pulse, the reader may wonder if some binarizations could be missed by not considering the rotations for which no onsets lie on a pulse and none lie halfway between two consecutive pulses. A rhythm with these properties is said to be in *general position*. By contrast, a configuration of a rhythm in which at least onset coincides with a lattice pulse will be called *anchored*.

First note that if the fume–fume is in general position, each onset will have its nearest pulse located either ahead of it (clockwise) or behind it (counterclockwise). Now assume that the rhythm is rotated in a clockwise direction by some distance d^*, which is small enough so that no onset crosses over a pulse. Then each onset will be rotated by that same distance d^*. Furthermore, the distances of those onsets with clockwise nearest pulses will decrease by d^*, and those of onsets with counterclockwise nearest pulses will increase by d^*. Therefore, for rhythms with an odd number of onsets, such as the fume–fume, the sum of the increases cannot cancel out the sum of the decreases. This means that rotating in one of the two directions will cause the distance between the rhythm and its binarization to decrease, implying that a rhythm in general position cannot realize the minimum distance to its binarization. This means that to search for the nearest binarization to the fume–fume it is sufficient to consider only anchored configurations. Consider for example the top-left rhythm in Figure 12.4. If the fume–fume is rotated clockwise by d^*, the distances will increase by d^* at pulses 16, 28, 36, and 0, and decrease by d^* at pulse 8, for an overall increase of $3d^*$. On the other hand, if the fume–fume is rotated in a counterclockwise direction by the same amount, then the distances will increase by d^* at pulses 0, 36, and 8, and decrease by d^* at pulses 28 and 16 for an overall increase of d^*.

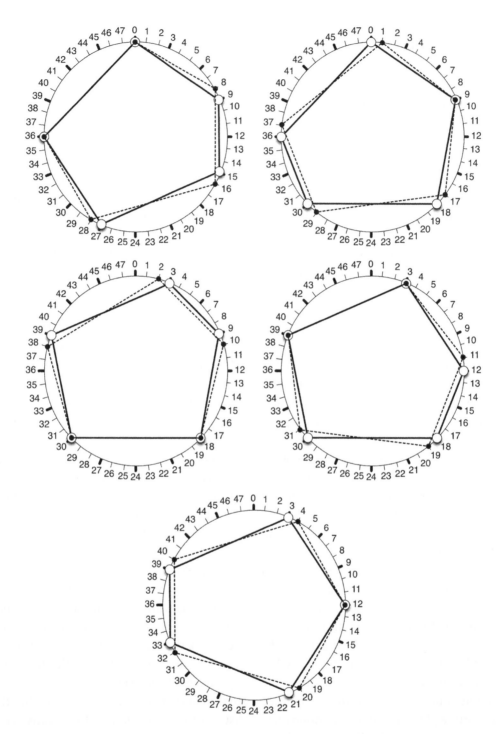

FIGURE 12.4 All the binarizations of the fume–fume (in general position) are obtained with the nearest-pulse snapping rule.

Since rotating the rhythm in both directions increases the distance, we conclude that this binarization yields a local minimum in the distance function. Turning to the binarization in the upper right of Figure 12.4, where the fume–fume has one onset coincident with the lattice pulse at position 9, it may be observed that rotating the fume–fume either clockwise or counterclockwise by d^* causes, in both directions, a distance change of $3d^* - 2d^*$ for a net increase of d^*. Therefore, the *clave son* also realizes a local minimum in the distance function. The smallest of all the local minima obtained in this way is the global minimum that we are looking for, and in this case, it does not correspond to the *clave son*.[8]

Therefore, we conclude that there exist only three different binarizations of the fume–fume, modulo rotation: [3-2-4-3-4], [3-3-4-2-4], and [2-3-4-3-4]. Furthermore, the first and third binarizations may be considered the same in the following sense. The first binarization starting at location zero has durations [3-2-4-3-4], but the third binarization has the same duration pattern [3-2-4-3-4] if it is started at position 18, and is played backwards, that is, in a counterclockwise direction. Therefore, the two binarizations obtained, aside from the *clave son*, are the same modulo rotations and mirror image reflections. The topic of rhythm equivalence with respect to rotations and mirror image reflections will be explored further in Chapter 14.

Given that for the fume–fume rhythm, the nearest-pulse snapping rule yields three binarizations (modulo rotations), it remains to explore geometrical explanations of how the *clave son* came to be chosen as the "authentic" binarization, as claimed by Rolando Pérez Fernández. One compelling argument would be obtained by showing that, of these three binarizations, the *clave son* is the most similar to the fume–fume. This leads us to broach the vast field concerned with measuring the similarity between rhythms, a topic about which volumes has been written, and which we will revisit in more detail in Chapter 36.[9] For now let us consider the simple measure suggested earlier: the sum of absolute values of the differences between the location coordinates of the attacks of the fume–fume and the corresponding snapped attacks of its binarizations.

As pointed out earlier, this measure of distance does not realize its minimum value for the *clave son*, but rather for the two other binarizations, each of which has a distance of three from the fume–fume, whereas the distance to the son is four. One might suspect that the reason the hoped-for answer is not obtained is because this distance measure is not the "right" one for the task at hand. This distance measure is after all a special case of a large general family of measures called Minkowski metrics.[10] Consider two rhythms X and Y consisting of d attacks each. Let the circular arc coordinates of the attacks of rhythm X be $x_1, x_2, \ldots, x_d$, and of rhythm Y be $y_1, y_2, \ldots, y_d$. Then the Minkowski metric of order p between rhythms X and Y is given by

$$d_p(X, Y) = \left(\left| x_1 - y_1 \right|^p + \left| x_2 - y_2 \right|^p + \cdots + \left| x_d - y_d \right|^p \right)^{1/p}$$

where $1 \leq p \leq \infty$. In other words, the discrepancies between each corresponding pair of attack points are raised to the power of p, added together, and finally the pth root of the resulting sum is taken. For the case when $p = 1$, and $d = 5$, the Minkowski metric reduces to the distance measure used above, i.e., the sum of the absolute values of the five differences. When $p = 1$, the Minkowski metric is known by several popular names including *Manhattan* metric, *city-block* distance, and *taxi-cab* distance.[11] One might wonder if for a value of p different from one, the Minkowski measure of distance would favor the *clave son* binarization over the other two. The two most popular alternate values of p used in practice are 2 and ∞. When $p = 2$, the Minkowski metric becomes the ubiquitous Euclidean distance, and for $p = \infty$, it is called the *sup* metric[12] because it can be shown[13] that as p becomes infinitely large

$$\left(\left| x_1 - y_1 \right|^p + \left| x_2 - y_2 \right|^p + \cdots + \left| x_d - y_d \right|^p \right)^{1/p}$$
$$= \max \left\{ \left| x_1 - y_1 \right|, \left| x_2 - y_2 \right|, \ldots, \left| x_d - y_d \right| \right\}.$$

These three Minkowski metrics have the following natural geometric interpretations illustrated in Figure 12.5, where it is desired to measure the distance between two points A and B. The popular Euclidean distance between A and B is of course the length of the straight line from A to B. If A and B were two street corners in Manhattan, then with a helicopter one could travel such a straight line over the buildings. The city-block distance between A and B is the sum of the horizontal length AC plus the vertical length CB. This corresponds to the minimum distance a car would travel along two-way streets and avenues, to get from A to B. The *sup metric* distance with $p = \infty$ is the maximum of the horizontal length AC and the vertical length CB. In this case, the maximum is AC. There exist applications where this measure of distance is preferred over the others for different reasons. Examples include classification in data mining,[14] efficient warehousing in operations research, a class of problems similar to computing the shortest time for a

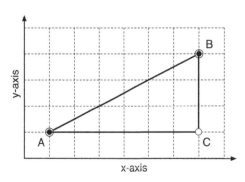

FIGURE 12.5 The Minkowski distances between points A and B with $p = 1, 2,$ and ∞.

mechanical plotter to draw a figure,[15] and optimizing geometric reconstruction problems in computer vision.[16]

Some insight concerning the nature of Minkowski metrics may be obtained by examining the shape of their unit "circles." Recall that a unit circle with center O is the locus of points that are at a distance one from O. With the Euclidean distance, the unit circle is the round circle we are so familiar with. Some unit "circles" for several Minkowski metrics are shown in Figure 12.6. The unit circle for $p = \infty$ is a square; for $p = 1$, it is a diamond; and for other values of $p > 1$ it is a convex curve that lies somewhere in between the square and diamond.

Tenney and Polansky (1980) experimented with the Euclidean and city-block distance measures for combining different features of music, and write that: "A definitive answer to the question as to which of these metrics is the most appropriate to our musical 'space' would depend on the results of psychoacoustic experiments."[17] However, in the absence of such experimental knowledge, they argue that the city-block distance is preferable since it treats all dimensions equally. Furthermore, the results they obtained with the city-block distance were superior to those observed with the Euclidean distance. Similar results favoring the city-block distance have been obtained in the visual domain.[18] Indeed, the "Householder-Landahl" hypothesis states that the city-block distance captures the dissimilarity judgments that human subjects make when comparing physical shapes.[19]

So how do these other Minkowski metrics compare in the case of our problem of finding which of the three anchored binarizations of fume–fume is closest to the *clave son*? From Figure 12.4, we can observe that the Euclidean distance between the fume–fume and the *clave son* is $[(0)^2 + (1)^2 + (1)^2 + (1)^2 + (1)^2]^{1/2} =$

$[4]^{1/2} = 2$. On the other hand, the Euclidean distance between fume–fume and the other two binarizations is $[(0)^2 + (1)^2 + (1)^2 + (1)^2 + (0)^2]^{1/2} = [3]^{1/2} = 1.732$. So here again, the *clave son* loses out over other binarizations.[20]

It remains to examine the Minkowski metric for $p = \infty$ the sup metric. Again, referring to Figure 12.4, we observe that all the three binarizations are equally distant from fume–fume, since max{0, 1, 1, 1, 1} = max{0, 1, 1, 1, 0} = 1. So, for this metric, the three binarizations are tied, showing some promise. Although the son is not favored over the two other binarizations, it is also not excluded, and we can break the tie by comparing their rhythmic contours. Both fume–fume with duration pattern [2-2-3-2-3] and the *clave son* with duration pattern [3-3-4-2-4] have the same rhythmic contour [0, +, −, +, −]. On the other hand, the two other binarizations of the fume–fume [3-2-4-3-4] and [2-3-4-3-4] have rhythmic contours [−, +, −, +, −] and [+, +, −, +, −], respectively, both of which differ on their first symbol with the son and fume–fume. Furthermore, this first symbol is arguably the most important symbol, since it determines the first two interonset intervals that we hear, and it is crucial that these two intervals have the same duration.

To summarize, in attempting to model the perceptual mechanism by which ternary rhythms could be converted to their binary counterparts, two different geometric mechanisms have been uncovered for converting the ternary fume–fume to the binary *clave son*. The first model (Figure 12.2) assumes that the ternary rhythm starts at its normal position, with the first onset at pulse zero, and snaps the remaining onsets, as they unfold, to their anticipated (clockwise) binary pulse positions, to obtain the *clave son* in a direct manner. The second model (Figure 12.4) assumes that the ternary rhythm starts a little later, and snaps the remaining onsets, as they unfold, to their nearest binary pulse positions, by using the sup metric as the distance, leading to an ambiguity that is resolved by comparing the durations of the first two interonset intervals. Which of these geometric models best fits the perceptual mechanism at work will, in the end, have to be determined by psychological experiments. However, if the rule of Occam's razor is invoked, then the simpler first model should be selected. The second model, in addition to being rather complex, has an awkward feature such that in order to compute the maximum of five onset discrepancies the listener has to wait until the entire rhythm is heard. By contrast, the first model allows the listener to project the anticipated binary locations on the fly.

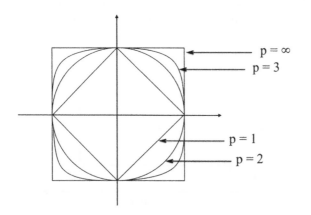

FIGURE 12.6 The Minkowski metric unit disks for four values of p.

NOTES

1 Jacoby, N. & McDermott, J. H., (2017) performed experiments in which subjects were asked to reproduce random rhythms that were fed back to them, and which mutated over time to reflect their internal cultural and experiential biases, suggesting that "priors on musical rhythm are substantially modulated by experience." See also Desain, P. & Honing, H., (2003), which provides psychological evidence that metric priming with old rhythms will cause perception of new unfamiliar rhythms to be perceived as familiar rhythms.

2 Johnson, H. S. F. & Chernoff, J. M., (1991), p. 68.

3 Pérez-Fernández, R. A., (1986), p. 105. That the fume–fume was binarized to the *clave son* when it emigrated to the new world from West Africa is only one possible historical scenario. It is also possible that the *clave son*, already established in Bagdad in the thirteenth century, was exported to West Africa, where it mutated to the ternary fume–fume. Yet, a third possibility is that the binary and ternary forms of this pattern were born independently in different places.

4 In the context of music transcription systems, Nauert, P., (1994), p. 229, calls this type of snapping *fixed quantization*. For a completely different approach to quantization that uses Bayesian decision theory, see Cemgil, A. T., Desain, P., & Kappen, B., (2000). For a recent original and promising quantization algorithm that uses near division see: Murphy, D., (2011).

5 Loy, G., (2007), p. 31. The snapping operation is a special case of the more general concept of rhythm quantization. See Gómez, F., Khoury, I., Kienzle, J., McLeish, E., Melvin, A., Pérez-Fernandez, R., Rappaport, D., & Toussaint, G. T., (2007), for a more detailed analysis of snapping rules applied to binarization of ternary rhythms and ternarization of binary rhythms.

6 Cargal, J. M., (1988a), Chapter 3, p. 2.

7 Any smaller but positive rotation will also yield the *clave son* as a binarization.

8 The general problem of finding such a matching between two sets of points on a line or circle is called a bijection in computer science, and has received a lot of attention in the field of operations research, and more recently, music theory as well. Werman, M., Peleg, S., Melter, R., & Kong, T. Y., (1986), were the first to find the optimal solution described here, i.e., the rotation and matching that minimize the sum of absolute values of the differences between all pairs of matched points may be obtained by restricting the search to configurations in which pairs of points from both sets coincide. Chen, H. C. & Wong, A. K. C., (1983), actually used the same procedure a few years earlier, but believed it was a suboptimal approximation. There is a difference worth pointing out between the minimum matching sought by these authors and the problem in this chapter. In their problem, the set that is shifted never changes. However, in the binarization problem considered here, the set of points for which a matching is sought changes during the shift according to how the nearest pulses change. Therefore, their problem applies to those rotation intervals for which the nearest pulses remain fixed. For the application of these concepts to the theory of musical chords and voice leading, the reader is referred to Tymoczko, D., (2006), and the references therein.

9 For a sampling of literature on measuring rhythm similarity, refer to the following papers. Antonopoulos, I., Pikrakis, A., Theodoridis, S., Cornelis, O., Moelants, D., & Leman, M., (2007), Bello, J. P., (2011), Berenzweig, A., Logan, B., Ellis, D. P. W., & Whitman, B., (2004), Gabrielson, A., (1973a, 1973b), Guastavino, C., Gómez, F., Toussaint, G. T., Marandola, F., & Gómez, E., (2009), Hofmann-Engl, L., (2002), Orpen, K. S. & Huron, D., (1992), Post, O. & Toussaint, G. T., (2011), Takeda, M., (2001), Toussaint, G. T., (2006b), and Toussaint, G. T., Campbell, M., & Brown, N., (2011). Polansky, L., (1996), provides a survey of a plethora of similarity metrics for use in music. A general theory of similarity of chords based on submajorization, recently developed by Hall, R. W. & Tymoczko, D., (2012), may have interesting consequences for measuring rhythm similarity as well. Tversky, A., (1977), considers the broader issue of similarity as a psychological construct.

10 Beckenbach, E. & Bellman, R., (1961), p. 103 and Toussaint, G. T., (1970).

11 Krause, E. F., (1975) and Reinhardt, C., (2005).

12 Schönemann, P. H., (1983), p. 314.

13 Beckenbach, E. & Bellman, R., (1961), p. 105.

14 Anand, A., Wilkinson, L., & Tuan, D. N., (2009).

15 Langevin, A. & Riopel, D., Eds., (2005), p. 96. Consider a computer plotter that has to make a large complicated technical drawing with many lines consisting of start and end points. The ink-head must travel to all start points and trace the lines until the end points are reached. The ink-head is attached to motors on the sides of the table by cables. The motors pull the ink-head at a constant speed along both horizontal and vertical directions. Therefore, the time taken for the ink-head to travel from point A to point B is determined by the maximum of the horizontal and vertical directions that the ink-head must travel.

16 Hartley, R. I. & Schaffalitzky, F., (2004), compare the sup metric with the Euclidean distance to solve the computer vision problem of motion recovery from omnidirectional cameras. They found that the sup metric had the advantage of lower computational cost, but the drawback of being too sensitive to outliers in the data.

17 Tenney, J. & Polansky, L., (1980), p. 213.

18 Sinha, P. & Russell, R., (2011), tested perceptual image similarity judgments of human subjects and found a consistent preference for images matched with the city-block distance over the Euclidean distance, leading them to conclude that this "metric may better capture human notions of image similarity."

19 Schönemann, P. H., (1983), p. 312.
20 Note that the Euclidean and sup metrics are calculated here with respect to the anchored binarizations which realize the local minima of the city-block distance function. It has not yet been determined whether using the Euclidean and sup metrics in the quantization algorithm will generate binarizations of the fume–fume that are not generated by city-block distance quantization.

CHAPTER 13

Syncopated Rhythms

METRICAL COMPLEXITY

"SYNCOPATION IS THE PIQUANT IN RHYTHM."[1] It adds surprise to an otherwise bland rhythm, and has been shown to elicit pleasure and the desire to move, a behavior that has been called "groove."[2] We can quite easily feel when a rhythm has syncopation, but translating that feeling to mathematical terms (my goal with *all* musical properties explored in this book) is easier said than done. A prerequisite for making progress in this direction is a precise constructive definition of the musical properties we are exploring. Consider how some dictionaries explain what syncopation is. The *New Oxford American Dictionary* defines a *syncopated* rhythm as one in which the "beats or accents" are displaced "so that strong beats become weak beats and vice versa." From the mathematical point of view, this definition is not satisfactory because the notions of "strong" and "weak" beats have not been defined. The Oxford Grove Music Online dictionary defines syncopation as "The regular shifting of each beat in a measured pattern by the same amount ahead of or behind its normal position in that pattern." This definition also lacks mathematical rigor because the notion of "normal" has not been specified. The Harvard Dictionary of Music defines syncopation as "A momentary contradiction of the prevailing meter."[3] This definition assumes we know what meter is, but more problematically, how do we interpret the words 'momentary' and 'prevailing'? As a final example, consider the lesser known online Virginia Tech Multimedia Music dictionary; it defines syncopation as the "deliberate upsetting of the meter or pulse of a composition by means of a temporary shifting of the accent to a weak beat or an offbeat." Does this mean that if the shifting of the weak beat is not carried out deliberately, there is

no syncopation? Furthermore, what is the difference between a weak beat and an offbeat? Such imprecise definitions hinder progress in a mathematical direction.

There must be more than 50 traditional definitions of syncopation adorning the pages of dictionaries, books, and Internet sites. Like the definitions offered here, most have their own particularities. However, to a mathematician, they all have one thing in common: vagueness. Of course, this is not surprising considering that we are trying to define with precise mathematical tools a slippery human perceptual judgment of an imprecise concept. We can all read text without effort, which implies we recognize characters such as A, B, C, etc., without difficulty. However, we do not know how to define what an "A" or a "B" is. This is a major problem in artificial intelligence.[4] Indeed, in the words of Michael Keith, "although syncopation in music is relatively easy to perceive, it is more than a little difficult to define precisely."[5] Some readers may believe that the desire for mathematical precision is completely inappropriate here, and that such demands lead inevitably to irrelevance regarding the psychological aspects of music. On the contrary, I concur with the philosopher Mario Bunge that we should mathematize everything we can, and the only way to know if a fuzzy concept can be successfully modeled mathematically is to try.[6] These are the *sine qua non* and hallmarks of artificial intelligence, which push the boundaries of the relevant psychology of music. In spite of the difficulties that such a task may pose, it is possible to construct unambiguous mathematical definitions of notions that may be used as useful models that bolster or even supplant the traditional concept of syncopation. Furthermore, in due time, as the scientific and technological approaches of the study of music continue to expand, some of these mathematical versions of syncopation may completely replace the traditional notion. Indeed, it has also been suggested that it might be advantageous to replace the notion of *syncopation* with that of *rhythm complexity*.[7] Syncopation is very much a Western concept, and for some types of music, new mathematical substitutes for syncopation, which are independent of culture, may be more appropriate and useful.[8] Referring to sub-Saharan music, Simha Arom states the case more bluntly: "terms such as … syncope … should be dispensed with as foreign to it."[9] Jay Rahn reflects this evolving terminology by offering two definitions of syncopated: one descriptive of the Western culture and another mathematically inspired.[10] His first definition of syncopated is "deviating from an oriental metrical organization in that one or both of the immediately adjacent presented moments to a given moment is not resolved." Here the term *moment* is used to mean an "irreducible portion of time." His second definition of syncopated is "not commetric." The term commetric here is synonymous with *regular*, a simple and well-defined mathematical notion. Thus, Rahn's second definition of a syncopated rhythm is one that is irregular.

In 1996, Fred Lerdahl and Ray Jackendoff published a book titled *A Generative Theory of Tonal Music* (GTTM), in which they proposed a hierarchy of accents for musical rhythm inspired by research work in linguistics.[11] For a timeline of 16 pulses, their hierarchy of accents or metrical weights may be expressed using the graph shown in Figure 13.1. One way to construct this graph is as follows. First, starting at pulse zero, and proceeding from left to right, assign a weight of one to every pulse (shown as shaded boxes). Second, in a similar manner, increment by one the weight of every second pulse. Third, increment by one every fourth pulse. Next, increment by one every eighth pulse, and finally every

sixteenth pulse. The resulting height of the column at any pulse location gives the weight or degree of accent given to an onset that occurs at that pulse location. In other words, the pulse emphasized most strongly is pulse zero with a weight of five. The next most salient pulse is number eight with a weight of four. Pulses 4 and 12 have a weight of three; pulses 2, 6, 10, and 14 have a weight of two; and all the remaining (odd-numbered) pulses have a weight of one.

This metrical hierarchy may be used to design a precise mathematical definition of syncopation, that we shall call *metrical complexity*, as follows.[12] Consider the *clave son* timeline shown in Figure 13.2 in box notation directly below the metrical hierarchy. The *clave son* consists of onsets at pulses 0, 3, 6, 10, and 12, with metrical weights equal to five, one, two, two, and three, respectively. These metrical weights express how normal or typical it is for a beat to occur at that pulse location according to the theory of Western music practice expressed by Lerdahl and Jackendoff. Therefore, the lower the weight is for an onset, the more unexpected the onset is, and thus the more syncopated it is as well. For the *clave son*, the onset with the lowest weight is the second onset occurring at pulse three that has a weight of one. Therefore, this onset is considered to be the most syncopated of the five onsets. Interestingly, in some Latin music such as *salsa*, a rhythm that accentuates this second onset of the *clave son* is called *bombó*, also the name of a bass drum used in the Afro-Cuban *comparsa* music that is played in carnivals.[13]

To measure the total metrical expectedness (or *simplicity*) of the rhythm, we may add the metrical weights of all its onsets. Thus, for the *clave son*, the metrical expectedness is equal to 13. To convert this measure to a measure of *metrical complexity* or *syncopation*, it suffices to subtract the metrical expectedness value of a given rhythm with k onsets and n pulses from the maximum

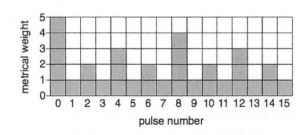

FIGURE 13.1 The GTTM metrical hierarchy of Lerdahl and Jackendoff.

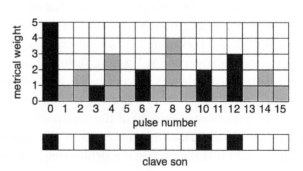

FIGURE 13.2 The GTTM metrical complexity of the *clave son*.

possible value that any rhythm with k onsets and n pulses may have. For a rhythm with 5 onsets and 16 pulses, the maximum expectedness value is 17, obtained by summing the column heights at pulses 0, 8, 4, 12, and any one of 2, 6, 10, and 14. This value is realized by several rhythms, including the popular classical music ostinato rhythm [4-4-2-2-4] with onsets at pulses 0, 4, 8, 10, and 12. Thus, the metrical complexity of the *clave son* is $17 - 13 = 4$. For comparison, the more syncopated clave rumba that has its third onset at pulse number 7 has a metrical complexity equal to $17 - 12 = 5$.[14]

KEITH'S MEASURE OF SYNCOPATION

Michael Keith proposed a mathematical measure of syncopation in the context of sustained musical notes that is defined by onsets as well as offsets.[15] Recall that the onset is the point in time at which a note starts sounding, and an offset is the point at which it stops sounding. Keith's definition of syncopation is based on three types of events he calls *hesitation*, *anticipation*, and *syncopation*.[16] Although in the strict sense he reserves the term *syncopation* for the more limited situation in which a note exhibits both hesitation and anticipation, his general measure of syncopation encompasses a weighted combination of all three events. Figure 13.3 illustrates these events for the special case in which a rhythmic cycle (or measure) contains four fundamental beats at pulses zero, two, four, and six. The leftmost illustration shows an example without syncopation: the note starts and ends at two fundamental beats, in this case, zero and two, respectively. The second diagram shows an example of *hesitation*: a note starts on a fundamental beat and ends *off the beat*, or in between two fundamental beats, here pulses zero and three, respectively. The third diagram, in which the note starts in between two beats (*off the beat*) and ends on a beat, here pulses one and four, respectively, is called *anticipation*. Finally, the rightmost diagram exhibits an example of genuine *syncopation*: both the start and end points of the note occur in between two beats, here at pulses one and three, respectively.

To construct his weighted general measure of syncopation, Keith assigns to *hesitation* a weight of one. He considers anticipation to be a stronger form of syncopation than hesitation, and therefore gives *anticipation* a weight of two. Finally, since syncopation combines both hesitation and anticipation, he adds these two weights together to obtain a weight of three for *syncopation*. It remains to define precisely what Keith means for an onset or offset to be "off the beat." To keep things simple, Keith restricts the definition of "off the beat" to hold only for cycles in which the total number of pulses n is a power of two, such as $n^2 = 4$, $n^3 = 8$, $n^4 = 16$, and $n^5 = 32$. This power is a parameter called d, and it is chosen so that n^d is small enough to be able to identify the smallest interonset interval (IOI) necessary to specify the granularity (resolution, elementary pulse) of the rhythm. For example, the tresillo has intervals [3-3-2], which makes $n = 8$ and $d = 3$. On the other hand, the *clave son* has intervals [3-3-4-2-4], making $n = 16$ and $d = 4$.

Let δ denote the duration of a note (in terms of the number of pulses that occur from onset to offset) as a multiple of $1/2^d$, and let S be the time coordinate at which the note starts (the onset). Furthermore, let D denote the value of δ rounded down to the nearest power of two. Then the onset of the note is defined to be "off the beat" if S is not a multiple of D. Similarly, the offset of the note is defined to be "off the beat" if $(S + \delta)$ is not a multiple of D. The syncopation value for the ith note in the rhythmic pattern, denoted by s_i is defined as: $s_i = 2$ (if the onset is off the beat) $+ 1$ (if the offset is off the beat). Finally, the overall measure of syncopation of the rhythmic pattern is the sum of the syncopation values s_i summed over all i.

Keith's measure of syncopation is defined in the context of sustained notes that start and end at positions

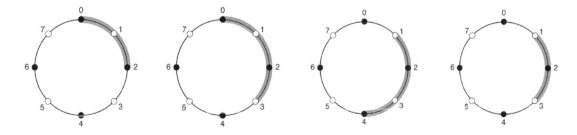

FIGURE 13.3   From left to right, no syncopation, *hesitation*, *anticipation*, and *syncopation*.

anywhere in the cycle. In this book, on the other hand, the rhythmic patterns consist of sounds that have extremely short durations, and that for all practical purposes, as well as theoretical analysis, are considered as attacks with zero duration. Therefore, it is assumed here that the offsets do not exist, and as a consequence, no offset can be "off the beat". This implies that the second term in Keith's syncopation value for a "note" may be dropped altogether. Furthermore, the weight of two for the first term may now be changed to one, since it is no longer necessary to emphasize that an *onset* that is "off the beat" is twice as important as an *offset* that is "off the beat." This simplifies the computation of the syncopation value of a rhythmic onset: it is one if the onset is "off the beat," and zero otherwise.

It is instructive as an example to walk through these computations for two ubiquitous rhythms, the distinguished and syncopated *clave son* introduced in Chapter 6, with interval structure [3-3-4-2-4], and the simpler (less syncopated) rhythm with interval structure [4-4-2-2-4], a classical music ostinato also used in traditional and popular music. It is also the rhythm of a prominent protest rallying call chanted in demonstration marches in several countries. Before diving into the mathematical exercise involved in the computation of Keith's syncopation measure, let us slow down and get to know this second rhythm [4-4-2-2-4] a little better. Figure 13.4 (left) shows the rhythm in polygon notation. This pattern of IOIs is a special case of the more general rhythm with interval structure [2d-2d-d-d-2d], where d is an integer that represents an IOI of duration d, which in effect determines the tempo or the type of the resulting rhythm (such as binary or ternary). For the case d = 1, the rhythm reduces to the energetic binary eight-pulse rhythm [2-2-1-1-2], a rhythm played in Rumanian folk dance[17] and Peking Opera.[18] This rhythm is also a popular accent pattern of protest chants, perhaps rising to prominence during the protest marches against the Vietnam War during the 1960s, with the chant: "Hell, no, we won't go!" The first half [2-2] and second half [1-1-2] are the ancient Greek rhythmic forms called the *spondee* and *anapaest*. According to C. F. Abdy Williams, "the spondee was suitable for solemn hymns to the gods; the anapaest, used especially for marches, induced energy and vigour."[19] Perhaps then it is not surprising that the [2-2-1-1-2] rhythm was adopted for demonstration marches. For the case d = 3, the pattern is the slow ternary 24-pulse rhythm [6-6-3-3-6], which is the rhythm timeline of the song *Mujiba* performed by the South African band *Amampondo*, played on the *shekere*.[20] The shekere is a percussion instrument made from a dried calabash gourd enveloped in a beaded fishnet (an illustration of the instrument is given in Figure 17.4 of Chapter 17). For the case d = 2, the pattern is the medium-tempo binary 16-pulse rhythm with duration structure [4-4-2-2-4], represented by bold-line polygons connecting pulses 0, 4, 8, 10, and 12, in all three diagrams of Figure 13.4. The rhythm on the left corresponds to the *Reggae* protest song "*Get Up, Stand Up*" written by Bob Marley and Peter Tosh in 1973, punctuating it with the phrase: "**get** up, **stand** up, **stand** up **for** your **right**," where the words in bold correspond to the beats at positions 0, 4, 8, 10, and 12.[21] The rhythm in the center diagram has two pickup grace notes before the downbeats at positions 0, 4, and 8, shown as gray-filled circles. This is the classical music accent pattern: *da da **dum**, da da **dum**, da da **dum dum dum**,* (the *da* being the gray-filled pickup grace notes, and the ***dum*** being the downbeats), of the *Finale* of the *William Tell Overture* by Giaochino Rossini.[22] The rhythm in the right diagram shows grace notes (gray-filled circles)

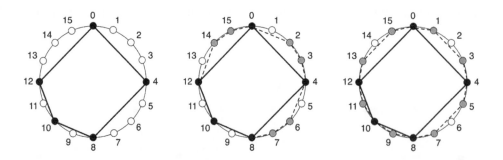

FIGURE 13.4   The ubiquitous [4-4-2-2-4] rhythm (left), with double pickup notes (center), and with grace notes on both sides of the downbeats (right).

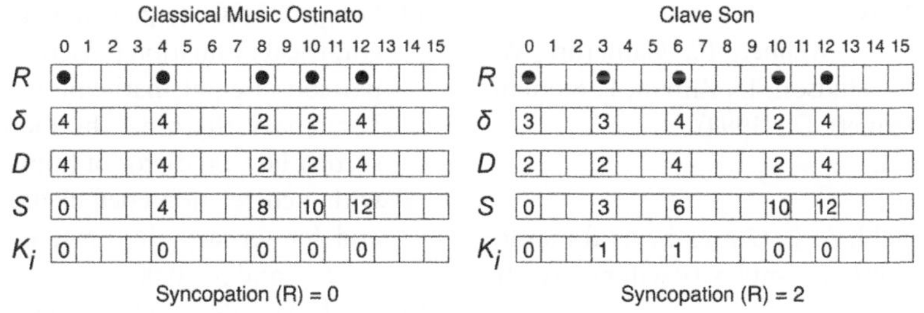

Classical Music Ostinato

	0	1	2	3	4	5	6	7	8	9	10	11	12	13	14	15
R	●				●				●		●		●			
δ	4				4				2		2		4			
D	4				4				2		2		4			
S	0				4				8		10		12			
K_i	0				0				0		0		0			

Syncopation (R) = 0

Clave Son

	0	1	2	3	4	5	6	7	8	9	10	11	12	13	14	15
R	●			●			●				●		●			
δ	3			3			4				2		4			
D	2			2			4				2		4			
S	0			3			6				10		12			
K_i	0			1			1				0		0			

Syncopation (R) = 2

FIGURE 13.5 Illustrating the calculation of Keith's syncopation measure for the classical music ostinato [4-4-2-2-4] and the *clave son* timeline [3-3-4-2-4].

	Shiko	Son	Rumba	Gahu	Bossa-Nova	Soukous
Keith Syncopation	1	2	2	3	3	4
Metrical Complexity	2	4	5	5	6	6

FIGURE 13.6 Values of Keith's syncopation measure and the GTTM metrical complexity for the six distinguished timelines. Note that both measures are functions that are monotonically nondecreasing.

just before and just after each of the downbeats at positions 0, 4, and 8, 10, and 12. It is the rhythm of the Spanish protest chant: "El pueblo, unido, jamás será vencido." This chant translates to English as: ("The people, united, will never be defeated.").[23] The accents corresponding to the beats at positions 0, 4, 8, 10, and 12, can be seen by splitting up the words into syllables, and highlighting the accents in bold as follows: "El-**pue**-blo, u-**ni**-do, ja-**más** se-**rá** ven-**ci**-do."

Let us now return to the calculations of Keith's syncopation measure for the two rhythms in question. The terms S, D, and δ in Figure 13.5 have already been defined in the preceding. The rhythmic pattern is denoted by R, and K_i represents the syncopation value of the onset present at coordinate value i. First consider the classical music ostinato on the left. Since for all five onsets the value of S is a multiple of D, by zero, one, four, five, and three, respectively, no onset is considered to be "off the beat." Therefore, the syncopation value for each onset is zero, and the overall syncopation value of the rhythm is also zero. On the other hand, in the case of the *clave son* on the right, for the second onset, three is not a multiple of two, and for the third onset, six is not a multiple of four. Therefore, both K_3 and K_6 take on a value of one, and the overall syncopation value of the *clave son* is two. Musicologists would unanimously agree that, in the context of a four-beat underlying meter, the *clave son* is more syncopated that the classical music ostinato, and thus in this particular case Keith's measure correctly captures human judgments.

The values of GTTM metrical complexity and Keith's measure of syncopation for the six distinguished timelines are given in increasing order of complexity from left to right in Figure 13.6. The Spearman rank correlation between these two orderings (with some ties for both measures) is high and statistically significant: $r = 0.894$ with $p < 0.05$. There is also evidence that these measures agree well with human judgments of rhythm complexity.[24] However, these measures have been tested on *monophonic* rhythms or rhythms with only one voice. Real world music by contrast has, most of the time, more than one layer of interacting rhythms. It remains an open problem to find measures of complexity or syncopation that predict human perception of syncopation or complexity for the case of multiple rhythms played simultaneously. Some progress with small datasets has been made in this direction for popular drum rhythms composed of two voices: bass drum and snare drum.[25]

In closing this chapter, it is worth noting that an open-source software toolkit in the Python programming language is now available for computing seven widely used syncopation models, including the two measures described here: metrical complexity and Keith's syncopation measure.[26]

NOTES

1 Gow, G. C., (1915), p. 649.
2 Witek, M. A. G., Clarke, E. F., Wallentin, M., Kringelbach, M. L., & Vuust, P., (2014).

3 Randel, D. M., Ed., (2003).

4 Longuet-Higgins, H. C., Webber, B., Cameron, W., Bundy, A., Hudson, R., Hudson, L., Ziman, J., Sloman, A., Sharples, M., & Dennett, D., (1994).

5 Keith, M., (1991), p. 133.

6 Mahner, M., Ed., (2001).

7 Fitch, W. T. & Rosenfeld, A. J., (2007). Several mathematical measures of rhythm complexity will be investigated in Chapter 17.

8 Vurkaç, M., (2012), for example, finds the mathematical definition of *off-beatness* (to be explored in Chapter 16) more useful than syncopation, for the analysis of rhythms in traditional contexts.

9 Arom, S., (1991), p. 183.

10 Rahn, J., (1983), p. 248.

11 Lerdahl, F. & Jackendoff, R., (1983).

12 This measure of syncopation was first proposed in Toussaint, G. T., (2002). In a subsequent study by Thul, E. & Toussaint, G. T., (2008a), it was compared with a large group of measures of rhythm complexity, irregularity, and syncopation, against human judgments of performance and perceptual complexity, and gave superlative performance. Flanagan, P., (2008), proposes a mathematical measure of syncopation that computes an average with respect to many possible underlying meters.

13 Uribe, E., (1996), p. 49. Therefore, in this music as well as *rumba* styles, it is not uncommon to consider that the rhythmic phrase starts on the second attack of the *clave son*, rather than the first. Morales, E., (2003), p. 174, quotes from I. Leymaire's book *Cuban Fire* that the *bombó* attack "falls on the second quarter note of the second bar." This would seem to contradict Uribe. Morales does not explicitly notate the rhythms, but Leymaire is probably referring to 2–3 versions of *clave son* in which the two bars are reversed as in [. . x . x . . . x . . **x** . . x .], where the *bombó* attack is shown in bold.

14 Song, C., (2014), provides a rigorous theoretical and empirical evaluation (based on listening tests) and comparison of eight mathematical models of syncopation. For binary rhythms, the metrical complexity was the second best predictor of human ratings of syncopation, with a correlation coefficient of $r = 0.92$ and $p < 0.001$, and for ternary rhythms, it was the third best with $r = 0.67$ and $p < 0.001$.

15 Keith, M., (1991), pp. 134–135.

16 For additional types of syncopations and extensions, see Gatty, R., (1912), p. 370, and Smith, L. & Honing, H., (2006).

17 Proca-Ciortea, V., (1969), p. 183.

18 Srinivasamurthy, A., Repetto, R. C., Sundar, H., & Serra, X., (2014). http://compmusic.upf.edu/bo-perc-patterns Accessed December 30, 2017.

19 Abdy Williams, C. F., (2009), p. 27.

20 The song "*Mujiba*" appears in the *Amampondo* album titled "An Image of Africa," released in 1992 by EWM Records (CD AM 24).

21 The song "Get Up, Stand Up" first appeared on The Wailers' 1973 album *Burnin'*. See the May 22, 2017 article "Stand Up for Your Rights!" by M. Romer in ThoughtCo: www.thoughtco.com/bob-marleys-best-protest-songs-3552848. (Accessed July 27, 2017).

22 Grahn, J. A. & Brett, M., (2007), p. 893. The *Finale* in the *William Tell Overture* is also known as the "March of the Swiss Soldiers."

23 "The Rhythm of Revolution: Protest Chants from Egypt to Ecuador," Center for Strategic and International Studies, *Newsletter*, May 17, 2011. www.csis.org/analysis/rhythm-revolution-protest-chants-egypt-ecuador (Accessed July 27, 2017).

24 Thul, E. & Toussaint, G. T., (2008a).

25 Hoesl, F. & Senn, O., (2018).

26 Song, C., Pearce, M., & Harte, C., (2015).

CHAPTER 14

Necklaces and Bracelets

Consider the *bembé* timeline played in the usual sub-Saharan African context of an underlying meter that places an accent at every third pulse starting with pulse zero. In Figure 14.1, these four metrically strong pulses {0, 3, 6, 9} are indicated by the vertical and horizontal lines (a four-beat measure). Note that the *bembé* (left) has attacks on the first and last metric accents at positions zero and nine. Examine what happens when this rhythm is rotated clockwise by one pulse so that the new rhythm starts on the last onset of the *bembé*, as shown in Figure 14.1 (right). The new rhythm contains attacks on the first, second, and third metrically strong pulses at positions zero, three, and six. This is a considerable change, and not surprisingly, if I play this rhythm on a bell with my hands, while playing a bass drum with my foot on pulses {0, 3, 6, 9}, the new rhythm sounds and feels quite different from the *bembé*. Indeed, it still feels considerably different even if I don't play the bass drum, and just mentally partition the cycle into a [3-3-3-3] metric subdivision. Therefore, from the point of view of music making, we may consider that

these two rhythms are different. However, it is obvious that the interval contents of these two durational patterns, and their resulting histograms are identical, since the interval content of a rhythm is invariant to a rhythm's rotation. Therefore, from certain analytical perspectives, the two rhythms may be considered to be the same. In the mathematical field of combinatorics, the two rhythms in Figure 14.1 are said to be instances of the same *necklace*.[1] In the pitch domain in music theory, a necklace corresponds to a *chord type*.[2]

A necklace is a closed string of beads (or pearls) of different colors, such as one might wear around one's neck. We are interested here in *binary* necklaces, i.e., necklaces with pearls of two colors: black and white. Two necklaces are considered to be the same if one can be *rotated* so that the colors of its beads correspond, one-to-one, with the colors of the beads of the other necklace. Figure 14.2 shows two more instances of identical necklaces. The rhythm on the right is obtained by rotating the one on the left clockwise by three pulses.

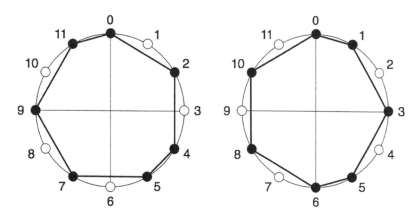

FIGURE 14.1   The *bembé* timeline (left) and its clockwise rotation by one pulse (right).

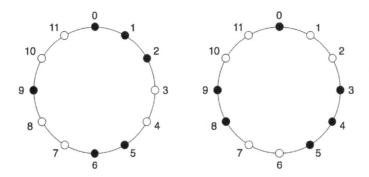

FIGURE 14.2 Two instances of the same *necklace.*

Obviously, it is possible to have two rhythms that are not instances of the same necklace and that still have the same interval content, namely, if one rhythm is the mirror image of the other. To include such cases, we use the mathematical term: *bracelet.* In other words, two bracelets are considered to be the same if one of them can be *rotated or turned over* so that the colors of their beads are brought into one-to-one correspondence. Figure 14.3 shows two rhythms that are not the same necklace, but they are the same bracelet: the rhythm on the right is a mirror image reflection about a vertical axis through pulses zero and six, of the rhythm on the left. In the pitch domain in music theory, a bracelet corresponds to a *chord.*[3] This chapter presents some of the most well-known rhythmic necklaces and bracelets used as timelines in music around the world.

One way to measure the robustness of the effectiveness of a necklace as a template for the design of rhythm timelines is by the number of its rotations that are actually used in practice. If many rotations are used, it suggests that the effectiveness of the rhythms it generates does not depend crucially on the starting onset, even though the result may sound quite different. A rhythm necklace that has the property that all its onset rotations are used as timelines in practice will be called a *robust* rhythm necklace.

The tresillo timeline, already introduced in Chapter 3, with durational pattern [3-3-2] shown in Figure 14.4 (Rotation 0 on the left) is one instance of a robust rhythm necklace. Note that Rotation 2 is also the mirror image reflection of Rotation 0 about a vertical line through pulses zero and four. The tresillo timeline is used so often in traditional music all over the world that it may be regarded as a *universal* rhythm or cultural *meme.*[4] In India it is one of the *talas* used in Carnatic music; in Central Africa, it is played by the Aka Pygmies with blades of metal, and in the United States it is a common rhythm played on the banjo in *bluegrass* music. Musicologists consider it to be a signature rhythm of Renaissance music. Historically, it can be traced back to at least classical Greece under the name of *dochmiac* rhythm. Its Rotation 1 in the center of Figure 14.4 is common in Bulgaria, Turkey, and Korea, and its Rotation 2 (right) is the Nandon Bawaa bell pattern of the Dagarti people of Ghana. The Adowa rhythm of the Ashanti people of Ghana incorporates both of these rotations, played on two hourglass-shaped talking drums (*donno*). Rotation 0 with intervals [3-3-2] is played on Donno-1

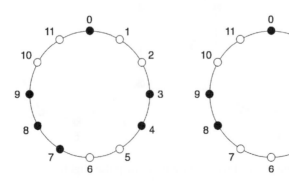

FIGURE 14.3 Two instances of the same *bracelet.*

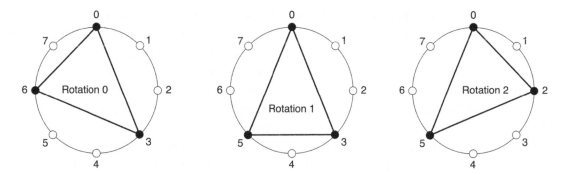

FIGURE 14.4 The *tresillo* timeline and its two onset rotations.

and Rotation 2 is played on Donno-2.[5] The resultant rhythm of the two donno rhythms played in unison is the cinquillo rhythm given by [x . x x . x x .]. It is also found in other places such as Namibia in Southern Africa and Bulgaria in Europe. By far, the most preferred of these rotations is Rotation 0. This can be explained by the fact that, unlike the other two, it creates in the listener an interesting broken expectation of a regular rhythm by starting with the pattern [3-3]. Although this necklace pattern is not as common in Southeast Asia as in the rest of the world, Rotation 0 and Rotation 2 are found as drum patterns in Burma (Myanmar) and Cambodia.[6]

A robust rhythm necklace with four onsets among nine pulses is pictured in Figure 14.5. Rotation 0 is the Turkish *aksak* rhythm also found in Greece, Macedonia, and Bulgaria. Simha Arom made the surprising discovery, in one of his many excursions to Africa, that this rhythm is the timeline of a lullaby used in southwestern Zaïre. It is rather unusual to find nine-pulse rhythm timelines in sub-Saharan Africa. This traditional rhythm necklace has been incorporated into jazz as well as modern art music in the twentieth century. Dave Brubeck used this pattern as the meter in one of his best-selling compositions *Rondo a la Turk*.[7] Rotation 1 is used in Serbia as well as Bulgaria. Rotation 2 is common in

Bulgaria, Macedonia, and Greece. Rotation 3 is used in the traditional music of Turkey, and in modern Western music as the meter in *Strawberry Soup* composed by Don Ellis.

Although in the traditional music of Turkey, all four of these rhythms are employed as timelines, there exists nevertheless a marked order of preference in terms of the frequency with which each pattern is used in practice that has been statistically observed and documented by ethnomusicologists. The most frequent of these is Rotation 0, followed in decreasing order by Rotations 3, 1, and 2. This preference may be explained in terms of Gestalt psychology principles. Rotations 0 and 3 are preferred over the other two, perhaps from the fact that rhythms are most easily perceived as starting or ending with the longest gap, in this case, three pulses. Furthermore, Rotation 0 has a greater surprise value, or in technical terms, a more pronounced Gestalt *despatialization* effect,[8] due to the fact that the initial regular pattern [2-2-2] creates the expectation of the complete cycle [2-2-2-2], which is suddenly broken by the introduction of a three-pulse interval to yield the irregular rhythm [2-2-2-3]. This expectation of a regular rhythm is not induced with Rotation 3, which starts with the irregular pattern [3-2]. This Gestalt despatialization

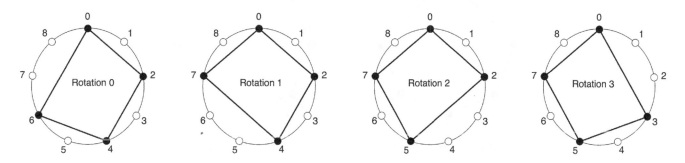

FIGURE 14.5 The *aksak* timeline and its three onset rotations.

effect also happens at the start of Rotation 1 with the pattern [2-2-3], but not with Rotation 2, thus perhaps explaining why the former is preferred over the latter.

Another well-known robust rhythm necklace is the five-onset, eight-pulse pattern shown in Figure 14.6. Rotation zero is the well-known Cuban cinquillo pattern[9], and like the tresillo, it is found in traditional music all over the world. Rotation 1 is a Middle Eastern popular rhythm, also called the *timini* in Senegal, the *adzogbo* in Benin, the *tango* in Spain, and the *maksum* in Egypt. Historically it can be traced back to thirteenth-century Persia, where it went by the name *al-saghil-al-sani*. Rotation 2 is known as the *müsemmen* rhythm in Turkey. Rotation 3 is the *kromanti* timeline, popular in Surinam. Finally, Rotation 4 is the *lolo* timeline played in Guinea.

The cinquillo rhythm in Figure 14.7 (Rotation 0) has syncopation at pulse 4 by virtue that it has a silent beat at

this position. If an onset is added to the cinquillo at pulse 4, we obtain the *Bangu* (*Man-changchui*) rhythm played with a combination of the clappers and a high-pitched drum, in Peking Opera. A rotation of this rhythm by 90° in a clockwise orientation yields the *Bangu* (*Duotuo*) rhythm played with the clappers and drum in Peking Opera.[10]

Figure 14.8 shows one of the most important families of ternary timelines used in sub-Saharan Africa. It consists of five onsets distributed among 12 pulses. Its most well-known representative is Rotation 0, common in West and Central Africa[11] but also used in the former Yugoslavia. In some places, it is called the *fume–fume*, and in others, the *standard short pattern* or the African *signature* rhythm.[12] It is the ternary version of the *clave son* with interonset intervals [2-2-3-2-3]. Rotation 1 is a bell pattern used in the Dominican Republic as well as Morocco. Rotation 2 (a vertical mirror image

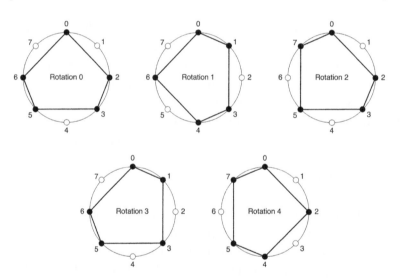

FIGURE 14.6 The cinquillo timeline (Rotation 0) and its four onset rotations.

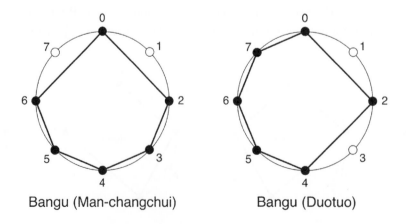

Bangu (Man-changchui)   Bangu (Duotuo)

FIGURE 14.7 Two Peking opera rhythms that are onset rotations of each other.

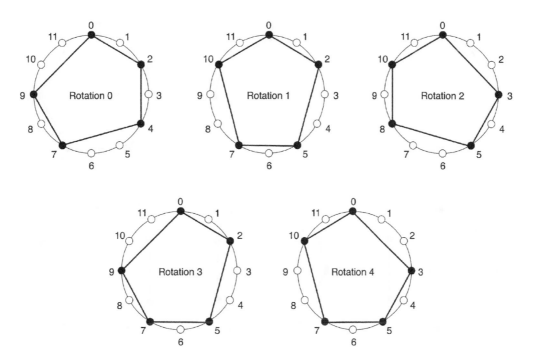

FIGURE 14.8   The ternary version of the *clave son* (fume–fume) and its four onset rotations.

of Rotation 0) is used as a metal blade timeline by the Aka Pygmies of Central Africa. It is also a metric pattern used in Macedonian music, and employed as a hand-clapping rhythm in a children's song by the Venda people of South Africa. Rotation 3 is the *columbia* bell pattern popular in Cuba. It is also the *abakuá* timeline in West Africa,[13] as well as a timeline of the Swahili in Tanzania.[14] Finally, Rotation 4 (a vertical mirror image of Rotation 3) is a bell pattern used by the Bemba people of Zimbabwe as well as a Macedonian dance rhythm.

Of these five rhythms, the most written about in the literature are Rotations 0 and 3.[15] This is probably because these are the only rotations in which their first and last onsets coincide with the first and last beats of a regular four-beat meter [3-3-3-3] at pulses zero and nine, which makes them have closure, be more stable, and easier to mark time for a performance. Furthermore, of these two, Rotation 0 appears to be more popular. This may be due to the surprise value caused by establishing the regular pattern [2-2] at the beginning, causing the listener to anticipate the fourth onset to occur at pulse 6, and then breaking the expectation by delaying this onset to pulse 7, thus introducing an interval of duration three.

Perhaps the most salient necklace in Sub-Saharan Africa is the seven-onset, 12-pulse group of bell rhythms pictured in Figure 14.9. All seven of its onsets are used as starting points for timelines.[16] By far the most important

rhythm is Rotation 0, which corresponds to the major scale or the ionian mode of the diatonic scale. As we have already remarked in previous chapters, this rhythm, denoted by [x . x . x x . x . x . x], is (internationally) the most well-known of all the African timelines. Indeed, the master drummer Desmond K. Tai has dubbed it the *standard pattern*, and it also goes by the name African *signature tune*. In West Africa, it is found under various names among the Ewe and Yoruba peoples. In Ghana, it is the timeline played in the *agbekor* dance rhythm found along the southern coast of Ghana,[17] and in the *agbadza*, as well as the *bintin* rhythms. Among the *Ewe* people, this rhythm is a bell pattern used in the *adzogbo* dance music. This standard pattern is one of the five Gamamla bell patterns played on the *gankogui*, with the first note played on the low-pitched bell, and the other six on the high-pitched bell. The same is done in the *sogba* and *sogo* rhythms. It is played in the *zebola* rhythm of the *Mongo* people of Congo, and in the *tiriba* and *liberté* rhythms of Guinea. It is equally widespread in America. In Cuba, it is the principal bell pattern played on the *guataca* or hoe blade, in the *batá* rhythms, such as the *columbia de La Habana*, the *bembé*, the *chango*, the *eleggua*, the *imba-loke*, and the *palo*.[18] The pattern is also used in the *guiro*, a Cuban folkloric rhythm. In Haiti, it is called the *ibo*. In Brazil, it goes by the name of *behavento*. In North America, this rhythm is sometimes called the *short*

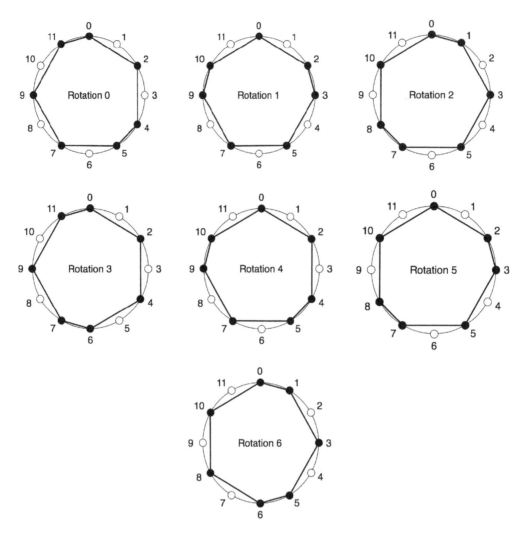

FIGURE 14.9 The standard seven-onset bell pattern and its six onset rotations.

African bell pattern.[19] Rotation 1 is a rhythm found in Northern Zimbabwe called the *bemba* (not to be confused with the *bembé* from Cuba), and played using axe blades. In Cuba, it is the bell pattern of the *sarabanda* rhythm associated with the Palo Monte cult. Rotation 2 is the *bondo* bell pattern played with metal strips by the *Aka* pygmies of Central Africa. Rotation 3 is a bell pattern found in several places in the Caribbean, including Curaçao, where it is used in a rhythm called the *tambú*, and where originally it was played with only two instruments: a drum and a metallophone called the *heru*.[20] Note that the word tambú sounds like *tambor*, the Spanish word for drum, and heru sounds like *hierro*, the Spanish word for iron. This bell pattern is also common in West Africa and Haiti. In Central Africa, it is called the *muselemeka* timeline, and in North America, it is sometimes called the *long* African bell pattern.[21] Strangely enough, Changuito uses this pattern in what

he calls the *bembé*, thus at odds with what everyone else calls *bembé*, namely the pattern [x . x . x x . x . x . x]. Rotation 4 is a *Yoruba* bell pattern of Nigeria, a *Babenzele* pattern of Central Africa, and a *Mende* pattern of Sierra Leone.[22] Among the Yoruba people, it is also called the *konkonkolo* or *kànàngó* pattern.[23] Rotation 5 is used in Ghana by the *Ashanti* people in several rhythms and by the *Akan* people as a juvenile song rhythm. In Guinea, it is used in the *dunumba* rhythm. It is also a pattern used by the *Bemba* people of Northern Zimbabwe, where it is either a hand-clapping pattern or played by chinking two axe blades together. Rotation 6 is a hand-clapping pattern used in Ghana, South Africa, and Tanzania. It is sometimes played on a secondary low-pitched bell in the Cuban *bembé* rhythm.

Note that Rotation 2 is a mirror image reflection of Rotation 0 about a vertical line through pulses zero and six. The same relation holds for Rotations 3 and 6 as well as Rotations 4 and 5. These rotations are equivalent to playing the rhythms backwards.

In the earlier examples, the numbers of onsets and pulses are relatively small. This appears to be a requirement for a timeline necklace to be robust. As these values become large, the number of rotations also grows, reducing the fraction of these that remain salient.

The rotations discussed in the preceding were called *onset*-rotations because all the rotations had an onset at pulse zero. However, there exist examples of timelines that do not have an onset at pulse zero. Indeed, one not uncommon property of African musical culture is the absence of the first pulse or down beat. A unique example is afforded by the eight *pulse*-rotations of the tresillo timeline [3-3-2] shown in Figure 14.10. This eight-pulse, three-onset necklace is so robust that all pulse rotations (except one) have been documented in African and Diasporic music.[24] Note that the danmyé from Martinique makes use of two rotations of the tresillo:

the pattern second from left in the top row and the leftmost pattern in the bottom row.

Some additional examples of timelines that start on a silent pulse are shown in Figure 14.11. The first on the left is the timeline from the highlife music[25] of West Africa. All three onset rotations of this duration pattern [2-2-4] or [1-1-2] are popular in Afro-Cuban music. A second example is the rotation of the cinquillo pattern with durational pattern [4-2-4-2-4] in a counterclockwise direction by one pulse, as shown in the second diagram of Figure 14.11, which is a rhythm used in a Rumanian dance. The last example is the rotation of the *bembé* rhythm by six pulses as shown in the rightmost diagram of Figure 14.11. This is equivalent to a reflection of the *bembé* about the line through pulses 5 and 11. Note that this rhythm is the complementary rhythm of the five-onset fume–fume. This timeline is a *palitos* rhythm used in the *columbia* style of Cuban *rumba* dance music.

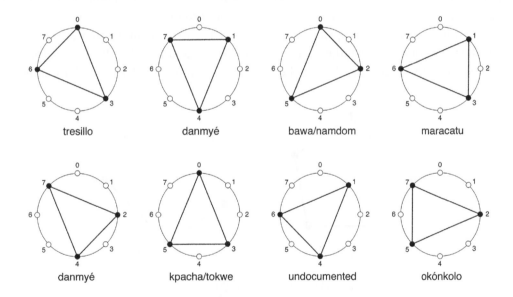

FIGURE 14.10 All eight *pulse*-rotations of the tresillo timeline.

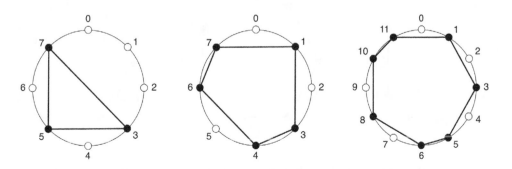

FIGURE 14.11 Some timelines that do not have an onset at pulse zero (anacrusis).

The *clave son* (Figure 14.12 left) is sometimes changed, to the silent first beat version (right), by rotating it by half a measure. These two claves are characterized as having a different *direction*.[26] Such a rotation is equivalent to two mirror image reflections, one about the vertical line through pulses 0 and 8, and the other about a horizontal line through pulses 4 and 12. The result is that both versions retain mirror symmetry about the line through pulses 3 and 11. The left and right versions of the *clave son* are often called the *three-two* and *two-three* claves,[27] as well as the *forward* and *reverse* claves, respectively.[28]

To close this chapter, let us return to the problem of listening to, and perceiving, rhythms in the context of an underlying meter, whether sounded or internalized. At the start of the chapter, we saw an example in Figure 14.1 in which the *bembé* rhythm was rotated in a clockwise direction by one pulse, and the underlying four-beat meter remained constant at positions {0, 3, 6, 9}. The result was that the rotated rhythm sounded very different from the *bembé*. Perhaps you were not too surprised by this effect since the rhythm was rotated. Therefore, consider now the more compelling case in which the *bembé* is not rotated, but rather is heard against three different underlying meters, the four-beat

meter [3-3-3-3], the six-beat meter [2-2-2-2-2-2], and the three-beat meter [4-4-4], as illustrated in Figure 14.13. In the figure, the metric beats are highlighted with a circle, and connected with thin line segments with labels denoting their duration. In this situation, ignoring the meters, the three rhythms are identical. However, if we play the meters on a bass drum while playing the *bembé* rhythm on a bell, the difference in sound and feel between the three renditions of *bembé* is considerable. Note that the resultant patterns obtained by taking the unions of the rhythm and meter attacks are very different. The resultant rhythm for the four-beat meter is [x . x x x x x x . x . x], for the six-beat meter is [x . x . x x x x x x x x], and for the three-beat meter is [x . x . x x . x x x . x]. The first has nine attacks, the second has ten attacks, and the third has eight. Furthermore, the groupings are quite different. The first resultant rhythm has one large group of size 6, the second has a large group of size 9, and the largest group of the third resultant has size 3. Furthermore, the third resultant rhythm has four groups: one of size one, two of size two, and one of size three. Incidentally this latter resultant pattern of the *bembé* and the [4-4-4] meter is the rhythmic necklace pattern used by Steve Reich in *Clapping Music*.

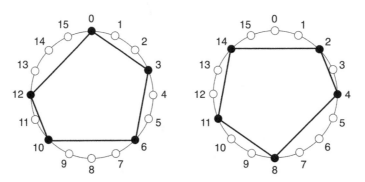

FIGURE 14.12   The *clave son* (left) and its rotation by eight pulses (right).

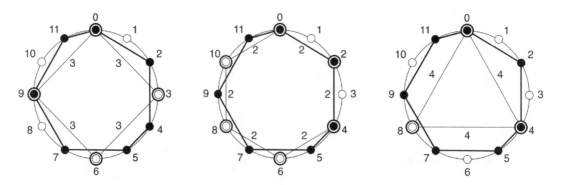

FIGURE 14.13   The *bembé* rhythm in articulated four-beat (left), six-beat (center), and three-beat (right) regular (isochronous) meters.

NOTES

1 Keith, M., (1991), p. 15. A useful computational tool for studying rhythm necklaces is an algorithm for generating them. Ruskey, F. & Sawada, J., (1999) describe an efficient algorithm that, when given the number of pulses and attacks, generates all possible necklaces, in execution time proportional to the number of necklaces generated.

2 Tymoczko, D., (2011), p. 38.

3 *Ibid*, p. 39.

4 Jan, S., (2007). Universal rhythms that exhibit symmetries such as mirror symmetry are instances of the principle of symmetry, a candidate for a more general *grand rhythmic universal*. Voloshinov, A. V., (1996), p. 111, puts it this way: "Symmetry is a universal genetic constant, collectivizing each and every rhythm into the Rhythm par excellence."

5 Kauffman, R., (1980), p. 396.

6 Becker, J., (1968), p. 186.

7 London, J., (1995), p. 67.

8 McLachlan, N., (2000).

9 Manuel, P. with Bilby, K. & Largey, M., (2006), p. 40.

10 Srinivasamurthy, A., Repetto, R. C., Sundar, H., & Serra, X., (2014). http://compmusic.upf.edu/bo-perc-patterns Accessed December 30, 2017.

11 Poole, A., (2018).

12 Agawu, K., (2006).

13 Pérez Fernández, R. A., (2007), p. 7.

14 Stone, R. M., (2005), p. 82.

15 These rhythms are played in the Afro-Cuban religious *batá* drumming, Moore, R. & Sayre, E., (2006), p. 129.

16 Since this rhythm necklace in the time domain is isomorphic to the diatonic scale in the pitch domain, these seven starting points correspond to the seven modes of the diatonic scale. See Anku, W., (2007), p. 11, Ashton, A., (2007), p. 52, and Loy, G., (2006), p. 20. See also Leake, J. (2007), Pressing, J. (1997), and Rahn, J. (1996).

17 Chernoff, J. M., (1979), p. 119.

18 The word *palo* in Spanish means *stick* but refers also to sugarcane. The rhythm acquired the name because it was played during the cutting of sugarcane.

19 Dworsky, A. & Sansby, B., (1999), p. 111.

20 de Jong, N., (2010), p. 202 and Rosalia, R. V., (2002).

21 Dworsky, A. & Sansby, B., (1999), p. 111.

22 Stone, R. M., (2005), p. 82.

23 King, A., (1960), p. 52, considers this rhythm with duration pattern [2-2-1-2-2-1-2] to be a variant of the standard pattern [2-2-3-2-3].

24 Gerstin, J., (2017), p. 33, has documented seven of the eight possible rotations of the tresillo duration pattern [3-3-2].

25 Agawu, K., (1995a), p. 129. This rhythm is also called the *sichi* rhythm (from Ghana) by Dworsky, A. & Sansby, B., (1999), p. 84.

26 Vurkaç, M., (2011), p. 27.

27 Mauleón, R., (1997), p. 24.

28 Traditionally, the contextual rhythmic analysis of the *clave son* is based on dividing the 16-pulse cycle into two 8-pulse half cycles corresponding to the *three-attack* and *two-attack* portions, and subjecting the two parts to further analysis based on syncopation. However, Vurkaç, M., (2012), finds it more useful to partition the 16-pulse cycle into an inner part flanked by two outer parts, and analyzing the parts by means of *off-beatness* rather than syncopation. We shall consider the notion of off-beatness in Chapter 16.

CHAPTER 15

Rhythmic Oddity

Do TWENTIETH CENTURY, East London, acid jazz music, and the ancient Aka Pygmy music of Central Africa have anything noteworthy in common? Yes, they do. There exist pieces of music in both domains that use rhythmic timelines that possess the *rhythmic oddity* property. But that is getting ahead of our story. First, we must backtrack more than half a century to 1963, when a 33-year old horn player with the symphony orchestra of an Israeli radio station received an invitation to work on a project spearheaded by the Israeli Ministry of Foreign Affairs: the setting up of a youth orchestra in the Central African Republic. The horn player's name was Simha Arom, and although he was not overly enthusiastic about the project itself, he was excited by the possibility of discovering a world of music unknown to him. Besides, he was ready to break up the routine that had enveloped his life. When he first heard the music of the Aka Pygmies he was instantly overwhelmed. He felt that their music not only had ancient roots, but that it also touched roots deep inside him.[1] The rest is history. Arom went on to develop original methods of musicological research, and new tools with which to collect data. He made multiple recordings of African traditional music and created a museum of arts and popular traditions. He studied the music of the Aka Pygmies for decades, becoming one of the foremost systematic ethnomusicologists in the world.

While studying the music of the Aka Pygmies of Central Africa, Arom noticed that their music contained rhythmic timelines that exhibited a property that he christened *rhythmic oddity*. A rhythm with an *even* number of pulses in its cycle has this property if no two of its onsets divide the rhythmic cycle into two half cycles, i.e., two segments of equal duration.[2] This property is not

defined for rhythms with an odd number of pulses, since it is impossible for two pulses to lie diametrically opposite each other on the rhythm circle (an odd number is not evenly divisible by two). It is quite easy in theory to construct examples of rhythms that have this property.[3] Figure 15.1 shows two such examples: the rhythm on the left has onsets on the first four of its eight pulses, and the one on the right has onsets on the first 6 of its 12 pulses. It is obvious that, in general, for any even number n of pulses, a rhythm that contains fewer than $n/2$ consecutively adjacent onsets, has the rhythmic oddity property. However, rhythms constructed in this way are not particularly interesting musically and are not used as timelines in world music, except when the number of pulses in the cycle is a small number such as $n = 4$ or $n = 6$, in which case we may obtain for example the two-onset and three-onset rhythms with interonset intervals [1-3] and [1-1-4], respectively, which can be heard as ostinatos in several musical traditions. Its drawbacks notwithstanding, we shall identify this procedure as the *Walk* Algorithm, since we can think of starting a walk at pulse zero, taking k short steps of one pulse durations each, where k is less than $n/2$.

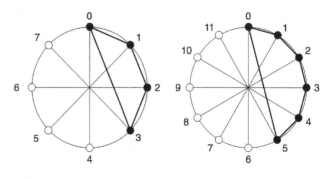

FIGURE 15.1 Two humdrum rhythm timelines that possess the *rhythmic oddity* property.

In spite of their easy construction from the mathematical point of view, in practice, timelines that possess the rhythmic oddity property are unusual in world music. For rhythms to be effective as timelines, they should in general not contain silent gaps longer than half of their cycle, and they should exhibit a certain degree of regularity. These two constraints are often enough to inadvertently prevent the rhythmic oddity property from being satisfied. Figure 15.2 illustrates five such examples of traditional rhythm timelines that satisfy these two conditions but lack the rhythmic oddity property. In the rhythm on the top left with interonset intervals [1-1-2], the first and last onsets violate the rhythmic oddity property. This simple pattern is used almost universally. For example, it is the *baiaó* rhythm of Brazil as well as the *polos* rhythm of Bali. When it is started on the second onset, it turns into the *catarete* rhythm of the indigenous people of Brazil. Started on the third onset, it becomes an archetypal pattern of the Persian Gulf region,[4] the *cumbia* from Colombia, and the *calypso* from Trinidad. It is also a thirteenth-century Persian rhythm called *khalif-e saghil*, as well as the *trochoid choreic* rhythmic pattern of ancient Greece. Starting it on the silent pulse (anacrusis) yields a popular flamenco hand-clapping pattern (also *compás*) used in the flamenco styles called the *taranto*, the *tiento*, the *tango*, and the *tanguillo*. It is also the *rumba* clapping pattern in flamenco as well as another pattern used in

the *baiaó* rhythm of Brazil.[5] In the top-center rhythm with interonset intervals [3-3-3-3], the rhythmic oddity property is violated twice, once with the first and third onsets, and again with the second and fourth onsets. This rhythm is the meter or *compás* of the *fandango* music of Spain. It is often accompanied by hand clapping every pulse, but with loud claps at pulses 0, 3, 6, and 9. The top-right rhythm with interonset intervals [1-1-1-1-2] contain two violations of the rhythmic oddity property at pulses zero and three as well as one and four. It is the *york-samai* pattern, a popular Arabic rhythm as well as a hand-clapping rhythm used in the *al-medemi* songs of Oman. The bottom-left rhythm with interonset intervals [2-1-1-1-1-1-1] contains three violations of the property at pulses 0 and 4, 2 and 6, and 3 and 7. It is a typical rhythm played on the *bendir* (frame drum) and used for the accompaniment of women's songs of the Tuareg people of Libya.[6] Finally, the rhythm at the bottom right with interonset duration pattern [1-1-1-1-1-2-3-2] contains two violations at pulses one and seven as well as pulses four and ten. This rhythm, played on the cajón, is the *samba malató* from the Afro-Peruvian repertoire.[7]

Let us turn to rhythms that contain the rhythmic oddity property and that satisfy the earlier constraints. Two examples are the 24-pulse timeline rhythms used by the Aka Pygmies pictured in Figure 15.3. The rhythm on the left has nine onsets with interonset intervals

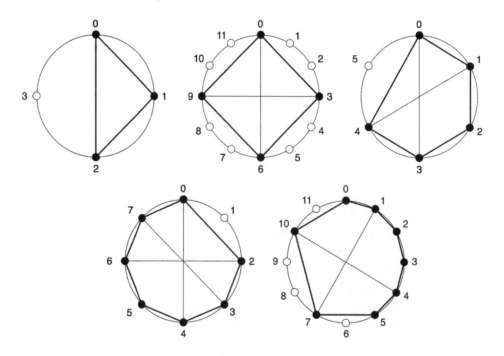

FIGURE 15.2 Five examples of traditional rhythm timelines without the rhythmic oddity property.

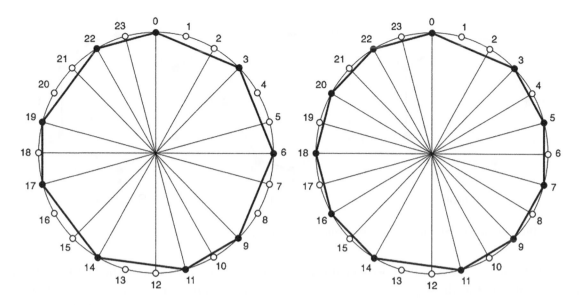

FIGURE 15.3   Two timelines that possess rhythmic oddity used by the Aka Pygmies.

[3-3-3-2-3-3-2-3-2], whereas the one on the right has 11 onsets with interonset intervals [3-2-2-2-2-3-2-2-2-2-2].

The Aka Pygmies also use the five-onset, 12-pulse timeline shown in Figure 15.4 (left). It has interonset intervals [3-2-3-2-2] and possesses the desired properties. As an aside, it is interesting to note that if the five intervals are permuted to yield [2-2-3-3-2], we obtain the hand-clapping pattern and meter (*compás*) used in the *seguiriya* style of the flamenco music of southern Spain shown on the right.[8] With the two interonset intervals of length three adjacent to each other, the rhythmic oddity property is violated at pulses four and ten.

Given that the music of the Aka Pygmies is characterized by having rhythmic timelines that possess the rhythmic oddity property, a natural ethnomusicological question arises: to what extent does this property manifest itself in other cultures, such as for example

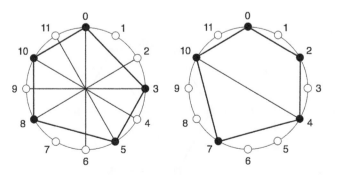

FIGURE 15.4   A 12-pulse timeline used by the Aka Pygmies (left) and the seguiriya *compás of the flamenco music of southern Spain (right).*

West Africa, South Africa, or Cuba? To try to answer this question, consider an archetype timeline structure used extensively in these three geographical regions that consists of seven onsets in a cycle of 12 pulses, with the constraint that all the interonset intervals must be of only two distinct durations: either one or two units. Three examples of such bell-pattern timelines are *bembé*, tonada,[9] and sorsonet pictured in Figure 15.5. Note that none of them possess the rhythmic oddity property. However, before we dismiss the usefulness of this property altogether as a discriminating feature of these rhythms, it is worth noting that the *bembé* contains one violation, the tonada[10] contains two and the sorsonet has three. This observation suggests a way to generalize the rhythmic oddity property as described in the following.

Arom defined the rhythmic oddity property in the form of a strict binary, all or none, category, i.e., a rhythm either has or does not have the rhythmic oddity property. This concept may be extended to a multivalued function that measures the *amount* of rhythmic oddity that a rhythm possesses. This function, which will be called *rhythmic oddity*, depends on the number of violations of the rhythmic oddity property present in a rhythm. Stated another way, a violation of the rhythmic oddity property yields a partition of the rhythmic cycle into two half cycles by pairs of its antipodal onsets. Let us call such a partition of the cycle an *equal bipartition*. Then the fewer equal bipartitions a rhythm admits, the more rhythmic oddity it possesses. The timeline in

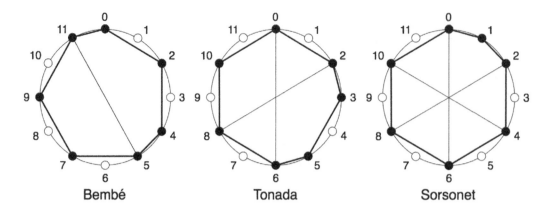

FIGURE 15.5 Three archetypal rhythmic timelines from West Africa.

Figure 15.4 (left) used by the Aka Pygmies contains no equal bipartitions, whereas the seguiriya *compás* of the flamenco music (right) contains one. Regular rhythms with an even number of pulses have the maximum number of equal bipartitions, since in a rhythm of $2n$ pulses, every onset has an antipodal onset, and therefore there are n bipartitions. Hence, regular polygons have a minimum amount of rhythmic oddity (indeed, no rhythmic oddity at all). But rhythms need not be regular, in order for them to have zero rhythmic oddity. It suffices for their polygonal representations to have parallel opposite sides. Such rhythms called *zonogon* rhythms will be revisited in more detail in Chapter 25.

Let us return to the three archetypal timeline necklaces from West Africa shown in Figure 15.6. Recall that if we disregard the rotations of a rhythm so that all its rotations form an equivalence class, we call such an object a necklace. The relevance of necklaces here comes from the fact that the rhythmic oddity function is independent of the rotations of a rhythm; it is a property of the necklace. Figure 15.6 depicts three distinct necklaces, and each necklace determines seven different rhythms

depending on which onset of the rhythm is taken as pulse zero (not counting the rotations that yield rhythms with anacrusis that start on a silent pulse). As it turns out, if the interonset intervals are restricted to the values one and two, these three necklaces are the only mathematical possibilities. The two short intervals of length one may be separated by two, one, or zero long intervals of length two, in the *bembé*, the tonada, or the sorsonet necklaces, respectively. Our original question concerning the postulated preference of timelines is as follows: which of the three necklaces in Figure 15.6 are preferred in West African music? This question is not easy to answer without first agreeing on the definition of "preference" and spending time in the field performing listening experiments. In the absence of all these requirements, we may attempt to answer this question as an arm-chair musicologist by counting, for each necklace, how many of its rotations are used in musical practice. However, this definition of preference still needs elaboration since it is conceivable that one necklace appears frequently, but in only one of its rotations, whereas the other necklaces appear infrequently, but in all their rotations. However,

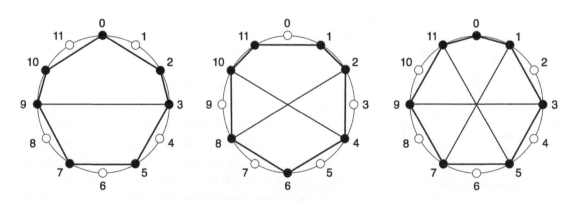

FIGURE 15.6 The three necklaces with two short and five long durations.

if only one rotation of a necklace is used, then it is that rhythm that is preferred, and not the necklace that gives rise to the rhythm. What is intended by preference here is precisely the necklace. Which necklace is preferred by nature, which has the greatest fecundity.

It turns out that in West African music, the sorsonet necklace is one of the least preferred of the three, yielding one timeline used in traditional music, the sorsonet rhythm of Figure 15.5. However, the rotation with durational pattern [2-2-2-2-1-1-2] is the Persian rhythm *kitāb al-Adwār*,[11] and the rotation [2-1-1-2-2-2-2] is the rhythm of the Polish *polonaise*.[12] Rotations of the tonada necklace are encountered more frequently, yielding two West African rhythms, the tonada with intervals [2-1-2-1-2-2-2][13] and the asaadua given by [2-2-2-1-2-1-2], and one Persian rhythm, the al-ramal with intervals [2-2-2-2-1-2-1]. The *bembé* necklace is overwhelmingly preferred over the other two necklaces. Indeed, all seven rhythms obtained by starting the cycle at every one of its seven onsets are heavily used. It is evident then that among this family of rhythms, there may have been an evolutionary preference for those that admit as few as possible equal bipartitions and thus a higher degree of rhythmic oddity. This is not to imply that there are no other mathematical properties that can produce the same preference ranking of these three necklaces. Perhaps the most obvious one is the separation distance between the two short intervals in the cycle, which are separated either by a minimum of two, one, or zero long intervals. This implies that the same preference ranking may also be obtained by measuring how evenly the seven attacks are spaced out in the circle. Yet another method for obtaining the same preference ranking is by calculating the minimum number of elementary mutations required for each of the necklaces to become a regular hexagon, which is another measure of evenness of these necklaces. An elementary mutation here either deletes an attack or inserts an attack. In the leftmost necklace of Figure 15.6 deleting the three attacks at pulses 10, 0, and 2, and inserting two attacks at pulses 11 and 1, does the job, yielding a total of five mutations. The necklace in the middle may be transformed into a regular hexagon by deleting the attacks at pulses 11 and 1, and inserting an attack at pulse zero, for a total of three mutations. Finally, the necklace on the right requires only the deletion of one attack at pulse zero. We will return to such mutation operations in more depth later in the book. All these mathematical methods are in effect theoretical explanations that fit the data. Whether any of these methods actually guided the evolutionary selection process is another matter altogether. It would be interesting to test experimentally which of these properties has the most perceptual reality. Under what circumstances, if any, is the degree of rhythmic oddity possessed by a rhythm, more easily perceived by humans than the amount of evenness? A dancing culture might have selected a timeline on the basis of rhythmic oddity, in as much as this property has a marked effect on the order of upbeats and downbeats of the feet, thus rendering evenness as a byproduct.

In the pitch domain, the three necklaces in Figure 15.6 are the three well-known scales called (from left to right) the diatonic scale, the ascending melodic minor scale, and the Neapolitan major scale. Michael Keith proposes measuring the evenness of scales by a suitable distance function between each note and the ideal note. The ideal notes are located at multiples of 12/7 on the circle, yielding the coordinate values along the circle: 0.0, 1.714, 3.428, 5.142, 6.856, 8.570, and 10.284. His measure called the *scale-idealness* also ranks the three necklaces of Figure 15.6 in decreasing order from left to right.[14] If we compute the sum of the absolute values of the differences between these coordinates, and those of the attacks of the *bembé*, tonada, and sorsonet rhythms of Figure 15.5, we obtain the distances: *bembé* = 2.290, tonada = 2.566, and sorsonet = 4.994. Thus, the *bembé* is slightly more even than the tonada, and both are much more even than the sorsonet.

Let us return to the topic of generating rhythms that exhibit the rhythmic oddity property. At the start of this chapter, the *Walk* Algorithm was presented that constructs rhythms that have the rhythmic oddity property but place all the onsets within a total duration region that spanned less than one half cycle, thus producing not the best of timelines. We close this chapter with a demonstration of a modification of the procedure that yields timelines that satisfy the rhythmic oddity property, such that every half cycle contains at least one onset. Furthermore, the timelines obtained in this way turn out to be better. This algorithm will be called the *Hop-and-Jump* Algorithm. It falls in the general category of algorithms for obtaining *generated* rhythms, and in Chapter 27, we shall see its relation to other generative methods for producing *deep* rhythms. Thus, one application of the rhythmic oddity property is to the algorithmic generation of "good" rhythms.

Let us assume we want to generate a rhythm with five onsets in a cycle of 12 pulses. The algorithm is illustrated with five clock diagrams (left to right) in Figure 15.7. The first onset is placed at pulse zero. This implies that the diametrically opposite pulse six is now unavailable for placing an onset, since we want the rhythmic oddity property to be satisfied. To place the next onset, we *hop* to pulse two, making pulse eight unavailable. This process is continued always advancing by hopping a distance of two units if this is possible. When this is not possible, as is the case when we want to hop to onset number four at pulse six (which is unavailable), we try the next pulse (here pulse seven). If it is available (as it is in this example), we take it. Otherwise, we continue skipping pulses until an available pulse is found. Since in this case we advanced by a distance of more than two pulses, we call this a *jump*. Following a jump, we continue as before, making hops of distance two if possible (or jumps otherwise), yielding the fifth onset at pulse nine.

The *Hop-and-Jump* algorithm is obviously guaranteed to yield rhythms with the rhythmic oddity property, since it never places an onset on an unavailable pulse location. A formal theoretical mathematical characterization of the *Hop-and-Jump* algorithm was shown by André Bouchet.[15] Furthermore, by choosing the number of onsets and *hop* distance appropriately, we may guarantee that there are no silent gaps longer than a half cycle. Finally, note that the resulting five-onset rhythm obtained in Figure 15.7 is the *fume–fume* bell pattern (also the *standard* pattern) widely used in West Africa, and is the same as the Aka Pygmie rhythm of Figure 15.4 either played backwards or rotated in a counterclockwise direction by four pulses.

Let us consider a few more examples with different numbers of onsets and pulses, and different sizes of hops, to substantiate our claim that the *Hop-and-Jump* Algorithm is successful at generating good timelines.

For three onsets out of eight pulses, and hop-size two, the algorithm generates the rhythm with interonset intervals [2-3-3] as shown in Figure 15.8. Recall that this rhythm is the *nandon bawaa* bell pattern of the Dagarti people of northwest Ghana, and is also found in Namibia and Bulgaria.[16] It is a rotation of the rhythm with interonset intervals [3-3-2] (the Cuban *tresillo*), which, as pointed out earlier, is the most important traditional bluegrass banjo rhythm, as well as a metalblade pattern of the Aka Pygmies. The latter is common in West Africa and many other parts of the world such as Greece and Northern Sudan. Some scholars consider it to be one of the most important rhythms

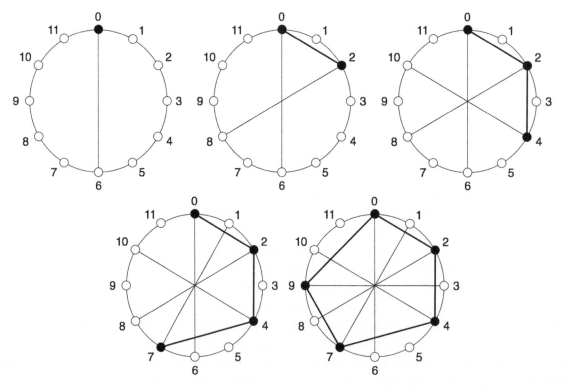

FIGURE 15.7 The *Hop-and-Jump* algorithm for generating good rhythms that have the rhythmic oddity property: five onsets among 12 pulses.

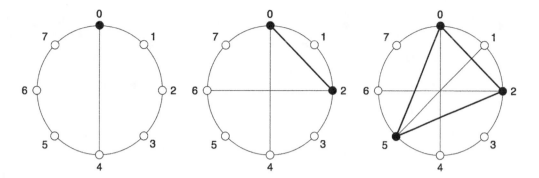

FIGURE 15.8   The *Hop-and-Jump* algorithm with three onsets among eight pulses.

in Renaissance music. Indeed, the pattern [3-3-2] dates back to the Ancient Greeks who called it the *dochmiac* pattern. In India, it is one of the talas of Carnatic music. The rotation with intervals [3-2-3] is a drum pattern used in Korean instrumental music and is also found in Bulgaria and Turkey.

Figure 15.9 illustrates the algorithm at work with five onsets out of 16 pulses, and hop-size three. It generates the rhythm with interonset intervals [3-3-3-3-4] that is a rotation of the bossa-nova clave rhythm of Brazil. The actual bossa-nova rhythm usually starts on the third onset. It is also a maximally even rhythm, since this is the most even manner in which one may distribute five onsets among 16 pulses.

The last example in Figure 15.10 shows the generation of a rhythm with seven onsets among 16 pulses using a hop-size of two pulses. It has interonset intervals [2-2-2-3-2-2-3] and is a rotation of a Samba rhythm from Brazil. The actual Samba rhythm starts on pulse four and coincides with a Macedonian rhythm. Other rotations of this rhythm are found in the music of Ghana as well as former Yugoslavia.

All the examples of timelines containing the rhythmic oddity property discussed in the preceding emerge from the traditional drumming music of West and Central Africa as well as the African Diaspora. With the modern and commercial preoccupation of twentieth century music, and the ubiquitous "square" divisive timelines

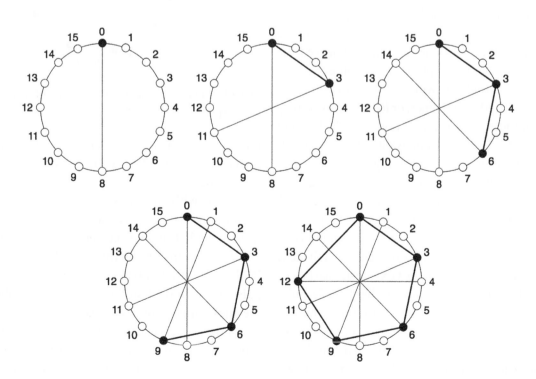

FIGURE 15.9   The *Hop-and-Jump* algorithm with five onsets among 16 pulses.

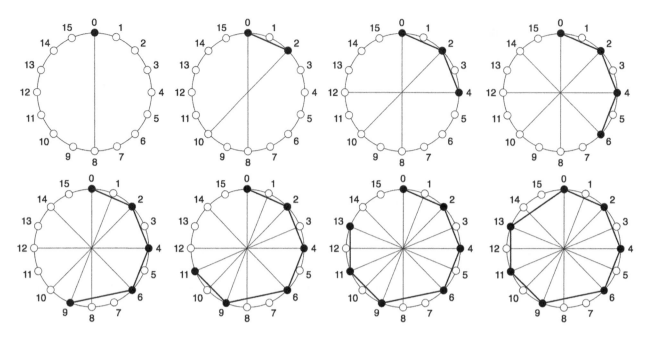

FIGURE 15.10 The *Hop-and-Jump* algorithm with seven onsets among 16 pulses.

that dominate much of pop music, one may wonder if there are any contemporary newly composed timelines out there that exhibit the rhythmic oddity property. A fascinating example of one such timeline that uses a highly syncopated ten-onset, 32-pulse cycle may be found in the song "*Cosmic Girl*" released in the United Kingdom in 1996 by the acid jazz band *Jamiroquai*.[17] Acid Jazz is both an East London recording company as well as a music genre, and there is an ongoing debate about which of the two came first and influenced the other. As a genre, acid jazz appears to combine elements of hip-hop, funk, and jazz with a strong rhythmic element that uses rhythm timelines, or *looped beats*, as they are called in the electronic music world. As the lyrics of "*Cosmic Girl*" testify, *Jamiroquai* wanted to create a feeling of outer space, of distance, of strangeness and science fiction, by using words such as "hyperspace," "galaxy," "quasar," "teleport," and the lightness of "zero gravity." To create these feelings, the band composed a timeline that does the job quite well. The electronic timbre is no doubt appropriate, but what really made this psychedelic song blast off to number six on the U.K. music charts is its unique timeline. The length of this rather long 32-pulse timeline gives the cosmic feeling time to sink in, but the timeline's main power comes from its long sequence of three-pulse interonset intervals that twist and turn around the regular four-pulse underlying beats shown with thin solid lines in Figure 15.11, and the property

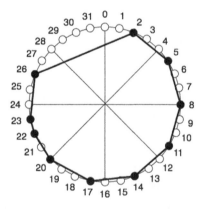

FIGURE 15.11 The opening timeline of acid jazz band Jamiroquai's "*Cosmic Girl*."

that even with as many as ten onsets, the timeline still manages to exhibit the rhythmic oddity property.

A straightforward application of the *Hop-and-Jump* algorithm with nine onsets and 32 pulses yields the rhythm in Figure 15.12 (left). If Jamiroquai experimented with this version before adopting their final one, it is easy to see why they would have abandoned it. With three of its onsets coinciding with downbeats at pulses 0, 12, and 24, and two of these downbeats being the most important downbeats, namely the first and last of the sequence, it lacks the element of surprise, and provides little energy in the margins. Furthermore, the onset on the downbeat at pulse 12 falls squarely halfway between those at pulses 0 and 24, providing too much

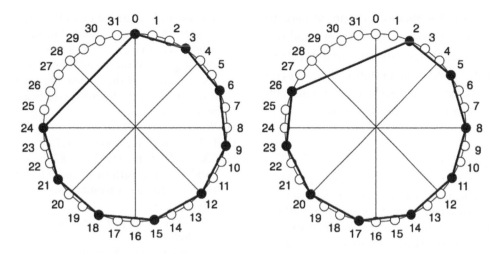

FIGURE 15.12   Rhythm obtained with the *Hop-and-Jump* algorithm (left), and its forward rotation by two pulses (right).

symmetry. Consider what results when this rhythm is rotated in a clockwise direction by two pulses to yield the rhythm on the right. Now there are onsets on only two downbeats, and they are not the first or last in the sequence, but rather the third on pulse eight and the sixth on pulse 20, more unexpected locations to be sure. Furthermore, since "*Cosmic Girl*" starts with only this timeline for a while, we are totally surprised when the beats start coming in with the remaining percussion instruments. Finally, add a tenth onset on pulse number 22, as in Figure 15.11, and now the timeline also has closure just before the end of the cycle; it is the icing on the cake. Besides, it is interesting to note that the first four attacks of this timeline are the same as the repeating rhythmic riff in George Gershwin's "I Got Rhythm," given by [..x..x..x..x....], the source for the most common chord progression in jazz.[18]

The methods described in this chapter are radically different from those used for generating rhythms automatically, that are described in the artificial intelligence and music information retrieval literatures. The latter approaches are inspired either by models of biological processes, such as neural networks that learn from experience, by genetic programming methods that model the evolutionary laws of natural selection, or by statistical models such as Markov processes.[19] Many of these techniques are based on guided random search of the space of all possible rhythms. Typically, genetic methods first define a measure of rhythmic "goodness" generally termed a *fitness function*, and then use simple rules for transforming a given collection of rhythms in such a way as to improve their fitness. These rules are usually described

in general terms as *reproduction*, *crossover*, and *mutation*, and applied in this order. Reproduction selects a pair of rhythms, say A and B, at random from the collection. Crossover involves creating new offspring rhythms of A and B by swapping some elements from A to B and vice versa. Mutation involves changing one of the elements of a new offspring of A or B at random, and usually with low probability. Finally, the algorithm is programmed to stop (or is stopped by the user) when the fitness function has (or seems to have) reached a maximum value.[20]

Gibson and Byrne (1991), incorporate a neural network in their genetic approach. First, they use humans to label a collection of training rhythms as either "good" or "bad." Then they use the trained neural network to classify new rhythms generated by the genetic algorithm as either "good" or "bad," thus serving as the fitness function.[21] Horowitz (1994) describes an interactive approach that allows the user to "simply execute fitness functions (that is, to choose which rhythms or features of rhythms the user likes) without necessarily understanding the details or parameters of these functions."[22] This "ostrich-head-in-the-sand" approach may be attractive and useful to those composers and other users that are satisfied with only the end product. By contrast, the methods proposed in this chapter and the book in general for generating "good" rhythms are *structural* in nature, and guided by musicological and empirical knowledge of rhythms that humanity has come to cherish over thousands, if not millions, of years of evolution. The crux in these methods is precisely the understanding of details and the elimination of the parameters in neural networks that must be tweaked ∎ to obtain good rhythms. The methods proposed here are closer in spirit to computational music theory and represent an attempt to understand the temporal structures that make a rhythm "good." Furthermore, if desired, the properties discussed here may also be incorporated into fitness functions for use in genetic algorithms.

NOTES

1 Arom, S., (2009), p. 7.

2 Chemillier, M., (2002), p. 176 and Chemillier, M. & Truchet, C., (2003). The convex polygons inscribed in a circle that correspond to rhythms with the rhythmic oddity property have inspired research in mathematics and computer science, where they are called *antipodal* polygons. Aichholzer, O., Caraballo, L. E., Díaz-Báñez, J. M., Fabila-Monroy, R., Ochoa, C., & Nigsch, P., (2015) prove a variety of mathematical properties of such polygons.

3 It is more difficult to enumerate *all* rhythms that have the rhythmic oddity property. Chemillier, M., (2004), p. 615, shows how this can be done using Lyndon words.

4 Olsen, P. R., (1967), p. 31.

5 The song "Baião" by Luiz Gonzaga uses the rhythms [x . . x x . . .] and [. . x . . . x]. See Murphy, J. P., (2006), p. 97.

6 Standifer, J. A., (1988), p. 50.

7 Miranda-Medina, J. F. & Tro, J., (2014), p. 217 and Feldman, H. C., (2005), p. 215.

8 Fernández, L., (2004), p. 35.

9 The Cuban tonada is called *djouba* in Haiti, and *ternary tibwa* in Martinique (see Gerstin, J., (2017), pp. 71–72).

10 The tonada has one less onset than a popular traditional nineteenth-century Cuban timeline called the *clave campesina* given by [x . x x . x x . x x x .] (see Mauleón, R., (1997), p. 10). With the additional onset in between the last two onsets of the tonada, the clave campesina has a third violation of rhythmic oddity at pulses three and nine.

11 Wright, O., (1995).

12 Dahlig-Turek, E., (2009), p. 127.

13 Nketia, J. H. K., (1962), p. 85, lists this rhythm as a hand-clapping pattern of the Akan people of Ghana. However, the term *tonada* is used in Cuba to describe this rhythm.

14 Keith, M., (1991), p. 97. See also Tymoczko, D., (2011).

15 Bouchet, A., (2010). Additional mathematical properties and characterizations of rhythms with the rhythmic oddity property may be found in Jedrzejewski, F., (2017).

16 Nketia, J. H. K., (1962), p. 123, includes this rhythm as hand-clapping pattern used in *Nayalamu*, a recreational maiden song of the Gonja people of Ghana.

17 I am indebted to mathematician Ben Green of the University of Cambridge for bringing this music to my attention.

18 Crawford, R., (2004), p. 163. I am indebted to Dmitri Tymoczko for pointing out this connection.

19 Paiement, J.-F., Bengio, S., Grandvalet, Y., & Eck, D., (2008).

20 Burton, A. R. & Vladimirova, T., (1999).

21 Gibson, M. & Byrne, J., (1991).

22 Horowitz, D., (1994).

CHAPTER 16

Offbeat Rhythms

THE MUSIC OF THE SAN PEOPLE, a group of hunters and gatherers that live in the Southern African countries of Angola, Botswana, and Namibia, is characterized by the use of an instrument called the *musical bow*, illustrated in Figure 16.1. This instrument consists of a bow, such as one might use for hunting, with a tight steel string fastened to the two ends. In addition, a gourd attached to the bow is used to create a resonating cavity to produce a particular tone and timbre. One rather unique tradition of these hunter-gatherers, called *kambulumbumba*, involves three individuals playing one bow simultaneously.[1]

One player, while securing the bow with his feet and mouth, plays the leftmost rhythm in Figure 16.2 by striking the string with a stick. This rhythm sets up an isochronous steady regular rhythm [3-3-3-3] with four beats per cycle. Another musician plays the rhythm in the center, also with a stick, but on the upper end of the bow. This rhythm has five onsets with intervals [3-3-2-2-2]. The first three onsets of the latter rhythm coincide with the first three onsets of the regular four-beat rhythm. However, the last two onsets at pulses 8 and 10 fall in between the regular beat. These onsets are said to be *offbeat*. The third performer plays a regular six-beat rhythm [2-2-2-2-2-2] (rightmost diagram in Figure 16.2) also with a stick, that contains two onsets at pulses two and four that are offbeat with respect to the other two rhythms.

The diagram in the middle of Figure 16.2 shows how the two regular rhythms (dotted lines) interact with the irregular rhythm. Interestingly enough, this irregular

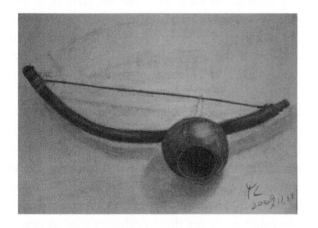

FIGURE 16.1   Musical bow. (Courtesy of Yang Liu.)

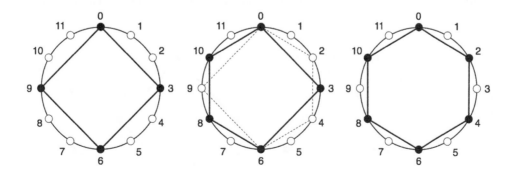

FIGURE 16.2   The three musical bow rhythms employed by the *San* people.

rhythm is also the meter (compás) of the guajira style of the flamenco music of Southern Spain.

Although the multivalued measure of the quantity of rhythmic oddity, discussed in the previous chapter, is more successful than the binary-valued rhythmic oddity property, at discriminating rhythmic preference in West Africa, it still has its limitations. For instance, among the seven rhythms determined by the rotations of the *bembé* necklace, some are preferred over others. In fact, one of these, the *bembé* itself, with intervals [2-2-1-2-2-2-1] is by far the most favored of the seven, as it is considered to be the African signature bell pattern. Afro-Cuban music has escorted it across the planet, and it is used frequently on the ride cymbal in jazz. Since all seven rhythms belonging to this necklace obviously have exactly one equal bipartition, even the multivalued rhythmic-oddity measure does not discriminate among these seven, and thus does not favor the *bembé* rhythm over its six other rotations. To resolve this conundrum, we recruit another mathematical measure of syncopation or irregularity termed *offbeatness*. To illustrate how this measure works, consider a cycle of 12 pulses. Such a cycle may be evenly divided (without remainder) by the integers two, three, four, and six, to yield the four regular rhythms with interonset intervals [6-6], [4-4-4], [3-3-3-3], and [2-2-2-2-2-2], respectively, pictured in Figure 16.3. If a piece of music uses a particular regular meter that has strong beats at say pulses 0, 3, 6, and 9, as in the third diagram from the left, then the notes that are played on the other eight pulses are considered to be offbeat relative to such a meter. Sub-Saharan African drum ensemble music is polyrhythmic and most music that uses a 12-pulse cycle incorporates most if not all four rhythms shown in Figure 16.3, played either on different types of drums, other percussion instruments, or clapping.[2] If we superimpose all four rhythms on one

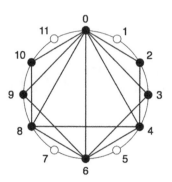

FIGURE 16.4 In a 12-pulse clock, the *offbeat* onset positions are {1, 5, 7, 11}.

circle. we obtain the diagram in Figure 16.4 that reveals four pulses that remain without any onset; these occur at positions 1, 5, 7, and 11. If beats were played at any of these four positions in this context, they would be considered as being *strongly* off the beat, in the sense that they are offbeat relative to all possible regular meters. Therefore, we define the *offbeatness* measure of a rhythm as the number of onsets that the rhythm contains at these four distinguished locations.[3]

The offbeatness measure is the converse of Stephen Handel's measure of *metrical strength*, which is defined as the number of cooccurrences of the onsets of the rhythm with the metrically strong beats, which in a 16-pulse cycle as in Figure 16.6 are {0, 4, 8, 12}.[4]

Armed with this new measure of irregularity, let us reconsider the *bembé*, *tonada*, and *sorsonet* rhythm necklaces shown in Figure 15.6. It is noteworthy that both tonada and sorsonet rhythms take on offbeatness values equal to 1, due to their onset at positions five and one, respectively, whereas the *bembé* has an offbeatness value of 3, due to its onset at positions 5, 7, and 11. Indeed, all the rotations of the three necklaces used in practice have offbeatness values equal to 1 and 2, except for the *bembé*. Therefore, the offbeatness measure may provide

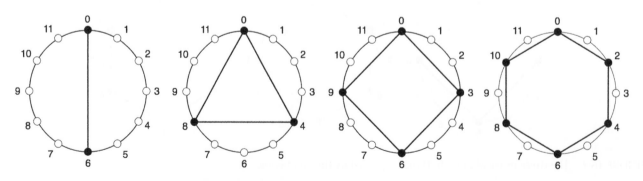

FIGURE 16.3 The four divisors of 12 other than 1 and 12.

a mathematical formula for rating the preference of the *bembé* timeline among this family of timelines.

The set of four offbeat pulse positions {1, 5, 7, 11} has an interesting mathematical interpretation as well. These numbers are the numbers between 0 and 12 that generate (visit) all 12 pulses when we travel along the circle starting at zero and advance in steps of size equal to the numbers. Assume for example that we travel in a clockwise direction starting at pulse zero, and refer to Figure 16.5. If the steps are of size 1, the sequence (0, 1, 2, 3, 4, 5, 6, 7, 8, 9, 10, 11, 12) that determines a convex polygon is generated. If the steps are of size five, the sequence (0, 5, 10, 3, 8, 1, 6, 11, 4, 9, 2, 7) determining a star polygon is obtained. Steps of size seven realize the sequence (7, 2, 9, 4, 11, 6, 1, 8, 3, 10, 5, 0) of the same star polygon. Finally, steps of size 11 produce the sequence (0, 11, 10, 9, 8, 7, 6, 5, 4, 3, 2, 1), the same as the previous convex polygon.

The offbeatness measure is easily generalized to other even values of the number of pulses. For 16-pulse cycles, the offbeat onset positions are {1, 3, 5, 7, 9, 11, 13, 15} as illustrated in Figure 16.6,[5] and for 24-pulse cycles the offbeat onset positions are {1, 5, 7, 11, 13, 17, 19} as shown in Figure 16.7.

The offbeatness property provides a tool for categorizing rhythms, as well as for illuminating musicological discourse, as the following examples illustrate. For the first example, consider the *clave son* and the clave rumba illustrated in Figure 16.8. In Cuba, the *clave son* is associated with secular music and dance heavily influenced by Spanish Christian sensibilities, whereas the rumba is considered to be less commercial and closer to traditional folkloric African religious roots. Whereas the Christian church has had a long history of vilifying syncopated music, the African religions venerated it. As a consequence, one might expect the more traditional rumba to be more syncopated than the son. From Figure 16.8, we see that the offbeatness value of the son is 1 since it has only one onset at pulses {1, 3, 5, 7, 9, 11, 13, 15}. On the other hand, the rumba has an offbeatness value of 2, due to the onsets at pulses 3 and 7. Since offbeatness measures a type of mathematical syncopation, it confirms our expectation. In contrast to the offbeatness measure, Handel's metrical strength yields a value of two for both rhythms determined by pulses {0, 12}, and thus does not discriminate between the son and the rumba.

For a second example, consider the five ternary meters (compás) used in the more than 70 styles of flamenco music of southern Spain and refer to Figure 16.9. First, if we compare the offbeatness measure with the rhythmic-oddity

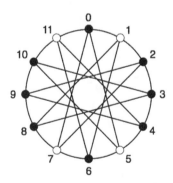

FIGURE 16.5 The four generators of all the pulses are the *offbeat* pulse numbers.

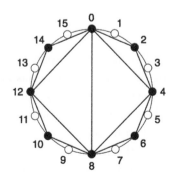

FIGURE 16.6 In a 16-pulse clock, the *offbeat* positions are {1, 3, 5, 7, 9, 11, 13, 15}.

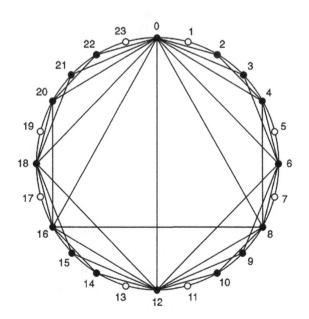

FIGURE 16.7 In a 24-pulse clock, the *offbeat* positions are {1, 5, 7, 11, 13, 17, 19}.

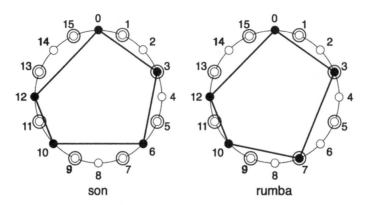

FIGURE 16.8   The *offbeatness* of the *clave son* and the clave rumba.

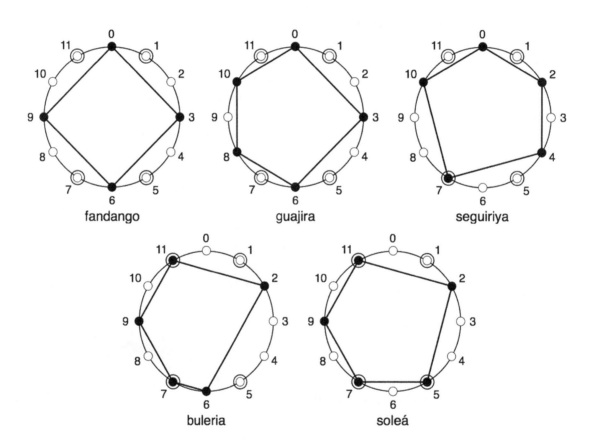

FIGURE 16.9   Calculation of the *offbeatness* values of the five flamenco ternary meters.

property, it is interesting to note that of the five meters, the bulería is the only one that has the rhythmic-oddity property, and thus it is not a very discriminating property. While it is true that the bulería is the only rhythm among these five that contains intervals of lengths 1, 2, 3, and 4 (the other rhythms have intervals of lengths 2 and 3 only), it would be nice to be able to discriminate between the remaining rhythms based on some measure of syncopation. The offbeatness value goes further in this direction.

The fandango and guajira are the only rhythms with an offbeatness value of 0. The seguiriya has an offbeatness value of 1, the buleria an offbeatness value of 2, and the soleá has the highest value of 3. It is worth noting that the soleá is considered to be one of the most paradigmatic and genuine styles of flamenco music. In the words of Nan Mercader, "*la soleá es uno de los palos más jondos del flamenco.*" Might this be explained by the fact that it has such a high offbeatness value?

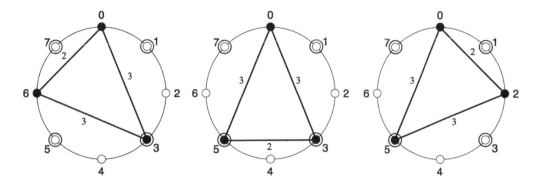

FIGURE 16.10   The *offbeatness* of the three most common rhythms of the [3-3-2] necklace.

As a final example, consider the three rhythms belonging to the [3-3-2] necklace that start on the three onsets (refer to Figure 16.10). The rhythms on the left-most and the rightmost diagrams have an offbeatness value of 1, whereas the one in the center has a value of 2. Of these three rhythms, the leftmost rhythm [3-3-2] is preferred over the other two, in the sense that it is encountered in the musical practice of more cultures around the world. Note that only this rhythm has two of its onsets on the first and last fundamental beats at pulses 0 and 6, thus introducing closure. Furthermore, only this rhythm engenders a cognitive surprise, due to the fact that the first two interonset intervals are equal. The listener expects a regular rhythm with interon-set interval durations of three pulses, which is broken by the third interval of two pulses. Therefore, in this example, these properties appear to override the pos-sible desirability of a higher offbeatness value of the rhythm in the middle.

NOTES

1 Kaemmer, J. E., (2000), p. 314. See Kubik, G., (1975–1976), for an account of musical bows in Angola.

2 Nketia, J. H. K., (1962), p. 83.

3 The offbeatness measure is a precise objective math-ematical measure. Syncopation, on the other hand, has many definitions in the literature, most of which are subjective. Stone, R. M., (1985), p. 140, however, equates offbeat with syncopation. Locke, D., (1982), provides a discussion on how important the principle of offbeat timing is in much of the sub-Saharan music. Vurkaç, M., (2011, 2012), uses the offbeatness measure to analyze the directionality of timelines in a variety of Afro-Latin musics.

4 Handel, S., (1992). As a measure of "irregularity" the off-beatness measure as defined here attributes a lot of weight on only four metric positions 1, 11, 5, 7. However, these are the most important positions in a polymetric context, and it is interesting to determine how useful such a streamlined measure can be. The measure can be generalized to a weighted version that does not put so much weight on these four positions. One way to do this is to create an offbeatness weight for every pulse in the cycle that depends on how many meters render the pulse offbeat. Thus, for a 12-pulse cycle with meters [12], [6-6], [4-4-4], [3-3-3-3], and [2-2-2-2-2-2], we would obtain, for pulses 0–11, the weights {0, 5, 4, 4, 3, 5, 2, 5, 3, 4, 4, 5}. The generalized offbeatness value would then be calculated by sum-ming these weights for all pulses that have an onset, and normalizing by dividing the sum by the number of attacks in the rhythm. It would be interesting to compare such a weighted offbeatness measure with the unweighted version.

5 Flatischler, R., (1992), p. 120, calls these pulse positions *double-time offbeat*, and reserves the term *offbeat* for pulses {2, 6, 10, 14}. See Hennig, H., Fleischmann, R., & Geisel, T., (2012) for the science of being slightly off.

CHAPTER 17

Rhythm Complexity

ARE WEST-AFRICAN TRADITIONAL RHYTHM *TIME-LINES* more complex than North Indian *talas*? Can the choice of the ostinato rhythmic pattern in Steve Reich's *Clapping Music* be informed in terms of the complexity of its rhythm? Has the evolution of the popular rhythms of the world favored an increase in their complexity? Can the difficulty of learning to perform a rhythm be predicted with a simple and elegant mathematical formula? How similar is the rhythmic oddity property prevalent in the Aka Pygmy music to the Western concept of syncopation? How powerful is rhythm complexity as a feature for music genre classification and music information retrieval? Leaving music aside, can the complexity of heartbeat rhythms and neural spike trains be used to aid in heart and brain disease diagnosis, respectively? An introduction to the search for answers to these questions is the focus of this chapter.

Rhythm is arguably the most fundamental aspect of music,[1] and complexity is one of its most salient features.[2] Musicologists routinely comment on the complexity of rhythm present in music from different cultures. In his analysis of African rhythmic systems, Simha Arom writes that they are "the most complex of all those which are known all over the world."[3] According to Reverend Arthur Morris Jones: "No European musician could clap and sing any but the simpler examples of African music."[4] Yet the formal investigation of the complexity of rhythm has been largely overlooked in the literature. A musical concept closely related to rhythm complexity is syncopation, a topic already explored in Chapter 13. However, as we saw there, formal definitions of syncopation are lacking. A typical definition of syncopation is the one found in Collins English Dictionary: "The displacement of the usual rhythmical accent away from a strong

beat onto a weak beat." A mathematician would not only demand formal definitions of "strong" and "weak" beats, but would be baffled by how to interpret the term "usual." On the other hand, many formal (mathematical) definitions of complexity do exist, mostly from domains other than music, but some from music itself. A typical example of the former is the Lempel-Ziv complexity of a binary sequence,[5] and two representatives of the latter category are the *rhythmic oddity* property[6] and the *offbeatness*, as discussed in Chapter 16.[7]

Concerning the rhythms of India, the journalist and producer Joachim-Ernst Berendt writes: "It is necessary ... to say a few words about the mysteries of Indian music. Its talas, its rhythmic sequences—incomprehensible for Western listeners—can be as long as 108 beats; yet the Indian ear is constantly aware of where the sam falls."[8] Kofi Agawu reviews a plethora of published claims about the purported complexity of African rhythms.[9] Comparing African and Indian music with European music, Benjamin I. Gilman writes: "Hindu and African music is notably distinguished from our own by the greater complication of its rhythms. This often defies notation."[10] Concerning the measurement of rhythmic complexity, Martin Clayton writes: "I can think of no objective criteria for judging the relative complexity or sophistication of rhythm in, for example, Indian rag music, Western tonal art music, and that of African drum ensembles."[11]

The concept of complexity is extremely fluid. Its definition depends to a great extent on the context[12] and the purpose to which it is put.[13] In an information theory setting, a metronomic pulsation is least complex, and random noise is most complex. However, in a musical context completely random (disorganized) music is not complex at all. The most complex musical rhythms exhibit a degree of complexity that lies somewhere between complete order and complete disorder.[14] However, determining the exact location within this continuum that maximizes the complexity is easier said than done. As a consequence, numerous definitions of complexity have been proposed. Complexity is also multidimensional, and there are many ways of measuring and combining these dimensions.[15] Ilya Schmulevitch and Dirk-Jan Povel distinguish between three broad categories of complexity measures for musical rhythms: *hierarchical*, *dynamic*, and *generative*.[16] Hierarchical measures refer to structure at several levels simultaneously, dynamic measures refer to the nonstationare of the input over time, and generative measures depend on the amount of effort required to generate rhythms. Rhythm complexity may also be measured with respect to perception and performance (also called production[17]). Furthermore, these complexities depend on additional factors such as tempo and the underlying meter.[18] Experiments by D. J. Povel demonstrated that "changing the tempo of temporal sequences may cause dramatic changes in the perceived rhythmical characteristics."[19] In some contexts such as music transcription, it is desirable to determine the notation of a rhythm that minimizes the performance complexity while affecting the perceptual complexity as little as possible.[20]

In this chapter, several definitions and measures of rhythm complexity are compared.[21] Some of these are better than others, and the reader may wonder: why not just describe the best one? The answer is that there is no best. The usefulness of a measure depends on its intended application. Furthermore, it is hoped that some readers may be inspired by these concepts to invent new measures that may perhaps combine features of the measures described here. It has been my experience during many years of teaching at universities that just presenting the best correct algorithm is not necessarily the best way to teach. It is sometimes better to teach inferior or incorrect algorithms first. Even better is to teach incorrect algorithms that students instinctively believe to be correct. Then, after seeing counterexamples, students have the opportunity to learn the reasons for their failure and to attempt to fix them. Following such an experience, students not only have greater appreciation for correct solutions but also acquire better skills at designing good algorithms to start with. The best learning does not happen when knowledge is served on a plate,

but when the learner has to construct that knowledge. There is an old proverb that goes something like this: give a man a fish and you feed him for a day; teach him how to fish and you feed him for a lifetime. The same applies to algorithms.

OBJECTIVE, COGNITIVE, AND PERFORMANCE COMPLEXITIES

Everyone can understand the principle behind juggling three balls as well as the written instructions in a book on how to juggle. On the other hand, picking up three balls and juggling them is another matter altogether. In other words, we all know very well that *perceptual* or *cognitive* complexity is not the same as *performance* complexity. It is easier to recognize a favorite song than to sing it. In the words of artificial intelligence, pioneer Marvin Minsky: "Learning to recognize is not the same as memorizing. A mind might build an agent that can sense a certain stimulus, yet build no agent that can reproduce it."[22] In the same way, there is no logical a priori reason why a formal mathematical measure of complexity should agree with either cognitive or performance complexities. The structures inherent in cognitive or performance complexities may not be adequately captured by a simple mathematical formula. In this section, several measures of complexity of rhythm are compared by means of illustration with respect to the distinguished five-onset, 16-pulse clave rhythms highlighted in the preceding chapters.

One of the oldest measures of complexity used in music analysis is defined in terms of the predictability of outcomes of a random process. To illustrate this idea, assume for the sake of a simple *gedanken* experiment that in the music from a fictitious planet called Alpha, songs use interonset interval durations of either one or two pulses, and that each of these two types of songs occurs with the same frequency, i.e., if we select at random a rhythm from a song from planet Alpha, it will have durations of one pulse with probability 0.5 and durations of two pulses with probability 0.5. The probability distribution characterizing this scenario is pictured in Figure 17.1 (left). Assume further that in another planet Zeta the inhabitants use only durations of two pulses. Then the probability distribution characterizing the songs from planet Zeta is given in Figure 17.1 (right). The difference between these two extreme distributions implies that in planet Alpha one cannot predict with certainty the durations used

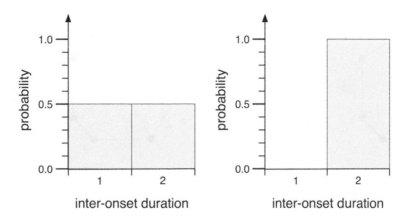

FIGURE 17.1   Two probability distributions: perfectly flat (left) and perfectly peaked (right).

in a song chosen at random, whereas in planet Zeta, one is certain that a song selected at random will have durations of two pulses. In this context, predictability implies there is no information obtained by selecting a song from planet Zeta. On the other hand, the nonpredictability of the outcome in planet Alpha implies that a maximum amount of information about the duration used is gained by selecting a song. Translating these ideas into the language of complexity, we obtain that predictability suggests simplicity, whereas nonpredictability or randomness suggests complexity.

From the geometrical point of view, nonpredictability, randomness, and therefore complexity may be characterized by the flatness of the underlying probability distribution. In this sense, a flatter distribution implies greater complexity, and thus the music in planet Alpha is more complex than the music in planet Zeta. Thus, the problem of measuring the complexity of a process has been converted to measuring the *flatness* of a probability distribution.

There are an uncountable number of ways to measure the flatness of a probability distribution, histogram, or by analogy, a geographical terrain. One measure is the smallness of the maximum height of the distribution. In Figure 17.1 the maximum height on the left is 0.5 and on the right is 1.0. Since 0.5 is smaller than 1.0, we would conclude that the distribution on the left is flatter than the one on the right. In general, a smaller maximum height implies a flatter distribution, but this is not necessarily so when the random variable can take on more than two values.

One very popular measure of the flatness of a distribution is *entropy*.[23] With respect to the probability distribution in Figure 17.1, let p_1 and p_2 denote the probabilities

of observing interonset durations of one and two pulses, respectively. Let P = (p_1, p_2) denote the probability distribution. It follows that $p_1 + p_2 = 1$. Then the entropy, usually denoted by $H(P)$, is given by the negative of the quantity ($p_1 \log p_1 + p_2 \log p_2$). This quantity takes a maximum value when the distribution is flat, that is when all the probabilities are equal, in this case when $p_1 = p_2$. It takes on its minimal value when one probability is equal to one and the other zero. In this case for the distribution on the left with $p_1 = p_2 = 0.5$ the entropy is one, and for the distribution on the right with $p_1 = 0$ and $p_2 = 1$ the entropy is zero (note that by convention $0 \log 0 = 0$).

In the more general case in which the random variable takes on N different values where $N > 2$, we have a probability distribution given by P = (p_1, p_2, ..., p_N), and the entropy is then given by

$$H(P) = -\sum p_i \log p_i$$

where the summation is over all $i = 1, 2, ..., N$.

One natural way to use the entropy as a measure of the flatness in the case of rhythm is to apply it to the histogram of all interonset intervals contained in a rhythm. Strictly speaking such a histogram is not a probability distribution that describes the behavior of a random variable, but rather a frequency or multiplicity count of the number of interonset durations of any given length that are present in the rhythm. Nevertheless, we may normalize the histogram so that its area is equal to one, and pretend that it is a probability distribution. The important point is not the faithfulness of the interonset interval histogram to probability theory, but rather the entropy's ability to measure the flatness of any histogram, no matter what its origin.

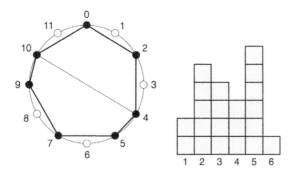
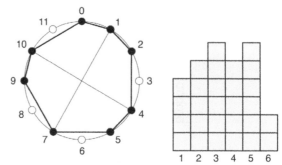

FIGURE 17.2   Entropy (left) = 2.398 and (right) = 2.513.

Consider the African bell pattern shown in Figure 17.2 (left). This is a rotation of the *bembé* timeline, and it is a deep rhythm, as can be seen from its histogram laid out next to it. Deep rhythms have relatively peaked histograms since they resemble one-sided Maya pyramids when the histogram bin heights are sorted by increasing height. The rhythm on the right in Figure 17.2 is the clapping pattern used by Steve Reich in his minimalist piece *Clapping Music*. Reich's pattern contains one additional onset at pulse one, compared with the African bell pattern. This additional onset introduces another antipodal pair of onsets with distance six between pulses one and seven, to the pair (4, 10) already present. However, as may be observed from the figure, this change also makes the histogram flatter. This is reflected by the increase in entropy from 2.398 to 2.513. In this sense, Reich's pattern transforms the African bell pattern into a more complex rhythm by including that onset. Alternately, consider the nine-onset *shekere* rhythm used in the *bembé* music of Cuba pictured in Figure 17.3. Removing the onset at pulse 11 converts this rhythm to a rotation of Reich's *Clapping Music* pattern (when started at pulse five). From this point of view, Reich's pattern is a transformation of the asymmetric *bembé* shekere rhythm to one that contains mirror symmetry.

In addition to the wooden claves and metal bells described in Chapters 4 and 5, a *shekere*, such as the one illustrated in Figure 17.4, is another widely used instrument for playing rhythmic timelines in African and Afro-Cuban traditional music. It is made from a hollowed-out gourd enveloped by a fishnet that holds a large quantity of beads (or seeds) loosely around the gourd. Rhythms are typically played by either pulling on the fish net or bouncing the gourd between one's thigh and free hand.

Steve Reich intended *Clapping Music* to be performed by two people clapping hands. Both performers clap the same rhythm shown in Figure 17.2 (right). One performer repeats the sequence continually throughout the piece, while the second player shifts the pattern by one time unit every time the pattern has been repeated 12 times. The piece ends when both performers play in unison again.

There has been speculative analysis about how Reich might have come to adopt this particular rhythmic

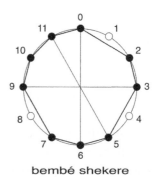

bembé shekere

FIGURE 17.3   A *shekere* rhythm used in the *bembé* rhythm ensemble.

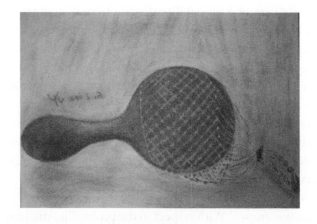

FIGURE 17.4   A *shekere*. (Courtesy of Yang Liu.)

pattern [x x x . x x . x . x x .] for this composition, and why this pattern is more successful than other possible candidates. One combinatorial analysis by Joel Haak proceeds by eliminating candidates while respecting several mathematical constraints.[24] His argument proceeds as follows. There are eight claps per cycle of 12 pulses in *Clapping Music*. The number of candidates or ways one can select 8 out of 12 pulses in which to clap is (12!)/(8!)(4!) = 495. His first constraint is that a pattern should begin with a clap rather than a silent pulse. His second constraint is that the silent interval between two consecutive claps should be short, and therefore two consecutive pauses (silent pulses) are not permitted. With these two constraints, the original 12 units, composed of eight claps and four rests, are reduced to eight units made up of four clap–rest patterns [x .] and four solitary claps [.]. In this setting, there are now only eight two-valued elements taken four at a time, and thus the formula for the total number of possible patterns becomes 8!/((4!)(4!)) = 70. Among these 70 patterns, there are some that are rotations of each other, and therefore are redundant, since they would yield the same composition from a different starting point. This observation leads to Haack's third constraint: the patterns should not be cyclic permutations of each other. With the addition of this third constraint, the 70 possible patterns are reduced to only 10 patterns. His fourth constraint is that during the execution of the entire piece the combined 12-pulse clapping patterns made by both performers should not repeat themselves before the ending of the piece. His fifth and last constraint is that consecutive repetitions of phrases consisting of the number of claps between consecutive pauses are not allowed. In other words, patterns such as [x x x . x x . x x . x .], [x x x . x . x x . x x .], and [x x x x .

x x . x . x .] are not permitted because of the presence of repetitive subunits such as [x x . x x .] = [x x .] [x x .] and [x . x .] = [x .] [x .]. With these five constraints, only two of the 495 patterns remain as possible candidates. One is the pattern Reich chose in Figure 17.2 (right), and the other is the pattern [x x x x . x . x x . x .] shown in polygon notation in Figure 17.5.

Haak does not speculate on the criteria that might be employed for choosing between the two finalist candidates that remain after the five rounds of constraint-satisfaction eliminations have been applied. Indeed, there are several arguments that emerge from musicology, geometry, and information theory that may be enlisted to come to the rescue here. One obvious solution is to pick the rhythm that minimizes the number of consecutive claps without gaps of silent pulses. Then we end up with Reich's pattern that starts with a group of three rather than four claps.

Other possible criteria for selecting Reich's pattern become evident by comparing the polygonal representations of these rhythms in Figures 17.5 and 17.2 (right). For one, although both patterns exhibit mirror symmetry, the mirror symmetry in Reich's pattern is with respect to a line that is incident to two antipodal onsets at pulses 1 and 7. Haak's rhythm is not symmetric about a pair of onsets but rather about a line that falls midway between the pairs (1, 2) and (7, 8). Whether this mathematical property has musicological capital is yet to be investigated. Another difference between the two rhythms with respect to their antipodal pairs of onsets is that although both rhythms contain two such pairs, the pairs in Haak's rhythm given by (1, 7) and (2, 8) are adjacent to each other, whereas the pairs in Reich's rhythm given by (1, 7) and (4, 10) are orthogonal to each other,

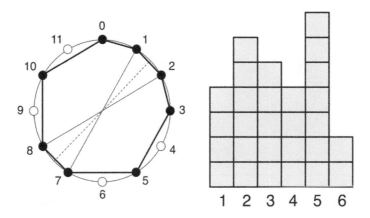

FIGURE 17.5   The second rhythm found by Joel Haak (left), and its interval histogram (right) with entropy = 2.49.

and thus form a regular four-beat underlying structure. This difference between the two candidates probably carries greater musicological weight, although the nature of this weight is also a topic that needs investigation.

Alternately, we may resort to the criterion of rhythmic evenness of the two patterns to arrive at Reich's pattern. Figure 17.6 shows the two rhythms with their onsets plotted in a two-dimensional *onset-pulse* plane in which the *x*-axis is the pulse number (time) and the *y*-axis is the onset number. The onsets are connected together to form a polygon (shaded). The longest edge of this polygon, from pulse 0 to 12 (zero), represents the location of onsets of perfectly even rhythms. Thus, the area of the shaded polygon is a measure of unevenness of the rhythm. A smaller area implies a more even rhythm.[25] Comparing the polygon on the left from Haak's second rhythm with the polygon on the right from Reich's rhythm, it is clear that Reich's rhythm is more even. Indeed, the reduction in area is the result of the movement of fourth and seventh onsets, by one pulse each, closer to the diagonal baseline of the polygon.

Finally, one could use the entropy of the interval content histograms to select Reich's pattern over Haak's second rhythm. Haak's second rhythm has entropy equal to 2.49, whereas the entropy of Reich's pattern is 2.513. The difference between the two is not large, but Reich's pattern still comes out ahead.

We have entertained several speculations regarding how Steve Reich might have come to adopt the pattern [x x x . x x . x . x x .] for his composition *Clapping Music*. Haack proposed musicological constraints that uniquely isolated this pattern from among the 495 possibilities of selecting the locations of eight claps from a cycle of 12 pulses. Then there are well-known timeline bell patterns of seven onsets for which inserting one additional onset in the right location yields Reich's pattern. There are also rhythms of nine onsets for which removing one judicial onset yields Reich's pattern.

In closing this exploration, it is fitting to recount what Steve Reich himself has said regarding the selection of this pattern for his composition. Russell Hartenberger, who initially performed *Clapping Music* with Steve Reich, has written a wonderful historical account of the development of *Clapping Music* (originally titled *Pulse Music*), that includes comments made by Reich during an interview. According to Hartenberger, Reich "devised the *Clapping Music* rhythm in an attempt to create a pattern that was a variation of the Atsiagbekor African bell pattern" (referred to as *bembé* in this book). To quote Steve Reich during the interview: "I didn't want to use the African pattern at that time; I wanted to make my own variation on it. Then there was the 3 2 1 2 that occurred to me."[26]

Finally it is worth noting that Reich's pattern [x x x . x x . x . x x .] is a hand-clapping resultant rhythm used in the *Beer Dance* of the *Lala* people of former Rhodesia. In this piece, three performers each clap one pattern that together yield Reich's pattern as a resultant rhythm, started at the seventh pulse.[27] The three clapping patterns and their resultant are illustrated in box notation in Figure 17.7. The patterns are the regular [4-4-4] pattern, the five-onset *fume–fume*, and the seven-onset *bembé*. By all accounts, Reich discovered this pattern independently, and considered this resultant pattern to be so successful that he went on to use it in several other compositions such as *Music for Eighteen Musicians*.[28]

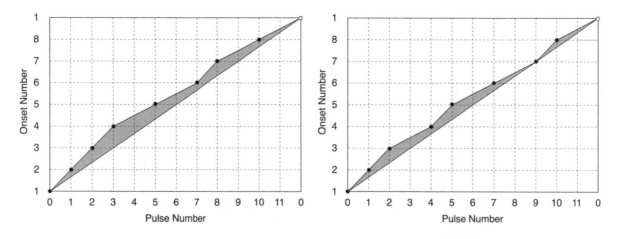

FIGURE 17.6 Deviation of Haak's (left) and Reich's (right) rhythms from a perfectly even regular rhythm.

	0	1	2	3	4	5	6	7	8	9	10	11
Clap 1	●				●				●			
Clap 2	●		●		●			●		●		
Clap 3	●		●		●	●		●		●		●
Resultant	●		●		●	●		●	●	●		●

FIGURE 17.7 Reich's pattern from *Clapping Music* is a rotation of the rhythm timeline that results from the three clapping patterns performed in the *Beer Dance* of the *Lala* people.

The entropy may also be used as a global feature for comparing and classifying rhythms. As an example, consider the six distinguished five-onset, 16-pulse timelines of Figure 7.1. The entropies of their full and adjacent interval histograms shown in Figure 8.3 are listed in the following table in Figure 17.8, in increasing order from left to right. One of the weaknesses of the entropy for measuring rhythm complexity is immediately evident from the table. Even though the bossa-nova is arguably more complex than the shiko, their full interval histograms look quite different, they have the same entropy value of 1.84. This is because the entropy function depends only on the height of the histogram bins and not on their location within the histogram. Since both histograms have four occupied bins of heights one, two, three, and four, their entropies are equal. Similarly,

the entropy cannot distinguish between the full interval histograms of the gahu and soukous. Interestingly enough, in these two cases, the entropies of the adjacent interval histograms disambiguate the shiko from the bossa-nova, and the gahu from the soukous, even though they cannot by themselves distinguish between the son, rumba, and gahu.

Figure 17.9 (left) shows a plot of the full interval histogram entropy along the abscissa and the adjacent interval histogram entropy along the ordinate. The six timelines fall naturally into three clusters: one cluster is made up of shiko and bossa-nova that share the same value of full interval entropy, another cluster consists of the son, rumba, and gahu, which have the same value of adjacent interval entropy, and soukous is off by itself.

Although either of the two entropies is unable to distinguish between all six rhythms, if a new measure is defined as the sum of both entropies, then a perfect distinguishing ordering is possible, as illustrated in Figure 17.9 (right). This diagram shows the diagonal lines that are loci of constant sum of the ordinate and abscissa values. This measure produces the ordering: bossa-nova, shiko, son, rumba, gahu, and soukous.

In Chapter 9 we described several popular methods used in music information retrieval to classify rhythms automatically. Here we take this opportunity to revisit

Entropy	Bossa-Nova	Shiko	Son	Rumba	Gahu	Soukous
Full Interval	1.84	1.84	2.24	2.44	2.72	2.72
Adjacent Interval	0.72	0.97	1.52	1.52	1.52	1.92

FIGURE 17.8 The full and adjacent interval entropies of the six distinguished timelines.

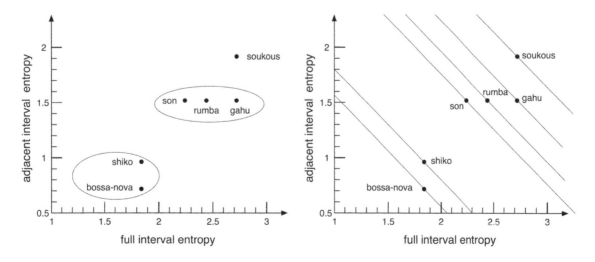

FIGURE 17.9 Clustering and ordering entropies of adjacent and full interval histograms.

the topic by introducing another widely used approach to classification that constructs decision trees by means of *space partitioning*.[29] The method is illustrated in Figure 17.10 with the toy example of the six distinguished time-lines used previously. The idea is to partition the space with vertical and horizontal lines in an alternating fashion (if possible) so that we can easily produce a decision tree afterwards. First the vertical line A is inserted at a coordinate value of 2.0. This produces two half-spaces that are partitioned next. Accordingly, horizontal line B is inserted on the left at coordinate 0.85, and horizontal line C on the right at coordinate 1.7. Next, vertical lines are inserted at coordinates 2.6 and 2.35 to separate son, rumba, and gahu rhythms.

The binary space partition of the six rhythms shown in Figure 17.10 yields the binary decision tree shown in Figure 17.11.

THE LEMPEL–ZIV COMPLEXITY

In 1976, Abraham Lempel and Jacob Ziv proposed an empirical information-theoretic measure of the complexity of a finite-length sequence of symbols in the context of data compression.[30] Their goal was to store a sequence of symbols in such a way as to use as little memory as possible. Their novel approach in fact yields a measure of the complexity of a given finite-length sequence by scanning it from left to right, looking for the shortest subsequences (patterns) that have not yet been encountered during the scan. Every time such a pattern is found, it is inserted in a growing dictionary

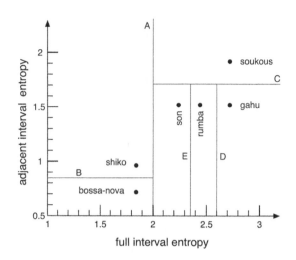

FIGURE 17.10   A binary space partition of the rhythms based on two entropies.

of patterns. When the scan is completed, the *size* of this dictionary is the measure of the complexity of the sequence. For the special case of cyclic sequences such as the rhythms considered here, a concatenation of two instances of the rhythm cycle is scanned. The application of this measure of sequence complexity to musical rhythm was explored by I. Shmulevich and D.-J. Povel.[31]

Since the publication of the original data compression algorithm of Lempel and Ziv, many variations on their theme have been proposed. To illustrate just one simple variant by means of an example, consider the *clave son* timeline shown in a binary box notation in Figure 17.12. Concatenating two copies of this rhythm

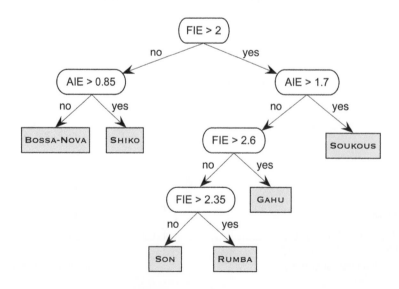

FIGURE 17.11   A binary decision tree based on space partitioning in Figure 17.9.

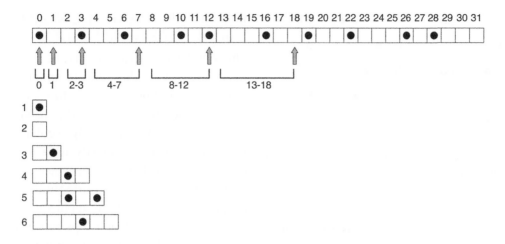

FIGURE 17.12 Illustrating the computation of the Lempel–Ziv complexity of the *clave son*.

yields the 32-pulse pattern shown at the top along with the pulse numbers. Figure 17.12 also shows each new subsequence encountered during the scan. The arrows underneath the sequence indicate the positions at which a new subsequence is discovered. The brackets with numbers underneath the arrows indicate the first and last pulses of each new subsequence discovered. A subsequence is considered newly discovered if it does not occur to the left of the previous arrow. Let us step through the algorithm to clarify the process. The scan is initialized at pulse zero, and of course this pattern is the first newly discovered sequence. The scan advances to pulse one, discovering another new sequence consisting of pulse one. The next newly discovered pattern consists of pulses two and three. The fourth pattern consists of pulses 4 –7, the fifth is made up of pulses 8–12, and the sixth of pulses 13–18. Starting at pulse 19, no new patterns are found because the sequence from pulse 19 to the last pulse 31 with intervals [3-4-2-4] already occurs between pulses 3 and 16. All the different subsequences discovered in this way are listed in a dictionary at the lower left and labeled in the order in which they are discovered. For the *clave son* rhythm, six subsequences are generated by this procedure, and therefore its complexity is equal to six. This measure is relatively simple to compute, and it is, like the entropy, completely objective in the sense that it is defined in pure mathematical terms without any explicit dependencies on psychological principles of perception.

The Lempel–Ziv measure has been compared experimentally with the complexity perceived by human subjects. The experiments used rhythms with a 16-beat measure typical of those found in Western music and yielded negative results. The comparison of this measure with the other measures of rhythm complexity discussed here, with respect to the six African, Cuban, and Brazilian clave patterns, indicates that this measure is also inferior for non-Western rhythms. Furthermore, looking at the scores obtained for the six rhythms in Figure 17.14 shows that this measure is deficient for other reasons as well. There is almost no variance in the scores: all values are six except for the bossa-nova, which receives a five. So, the measure does not discriminate well between short sequences such as these. Also, the scores do not make sense to anyone experienced in teaching or playing these rhythms. For example, shiko is the simplest of the six rhythms, and gahu more complex, both to recognize and to play, yet the Lempel–Ziv complexities are six for both of these rhythms. In conclusion, at present, it appears that information-theoretic measures per se are not able to capture well the human perceptual, cognitive, or performance complexities of short musical rhythms such as timelines. It is quite probable that the Lempel–Ziv measure may perform better for much longer rhythms or entire musical compositions.[32] Indeed, compression-based measures of complexity, akin to Lempel–Ziv complexity, have also been used to define measures of music (melodic) similarity and have been shown to correlate significantly with human perceptual judgments of similarity.[33]

THE COGNITIVE COMPLEXITY OF RHYTHMS

In contrast to the information-theoretic measures of complexity, Jeff Pressing proposed a measure of the cognitive complexity of musical rhythms based on

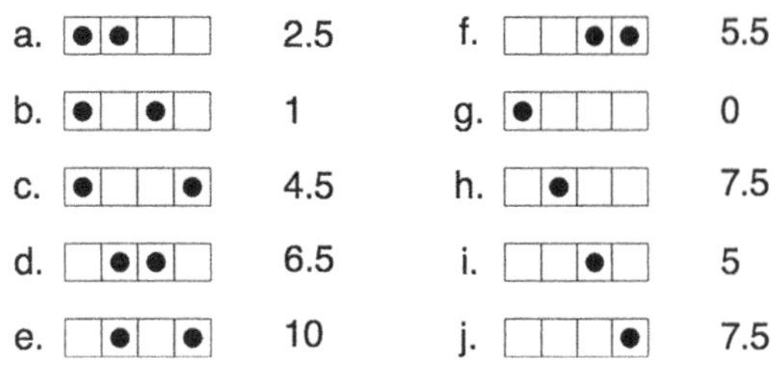

a. ● ● □ □ 2.5   f. □ □ ● ● 5.5
b. ● □ ● □ 1   g. ● □ □ □ 0
c. ● □ □ ● 4.5   h. □ ● □ □ 7.5
d. □ ● ● □ 6.5   i. □ □ ● □ 5
e. □ ● □ ● 10   j. □ □ □ ● 7.5

FIGURE 17.13   Pressing's cognitive complexities of ten basic four-pulse rhythmic units.

IRREGULARITY AND THE NORMALIZED PAIRWISE VARIABILITY INDEX

The complexity of a rhythm may be characterized by the irregularity of the durations of its interonset intervals. There are many possible ways to measure irregularity. One approach is to measure the distance between the given rhythm and a perfectly regular one.[35] A widely used measure of irregularity used in statistics is the classical standard deviation, which has been applied frequently to the analysis of rhythm in speech and language. However, in these applications, as in musical rhythm, the order relationships between adjacent intervals are important, and the standard deviation disregards them. To take order information into account, a measure should be sensitive to local change. The "normalized Pairwise Variability Index" (nPVI) is a measure that attempts to capture this notion of change. The nPVI for a rhythm is defined as

$$nPVI = \left(\frac{100}{m-1}\right)\sum_{k=1}^{m-1}\left|\frac{d_k - d_{k+1}}{\frac{(d_k + d_{k+1})}{2}}\right|$$

where m is the number of adjacent interonset intervals and d_k is the duration of the kth interval. Although the nPVI has a long history of application to language, its ramifications in the music domain are beginning to be explored.[36]

The values of the nPVI in Figure 17.14 show that irregularity does not necessarily translate monotonically to complexity. Shiko is clearly a much less complex rhythm than bossa-nova, as Pressing's complexity underscores. However, the nPVI is much greater for shiko (66.7) than for bossa-nova (14.3). These results serve to highlight the fact that measuring rhythmic complexity is a complex problem, and much work still remains to be done.

psychological properties of perception as well as musicological principles, such as the amount of syncopation present in the rhythm.[34] The cognitive complexities of the ten 4-pulse patterns containing one-onset and two-onset computed with Pressing's measure are given in Figure 17.13. One simple way to obtain a measure of cognitive complexity for longer rhythms such as the 16-pulse rhythms considered here is to first partition these rhythms into four units of four pulses each, then compute the complexities for each unit, and finally add these four complexity values. For example, the shiko pattern consists of the concatenation of patterns [x . . .], [x . x .], [. . x .], and [x . . .]. Referring to Figure 17.14, we find the corresponding complexity values 0, 1, 5, and 0 for a total of 6. On the other hand, rumba yields a Pressing cognitive complexity of 4.5 + 7.5 + 5 + 0 = 17. Examining the Pressing cognitive complexities of all six clave rhythms in the table of Figure 17.14 reveals more information than the Lempel–Ziv complexity. For one, all the scores are different and the variance is quite large ranging from 6 for the shiko to 22 for the bossa-nova. The scores are also in good agreement with my personal teaching and performing experience. Shiko is easy, rumba is more difficult than son, and bossa-nova is the most difficult to recognize and perform.

	Pressing	Lempel-Ziv	Entropy Adjacent	Entropy Full	Metric	Distinct Distances	nPVI
Shiko	6	6	0.97	1.84	2	4	66.7
Son	14.5	6	1.52	2.24	4	5	40.5
Soukous	15	6	1.92	2.72	6	7	70.5
Rumba	17	6	1.52	2.44	5	6	41.0
Gahu	19.5	6	1.52	2.72	5	7	23.8
Bossa	22	5	0.72	1.84	6	4	14.3

FIGURE 17.14   A comparison of seven measures of rhythm complexity.

NOTES

1 Although most musicologists argue for the supremacy of rhythm over other features of music, this thesis is not without its detractors. The composer Olivier Messiaen, for example, writes: "The melody is the point of departure. May it remain sovereign! And whatever may be the complexities of our rhythms and our harmonies, they shall not draw it along in their wake, but, on the contrary, shall obey it as faithful servants." See Messiaen, O., (1956), p. 13.

2 Gabrielson, A., (1973a, 1973b), used factor analysis and multidimensional scaling to uncover 15 perceptual features of rhythm, and complexity stood out among them. Conley, J. K., (1981), p. 69, experimented with ten physical features of music complexity calculated from Beethoven's *Eroica* Variations, Op. 35, and found that the rate of rhythmic activity (in terms of the number of rhythmic events) was the most powerful measure of complexity. Interestingly, Wang, H.-M., Lin, S.-H., Huang, Y.-C., Chen, I.-C., Chou, L.-C., Lai, Y.-L., Chen, Y.-F., Huang, S.-C., & Jan, M.-Y., (2009), showed that the complexity of rhythms can modify the interbeat duration patterns of the heart of the listener. See Diaz, J. D., (2017) for techniques that increase the complexity of rhythms in Afro-Bahian Jazz.

3 Arom, S., (1984), p. 51.

4 Jones, A. M., (1949), p. 295.

5 Lempel, A. & Ziv, J., (1976). See Coons, E. & Kraehenbuehl, D., (1958), for some early work on the application of information theory to the analysis of musical structure.

6 Chemillier, M., (2002) and Chemillier, M. & Truchet, C., (2003).

7 Toussaint, G. T., (2005b).

8 Berendt, J.-E., (1987), p. 202.

9 Agawu, K., (1995).

10 Gilman, B. I., (1909), p. 534.

11 Clayton, M., (2000), p. 6.

12 Repp, B. H., Windsor, W. L., & Desain, P., (2002). Toussaint, G. T., (1978), provides a tutorial survey on the dependence between the perception and recognition of patterns in spatial (visual) and temporary (auditory) modalities, and the context in which those patterns are perceived. See also Van der Sluis, F., Van den Broek, E., Glassey, R. J., Van Dijk, De Jong, E. M. A. G., (2014).

13 Wolpert, D. H. & Macready, W., (2007). See Crofts, A. R., (2007), p. 25, for the relevance of complexity to evolution.

14 Eglash, R., (2005), p. 154. See Akpabot, S., (1975) for the structure of random music performance of Birom in Nigeria.

15 Sioros, G. & Guedes, C., (2011), p. 385, propose a complexity measure that combines the density of events in a rhythm with its syncopation by means of the formula Complexity = $\{density^2 + syncopation^2\}^{1/2}$. See also Essens, P., (1995).

16 Shmulevich, I. & Povel, D.-J., (1998, 2000a, 2000b).

17 Fitch, W. T., (2005), p. 31.

18 Vinke, L. N., (2010), p. 41. Scheirer, E. D., Watson, R. B., & Vercoe, B. L., (2000). Palmer, C. & Krumhansl, C. L., (1990) have shown experimentally that rhythm perception (and therefore rhythm complexity) is influenced by the underlying meter. Rhythm perception also depends on rhythmic grouping. Music theorists, such as Lerdahl, F. & Jackendoff, R., (1983) have argued that meter and figural grouping are independent. However, psychologists have obtained experimental evidence that they are not only dependent but also that figural grouping may be even more important than meter in judging rhythm similarity. See Handel, S., (1992, 1998). Furthermore, in the field of music information retrieval, Chew, E., Volk, A., & Lee, C.-Y., (2005), have shown that a type of meter extracted from onset grouping is quite successful at classifying certain types of music. They call the measure-based definition of meter used in traditional Western music, the *outer meter*, and a meter extracted from the grouping of the note onsets (ignoring the measures and bar lines), the *inner meter*. This approach to solving music problems is referred to as *inner metric analysis*.

19 Povel, D. J., (1984), p. 330.

20 Nauert, P., (1994), p. 227.

21 See Thul, E. & Toussaint, G. T., (2008a) for a comparison of many more measures of rhythm complexity, and Thul, E. & Toussaint, G. T., (2008b, 2008c) for a comparison of the complexity between African timelines and North Indian talas. See also Ravignani, A. & Norton, P., (2017).

22 Minsky, M., (1981), p. 30.

23 Kulp, C. W. & Schlingmann, D., (2009), Cohen, J. E., (2007), p. 139, Gregory, B., (2005), p. 11, and Streich, S., (2006), p. 20. In his PhD thesis, Streich proposes algorithms to compute estimates of a variety of features of music complexity (based on rhythm, tonality, and timbre) from musical audio signals. Don, G. W., Muir, K. K., Volk, G. B., & Walker, J. S., (2010), p. 44, quantify the complexity of musical rhythms (represented as binary sequences) with the entropy function. De Fleurian, R., Blackwell, T., Ben-Tal, O., & Müllensiefen, D., (2016), report on experiments that support the hypothesis that entropy correlates with human judgments of complexity. In an early influential book, Moles, A., (1966), applies Shannon's entropy-based theory of information (uncertainty) to the analysis of expectancy and originality in music. Other information measures that have also been applied extensively to music (when *two* probability distributions are involved) include the *discrimination information* (also called the Kullback-Liebler distance). For example, Farbood, M. M. & Schoner, B., (2009), apply the discrimination information to determine the salience of several features of the acoustic signal for the perception of musical tension. That the discrimination information is appropriate for measuring the distance between two probability distributions is due to the fact

that the measure is closely related to the Bayes error probability (also Kolmogorov distance or simply *variation*); see Toussaint, G. T., (1975). Scheirer, E. D., (2000), p. 98, uses the variance of interonset durations as a measure of rhythm complexity.

24 Haack, J. K., (1991, 1998).

25 This is but one measure of rhythmic evenness, a topic to be explored deeper in Chapters 18–21.

26 Hartenberger, R., (2016), pp. 157–158.

27 Jones, A. M., (1954a), p. 44.

28 Potter, P., (2000), p. 225. See Cohn, R., (1992a), for an analysis of some of Steve Reich's other phase-shifting music.

29 Safavian, S. R. & Landgrebe, D., (1991).

30 Lempel, A. & Ziv, J., (1976).

31 Shmulevich, I. & Povel, D.-J., (1998).

32 See also Chaitin, G. J., (1974), for information measures based on the shortest possible description of a rhythm.

33 Pearce, M. & Müllensiefen, D., (2017), p. 137.

34 Pressing, J., (1997).

35 Toussaint, G. T., (2012b).

36 Toussaint, G. T., (2012c). See also the detailed comparison of the nPVI measure to other measures of complexity, and its application in characterizing families of rhythms from different cultures in: Toussaint, G. T., (2013c). The pairwise variability index as a measure of rhythm complexity. *Analytical Approaches to World Music*, 2/2:1–42. See also Condit-Schultz, N., (2016).

CHAPTER 18

Meter and Metric Complexity

WHAT IS METER?

Tʜᴇ ᴄᴏɴᴄᴇᴘᴛ ᴏꜰ ᴍᴇᴛᴇʀ ᴡᴀꜱ ʙʀɪᴇꜰʟʏ ɪɴᴛʀᴏᴅᴜᴄᴇᴅ in Chapter 3. In this chapter, we dig deeper into three specific aspects of meter in its two basic forms: *isochronous* meter and *nonisochronous* meter. The music literature is filled with a variety of definitions of meter, most of them deficient in one way or another. For instance, Richard Cohn, critical of today's paucity of teaching meter in music schools, lists four antiquated definitions of meter from the early chapters of recent American harmony textbooks, authored by well-known music theorists.[1] These four definitions are as follows.

1. "Beats are … grouped into a regular repeating pattern of strong and weak. This is the meter."

2. "This pattern of stressed and unstressed beats results in a sense of metrical grouping or meter."

3. "Meter provides the framework that organizes groups of beats and rhythms into larger patterns of accented and unaccented beats."

4. "Meter is the arrangement of rhythm into a pattern of strong and weak beats."

Upon reading these definitions, the reader may wonder what the difference is between *meter* and *rhythm*. Must rhythm without meter be devoid of strong and weak beats? Consider the definition of rhythm offered by G. Cooper and L. B. Meyer: "Rhythm may be defined as the way in which one or more unaccented beats are grouped in relation to an accented one."[2]

Since *accented* and *unaccented* beats are akin to *strong* and *weak* beats, respectively, there is in effect little difference between Cooper & Meyer's definition of *rhythm* and the last three definitions of *meter* listed by Cohn. Only the first definition of meter, with its inclusion of the word *regular*, hints at a possible difference between meter and rhythm, and is echoed approximately by Cooper & Meyer's definition of meter: "Meter is the measurement of the number of pulses between more or less regularly recurring accents. Therefore, in order for meter to exist, some of the pulses in a series must be accented—marked for consciousness—relative to others."[3] However, the specification of "regular" in the first definition listed by Cohn has been relaxed by Cooper & Meyer to "*more or less regular.*" In any case, the notion of regularity of beats appears to be one of the salient features in most definitions of meter.

In this chapter I will not attempt to answer the questions posed in the title of this section. The reader is directed to the literature that offers an abundance of attempts at precise definitions of *meter*.[4] Martin Clayton disentangles some of these definitions,[5] and Christopher Hasty illuminates the interface between meter and rhythm,[6] Justin London clarifies the distinction between meter and the concept of *grouping*,[7] and Johansson proposes a pattern recognition approach to meter, advocating that "*a top-down gestalt processing interacts with a bottom-up, additive mechanism.*"[8] Instead, in this chapter, I illustrate the application of three distinct but precise hierarchical mathematical definitions of meter to three problems: (1) the question of whether meter exists in African rhythm, (2) the measurement of the complexity of nonisochronous meter, and (3) the applicability of a model of the perception of meter.

DOES AFRICAN RHYTHM POSSESS METER?

Much has been written during the past century about the similarities and differences between African and Western music.[9] "What makes African rhythms sound so different from Western rhythms?" is a question often asked.[10] Some authors claim that African music has more complex rhythms[11] or that its rhythms are more developed.[12] The rhythms in African music have also been compared with those found in Indian music in terms of complexity[13] and their additive/divisive properties.[14] African music has also been compared with Western music using rhythm complexity measures of their melodies.[15] Comparing western (European) music to both African and Indian music, Benjamin I. Gilman writes: "Hindu and African music is notably distinguished from our own by the greater complication of its rhythms. This often defies notation."[16] Kofi Agawu chronicles a good deal of literature that focuses on the purported prominence of rhythm in African music, and its asserted complexity relative to that of Western music.[17] But is it indeed the case that African and Western musical rhythms are fundamentally different? In support of this view, John Miller Chernoff writes that "Western and African orientations to rhythm are almost opposite."[18] On the other hand, for David Temperley[19] "African and Western rhythms are profoundly similar." What is one to make of such antithetical pronouncements?

I will not take sides on this draconian dichotomous distinction of such a complicated issue. One reason for taking this stance is that there exists in the literature scores of different definitions of rhythm, as discussed in Chapter 1. By which definition then should African and Western rhythm be compared? As an example, consider one of these definitions penned by B. C. Wade, which stipulates that "A rhythm is a specific succession of durations." By this definition African and Western rhythms are more than "profoundly similar." They are in fact *identical*. Furthermore, although the analysis expounded here is quantitative in nature, the goal is not to pin down a number with which to characterize the amount of rhythmic similarity, that lies somewhere in between "almost opposite" and "profoundly similar." A possible way to attack such a problem quantitatively is to calculate a comprehensive list of scores of rhythmic features from both symbolic and acoustic samples of African and Western music, thus rendering the samples as points in a high-dimensional space. The distance between these

points according to a suitable metric might then yield a quantitative measure of the similarity of African and Western rhythm. Such an ambitious and difficult study is left for the future. Instead, I will provide an answer to Victor Grauer's question: "Are African rhythms actually based on an underlying meter, and if so, can such 'meters' be compared to the meters found so typically in European music?"[20] This more modest goal of the analysis presented in the first section of this chapter zooms in on a single mathematical property of rhythm that measures *hierarchical meter*, suggests a method of quantifying it by means of *pulse saliency histograms*, and calculates these histograms for some specific examples of African and Western music to determine how they can inform the issue of whether African rhythm exhibits hierarchical meter, and to provide some new quantitative data to illuminate the more general question of the similarity between African and Western rhythm.[21]

Before embarking further in this exploratory study, a word is in order concerning the samples of Western and African rhythms used, as well as the experimental methodology adopted. The word "African" here refers to the music indigenous in the region to the south of the Sahara, and thus excludes the Arabic rhythms of North Africa. However, it includes rhythms from the Caribbean and Brazil that are used by the sub-Saharan African communities there. On the methodological side, the statistical measures employed are descriptive rather than hypothesis-driven. No hypotheses are posited here, with regards to whether African and Western rhythm are similar or not. To properly carry out such a scientific study, random samples of all western and African music would have to be obtained, to be able to validly test such hypotheses. The study described here is rather a preliminary exploration (in the style of data mining) of some examples that may point the way to possible future more exhaustive analyses to test more rigorously specific hypotheses. In the absence of a random sample of African rhythms, an alternative approach is adopted in the form of a "worst-case" analysis, in which rather than obtaining a large *random* sample of African rhythms, a collection of *unique* special rhythms is selected for study. In particular, if the earlier claims are true that African rhythm does not have hierarchical meter, then these special rhythms (among all African rhythms) should be those least likely to possess hierarchical meter. Good candidates for this purpose are the asymmetric timelines, usually played with a variety of bells and high-pitched wooden sticks.[22] For the Western rhythms, samples were used for which the pulse saliency histograms were easily available or computable. These pieces span Renaissance and Common Practice music, and include Palestrina's *Pater Noster*,[23] German folk songs,[24] and compositions by J. S. Bach, Mozart, Brahms, and Shostakovich.[25] For comparison with music theory, the histogram determined by the mathematical *Generative Theory of Tonal Music* (GTTM) hierarchy is used, as explained in the following. Since for a 16-pulse time span (cycle, measure), the GTTM hierarchy is uniquely defined, the timelines and music samples selected for this study all had 16-pulse cycles, thus providing a sharper focus for the comparison.

At one end of the conceptual spectrum of definitions, meter is conceived as a pulsation of equally spaced (regular) beats (lacking any hierarchy) that may be sounded or merely felt, and that functions as the railing on which rhythms ride. At this level meter divides the time span cycle (measure, bar) into a specific number of regular beats such as 3, 4, 5, 6, 7, 8, 9, 12, 16, etc., without placing any emphases (accents) on any one beat. At the other end of the spectrum, the regular beats are hierarchically arranged according to their strength within an evenly divisible periodic cycle. Such is the view of Lerdahl and Jackendoff (1983) who define meter as a regular pattern of alternating strong and weak beats arranged in a specific hierarchical manner.[26] According to this definition, the meter furnishes the musician with a hierarchy of temporal reference points. This structure, referred to as the GTTM hierarchy, is illustrated in Figure 18.1 for the case of a 16-pulse periodic cycle. The height of the column in each pulse position reflects the relative strength of each pulse. The values of these heights should not be interpreted as absolute numerical quantities, but

rather as relative magnitudes with respect to each other. What is more important for characterizing the nature of GTTM hierarchy is the discrete ordering (rank) of these 16 magnitudes. This is the definition of meter that has been frequently invoked to contrast African with Western rhythm, and which is the focus of this exploration. For the purpose of the statistical analysis carried out, the magnitudes may be scaled so that they all sum to 1, and thus may be conveniently viewed as prescriptive probabilities of the occurrences of onsets at each of the 16 pulse positions.

Immediately following its publication, the GTTM model received considerable criticism from music psychologists and ethnomusicologists, regarding its applicability to nonwestern music.[27] One specific criticism has been that the GTTM hierarchy is based on intuition and music theory principles that are not supported by psychological experimental data.[28] This criticism spawned several empirical studies to evaluate GTTM's psychological reality.[29] In studies with Western music, it was found experimentally that the strength of a pulse location correlates well with the degree of the expectancy of occurrence of an onset at that particular location.[30] Even so, a second criticism of GTTM has been that it is a theory applicable only to western tonal music, and that its claims of universality have not been supported by intercultural research. Some writers contend that African rhythms in general, and timelines in particular, exhibit an additive structure rather than being hierarchically evenly divisible, as specified by the GTTM model.[31] Indeed, some evidence for this view suggests that under certain conditions the timeline-ground model achieves superiority over the Western pulse-ground model.[32]

In addition to the earlier criticisms of the GTTM hierarchy promulgated by music psychologists, some ethnomusicologists have derogated its applicability to African music. Regarding the role of meter in sub-Saharan African music, Simha Arom writes: "The pulsation is the only temporal reference the musicians have."[33] J. H. Kwabena Nketia emphasizes that "The African learns to play rhythms in patterns."[34] Such a sentiment is echoed by James Koetting, who writes that "African drummers do not think in terms of meter."[35] M. S. Eno Belinga dismisses meter outright: "In African music only one thing matters: the periodic repetition of a single rhythmic cell."[36] Of course, learning to play rhythms in patterns, repeating rhythmic cells, and drumming without thinking in terms of meter are not activities that *per se*

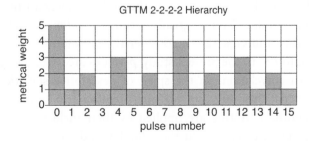

FIGURE 18.1 The GTTM metric hierarchy of Lerdahl and Jackendoff. (With permission from Toussaint, G. T. 2015. *World Music Journal*, 4(2):1–30.)

necessarily produce rhythms that lack meter. Meter may still slip in unconsciously. Nevertheless, the consensus of these and other authors is that African music does not possess meter in the hierarchical sense embodied by the Lerdahl–Jackendoff model.

The criticisms of GTTM described earlier, by both music psychologists and ethnomusicologists, are based on behavioral acts of perception and production of musical rhythms. However, instead of focusing only on the *subjective process* of generating and perceiving African rhythms, we can also seek answers to the aforementioned questions by analyzing the *objective product* instead, i.e., the rhythmic object or written score. In this section, the two questions outlined earlier concerning how similar African and Western rhythms are, and whether African rhythms possess hierarchical meter, are subjected to a mathematical analysis, using *pulse saliency histograms* that yield quantitative and qualitative objective measures that help to illuminate the structure and degree of this similarity, as well as how much and what type of hierarchical meter African rhythms possess. Unlike some previous more general comparisons of western with nonwestern music based on objective *acoustic tonal* features,[37] the analysis presented is based only on the mathematical features of *rhythm* and *meter*, and are restricted to *symbolic notated* music.

As intimated in the preceding, the study described here uses the methodology of descriptive statistics, rather than a formal hypothesis-driven approach. Since the GTTM hierarchy implies a ranking of the 16 pulse frequencies (expectations), with pulse 0 receiving the highest rank, pulse 8 the second highest rank, pulses 4 and 12 tied for the third highest rank, etc. (see Figure 18.1), a natural measure that may be used to compare the association between pulse and frequency histograms is the Spearman rank correlation coefficient,[38] which measures the degree to which the relation between two variables is a *monotonic* function. Another common measure of association between two variables is the Pearson correlation coefficient that measures how *linear* this relation is.[39] Which of the two correlations is a better description of the association for a particular problem at hand is part of an outstanding and ongoing debate.[40] Previous studies indicate that sometimes the Spearman correlation is higher than the Pearson correlation and vice versa.[41] The same behavior is observed with the rhythm data analyzed here, and for this reason, rather than listing only one of these two correlation

coefficients, both are included for comparison and completeness. A second issue related with reporting correlation coefficients in descriptive statistics concerns the reporting and meaning of the *p*-values (also called misleadingly, the levels of significance). This matter also has a long history of debate. Some researchers have asked the question "Should we stop using the *p*-value in descriptive studies?"[42] The purpose of *p*-value is to convince others that either a pattern discernible in data is real or it could plausibly have arisen by chance alone.[43] More specifically, the *p*-values are probabilities often used in a hypothesis-driven context to test how confident one can be that a statistic calculated from a sample of data taken from a larger dataset (the population) also applies to the population as a whole. It is usually assumed that the smaller this probability is, the more confident we can be in the veracity of the hypothesis. However, this assumption must be taken with a grain of salt since *p*-values are not fixed, but random variables, and thus have a probability distribution.[44] Concerning the 34 African asymmetric timelines used in this study, the question then arises as to whether they represent a random sample of some larger population. The purpose of this study is to utilize the 16-pulse asymmetric timelines adopted in music from the African Diaspora, as a worst-case litmus test. The 34 timelines represent all the 16-pulse timelines that I was able to collect from scholarly books and papers on the subject. Therefore, the "sample" may be considered to be the entire population, in which case the *p*-values would be considered to be meaningless in the traditional hypothesis-driven context. The reader may care to formulate various hypotheses in terms of larger populations, such as these 34 asymmetric timelines *plus* those I did not discover, or those asymmetric timelines consisting of all possible numbers of pulses (time spans, cycles, measures) such as 6, 8, 9, 12, 16, 18, and 24, or *all* rhythms used in the music of the African Diaspora. However, in these situations the 34 hand-picked timelines would diverge greatly from a *random* sample. Barber and Ogle (2014) consider the question "To *P* or not to *P*?" In the present study, *p*-values are listed along with correlations in the spirit of descriptive statistics and data mining, but no interpretations in terms of hypothesis testing are implied by their inclusion. The readers are referred to the complex and varied literature on the interpretation of *p*-values, and may interpret them as they see fit. To complement the numerical correlation coefficients and *p*-values reported, visual plots of the data and graphs of the histograms are provided, which may reflect the true nature of the relations better than abstracted numbers and purported statistical significance levels. After all, results may be statistically significant without being scientifically significant and vice versa.

Regarding the methodology for estimating correlations, an alternative approach to calculating the correlation coefficients between the pulse saliency histograms (frequency distributions) and the GTTM hierarchy (or other models) is to calculate the correlation value for each of the 34 individual timelines with the GTTM hierarchy, and then perform a *t*-test to determine whether the resulting correlation values are significantly greater than zero. This approach is also used to compare this procedure with the histogram method, and the results are described in the following.

PULSE SALIENCY HISTOGRAMS IN RENAISSANCE AND COMMON PRACTICE MUSIC

A *pulse saliency histogram* calculated from a given corpus (dataset) of symbolically notated music resembles the GTTM hierarchy shown in Figure 18.1. The difference is that the height of a column in the pulse saliency histogram corresponds to the *empirically observed frequency of occurrence* of an onset in that position of the rhythmic cycle. In this section, we compare the pulse saliency histograms of Renaissance and Common Practice music with the GTTM hierarchy and the 34 timelines of Figure 18.6. The pulse saliency histogram of the onsets in Palestrina's Sixteenth-Century motet, *Pater Noster*, compiled by Joshua Veltman, is shown in Figure 18.2.[45] The correspondence between this histogram and the

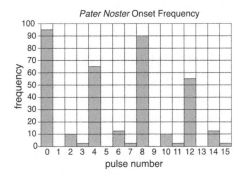

FIGURE 18.2 The pulse saliency histogram of Palestrina's *Pater Noster*. (With permission from Toussaint, G. T. 2015. *World Music Journal*, 4(2):1–30.)

GTTM hierarchy in terms of the ranks of saliencies of its pulses is visually striking. In both graphs, pulses 0 and 8 contain the highest and second highest columns, respectively. In both graphs, pulses 4 and 12 come next, and have approximately equal height. The same may be said of the third-level pulses 2, 6, 10, and 14, as well as the fourth-level pulses 1, 3, 5, 7, 9, 11, 13, and 15. This visual comparison is compelling enough in this case, but to obtain a quantitative measure of the relationship (similarity) that exists between these hierarchies (or ranks) of the two histograms, correlation coefficients are computed between the vectors determined by each of the 16 ordered heights. The resulting Spearman (r_s) and Pearson (r_p) correlation coefficients between the two histograms are $r_s = 0.935$ with $p < 0.00001$ and $r_p = 0.955$ with $p < 0.0001$, respectively (refer to Table 18.1 for this and other results). The remarkably high correlation between these two histograms provides additional quantitative evidence that the GTTM hierarchy is an accurate model of the metric hierarchy contained in the Sixteenth-Century Italian motets composed by Palestrina. Note that the Pearson correlation coefficient is higher than the Spearman correlation but has a lower *p*-value.

Palmer and Krumhansl (1990) calculated the pulse saliency histograms of a sample of Common Practice music. For their analysis they chose the following four composers and corresponding piano compositions: J. S. Bach, French Suites III (Allemande and Sarabande), IV (Gigue), and VI (Gavotte); Mozart, Piano Sonatas K310, 311, 545, and 576; Brahms, Piano Intermezzi Opus 118, No. 1, and 119, Nos. 2, 3, and 4; and Shostakovich, Piano Prelude III and Fugues III, VI, and XI. These pieces comprised music composed of a variety of meters, but all the pieces contained fragments with 16-pulse measures, which yielded the pulse saliency histogram shown in Figure 18.3. The resulting correlations between GTTM and the Palmer–Krumhansl histograms are $r_s = 0.937$ with $p < 0.000001$ and $r_p = 0.959$ with $p < 0.0001$. Again, the very high correlation between these two histograms provides quantitative evidence that the GTTM hierarchy is an accurate model of the metric hierarchy contained in the Common Practice music of these composers. Note also that again, the Pearson correlation is higher than the Spearman correlation but has lower *p*-value.

The pulse saliency histograms of the *Palestrina* piece (Figure 18.2) and the compositions by Bach, Mozart, Brahms, and Shostakovich (Figure 18.3) suggest the

TABLE 18.1 Spearman and Pearson correlation Coefficients for the GTTM Hierarchy and the Four Rhythm Corpora

	GTTM Hierarchy	Palestrina *Pater Noster*	Common Practice Music	German Folk Songs
African timelines	$r_s = 0.793$ $p < 0.0002$ $r_p = 0.774$ $p < 0.0005$	$r_s = 0.786$ $p < 0.0002$ $r_p = 0.588$ $p < 0.009$	$r_s = 0.785$ $p < 0.0002$ $r_p = 0.725$ $p < 0.0008$	$r_s = 0.730$ $p < 0.0007$ $r_p = 0.680$ $p < 0.002$
GTTM hierarchy		$r_s = 0.935$ $p < 0.00001$ $r_p = 0.955$ $p < 0.0001$	$r_s = 0.937$ $p < 0.000001$ $r_p = 0.959$ $p < 0.0001$	$r_s = 0.930$ $p < 0.00001$ $r_p = 0.973$ $p < 0.000001$
Palestrina *Pater Noster*			$r_s = 0.933$ $p < 0.000001$ $r_p = 0.934$ $p < 0.0001$	$r_s = 0.969$ $p < 0.000001$ $r_p = 0.971$ $p < 0.0001$
Common Practice music				$r_s = 0.943$ $p < 0.000001$ $r_p = 0.971$ $p < 0.0001$

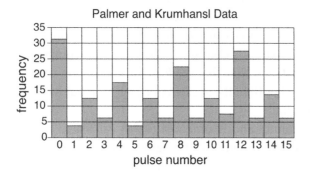

FIGURE 18.3 The pulse saliency histogram for 16-pulse time spans of the Common Practice music examples compiled by Palmer and Krumhansl (1990). (With permission from Toussaint, G. T. 2015. *World Music Journal*, 4(2):1–30.)

possibility of characterizing the difference between Renaissance music and Common Practice music. The Palestrina hierarchy follows the GTTM model almost perfectly in terms of the ranking of pulse positions; only the relative heights differ. This means that all the odd numbered pulses have almost equal heights, the pulses numbered 2, 6, 10, and 14 are almost the same, and pulses 4 and 12 have almost equal height. Even pulses 0 and 8 have almost equal heights. This hierarchy possesses several left–right mirror symmetries because all the levels of the hierarchy are flat. In particular, pulses 8, 4, and 12 show a mountain-peak structure with the central pulse 8 dominating. Examination of the Palmer–Krumhansl histogram (Figure 18.3) reveals that although similar symmetrical structures hold for the two lowest levels, the remaining levels tell a different story; pulses 4, 8, and

12 exhibit a linear increase, with pulse 12 dominating, and thus they lack left–right symmetry. This suggests the possibility of characterizing the difference between Renaissance and Common Practice music by the relative strengths of pulses 8 and 12. The meter of Renaissance music appears to be more symmetrical with respect to the midpoint of the measure than that of the Common Practice music. It would be interesting to confirm this hypothesis with larger samples of these two genres of music.

A feature that both histograms (Palestrina's *Pater Noster* and the Palmer–Krumhansl data) have in common with each other and with the GTTM hierarchy is that the relative magnitude of the ordered pulses are characterized by a perfect up-down-up-down alternating pattern. The Spearman and Pearson correlation coefficients between Palestrina's *Pater Noster* and the Palmer–Krumhansl histograms are $r_s = 0.933$ with $p < 0.000001$ and $r_p = 0.934$ with $p < 0.0001$, respectively.

AFRICAN RHYTHM TIMELINES AND WESTERN MUSIC

As already described in the preceding section, apart from a few exceptions, the sub-Saharan African rhythms that are considered to be the most different from rhythms employed in Western music (European, Renaissance, Common Practice, Classical) consist mainly of the *timelines* occurring in drum ensemble music. Much has been written about their uniqueness, complexity, and the amount of syncopation they possess (in relation to Western hierarchical meter).[46] Therefore, intuition dictates that these timelines provide a suitable dataset (a type of litmus test) to shed light on the claims that African and Western rhythms are fundamentally different, and that African music lacks GTTM hierarchical meter. To inform these claims, a corpus of 34 notated timelines used in African, Afro-Cuban, and Afro-American music was collected from published papers and books.[47] These timelines, shown in box notation in Figure 18.6, all have 16 pulses, and the number k of their onsets varies between 5 and 10. In this box notation, an empty box denotes a silent pulse, and a box containing an X denotes an onset (the strike of the bell, a wooden clave, or a drum). The histogram computed from these 34 rhythms, shown in Figure 18.4, looks quite different from the one in Figure 18.1. Apart from the dominant column at pulse zero, some structural properties of the GTTM hierarchy are not instantly perceivable. In GTTM and *Pater Noster* (Figure 18.2), the most salient pulses after pulse 0 are 4, 8, and 12, whereas in the African timelines, pulses 2 and 6 are preferred, indicating that the latter rhythms are more syncopated than the former. Another obvious difference is that the African timelines make use of almost all pulses with a relatively more similar frequency, noticeably largely disregarding pulses 1 and 15. However, before hastily concluding that African timelines are profoundly different from Western rhythms and lack hierarchical meter altogether, it is instructive to compare the hierarchies in these two histograms with the Spearman and Pearson correlations. The correlation coefficients between the African timelines and GTTM are $r_s = 0.793$ with $p < 0.0002$, and $r_p = 0.774$ with $p < 0.0005$. On the other hand, for African timelines and *Pater Noster,* the correlations are $r_s = 0.786$ with $p < 0.0002$, and $r_p = 0.588$ with $p < 0.009$. These are high correlations with low p-values. Therefore, although there are some illuminating and important differences between the histograms (discussed further later), what is being tested in addition to the similarity of the overall shape of the histograms, is whether they exhibit similar metric hierarchies as defined in GTTM: alternating strong and weak beats arranged in a hierarchy that exhibits a maximum number of left–right symmetries with the five-level height constraints of the histogram columns. The Spearman correlation coefficient measures the ordering of the relative heights of the histogram columns. It is therefore more sensitive to the contour of the histogram columns than to their absolute values. Recall that the contour of a histogram (or rhythm or melody) is a three-symbol string that takes on the symbols "+" if the value increases, "−" if it decreases, and "0" if it stays the same.[48] Note that all three histograms have exactly the same perfectly alternating contour: [− + − + − + − + − + − + − + − +]. Therefore, the correlation coefficient provides significant evidence that African timelines, as a family, do contain lower levels of hierarchical meter, and thus are similar in this regard to the rhythms of Western music. The shape of the histogram provides additional useful information as to how they differ.

The histogram for Palestrina's *Pater Noster* has an almost perfect resemblance to the GTTM hierarchy. The histogram of the Common Practice music in Figure 18.3 differs noticeably with the height of column 12, and one may wonder how the histogram might differ if obtained from yet another European music genre from another century. Thanks to the work of Huron and Ommen (2006), who calculated the onset frequencies for the well-known Essen corpus of Germanic folk songs, this question can be easily answered. The histogram for this corpus, shown in Figure 18.5, differs somewhat from the histogram of *Pater Noster,* but like the other four earlier, has exactly the same perfectly alternating contour: [− + − + − + − + − + − + − +]. The correlation coefficients between the German folk songs and GTTM are $r_s = 0.930$ with $p < 0.00001$ and $r_p = 0.973$ with $p < 0.000001$ (similar to *Pater Noster*), and the correlations between the German folk songs and the African timelines histogram are $r_s = 0.73$ with $p < 0.0007$ and $r_p = 0.680$ with $p < 0.002$. It is worth pointing out a discriminating feature that can be used to distinguish between the German folk songs and Palestrina's *Pater Noster,* which is that, in the latter,

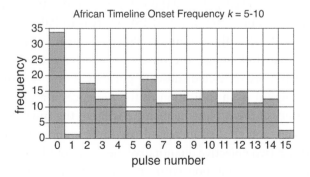

FIGURE 18.4 The onset histogram of the 34 African 16-pulse timelines for $k = 5 - 10$. (With permission from Toussaint, G. T. 2015. *World Music Journal,* 4(2):1–30.)

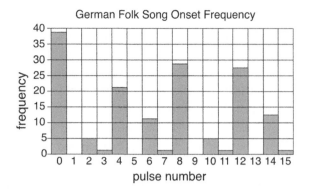

FIGURE 18.5 The onset frequencies of the German folk songs (Huron and Ommen 2006). (With permission from Toussaint, G. T. 2015. *World Music Journal*, 4(2):1–30.)

ignoring the column at pulse 0, the remaining histogram exhibits a strong mirror symmetry about pulse 8, so that for example, the heights at the fourth level in the hierarchy, at pulses 2, 6, 10, and 14 are all about the same. On the other hand, for the German folk songs, this mirror symmetry is absent: at the third level, pulse 12 is more prominent than pulse 4, and at the fourth level, pulses 6 and 14 are more prominent than pulses 2 and 10.

As pointed out in the preceding section, an alternate methodological approach to measuring the association between the African timelines and the GTTM model is to calculate the correlation coefficient for each individual timeline (rather than the frequency distribution of all the timelines) with the GTTM profile, and then perform a *t*-test to determine whether the resulting average correlations are significantly greater than zero.[49] The average of Spearman correlations (averaged over all timelines) is 0.296 with a standard deviation of 0.232, a standard error of 0.041, and a *t*-value equal to 7.219 with a *p*-value < 0.0001. In addition, the average Pearson correlation (averaged over all timelines) is 0.320 with a standard deviation of 0.198, a standard error of 0.035, and a *t*-value of 9.151 with a *p*-value < 0.0001. Therefore, this alternative method yields results in agreement with the process of comparing the *averaged* timelines in the form of the pulse saliency histograms (Figure 18.6). Indeed, Pearson correlations are even higher than that of Spearman correlations. From the theoretical point of view both methods, (1) correlating the averages or (2) averaging the correlations are valid methods for measuring the associations between groups, but correlated averages (the method of choice in this study) are more likely to make a point about the features of the timelines and how they relate to each other.[50]

These results provide some evidence that supports the hypothesis that African and Western rhythms, at least when compared in terms of the structure of their pulse saliency histograms, exhibit some similarities, although the term "profoundly similar" appears to be too strong a description. In order to make such a strong conclusion, additional study should be undertaken using ethnography, a wider all-encompassing set of musical features, and more detailed statistical performance models. Nevertheless, the correlation coefficients provide quantitative measures of the degree of the metric similarity. Furthermore, the structure of the histograms themselves provides insight into the types of similarities observed. Contrary to several musicology-based claims found in the literature, these results provide mathematical evidence that African rhythm does possess hierarchical meter, albeit to a limited extent, along the lines of the original GTTM model, thus lending mathematical support to the perspective of Kofi Agawu. The very high correlation between the Renaissance music, the Common Practice music, and the German folk songs with the African rhythm timelines also adds mathematical support to the view of Mark Hijleh (2008) that the "*practical theory of world rhythm* reveals that even older western art music can have more in common rhythmically with its Asian, African, and American counterparts than might be suspected."

The descriptive statistical comparisons of the histograms in terms of the Spearman and Pearson correlation coefficients are *global* comparisons. On the other hand, the *local* comparisons of the shapes of the histograms serve to focus on several specific differences between the corpora that are worth highlighting. From a visual inspection of the histograms in Figures 18.1–18.5, it is clear that the onsets of the African timelines are distributed more evenly among the 16 pulse locations (with the exception of pulses 0, 1, and 15), whereas in Palestrina's *Pater Noster*, the Common Practice music, and the German folk songs, the onsets are concentrated on a much more restricted subset of pulse locations. A flatter histogram is an indication of the presence of a greater variety of onset placements in relation to the time span (cycle), and thus of increased rhythmic complexity. To obtain two objective quantitative measures of histogram flatness, the Shannon entropy of histograms (normalized so that they resemble probabilities and sum to 1) and the standard deviations of the histogram bin heights may be calculated by viewing the histograms as discrete

Pulse Number

ID	Rhythm Name	0	1	2	3	4	5	6	7	8	9	10	11	12	13	14	15
1	Bossa-1	X			X			X				X			X		
2	Bossa-2	X			X			X		X			X				
3	Bossa-3	X			X			X			X				X		
4	Shiko	X				X		X				X		X			
5	Timini	X		X				X		X			X				
6	Kromanti	X		X				X				X		X			
7	Tuareg	X				X		X		X				X			
8	Tipitina	X		X		X					X					X	
9	Son	X			X			X				X		X			
10	Rumba	X			X				X			X		X			
11	Gahu	X			X			X				X				X	
12	Domba	X		X			X				X				X		
13	Kpatsa	X				X				X			X		X		
14	Rap-X	X				X			X		X			X			
15	Soukous	X			X			X			X	X					
16	Rap-2	X		X					X			X				X	
17	Mambo-1	X			X		X			X		X			X		
18	Mambo-2	X		X			X			X			X		X		
19	R-W-B	X		X			X			X			X			X	
20	Popcorn	X		X		X			X		X					X	
21	Funky	X		X		X			X		X			X			
22	Central-Africa	X		X		X		X		X			X			X	
23	Takoe	X			X		X			X		X		X		X	
24	Akom	X		X		X		X			X		X			X	
25	Adangme	X		X		X		X		X			X		X		
26	Samba	X		X			X		X		X			X		X	
27	Ghana	X		X		X		X			X		X		X		
28	Bembe-duple	X			X			X	X			X		X			X
29	Oyaa	X		X		X		X			X	X			X	X	
30	Ngbaka	X		X		X		X	X		X		X		X	X	
31	Ngbaka-Maibo	X		X		X		X	X		X		X		X		X
32	Kassa	X	X		X		X	X		X		X		X		X	
33	Mutuashi	X	X		X		X		X	X		X		X		X	
34	Rumba-palitos	X		X	X		X		X	X		X		X	X		X
	Histogram	34	2	18	13	14	9	19	11	13	12	15	11	15	12	13	3

FIGURE 18.6 The 34 rhythm timelines in box notation and the resulting histogram. (With permission from Toussaint, G. T. 2015. *World Music Journal*, 4(2):1–30.)

probability distributions.[51] The entropy has been enlisted before as a feature for comparing musical styles.[52] If we denote the height (probability) of the i-th histogram bin by p_i, the Shannon entropy H is given by the formula:

$$H = -\sum_{i=1}^{16} p_i \log_2 p_i$$

The resulting entropy calculations are shown in Table 18.2, in order of increasing entropy from left to right. The degree of flatness of the histograms is indicated by relatively high values of entropies, and relatively low values of standard deviations. Note that the standard deviations are decreasing monotonically from left to right, and thus both measures are in perfect

TABLE 18.2 The Shannon Entropies (in Bits) of the Histograms and the Standard Deviations of the Histogram Column Heights, for the GTTM Profile and the Pulse Histograms of the Four Rhythm Corpora

	Palestrina's *Pater Noster*	German Folk Songs	Common Practice Music	GTTM Hierarchy	African Timelines
Shannon entropy	$H = 2.67$	$H = 2.91$	$H = 3.70$	$H = 3.75$	$H = 3.81$
Standard deviation	$SD = 0.093$	$SD = 0.081$	$SD = 0.04$	$SD = 0.039$	$SD = 0.033$

agreement with respect to the relative complexities of the rhythms of these corpora. From these results, we may conclude that Palestrina's *Pater Noster* has the simplest rhythms, and the African timeline rhythms are the most complex (at least by these two measures of complexity). Furthermore, the GTTM profile is more complex than that of the Common Practice rhythms in the Palmer–Krumhansl samples, which is more complex than the profile of the rhythms of German folk songs compiled by Huron and Ommen, which in turn is more complex than the profile for the rhythms in Palestrina's *Pater Noster*. It is worth noting that the maximum possible value of the entropy, attained for a perfectly flat histogram, is 4.0, and that the entropy of the GTTM hierarchy ($H = 3.75$) is almost equal to that of the African timelines ($H = 3.81$). Indeed, both are close to 4.0, again suggesting another commonality between Western and African rhythms. The data also suggest that, on the whole, the meter of these samples of Western music has become more complex during the past four centuries, and has reached levels of complexity almost as high as that of the African timelines.

The shape of the pulse saliency histograms also allows the concept of meter to be broken down into different types. The definition of meter by Lerdahl and Jackendoff as a regular pattern of alternating strong and weak beats arranged in a specific hierarchical manner may be deconstructed into categories in terms of four distinct properties: (1) a regular pattern of alternating strong and weak beats, (2) a particular five-level hierarchical distribution of the relative strengths of these beats, in which the fifth

and fourth levels are realized by one pulse location each, the third level is realized by two pulse locations, the second level is realized by four pulse locations, and the first level is realized by eight locations, and (3) the placement of these histogram bins within the cycle so as to create a fractal pattern with a maximum number of subsymmetries among the histogram bins of different heights. These three categories will be referred to by the shorter terms: (1) *alternating property*, (2) *distribution property*, and (3) *fractal symmetry property*, respectively. Note that the third property is a combination of two additional distinct properties: (1) the presence of *subsymmetries* and (2) whether any existing subsymmetries form a *fractal* pattern. These concepts are illustrated with the GTTM profile in Figure 18.7. A *subsymmetry* of a one-dimensional pattern is a contiguous subpattern that possesses mirror symmetry, i.e., is palindromic. The number of subsymmetries contained in a pattern has been shown to be a good predictor of human perception of the complexity of both auditory and visual patterns.[53] The entire GTTM pattern ranging from pulses 0 to 15 is not mirror symmetric. However, the subpattern from pulses 1 to 15 contains mirror symmetry about pulse 8, and hence is a large subsymmetry of length 14. However, the GTTM profile contains several additional smaller subsymmetries. There are subsymmetries of length 3 involving levels 1 and 2, such as between pulses 1-3, 5-7, 9-11, and 13-15 (indicated by leaves of the tree in the left diagram), also involving levels 1 and 3, such as pulses 3-5 and 11-13 (shown in the right diagram). In the right diagram, there is also a subsymmetry of length 3 between

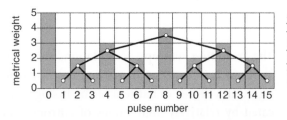
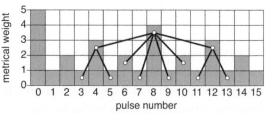

FIGURE 18.7 The subsymmetries present in the GTTM model and their fractal nature. The standard hierarchical fractal subsymmetries (left) and some additional nonfractal subsymmetries (right). (With permission from Toussaint, G. T. 2015. *World Music Journal*, 4(2):1–30.)

levels 1 and 4, between pulses 7 and 9, and a subsymmetry of length 5 between pulses 6 and 10, between levels 2 and 4. The reader is invited to discover the remaining subsymmetries present. What is even more interesting about the GTTM profile is that the subsymmetries in the left diagram (the standard GTTM hierarchy) have a *fractal* nature. Informally, a fractal pattern is one that is self-similar at different scales or "looks the same" at different levels of "zooming in." In the case of GTTM subsymmetries, the pattern from pulses 2 to 14 (levels 2–4) has the same structure as the two subpatterns from pulses 1 to 7 and 9 to 15 (levels 1–3). The subsymmetries highlighted in the right diagram on the other hand do not possess the fractal property.

In the comparison of African with Western rhythm, a natural question concerns the relative prominence of the preceding features of meter present in the families of rhythms under study. After all, it is logically possible for a histogram to reflect a pattern of alternating strong and weak beats without exhibiting any hierarchy other than at the two lowest levels sufficient for distinguishing between strong and weak beats (Figure 18.8, left). Such a profile represents one extreme of the GTTM definition: it contains the minimum amount of hierarchy (lacks the distribution and fractal symmetry properties) while still preserving the alternating property. It is also possible for a histogram to possess a hierarchy consisting of the same histogram bin heights present in GTTM (the distribution property) but arranged in such a manner as to have no alternations whatsoever between strong and weak beats (Figure 18.8, right). This profile represents another extreme of the GTTM properties inherent

in the definition. Note that there are only two possible profiles with this property: the one on the right, and its mirror symmetric version in which the tallest bin lies at pulse 15 rather than at pulse 0. The latter version is not explored in this study since it is the complete inverse of the GTTM profile. In general, a corpus of rhythms will contain a pulse saliency histogram that contains a mixture of all these properties. The correlations between these histograms and those of GTTM and the four corpora are shown in Table 18.3. For Palestrina's *Pater Noster*, Common Practice music, and the German folk songs, both the Spearman and Pearson correlations are higher for the GTTM hierarchy than for meters with only the alternation property. On the other hand, for the African timelines the reverse is true for the Spearman correlation: for pure alternations $r_s = 0.845$ with $p < 0.00002$, whereas for the GTTM hierarchy $r_s = 0.793$ with $p < 0.0002$. These results suggest that the meter in Western music tends to reflect the *distribution* property of GTTM, whereas the meter in African music expresses the *alternation* property of GTTM (Figure 18.8). These properties are also visually evident from the histogram in Figure 18.4.

Examination of the pulse saliency histogram of the African timelines in Figure 18.4 suggests the following hierarchy. Apart from pulse 0 which dominates all other pulses by far, six other pulses stand out from the remaining pulses: two pulses are almost nonexistent (pulses 1 and 15), and four are notably higher than the rest: pulses 2, 6, 10, and 12. Pulse 6 is the second highest, pulse 2 is the third highest, and pulses 10 and 12 are tied for fourth and fifth positions. These observations suggest

TABLE 18.3 Spearman and Pearson Correlation Coefficients for Alternations without the GTTM Hierarchy (Second Row), and for the GTTM Hierarchy (with the Distribution Property) without Alternations (Third Row)

	GTTM Hierarchy	Palestrina's *Pater Noster*	Common Practice Music	German Folk Songs	African Timelines
GTTM hierarchy	-	$r_s = 0.930$	$r_s = 0.937$	$r_s = 0.930$	$r_s = 0.793$
	-	$p < 0.00001$	$p < 0.000001$	$p < 0.00001$	$p < 0.0002$
	-	$r_p = 0.955$	$r_p = 0.959$	$r_p = 0.973$	$r_p = 0.774$
	-	$p < 0.0001$	$p < 0.0001$	$p < 0.000001$	$p < 0.0005$
Strong–weak alternations without GTTM hierarchy	$r_s = 0.935$	$r_s = 0.874$	$r_s = 0.881$	$r_s = 0.871$	$r_s = 0.845$
	$p < 0.000001$	$p < 0.000005$	$p < 0.000003$	$p < 0.000006$	$p < 0.00002$
	$r_p = 0.783$	$r_p = 0.663$	$r_p = 0.795$	$r_p = 0.740$	$r_p = 0.618$
	$p < 0.0002$	$p < 0.003$	$p < 0.0001$	$p < 0.0005$	$p < 0.006$
GTTM distribution property without alternations	$r_s = 0.069$	$r_s = 0.074$	$rs = -0.123$	$r_s = -0.034$	$r_s = 0.186$
	$p < 0.399$	$p < 0.392$	$p < 0.326$	$p < 0.450$	$p < 0.245$
	$r_p = 0.302$	$r_p = 0.210$	$r_p = 0.192$	$r_p = 0.236$	$r_p = 0.428$
	$p < 0.127$	$p < 0.216$	$p < 0.239$	$p < 0.190$	$p < 0.050$

For comparison, the first row provides the correlations with the original GTTM profile.

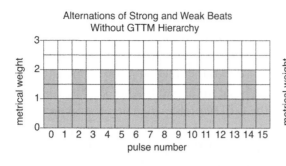
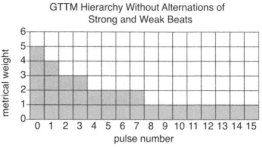

FIGURE 18.8 The weights for regular alternations between strong and weak beats (*alternation* property) lacking the GTTM distribution property (left), and the GTTM weights arranged to have the *distribution* property without any alternation (right). (With permission from Toussaint, G. T. 2015. *World Music Journal*, 4(2):1–30.)

two modifications of the GTTM model that may be more veridical as theoretical models of the metrical hierarchies of asymmetric African timelines. The first modification is obtained by shifting the columns of GTTM to match the heights of the African timelines as much as possible, while maintaining both the distribution and alternating properties of GTTM. Thus, column 8 of GTTM moves to position 6, column 4 of GTTM moves to position 2, and the two lowest levels of the hierarchy are placed so as to maintain the alternating property of GTTM. The resulting metric hierarchy is shown in Figure 18.9 (left). The correlations with the African timeline profile are shown in Table 18.4 with the title GTTM Distribution with Timeline Placements ($r_s = 0.908$ with $p < 0.000001$ and $r_p = 0.861$ with $p < 0.0001$). The second modification is obtained by relaxing the distribution constraint such that the heights of the columns may be altered by discrete units, but the total number of discrete units used to define the GTTM weights (which number 31) remains the same. The GTTM model may thus be viewed as a distribution of 31 unit squares among the 16 pulse positions distributed according to several constraints regarding alternations, height columns, and fractal symmetries.

This property is here dubbed the *numerosity* of the profile. The second modification of the GTTM profile makes the minimum number of changes to the distribution constraint with timeline placement while maintaining the numerosity and alternation properties and is shown in Figure 18.9 (right) titled GTTM Numerosity with Timeline Placement. Comparing the African timeline profile (Figure 18.4) with the profile of Figure 18.9 (left) shows that the main differences are that pulses 1 and 15 are almost zero, and pulse 0 is considerably higher than all other pulses. The profile in Figure 18.9 (right) is obtained from the profile on the left by moving the unit square from pulse 1 to pulse 0, to increase its domination, and moving the unit square from pulse 15 to pulse 10 to match the heights of pulses 10 and 12.

Correlations between the pulse saliency histograms of rhythm corpora and the GTTM (or any other model), as well as entropy values, are of course not the only ways to measure the similarity between metrical hierarchies. Furthermore, these measures do not assess the amount of symmetry or depth present in the metric hierarchies. Visual inspection of the histograms themselves provides additional complementary insights into the structure of

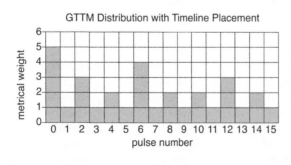
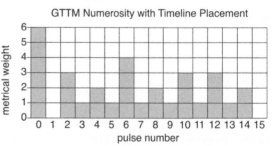

FIGURE 18.9 The weights for the GTTM distribution property with GTTM columns swapped to match the African timelines profile (left) and the weights adjusted to have the *numerosity* property to match the African timelines profile (right). (With permission from Toussaint, G. T. 2015. *World Music Journal*, 4(2):1–30.)

TABLE 18.4   Correlations of the African Timelines Profile with the GTTM Profile and Its Four Modifications

	GTTM Hierarchy without Alternations	Original GTTM Hierarchy	Strong–Weak Alternations without GTTM Hierarchy	GTTM Distribution with Timeline Placements	GTTM Numerosity with Timeline Placements
African timelines histogram	$r_s = 0.186$	$r_s = 0.793$	$r_s = 0.845$	$r_s = 0.908$	$r_s = 0.958$
	$p < 0.245$	$p < 0.0002$	$p < 0.00002$	$p < 0.000001$	$p < 0.000001$
	$r_p = 0.428$	$r_p = 0.774$	$r_p = 0.618$	$r_p = 0.861$	$r_p = 0.941$
	$p < 0.050$	$p < 0.0005$	$p < 0.006$	$p < 0.0001$	$p < 0.0001$

the African timeline hierarchy of Figure 18.4. The modification of GTTM in Figure 18.9 (right) that preserves the numerosity and alternation properties of GTTM, while maximizing the matching of the placements of the histogram columns at the expense of violating the distribution and fractal properties, is reproduced in enhanced form in Figure 18.10, with the addition of indications (in thick bold lines) of some of the remaining subsymmetries and the partial fractal structure remaining. From a comparison of the GTTM histogram with the histogram in Figure 18.10, it is clear that GTTM has many more subsymmetries. Table 18.5 lists the number of subsymmetries of each length. GTTM has a greater number of short subsymmetries of lengths 3 and 5, and the African timeline profile has no long subsymmetries of lengths 11, 13, and 15. Furthermore, the four-level fractal in Figure 18.6 (left) between pulses 1 and 15 has been reduced to the three-level fractal in Figure 18.10 (left) between pulses 3 and 9. It is well known that African timelines possess fewer symmetries than do Western

rhythms. Indeed, that is how they acquired the name "asymmetric" timelines. However, the number of subsymmetries and the size of the fractals present in the pulse saliency histograms are two features that quantify the asymmetries. Symmetry has been used extensively in the composition of music for centuries, but it is more difficult for the listener to perceive symmetries in the aural domain than for the composer to perceive them in the visual domain.[54] However, too much symmetry becomes uninteresting, and therefore much attention has been given to striking a balance between perfect symmetry and a more aesthetically pleasing asymmetry or "broken" and "crippled" symmetry as it is sometimes called.[55] For African rhythm, it appears that the GTTM model is too symmetric to correlate well with the timelines profile, and thus the "broken" symmetry in the profile of Figure 18.10 appears to be a better fit.

Another noteworthy feature of the African timeline profile is the almost complete absence of onsets at pulses 1 and 15. The reason for this may be that since one of

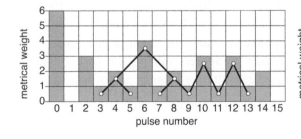
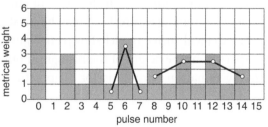

FIGURE 18.10   The idealized hierarchical meter model for the African timelines obtained by modifying GTTM while preserving the numerosity and alternating properties of GTTM while maximizing the timeline placements, at the expense of violating the distribution property and losing some subsymmetries and fractal structure. (With permission from Toussaint, G. T. 2015. *World Music Journal*, 4(2):1–30.)

TABLE 18.5   Number of Subsymmetries of Varying Lengths in GTTM and the Idealized Hierarchical Meter for the African Timelines

Length of subsymmetry	3	5	7	9	11	13	15
Number of subsymmetries in GTTM profile	7	3	2	1	1	1	1
Number of subsymmetries in African timeline profile	6	2	2	1	-	-	-

the main functions of timelines is their time-keeping role, the musicians need a clear signal as to where the beginning of the cycle (pulse zero) falls. This fundamental downbeat signal will be more perceivable (stand out from the crowd of beats) if it is surrounded on both sides by a silent pulse (pulse 1 just after the main onset, and pulse 15 just before). However, the same behavior is observed in the German folk songs. Indeed, only the Common Practice music examples make use of these two pulse positions. It may be that the same reasoning applies to the German folk songs, but this is a speculation that invites further research. It does not apply to Palestrina's *Pater Noster* because this piece tends to ignore almost completely all the odd-numbered pulses.

Examination of the various properties of the deconstructed GTTM definition helps to further characterize the differences between the Renaissance music of Palestrina, the Common Practice music samples compiled by Palmer and Krumhansl, and the German folk songs in the Essen collection studied by Huron and Ommen. The pulse saliency histogram of Palestrina's *Pater Noster* (Figure 18.2) is almost identical to the GTTM model in the sense of possessing all the fractal subsymmetries of GTTM. The main difference is in terms of the relative heights of histogram bins. For *Pater Noster,* the odd pulses are almost not used at all, and pulses 2, 6, 10, and 14 are much less used than pulses 4, 8, and 12. It would be interesting to determine whether this property holds for Palestrina's music in general, or even for Renaissance music as a whole.

The empirical pulse saliency histogram of the Common Practice samples (Figure 18.3) may be idealized to the hierarchical meter model shown in Figure 18.11 (left). Apart from the height of the histogram bin at pulse 12, the remaining profile is for all practical purposes identical to that of GTTM. However, the fact that the height of pulse 12 is higher than pulse 8 alters the number of subsymmetries present considerably, and destroys the fractal nature of the hierarchy, while preserving the alternation property. Interestingly the three heights of pulses 4, 8, and 12 are collinear, introducing a different type of regular structure that compensates for losing some of the symmetries. This pattern suggests that Common Practice music places more emphasis on the last (fourth) quarter note than on the third quarter note in the middle of the time span, as do Palestrina's *Pater Noster*, the German folk songs, and GTTM, where the quarter note at pulse 8 is used more frequently than the quarter notes at pulses 4 and 12. To speculate further, a tantalizing similarity is suggested between the frequency patterns of these three quarter notes and the accents (stresses) of three-syllable words in the English and French languages. English tends to stress the middle syllable (as in the English word "production"), whereas French tends to place the stress on the third syllables (as in the French word "production"). The metric hierarchy of the Common Practice music appears to match the French pattern of syllable stresses, whereas the metric hierarchies of *Pater Noster*, the German folk songs, and GTTM appear to match the English syllabic stress pattern. The commonalities of rhythm in language and music have been investigated previously.[56] However, these studies compared the complexity of English and French language with English and French music using the normalized pairwise variability index (nPVI) as a measure of rhythm complexity. The nPVI of the interonset intervals (IOIs) of musical notes has also been applied to study the changes in French, German, and Italian composers during the past 400 years, providing "quantitative support for historical musicology."[57]

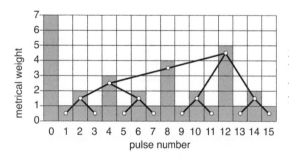
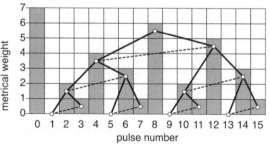

FIGURE 18.11 The idealized hierarchical meter models for the Common Practice music examples of Palmer and Krumhansl (left) and the German Folk Song data of Huron and Ommen (right). (With permission from Toussaint, G. T. 2015. *World Music Journal*, 4(2):1–30.)

It would be interesting to determine whether there also exist relations between music and language with respect to the location of stresses in musical meters and words, respectively. More specifically, has the French language influenced the *meter* of Common Practice music?

The empirical pulse saliency histogram of the Essen German Folk Song Collection (Figure 18.5) may be idealized to the hierarchical meter model shown in Figure 18.11 (right). This hierarchy is uniquely different from GTTM and the profiles of all the other pulse saliency histograms examined here, in that it contains none of the left–right mirror subsymmetries which are a hallmark property of GTTM, while still completely preserving the alternation and fractal properties of GTTM. The subsymmetries are broken by the fact that at every level of hierarchy that connects a "parent" pulse to its two "offspring" pulses, the height of the offspring on the right is higher than the offspring on the left (indicated in the figure by dashed lines). Thus, for pulses 1, 2, and 3, the height of column 3 is greater than the height of column 1. Similarly, for pulses 5, 6, and 7, the height of column 7 is greater than the height of column 5. For pulses 2, 4, and 6, the height of column 6 is greater than the height of column 2. At the highest level, for pulses 4, 8, and 12, the height of column 12 is greater than the height of column 4, and so on. It is as if the GTTM hierarchy has been rotated in a counterclockwise direction by some angle. Therefore, in spite of this "rotation," the hierarchy completely preserves the fractal property of GTTM. Furthermore, although the hierarchy does not possess the left–right mirror subsymmetries, it does possess more complex subsymmetries that involve simultaneous translations in a northeasterly direction accompanied by dilation. Thus, the triangle determined by pulses 5, 6, and 7 is larger than the triangle determined by pulses 1, 2, and 3. Similarly, the triangle determined by pulses 10, 12, and 14 is larger than the triangle determined by pulses 2, 4, and 6. These types of symmetries imply that the German folk songs are rhythmically more complex than the rhythms of Renaissance and Common Practice music, as well as those implied by the GTTM profile. They also suggest that the German folk songs tend to place more emphasis on the notes appearing at the ends of subsymmetries. It would be interesting to also explore the relationship between this structure in music and possible analogies to the syllable stress patterns in the German language.

Another property of rhythms that may be used as a feature for distinguishing between families of rhythms is the preference for utilizing pulse locations in the first half of the time span (cycle). This property can be measured by considering the normalized histograms as probability distributions, and calculating their expected values, which are equivalent to the centers of gravity of the normalized histograms. In a perfectly balanced meter, in which the number of onsets that occur in each half measure is the same, the expected value should be 8.5. The expected values, in increasing order, are (1) *Pater Noster* = 7.03, (2) German folk songs = 7.43, (3) African timelines = 7.61, (4) GTTM hierarchy = 7.71, and (5) Common Practice music = 8.09. Common Practice music is the closest to being perfectly balanced, *Pater Noster* has the most notes in the first half measure, and the African timelines are almost indistinguishable from the GTTM hierarchy. Hence, this is yet another feature of the pulse saliency histograms that the African timelines and GTTM have in common.

KEITH'S MATHEMATICAL MEASURE OF METER COMPLEXITY

In the introduction to this chapter, the distinction was made between isochronous and nonisochronous meters. The latter kind is also referred to as *complex* meters.[58] However, the complexity of meters also varies greatly among different meters that belong to the nonisochronous family. In this section, a particular mathematical measure of meter complexity is presented for a special class of nonisochronous meters.[59]

For a string of symbols S, Michael Keith proposed a measure of meter complexity, denoted here by $C(S)$, defined for the family of asymmetrical meters that consist of durational patterns made up of strings of 2's and 3's.[60] These meters include a vast collection of *aksak* [61] meters and rhythms, and a variety of additional examples such as the African signature *fume–fume* pattern [2-2-3-2-3], the *guajira* flamenco compás metric pattern [3-3-2-2-2], and the Bushmen *San* rhythm of the Northwest Kalahari [2-2-2-3]. However, even for the Western theoretical hierarchical underpinning of meter in the GTTM model, Lerdahl and Jackendoff stipulate that "at each metrical level, strong beats are spaced either two or three beats apart."[62]

Keith first partitions S into any string of disjoint *subunits*. At the lowest level, the first two 12-pulse meters are partitioned into subunits [2][2][3][2][3] and [3][3][2][2][2], respectively. At this level, the complexity of an individual [2]-unit is 2 and that of a [3]-unit is 3. For a given partition, the complexity of the string $C(S)$ is the sum of the complexities of subunits. Thus, at this level, the complexities of the fume–fume and guajira meters are the same: $2 + 2 + 3 + 2 + 3 = 3 + 3 + 2 + 2 + 2 = 12$. Keith defines a *unit* as one or more *identical contiguous* subunits. Thus, another possible pair of partitions for these two meters consists of [2-2][3][2][3] and [3-3][2-2-2], respectively. The complexity value of a unit U consisting of a number of identical subunits H is defined as $C(U) = \max\{\#subunits, C(H)\}$, where #subunits denotes the number of subunits. For example, the complexity of the unit [2-2] is max{2, 2} = 2, and that of the unit [2-2-2] is max{3, 2} = 3. If we denote a partition of S by S_U, then the complexity of a given partition of S into units, denoted by $C(S_U)$, is the sum of the complexities of the units, $C(U)$. Therefore, for this partition, $C(S_U)(\textit{fume–fume}) = 2 + 3 + 2 + 2 = 9$ and $C(S_U)(\textit{guajira}) = 3 + 3 = 6$. Finally, the complexity of the sequence $C(S)$ is the minimum complexity, minimized over all possible partitions *at all hierarchical levels*; $C(S) = \min\{C(S_U)\}$. In our example, the guajira admits several other partitions such as [3-3][2-2][2], [3-3][2][2-2], [3][3][2-2][2], and [3][3][2][2-2], the complexities of which are, respectively, 7, 7, 10, and 10. Therefore, the final complexities are $C(\textit{fume–fume}) = 9$ and $C(\textit{guajira}) = 6$.

It is important to note that the complexity must be computed hierarchically at all levels. For example, in the 16-pulse meter [2-2-2-2-2-2-2-2], the four units at the second level [2-2][2-2][2-2][2-2] can be further divided at the next higher level into two identical subunits of {[2-2]

[2-2]} and {[2-2][2-2]}. The correct Keith complexity of metric pattern [2-2-2-2-2-2-2-2] is 2, which is as it should be because the meter is so simple. For a second example, the meter [3-2-2-3-2-2-3] can be divided to include two identical subunits [3-2-2][3-2-2][3]. And from the complexity of the subunit [3-2-2], the final complexity of the meter [3-2-2-3-2-2-3] is 8, and not 13, because the repetition of subunit [3-2-2] simplifies the meter.

Measures of meter complexity find application in the design of automated composition systems, and are also a feature that may be incorporated into measures of rhythm similarity, a topic examined in Chapter 37.

THE INTERACTION BETWEEN METER AND RHYTHM PERCEPTION

In contrast to almost all the definitions of meter found in the literature, based on some formulation in terms of strong and weak beats, Christopher Hasty offers a unique original definition based on the perceptual concept of expectation or projection "as the potential of a duration to be immediately reproduced."[63] In this sense, meter and rhythm have a great deal in common, and their perception is inextricably intertwined. Hasty's treatise is philosophical in nature. However, it is possible to translate his idea of projection to a mathematical and computational model (algorithm). Indeed, Longuet-Higgins and Lee proposed such an elegant mathematical hierarchical meter-based model of rhythm perception, founded on two assumptions.[64] Their model is explained and illustrated using the six distinguished timelines as examples, in Figure 18.12. The first assumption is that after hearing

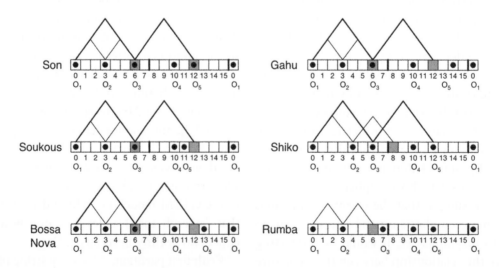

FIGURE 18.12 Illustration of the hierarchical meter-based model of rhythm perception proposed by Longuet-Higgins and Lee (1982) for the six distinguished timelines.

the first two onsets of a rhythm at times denoted by t_1 and t_2, respectively, the listener creates a hypothesis of an underlying regular (isochronous) meter and predicts that a third onset will occur at time $t_3 = t_2 + (t_2 - t_1)$, where $(t_2 - t_1)$ is designated to be the *beat length*. In other words, the third onset is predicted to occur such that the duration interval between the second and third onsets equals the duration interval between the first and second onsets. Consider, for instance, the *clave son*, in box notation, at the top left diagram of Figure 18.12. The five onsets of the *clave son* are denoted by O_1 to O_5. Here t_1 occurs at O_1 and t_2 occurs at O_2. Since the duration interval between O_1 and O_2 is three pulses, t_3 is predicted to be at pulse six. These two intervals are depicted by the two small triangles (thin lines). Now, if an onset occurs at the predicted time t_3, the hypothesis is confirmed, and the listener "attempts to move up the metrical hierarchy by uniting the two metrical units just established into one single metrical unit, doubling the beat length, and predicting a new value of t_3." This process of moving up the metrical hierarchy may be continued if the rhythm allows it. Referring to the *clave son* again, we see that onset O_3 does fall on the predicted time t_3. The new metrical unit now becomes the interval from O_1 to O_3 pictured by the larger triangle (bold lines), and the new value of t_3 is predicted to occur at pulse 12.

The second assumption is that under certain conditions relatively long IOIs may modify our estimate of the beat length. In particular, if an onset does not occur at the predicted time t_3, we determine whether an onset occurs between t_2 and t_3, such that its IOI is greater than that starting at t_2. If this is the case, t_2 is moved to this onset, and a new longer beat length is postulated. This case arises with the shiko timeline (right-middle diagram in Figure 18.12). For this timeline, t_1 occurs at O_1, as before, but t_2 occurs at O_2 in pulse 4, and therefore, t_3 is predicted to be at pulse 8. These two intervals are depicted by the two small triangles (thin lines), each spanning four pulses. Since there is no onset at pulse 8, we look for an onset in between onset O_2 at time t_2 and predicted time t_3 at pulse 8. Onset O_3 at pulse 6 is such an onset. Furthermore, the interval between O_3 and O_4 (four pulses) is longer than the interval between O_2 and O_3 (two pulses). Therefore, it satisfies the condition for the second assumption, so that t_2 is moved to O_3 at pulse 6, and a new beat length consisting of a duration interval of six pulses is adopted, indicated by the large

triangle (bold lines) spanning pulses 0 to 6. This new beat length predicts a new onset at time t_3 on pulse 12.

Applying the rhythm perception model of Longuet-Higgins and Lee (1982) to the six distinguished timelines yields the following observations. At the first level of metrical hierarchy, only the son, soukous, bossanova, and gahu have onsets O_3 at the predicted time t_3. Furthermore, of these four rhythms, at the second level of the metrical hierarchy, only the *clave son* has an onset O_5 at the new predicted value of t_3. At this second level in the hierarchy, the shiko also has an onset O_5 at the new predicted value of t_3, but only after failing to predict an onset at t_3, in the first level of the hierarchy. Therefore, regarding the simplest most accurate prediction of the onsets as the rhythm unfolds, the *clave son* is unique among this group of six rhythms.

The *clave son* has an additional characteristic feature that sets it apart from the other five distinguished rhythms. It is the only rhythm (among these six) with the property that the predicted onsets at both levels of the metrical hierarchy are in a *fractal* relationship with each other. Recall that a pattern has the fractal property if the pattern looks geometrically self-similar at different scales or levels of magnification.[65] Fractals have been studied extensively by mathematicians and computer scientists, and have been traditionally applied in visual arts, and used for modeling patterns in nature. However, they have also found application in the music domain.[66] A typical example of a visual two-dimensional fractal is the Sierpinski sieve (also gasket) pictured in the bottom-right diagram of Figure 18.13. Note that the three triangles surrounding the large white triangle look the same as the entire triangle at a different scale. It may be constructed from the leftmost pattern that consists of a black triangle with a smaller inverted white triangle superimposed in the middle. By iteratively superposing smaller white triangles within the black triangles (indefinitely), the fractal is obtained.[67]

Returning to the top-left diagram for the *clave son* in Figure 18.12, the fractal nature of its IOI duration pattern is evident. Onsets O_1, O_2, and O_3 form two small triangles in a fractal relationship to onsets O_1, O_3, and O_5 (the large triangles). Furthermore, the first 8 pulses of the 16-pulse *clave son* (the first half of the rhythmic cycle) consist of the IOI pattern [3-3-2], the *tresillo* timeline, at the first level of the fractal pattern. At the second level in the 16-pulse cycle (the large triangles), onsets O_1, O_3, and O_5 determine the IOI pattern [6-6-4],

FIGURE 18.13   The Sierpinski sieve (also gasket) fractal.

a *well-formed* expansion of the tresillo rhythm. Hence, in the *clave son*, we hear a fast tresillo embedded fractally in a slow tresillo at half the tempo. This fractal property may be a contributing factor of the worldwide popularity of the *clave son* rhythm, but this is speculation. It would be interesting to explore the psychological perceptual parameters of this fractal property of rhythm and meter.

NOTES

1   Cohn, R., (2015), p. 9.
2   Cooper, G. W. & Meyer, L. B., (1960), p. 6.
3   Cooper, G. & Meyer, L. B., (1960), p. 4.
4   Kvifte, T., (2007).
5   Clayton, M. R. L., (1996). See also Gotham, M., (2015).
6   Hasty, C. F., (1997).
7   London, J., (1995, 2012). Grouping will be covered in Chapter 19.
8   Johansson, M., (2017), p. 47.
9   Ward, W. E., (1927), Waterman, R. A., (1948), Koetting, J., (1970), and Eerola, T., Himberg, T., Toiviainen, P., & Louhivuori, J., (2006).
10 Tan, S.-L., Pfordresher, P., & Harré, R., (2010), p. 292.
11 Munyaradzi, G. & Zimidzi, W., (2012).
12 Muchimba, F., (2008), p. 113.
13 Thul, E. & Toussaint, G. T., (2008a).
14 Entsua-Mensah, T. E., (2015).
15 Eerola, T., Himberg, T., Toiviainen, P., & Louhivuori, J., (2006).
16 Gilman, B. I., (1909), p. 534.
17 Agawu, K., (1995).
18 Chernoff, J. M., (1979), p. 54.
19 Temperley, D., (2000), p. 289.
20 Grauer, V., (2017).

21 Toussaint, G. T., (2015c). Most of the material in this section of this chapter is taken (with permission) from "Quantifying musical meter: How similar are African and Western rhythm?" *Analytical Approaches to World Music Journal*, Vol. 4, No. 2, pp. 1–30. https://aawmjournal.com.
22 The timelines used were taken from the following papers: Agawu, K., (2006), Anku, W., (2002a, 2002b), Chernoff, J. M., (1979), Kubik, G., (2010a, 2010b), Leake, J., (2009), Locke, D., (1982), London, J., (1995), Pressing, J., (1983), Rahn, J., (1987, 1996), Temperley, D., (2000), & Toussaint, G. T., (2013a).
23 Veltman, J., (2006).
24 Huron, D. & Ommen, A., (2006).
25 Palmer, C. & Krumhansl, C. L., (1990).
26 Lerdahl, F. & Jackendoff, R., (1983).
27 Tan, S.-L., Pfordresher, P., & Harré, R., (2010), Magill, J. M. & Pressing, J. L., (1997), Arom, S., (1991), Nketia, J. H. K., (1963), Koetting, J., (1970), and Belinga, M. S. E., (1965).
28 Hansen, N. C., (2011).
29 Deliège, I., (1987), Palmer, C. & Krumhansl, C. L., (1990), Dibben, N., (1994), and Todd, N. P., (1994).
30 Palmer, C. & Krumhansl, C. L., (1990).
31 Tan, S.-L., Pfordresher, P., & Harré, R., (2010). For a discussion of the differences between additive and divisive representations of rhythm, see Agawu, K., (2003) and Entsua-Mensah, T. E., (2015).
32 Magill, J. M. & Pressing, J. L., (1997).
33 Arom, S., (1991), p. 206.
34 Kwabena Nketia, J. H., (1962), p. 10.
35 Koetting, J., (1970), p. 124.
36 Belinga, M. S. E., (1965), p. 18.
37 Gómez, E. & Herrera, P., (2008).
38 Spearman, C., (1904).
39 Pearson, K., (1920).

40 Maturi, T. A. & Elsayigh, A., (2010).

41 Hauke, J. & Kossowski, T., (2011).

42 Jekel, J. F., (1977).

43 de Valpine, P., (2014).

44 Murdock, D. J., Tsai, Y.-L., & Adcock, J., (2008).

45 Veltman, J., (2006).

46 Agawu, K., (2006, 2003, 1995), Arom, S., (1991), Cuthbert, M. C., (2006), Pressing, J., (1983), Rahn, J., (1987, 1996), Temperley, D., (2000), Thul, E. & Toussaint, G. T., (2008a,b), and Toussaint, G. T., (2011, 2013a,b).

47 Agawu, K., (2003, 1995), Arom, S., (1991), Chernoff, (1979), Locke, D., (1982), Pressing, J., (1983), Rahn, J., (1987, 1996), and Temperley, D., (2000).

48 Marvin, E. W., (1991), Morris, R. D., (1993), and Quinn, I., (1999).

49 Watkins, A., Scheaffer, R., & Cobb, G., (2004).

50 Monin, B. & Oppenheimer, D. M., (2005).

51 Vitz, P. C., (1968).

52 Knopoff, L. & Hutchinson, W., (1983) and Snyder, J. L., (1990).

53 Toussaint, G. T. & Beltran, J. F., (2013).

54 Handel, S., (2006).

55 Anderson, P. W., (1972), Feldman, M., (1981), and Don, G. W., Muir, K. K., Volk, G. B., & Walker, J. S., (2010).

56 Patel, A. D. & Daniele, J. R., (2003a, 2003b) and Huron, D. & Ollen, J., (2003). See also Patel, A. D., (2003) for a treatise on the similarities and differences between rhythm in language and music.

57 Hansen, N. C., Sadakata, M., & Pearce, M., (2016).

58 London, J., (1995).

59 Some of this material is taken (with permission) from Toussaint, G. T., (2013).

60 Keith, M., (1991).

61 Arom, S., (2004).

62 Lerdahl, F. & Jackendoff, R., (1983), p. 69.

63 Roeder, J., (1998), [2.2].

64 Longuet-Higgins, H. C. & Lee, C. S., (1982).

65 Mandelbrot, B., (1982).

66 Hodges, W., (2006). See also the work by Scherzinger, M., (2013), p. 62, on the harmonic fractals present in the music of Southern Africa.

67 Pickover, C. A., (1990).

CHAPTER 19

Rhythmic Grouping

WHEN ONE LOOKS AT THE PATTERN of 27 dots scattered in Figure 19.1 (left), one is most likely to perceive four separate groups of dots: two groups of six dots each at the top, one group of six dots in the middle of the bunch, and one group of nine dots at the bottom. When humans are presented with visual dot patterns such as these, or even more unstructured scattered dots, they tend to perceive the dots as distinct groups, according to psychological *gestalt* principles of perceptual organization.[1] One of the most fundamental gestalt principles of visual perception is the *law of proximity*. The computing science field of artificial intelligence is concerned with modeling human skills, such as the perception of dot patterns, using *algorithms*, the most consequential technological tool of the twenty-first century, that according to computing scientist Bernard Chazelle *"promises to be the most disruptive scientific development since quantum mechanics."*[2] When distance is the principal feature expressed in the law of proximity, then this gestalt principle is easily captured by a simple yet powerful algorithmic model. One algorithm that works admirably well for modeling human perception of dot patterns constructs the *nearest-neighbor graph*.[3] The

nearest-neighbor graph of a group of dots is obtained by joining each dot to its nearest neighbor with an edge (a dot may have more than one nearest neighbor if several distances are equal). The nearest-neighbor graph of the dot pattern in Figure 19.1 (left) is pictured in the diagram on the right. This graph consists of four disconnected components corresponding to the four groups perceived in the dot pattern on the left. Hence, for this group of dots, the nearest-neighbor graph models perfectly the human visual perceptual skill.

The gestalt principle of proximity also applies to the *figural organization* or grouping of temporal stimuli in the auditory domain.[4] Figural organization means that a sequence of tones is perceived as being composed of distinct clustered groups. Indeed, proximity is considered to be "the primary grouping force at the melodic and rhythmic level."[5] A duration pattern such as the ubiquitous *clave son* timeline [x . . x . . x . . . x . x . . .] will be perceived as consisting of two groups of onsets: one group of three onsets [x . . x . . x] followed by a second shorter group of two onsets [x . x]. The shorter group [x . x] is separated from the longer group by the longest silent interval [. . .]. More generally, rhythms are perceived as

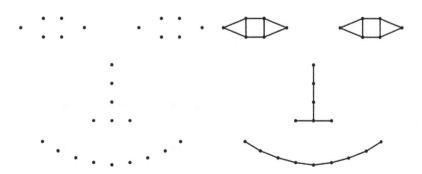

FIGURE 19.1   A dot pattern (left) and its nearest-neighbor graph (right).

consisting of a nested hierarchy of groups of onsets.[6] For over a century it has been widely believed that grouping was a universal basic principle of perception. However, recently some evidence has surfaced to support the hypothesis that grouping may depend on culture.[7] According to Cooper and Meyer (1960): "Rhythmic grouping is a mental fact, not a physical one. There are no hard and fast rules for calculating what in any particular instance the grouping is."[8] Nevertheless, today there exists a variety of rule-based physical formal models of grouping that predict the way humans perceive rhythmic groupings. The models of grouping examined here are based solely on durations, and hence apply to rhythms consisting of onsets that are identical to each other, and are not accented by any parameter other than time. What the different methods have in common is that they partition the onsets of a rhythm into groups by defining boundaries between the groups in terms of interonset intervals (IOIs) that possess specific proximity properties. However, the methods differ according to the properties that they enlist to determine which IOIs are designated as generators of group boundaries.

Tenney and Polansky (1980) and Temperley (2001) have proposed that an IOI will tend to be a grouping boundary if it is a *local maximum*, that is, if its duration is longer than that of the adjacent IOIs on *both* sides.[9] Consider the two common rhythms in box notation in Figure 19.2. According to the local maximum IOI rule, the top rhythm has one grouping boundary between the second and third onsets, and hence is designated as consisting of two groups: [x x .] and [x x . x .]. By contrast, the rhythm at the bottom is classified as consisting of just one group because it contains no local maxima among its IOIs. However, both rhythms can easily be perceived as consisting of three groups. The rhythm at the top consists of groups [1-2], [1-2], and [2], and the rhythm at the bottom of groups [2], [1-1-2], and [2]. Notwithstanding, Temperley writes regarding the *local*

maximum rule that "One nice effect of this is that it prevents having groups with only one note, something which clearly should generally be avoided." In rhythms such as these, however, it makes more sense to allow the existence of groups that consist of just one note (onset) if they are isolated. Therefore, the local maximum rule is inadequate as a general model for accurately predicting human perceptual grouping in this context.

Emilios Cambouropoulos acknowledges that the formalizations of rhythmic grouping mechanisms in terms of *local maxima* suffer from limitations and affirmed that although a local maximum IOI is sufficient to determine a grouping boundary, it is not a necessary condition.[10] To support this contention, Cambouropoulos provides the example rhythm in Figure 19.3 (top) which has no IOIs that are local maxima, but nevertheless induces perceptual grouping of its onsets. The rhythmic ostinato in Coldplay's *Viva la Vida* in Figure 19.3 (bottom), a member of Jay Rahn's family of *braided* rhythms, is another such example.[11] Indeed, a large subset of *braided* rhythms have this property that they contain no local maxima in their IOI sequences but nevertheless induce perceptual grouping.[12] In these rhythms the perception of grouping is induced by the pairs of closest onsets with unit IOI durations which can be felt as boundaries that produce a transition between consecutive downbeat onsets and consecutive upbeat onsets.

To remedy the limitations of the *local maximum* rule, Cambouropoulos proposed a local boundary detection model (LBDM) with the following *proximity rule*: "Amongst three successive objects that form different intervals between them a boundary may be introduced on the larger interval."[13] Application of the LBDM rule to the example rhythm in Figure 19.3 (top) yields the desired grouping into the three groups shown in Figure 19.4. Since the three onsets labeled *a*, *b*, and *c* form two different IOIs (two pulses from *a* to *b* and one pulse from *b* to *c*), a boundary is introduced between onsets *a* and *b*. The same symmetric situation occurs with onsets *b*, *c*, and *d*,

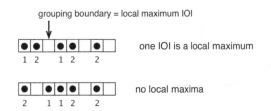

FIGURE 19.2 Onset groupings for two rhythms obtained with the *local maximum* IOI rule.

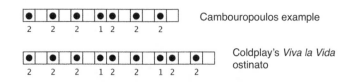

Cambouropoulos example

Coldplay's *Viva la Vida* ostinato

FIGURE 19.3 These two rhythms have no IOIs that are local maxima, but the rhythms still induce perceptual grouping of onsets.

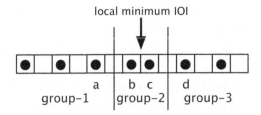

FIGURE 19.4 The LBDM rule applied to this rhythm yields three groups determined by the *local minimum* IOI.

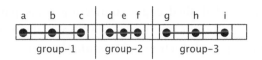

FIGURE 19.5 The *mutual nearest-neighbor* graph of this rhythm contains three disconnected components corresponding to the three onset groups (*a*, *b*, *c*), (*d*, *e*, *f*), and (*g*, *h*, *i*). Note that this rhythm has no *local maxima* and no *local minima* among its IOIs.

creating a second boundary between onsets *c* and *d*. These two boundaries partition the rhythm into the three groups shown. Note that in this case it is the *local minimum* IOI that determines the grouping, by introducing boundaries on either side of the *local minimum* IOI. Thus, the LBDM rule may introduce boundaries determined by IOIs that are either *local maxima* or *local minima*, or *step functions* such as [2-2-2-2-1-1-1-1], which contain three successive onsets that form different intervals between them.

The principle behind the *proximity rule* of the LBDM model may be epigrammatically captured by a structure from the discrete mathematics field of graph theory known as the *mutual nearest-neighbor* graph.[14] The mutual nearest-neighbor graph of the onsets of a rhythm is obtained by inserting an edge between two onsets if both onsets are nearest neighbors of each other. The resulting graph consists of connected components consisting of onsets connected to each other with edges. Each connected component (subgraph) in the graph determines one group in the grouping. The *mutual*

nearest-neighbor graph of the rhythm [2-2-2-1-1-2-2-2-2] is illustrated in Figure 19.5. The graph of this rhythm contains three disconnected components corresponding to the three onset groups (*a*, *b*, *c*), (*d*, *e*, *f*), and (*g*, *h*, *i*). Note that this rhythm has no *local maxima* and no *local minima* among its IOIs. Nevertheless, the three groups determined by the mutual nearest-neighbor graph correspond to the three perceived groups.

The preceding discussion considered the case of linear rhythms. However, the main focus in this book is on rhythm timelines which by definition are cyclic. Therefore, it is natural to extend the definition of *linear* mutual nearest-neighbor graphs to hold for geodesic distances measured on the circles of time used in the circular notation of rhythms.[15] These graphs may differ from their linear counterparts, since "rhythmic groups are not respecters of bar lines. They cross them more often than not."[16] The nearest-neighbor graphs of the six distinguished timelines introduced in Chapter 7 are shown in Figure 19.6. They consist

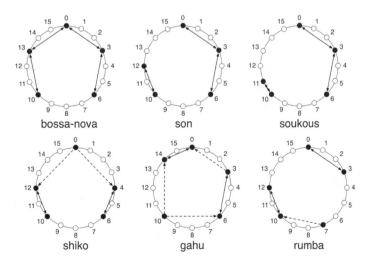

FIGURE 19.6 Onset groupings obtained with the mutual nearest-neighbor graph (solid lines) for the six distinguished timelines. The nearest neighbors that are not mutual nearest neighbors are indicated with dashed lines, with arrows pointing to their nearest neighbors.

of solid and dashed edges connecting some pairs of consecutive onsets (vertices of the graph). The nearest-neighbor pairs which are also *mutual* nearest neighbors are shown with solid lines. The nearest neighbors that are not mutual nearest neighbors are indicated with dashed lines, with arrows pointing to their nearest neighbors. The bossa-nova consists of one group, the son and soukous each contain two groups, and the shiko, gahu, and rumba each have three groups, one of which is a solitary group consisting of one onset. The graphs (groups) of the three rhythms at the top are made up of only mutual nearest-neighbor pairs, and they do not contain onset pairs that are asymmetric nearest neighbors. By contrast, the three rhythms at the bottom contain asymmetric nearest neighbors (nearest neighbors that are not mutual nearest neighbors) indicated by dashed lines, with arrows pointing to their nearest neighbors.

The nearest-neighbor graph of a rhythm's onsets represented in circular notation, as in Figure 19.6, provides a new tool for the visualization and analysis of the rhythm's grouping structure. Consider the property of "call-and-response," one of the rhythmic hallmarks of some of the world's cultures. In terms of purely durational patterns, it is reasonable to suppose that the "call" and "response" phrases should be perceived as two distinct groups. This perceptual property translates to the mathematical property of the corresponding nearest neighbor graphs having two disconnected components. From Figure 19.6 it follows that the son, soukous, and rumba are prime candidates for possessing the "call-and-response" property, with son and soukous having shorter "responses" to longer "calls," and the rumba exhibiting a longer "response" to a shorter "call." The nearest-neighbor graphs of the remaining three rhythms, bossa-nova, shiko, and gahu, each consist of only one connected component, suggesting that they do not have the "call-and-response" property.

One of the six rhythm timelines in Figure 19.6 stands out uniquely amongst the group. Only the gahu has a nearest-neighbor graph that consists of a single connected component that forms a *cycle* (a closed loop), a property in common with all regular rhythms. Every vertex (onset) of the graph has degree two (contains two incident edges). The single component of the nearest-neighbor graph of the two other rhythms (the bossa-nova and the shiko) consists of a *chain* rather than a

cycle, and therefore has two vertices of degree one, located at pulses 6 and 10. David Locke characterizes the gahu rhythm bell pattern as possessing a "tricky quality" due to its "spiraling effect," and its ability to "cause the beat to turn around" in the sense that the third, fourth, and fifth *upbeat* onsets at pulse positions 6, 10, and 14 can induce, during a performance, a "gestalt flip" and be perceived as *downbeats*.[17] It is tempting to speculate that these properties of the gahu bell pattern highlighted by David Locke are in part the result of the fact that the rhythm's nearest-neighbor graph consists of a connected cycle. Whether there is any reality behind this speculation is best left for psychological experimentation.

The nearest-neighbor and mutual nearest-neighbor graphs also provide features for new classifications of rhythms. In Chapter 9, a variety of decision trees were illustrated for classifying rhythms. We close this chapter with yet another example of decision trees applied to the classification of rhythms, which uses geometric information about the *groupings* of the onsets in a rhythm, obtained from these proximity graphs. One possible decision tree that uses grouping information obtained from the *mutual* nearest-neighbor graphs of the six distinguished timelines is shown in Figure 19.7. The root node of the decision tree partitions the rhythms according to the number of groups contained in each rhythm. The bossa-nova consists of a single group. Two rhythms have two groups: son and soukous. These two rhythms are distinguished by the durations of their short groups: the short group for the son, between pulses 10 and 12, has duration 2, whereas the short group for the soukous, between pulses 10 and 11, has duration 1. Three rhythms have three groups each. The shiko, gahu, and rumba, each contain one solitary group (single onset) and two groups of two onsets each. The shiko is distinguished from the other two rhythms in that its two nonsolitary groups have the same duration of two pulses. Finally, the gahu and rumba are discriminated by the position of the solitary group located in between the other two groups. The solitary group of the gahu at pulse 10 bisects the duration interval between pulses 6 and 14, whereas the solitary group of rumba is closer to pulse 10 than to pulse 3. Incorporating features of the nearest-neighbor graph, one can produce other similar decision trees that are left as exercises for the reader.

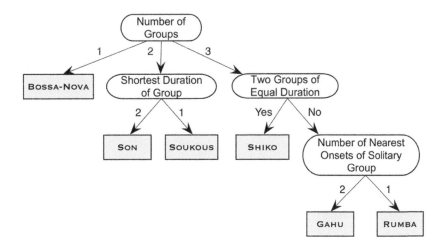

FIGURE 19.7   A decision tree classification of the six distinguished timelines using the mutual nearest-neighbor groupings of Figure 19.6.

NOTES

1 The concept of grouping in perception has received considerable attention in the domains of visual as well as auditory stimuli. See Papari, G. & Petkov, N., (2005), Temperley, D., (2000), Toussaint, G. T., (1988), Marr, D., (1982), and Vetterl, K., (1965).

2 Chazelle, B., (2006). "The algorithm: Idiom of modern science." Princeton University. www.cs.princeton.edu/~chazelle/pubs/algorithm-print.html.

3 Toussaint, G. T., (2005d). Note that the *nearest-neighbor graph* differs from the *sphere-of-influence* graph applied to modeling the concept of *streaming* in Chapter 30, in the sense that the former has no crossing edges. However, both graphs are mathematical–computational tools for grouping.

4 Deutsch, D., (1999) and Deliège, I., (1987).

5 Jensen, K., (2010), p. 331. Grouping the elements of a pattern into groups is also referred to as pattern *segmentation*. See Handel, S. & Todd, P., (1981) for segmentation strategies used by humans on dichotomous sequential visual, temporal, and motor patterns.

6 Bregman, A. S., (1990).

7 Iversen, J. R., Patel, A. D., & Ohgushi, K., (2008).

8 Cooper, G. W. and Meyer, L. B., (1960), p. 9.

9 Tenney, J. & Polansky, L., (1980), p. 211 and Temperley, D., (2001), p. 61.

10 Cambouropoulos, E., (2005), p. 281.

11 Rahn, J., (1987).

12 One subset of *braided* rhythms that have this property is the *sharp toggle* rhythms explored in Chapter 32.

13 Cambouropoulos, E., (2005), p. 282. In the present context, an "object" may be interpreted as an "onset."

14 Toussaint, G. T., (2016). Mutual nearest neighbors are sometimes referred to as *symmetric* nearest neighbors in computer science and *reflexive* nearest neighbors in biology. See Cox, T. F., (1981).

15 Toussaint, G. T., (2017). In Chapter 37, several formal models of grouping are compared, and features of the grouping structure are applied to the measurement of rhythm similarity.

16 Cooper, G. W. & Meyer, L. B., (1960), p. 6.

17 Locke, D., (1998), pp. 18–22.

CHAPTER 20

Dispersion Problems

Perfectly Even, Maximally Even, and Balanced Rhythms

ASSUME THAT A SATELLITE COMPANY wants to install six electronic communications transmission towers in the city where you live, and that the city government has given the company permission to build them in any of the 15 possible locations approved by its citizens, as indicated in the schematic map shown in Figure 20.1 (left). The company desires to select 6 out of 15 locations that minimize the total signal interference between all pairs of transmission towers. Clearly, to satisfy this criterion, the locations selected should be as far away from each other as possible. But how should we measure the concept: *as far as possible from each other*? There is a profusion of ways to measure this notion. However, in an idealized situation, and in the absence of additional knowledge that may affect our choice, it is reasonable to pick a natural and easily understood criterion: select the six locations that maximize the *average distance* between pairs of towers. Figure 20.1 (right)

shows six candidate locations along with the distance between every pair indicated by a straight edge. Since the average distance between a set of locations is the sum of all the pairwise distances divided by the number of distances, and the number of distances is fixed no matter which locations we select (in this case 6(5)/2 = 15), this criterion is equivalent to choosing the six locations that maximize the *sum* of their pairwise distances. This type of problem is called a *dispersion problem*[1] in the field of *obnoxious facility location*, within the broader research area of *operations research*.[2]

One of my main goals in writing this book is to build bridges between problems in music and other areas of science, mathematics, and engineering. It is hoped that such bridges will help both areas that are connected by such a bridge. To obtain some insight into the maximum dispersion problem and how it relates to the theory of musical rhythm, consider first the equivalent

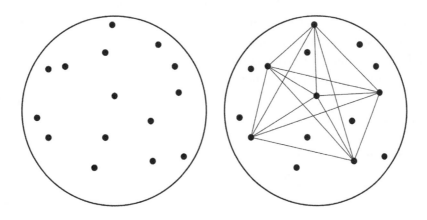

FIGURE 20.1   Selecting a subset of maximally dispersed locations in a region.

problem in a simpler situation where the locations are restricted to lie on a perfectly straight highway such as one might find in Kansas, Saskatchewan, or Abu Dhabi. For concreteness, and as a start, consider the example of selecting 2 of 10 possible locations illustrated in the top diagram of Figure 20.2. Obviously the two extreme locations at A and B, in the second diagram, maximize the distance between them. Consider now selecting the best three locations. We must select a third location C from among the eight remaining locations. However, in this situation it does not matter which location is selected (refer to the third diagram), because no matter where C is located in between A and B, the total sum of distances increases by AC + CB = AB, and hence the total sum of distances equals 2AB. This condition changes radically, however, when four locations are selected (refer to the fourth diagram). Having selected C anywhere between A and B, consider selecting the fourth location D anywhere between C and B. Clearly, no matter where D is located, its sum of distances to A and B remains fixed and equal to AD + DB. However, its distance to C varies depending on where D is located. Hence, D should be chosen to be as far as possible from C, i.e., in the location adjacent to B. Furthermore, by symmetry, given D's position, the total sum of distances can be increased further by selecting C to be as far as possible from D, i.e., in a location adjacent to A. This argument generalizes to any even number of locations selected from any number of allowable positions. In other words, the solution, in general, implies that half of the selected locations should be chosen from the leftmost positions, and the other half from the rightmost positions. The fourth diagram in Figure 20.2 shows the optimal solution for selecting four locations.

Since rhythm timelines are cyclic, let us turn our attention to the analogous dispersion problem of locating k points on a circle so as to maximize the sum of their pairwise distances. Since the highway is now a

beltway (circle) the straight-line distances of Figure 20.2 become arclengths or geodesic distances, as illustrated in Figure 20.3. The distance between A and B, for example, is the clockwise arclength starting at A and ending at B, or the counterclockwise arclength starting at B and ending at A, but not the clockwise arclength starting at B and ending at A, nor the counterclockwise arclength starting at A and ending at B.

The question now becomes: how should a set of k points on a circle be arranged so that the sum of their pairwise arclengths is maximized? It turns out that if the points are located such that they form the vertices of a regular k-sided polygon (all sides are of equal length), then the resulting sum is a maximum. Thus, the structure of this solution is completely different from the case in which the points fall on a straight line and might raise hopes that the sum of pairwise arclengths might be used either as a criterion for generating regular rhythms or perhaps maximally even rhythms, or for measuring how regular they are for the purpose of their automatic classification. A rhythm is maximally even if its attack points (onsets) are distributed in time as evenly as possible.[3] Unfortunately, there is little cause for celebration here because it turns out that the converse is not true; rhythms other than maximally even or regular, and even highly irregular rhythms, can also maximize the sum of pairwise arclengths.

Before proceeding further, however, the concept of maximally even rhythms should be made more precise, and before that, it is instructive to present their simpler offspring, perfectly even rhythms, and their opponents, perfectly uneven rhythms. Figure 20.4 (top) shows four regular rhythms. The first has two onsets in a cycle of four pulses, and the other three have four onsets

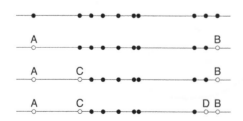

FIGURE 20.2 Selecting a subset of maximally dispersed locations on a line.

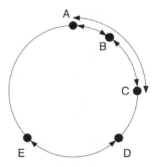

FIGURE 20.3 The distance between two points on a circle is the geodesic distance or shortest arclength between them.

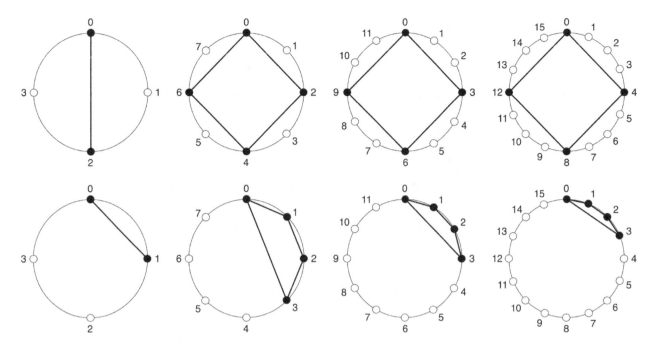

FIGURE 20.4 Four perfectly even rhythms (top) and their perfectly uneven counterparts (bottom).

in cycles of 8, 12, and 16 pulses, respectively. Regular rhythms are perfectly even rhythms. The four rhythms at the bottom of Figure 20.4 are the perfectly uneven counterparts of the rhythms at the top: their onsets are clustered together as tightly as the underlying discrete pulse structure permits.

To emphasize the point that the sum of pairwise arclengths between the onsets of rhythms is not an acceptable measure of rhythm evenness, consider the six distinguished Afro-Cuban timelines in Figure 20.5. Each rhythm is shown with straight-line segments joining every pair of onsets to clearly identify all the arclengths. First note that for all the rhythms the arclengths corresponding to the heavy edges that make up the convex polygon add up to 2π since they determine one entire circumference. This follows from the fact that no arclength is greater than a half circle, and so all the arclengths are traversed in a clockwise order around the circle. For example, the *clave son* (second diagram in Figure 20.5) is traversed in the pulse sequence (0, 3, 6, 10, 12, 0). Next, consider the five-sided star polygons drawn with thin lines that start at pulse zero and connect all the remaining onsets by skipping an onset at each step. For the *clave son*, this traversal consists of the pulse sequence (0, 6, 12, 3, 10, 0). None of these arclengths is greater than a half circle either, and, as we traverse the edges of the star polygons in clockwise

order until we return to pulse zero, we have gone around the circle twice in every case, so that the sum of the arclengths corresponding to the light edges is 4π. This means that for all six rhythms the sum of all pairwise arclengths between the five onsets is 6π. This implies that this distance is completely ineffective for discriminating between these six rhythms, and thus for measuring the evenness of rhythms. Intuitively, the bossa-nova appears to be more evenly spread than the son, which is in turn more even than the soukous, but this measure fails to capture these significant differences.

To understand the anomalous behavior of the sum of arclengths measure, the concept of a *balanced* rhythm is convenient. Consider a rhythm such as the *clave son* in Figure 20.6 (left) represented by its onsets as points on the circle. Draw any straight line through the center of the circle such as line *a*. Now imagine rotating this line about this center (say in a clockwise direction) by 180° until it returns to its starting orientation. If for every position of the line such that it is not incident on an onset, the number of onsets on one side of the line differs by at most one from the number of onsets on the other side, then the rhythm has been called *balanced*. Since the notion of a balanced rhythm has been defined independently in different ways[4] and additional definitions will be encountered later in this chapter, let us call this type of balanced rhythm *semicircle-balanced*.

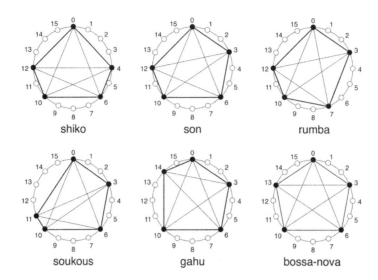

FIGURE 20.5 The six distinguished timelines all have the same sum of pairwise arclengths.

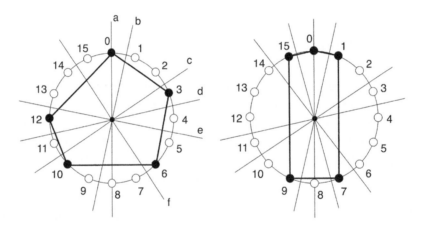

FIGURE 20.6 The *clave son* (left) is a *semicircle-balanced* rhythm, but so is the highly uneven rhythm (right).

In 2008, Minghui Jiang proved that the sum of all the pairwise arclengths for a rhythm is a maximum if, and only if, the rhythm is semicircle-balanced.[5] Examination of the six rhythms in Figure 20.5 reveals that they are all semicircle-balanced, and hence they all yield the same maximum value for the sum. This characterization of the rhythms that maximize the sum of pairwise arclengths in terms of semicircle-balance, clarifies why this sum is not an effective measure of the evenness of a rhythm. As the diagram on the right in Figure 20.6 makes clear, a rhythm can be extremely uneven and still be semicircle-balanced.

Fortunately, there exist other methods to measure the evenness of the distribution of points on a circle that work well for musical rhythm applications. Rather than calculate the distance between two onsets using arclength (which corresponds quantitatively to duration of time), use instead the length of the straight line or chord of the circle that connects the two points. This seemingly innocent change between the straight-line segments connecting a pair of points on a circle, and the circular arcs they subtend, makes all the difference in the world, in spite of signaling anathema to some music theorists. One might argue intuitively that the rhythms displayed on the circle of time are one-dimensional objects that are not embedded in the two-dimensional plane, and that the durations should correspond to arclengths. However, from the mathematical point of view there is nothing inherently wrong with such a mapping, if it serves some fruitful purpose, and here it does. Whereas the arclengths fail to measure maximal evenness, the straight-line distances between the vertices of the two-dimensional polygons work admirably well for this purpose. Furthermore, it is plausible that,

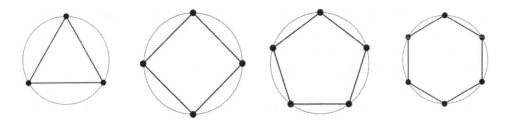

FIGURE 20.7   Perfectly even rhythms (regular polygons) with $k = 3, 4, 5, 6$ inscribed on the *continuous* circle of time.

in the case of the repeating nature of a cyclic rhythm timeline, the human brain may produce a semblance of a circular two-dimensional representation of the rhythm being perceived. It may be that the straight-line distance between two onsets on this circular representation has a perceptual reality. Psychological experiments to determine the validity of such speculations would be welcome.

In 1956, László Fejes Tóth proved that given k points anywhere on a circle (or its interior) the sum of all the pairwise Euclidean (straight-line) distances between the points is maximized if, and only if, the points form the vertices of a regular k-sided polygon inscribed in the circle.[6] This result immediately provides a definition of *perfectly even rhythms* as those consisting of k onsets placed so that they correspond to vertices of regular k-sided polygons on the continuous circle of time. Figure 20.7 illustrates four perfectly even rhythms for $k = 3, 4, 5,$ and 6.

The regular rhythms notated in Figure 20.7 without the specification of the number of pulses in the cycle are fine for certain types of music, such as electronic music or music that contains no meter. However, most music around the world, and especially dance music, depends heavily on an underlying cycle consisting of n pulses, and furthermore, the onsets of the rhythm are required to fall on a subset of these pulses.[7] If k divides evenly into n, then perfectly even rhythms corresponding to regular polygons are trivial to construct. For example, the rhythms in Figure 20.7 with three, four, and six onsets may be realized in a cycle of 12 pulses, but not the third rhythm with five onsets. The five-onset rhythm exists in a cycle of 10 pulses, but not the other three, and all four rhythms may be embedded in a cycle of 60 pulses. Therefore, what is to be done with regular rhythms that have onsets with cardinalities such as five and seven if the cycle of interest has a total of 8, 12, or 16 pulses? The answer to this conundrum is to relax the notion of *perfectly even* rhythms that correspond to *regular* polygons

to that of *maximally even* rhythms that correspond to polygons that are as regular as possible or *almost regular*. There are several ways to proceed in this direction. One natural solution is to apply the same snapping (quantization) process that was used in Chapter 12 to solve the problem of binarization of ternary rhythms, as illustrated in Figure 20.8 using a cycle of 12 pulses. If a maximally even rhythm with k onsets is desired, the idea is to first place a regular k-sided regular polygon anywhere on the circle and then to snap each vertex either to its nearest pulse, its nearest pulse in a clockwise direction, or its nearest pulse in a counterclockwise direction. It is to be understood that if a vertex of the regular polygon lies on a pulse location, then it stays where it is. Alternately, the regular polygon may be positioned so that its vertices do not lie on pulse positions or halfway between two pulse positions. In any case, the snapped positions then become the onsets of the maximally even rhythm.[8] For example, the left diagram of Figure 20.8 shows an arbitrarily placed square (solid lines with small black vertices) with its vertices snapped to its nearest clockwise pulses (dotted lines with gray onsets). A more interesting example is the pentagon on the right, in which the number of onsets does not divide evenly into the number of pulses. Here the initial placement has a vertex on pulse four, which stays there. The other four vertices snap to pulses 7, 9, 0, and 2, yielding

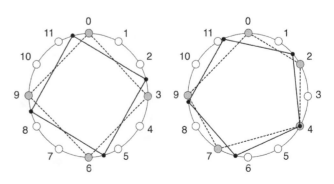

FIGURE 20.8   Maximally even rhythms via *snapping* regular polygons to lattice points.

the well-known African standard ternary bell pattern [2-2-3-2-3]. Note that if the vertices are rounded to their nearest pulses the same necklace pattern is obtained at pulses 2, 4, 6, 9, and 11. If the number of onsets divides evenly into the number of pulses, then the resulting maximally even rhythm is also perfectly even, as the example in Figure 20.8 shows.

Let us return to the notion of balanced rhythms. It was shown with the example of Figure 20.6 that a rhythm that is semicircle-balanced can be highly uneven in its distribution of points along the circle. Therefore, although good rhythm timelines tend to be semicircle-balanced, rhythms that are semicircle-balanced need not be good timelines by virtue of this property alone. However, another definition of balanced rhythms inspired by the physical property of *gravity* has more potential for informing music theory, the psychology of rhythm perception, and musical practice. Consider the circle of time embedded in a two-dimensional Cartesian coordinate system with the origin of this system as the center of a unit-radius circle. Furthermore, let the location of every onset of a rhythm be specified by its x and y coordinates. If the onset points are considered to be physical point-masses of unit weight, then a rhythm may be represented by its center of gravity (centroid). A rhythm will be defined as *perfectly centroid-balanced* if its center of gravity coincides with the center of the circle (the origin of the space). The distance between the centroid of a rhythm and the origin of the circle may then be used as a measure of how centroid-balanced a rhythm is.[9]

To compare the two notions of rhythm balance introduced in this chapter, it is useful to extend Minghui Jiang's binary (all-or-none) definition of semicircle-balance to allow it to measure the degree to which a rhythm is semicircle-balanced. For any nonnegative integer I, define a rhythm to be I-semicircle-balanced if all lines through the center of the circle partition the onsets of the rhythm such that the number of onsets lying strictly on each side of the line differs by at most I. For $I = 1$, this definition reduces to Jiang's definition. A rhythm that is 0-semicircle-balanced will have the same number of onsets on both sides of every center line and will be called *perfectly semicircle-balanced*. All perfectly even rhythms (regular polygons) with an even number of onsets are *perfectly semicircle-balanced* because if the center line does not intersect any onsets, then $n/2$ onsets lie on each side of the line, and if the line intersects one onset it must also intersect the onset's antipodal onset, and thus there are $(n − 2)/2$ onsets on each side of the line. If the number of onsets is odd, a perfectly even rhythm (regular polygon) is *1-semicircle-balanced* because if the center-line intersects one onset, then $(n − 1)/2$ onsets lie on each side of the line, and if the line does not intersect any onset, then one side of the line will contain $(n − 1)/2$ onsets and the other side will contain $(n − 1)/2 + 1$ onsets. On the other hand, *perfectly semicircle-balanced* rhythms need not be perfectly even, as the sub-Saharan African 12-pulse *kéné foli* rhythm and the 16-pulse Persian rhythm *al-thaqil al-thani* in Figure 20.9 illustrate. Note that these two rhythms have the property that for every onset there is an antipodal onset located diametrically opposite on the circle. For this reason, they are called *antipodal-symmetric* rhythms.[10] Thus, all antipodal-symmetric rhythms are also *perfectly semicircle-balanced*.

The center of gravity of a pair of antipodal onsets is the center of the line joining the two onsets, which is a diameter of the circle. Therefore, the center of gravity of an antipodal pair of onsets coincides with the center

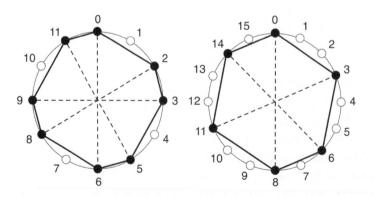

FIGURE 20.9 The *kéné foli* rhythm (left) and the *al-thaqil al-thani* (right).

of the circle, and the pair is therefore *perfectly centroid-balanced*. The Archimedes Lemma states: "If an object is divided into two smaller objects, the center of mass of the compound object lies on the line segment joining the centers of mass of the two smaller objects."[11] Therefore if two rhythms each have their centers of mass at the same point, the resultant rhythm will have its center of mass at that same point. It follows that the union of distinct pairs of antipodal onsets is therefore also *perfectly centroid-balanced*, which in turn implies that all *antipodal-symmetric* rhythms are also *perfectly centroid-balanced*. However, *perfectly centroid-balanced* rhythms need not be antipodal-symmetric, nor perfectly even, as we demonstrate in the following.

In the combined clapping pattern used by the *Lala* tribe in the *Icitelele* play song for girls, one person claps the rhythm [3-3-3-3] and the second person claps [4-4-4].[12] The polygons corresponding to these two perfectly centroid-balanced rhythms are shown in the top-left and top-center diagrams of Figure 20.10. Both rhythms are perfectly even. The first is also antipodal-symmetric and perfectly semicircle-balanced. The second rhythm is not antipodal-symmetric but is 1-semicircle-balanced. The resultant clapping pattern is the typical six-onset four-against-three rhythm shown in the top-right diagram. If the onset at pulse six is

deleted to create additional syncopation, one obtains the five-onset rhythm at the bottom-left diagram. Although this rhythm is not perfectly even and not perfectly semicircle-balanced, it is perfectly centroid-balanced. To see this, decompose the rhythm into two rhythms shown in the bottom-right diagram. One consists of the antipodal pair of onsets at pulses 3 and 9, and the other is a regular triangle with onsets at pulses 0, 4, and 8. Each of these two rhythms has their center of gravity at the center of the circle. Therefore, by the Archimedes Lemma, it follows that the resultant five-onset rhythm also has its center of gravity at the center of the circle, and is thus perfectly centroid-balanced.

It remains to elucidate the relationship between semicircle-balanced and centroid-balanced rhythms. The five-onset rhythm in Figure 20.10 (lower left) establishes that perfectly centroid-balanced rhythms are not necessarily perfectly semicircle-balanced. For example, the line through pulses 1 and 7 leaves two onsets on one side and three on the other. That *1-semicircle-balanced* rhythms need not be perfectly centroid-balanced follows from the rhythm in Figure 20.6 (right).

With this new tool in hand, let us return to the analysis of the six distinguished timelines in Figure 20.5 by calculating their centers of gravity. In the diagram on the left of Figure 20.11, the centers of gravity are shown

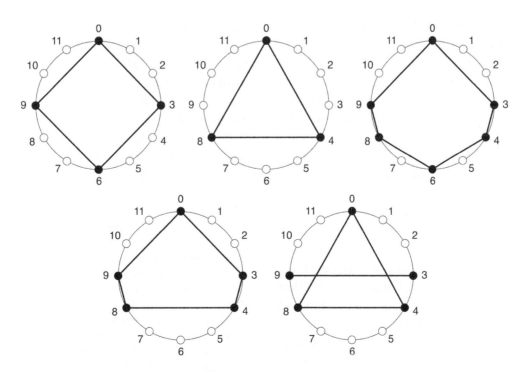

FIGURE 20.10 The five-onset rhythm on the lower left diagram is *perfectly centroid-balanced* but is not *perfectly even*, not *maximally even*, and not *antipodal-symmetric*.

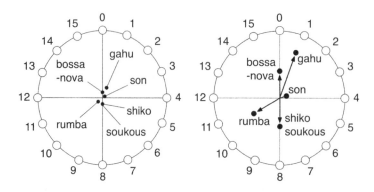

FIGURE 20.11 The centers of gravity of the six distinguished timelines (left) and the same center locations with distances from the origin magnified for the sake of clarity (right).

Rhythm	Son	Shiko	Bossa-Nova	Soukous	Rumba	Gahu
Distance	1.6	8.0	8.0	8.0	9.0	14.7

FIGURE 20.12 The approximate distances between the centers of gravity of the six distinguished timeline from the center of the circle of radius 100 units.

to scale and are clustered together quite closely. For the sake of clarity, the diagram on the right displays the locations of the centers of gravity with their distances from the origin (center of the circle) magnified. The shiko and soukous have identical centers of gravity. The center of gravity of the gahu is the furthest from the origin, and thus gahu is the most unbalanced rhythm. On the other hand, the son is extremely centroid-balanced compared with the other five rhythms. Indeed, it is almost perfectly centroid-balanced, and the most balanced of the six distinguished rhythms. Furthermore, recent exhaustive computer calculations performed by Zak Abel and Tom C. van der Zanden have verified my conjecture that among *all* possible rhythms with five onsets and 16 pulses, the *clave son* is the most balanced of them all.

The approximate distances of the centers of gravity of each rhythm from the origin (center of the circle) are shown in Figure 20.12, in increasing value from left to right, as a fraction of the radius of the circle that ranges from 0 to 100 units.

The distance between the center of gravity of a rhythm and the center of the circle is not the only information available as a descriptive feature of the rhythm. The location of the center of gravity of a rhythm with respect to the center of the circle provides additional descriptive information, as Figure 20.11 makes clear. The center of gravity of the gahu lies in the first half cycle of the rhythm, whereas the center of gravity of the rumba lies in the second half cycle. It is tempting to

speculate on whether the location of the center of gravity of a rhythm can be perceived while listening to the rhythm. However, exploring this question is best left for psychological experimentation.

The fact that the *clave son* has been such a popular rhythm all over the world for hundreds of years, and that it is almost perfectly balanced, suggests that perhaps centroid-balance is a feature that contributes to making a "good" rhythm good. It remains to determine by means of psychological experiments what role the degree of centroid-balance plays in the perception of rhythmic goodness.

The concept of elements in a set being distributed close to evenly, such as the maximally even and balanced rhythms considered here, is important also in the pitch domain, because chords which are highly consonant tend to divide the octave nearly evenly.[13]

NOTES

1 Ravi, S. S., Rosenkrants, D. J., & Tayi, G. K., (1994) and Tamir, A., (1998).
2 Herremans, D. & Chew, E., (2016), provide a review of an operations research approach to music generation with structural constraints.
3 Clough, J. & Douthett, J., (1991), in their seminal award-winning paper, originally developed the concept of maximally even sets in the context of pitch class sets. See also Block, S. & Douthett, J., (1994).
4 MacGregor, J. N., (1985), defines a temporal pattern as balanced if the longest runs (interonset intervals (IOIs)) occur at the beginning and ending of the pattern.

For this measure, the shiko rhythm with IOI pattern [4-2-4-2-4] is balanced, whereas the son timeline with IOI pattern [3-3-4-2-4] is not balanced.

5 Jiang, M., (2008).

6 Tóth, L. F., (1956, 1959). While it is true that almost anyone's intuition would dictate that the sum of pairwise (straight) distances of points on a continuous circle should be realized only by points that are equally spaced, it is a different matter altogether to prove it mathematically, and dangerous to take it as a postulate before a proof is established. After all, almost anyone's intuition would dictate that the sum of pairwise (arc) distances of points on a continuous circle should also be realized only by points that are equally spaced, but this is very far from the truth.

7 In practice, the onsets do not fall precisely on the pulses for a variety of reasons that include microtiming, expressive timing, and physical constraints such as the distance of some drums from the hands of the performer. Here we idealize the situation for theoretical purposes and to simplify the resulting analyses.

8 For an alternate definition of maximally even sets, see Douthett, J. & Krantz, K., (2008) and Johnson, T. A., (2003), p. 27. Amiot, E., (2007), gives a short history of maximally even sets.

9 This notion of *balance* was introduced by Milne, A. J., Bulger, D., Herff, S., & Sethares, W. A., (2015). However, these authors use the single-word term *balance*, which here is extended to *centroid-balance* to distinguish it from *circumcircle-balance*. See also Milne, A. J., Bulger, D., & Herff, S. A., (2017) and Milne, A. J. & Dean, R. T., (2016).

10 *Antipodal-symmetric* rhythms should not be confused with the *antipodal* rhythms studied in Aichholzer, O., Caraballo, L. E., Díaz-Báñez, J. M., Fabila-Monroy, R., Ochoa, C., & Nigsch, P., (2015). The rhythms these authors call antipodal consist of rhythms that do not contain pairs of onsets that are antipodal. Such rhythms here are said to possess the rhythmic oddity property, a topic covered in Chapter 15. Aichholzer, O., Caraballo, L. E., Díaz-Báñez, J. M., Fabila-Monroy, R., Ochoa, C., & Nigsch, P., (2015), provide a multitude of mathematical properties of such antipodal rhythms.

11 Apostol, T. M. & Mnatsakanian, M. A., (2000), p. 8.

12 Jones, A. M., (1954a), p. 35.

13 Tymoczko, D., (2011), p. 63.

CHAPTER 21

Euclidean Rhythms, Euclidean Strings, and Well-Formed Rhythms

I N THE YEAR 300 BC, the city of Alexandria in present-day Egypt was endowed with a magnificent library similar in spirit to the modern institution we call a university. In this Royal Library, which contained reading rooms and a large quantity of books in the form of papyrus scrolls, scholars from diverse parts of the neighboring world, financially supported by the government in Alexandria, gathered together to carry out research and write about a wide variety of topics including astronomy, geometry, and music.[1] One scholar in particular, named Euclid, wrote a book that became one of the best sellers of all time for more than 2,000 years. This book, titled *The Elements*, contains a wonderful compilation of algorithms (recipes) that were known at that time for solving an extensive variety of *geometric* problems that many of us have studied in secondary school.[2] Aside from geometry, Euclid also described a compelling algorithm for solving a fundamental problem concerned with the arithmetic of *numbers*. This algorithm has found many mathematical and computational applications since then, and is still actively investigated today.[3] Little did Euclid know that 2,300 years later this numerical algorithm would be shown to generate traditional musical rhythms used throughout the world. Although some readers may be surprised to learn that numbers and music are intimately related, the German philosopher and mathematician Gottfried Leibniz wrote: "*The pleasure we obtain from music comes from counting, but counting unconsciously. Music is nothing but unconscious arithmetic.*"[4]

The algorithm in question is one of the oldest and well-known algorithms, described in Propositions 1 and 2 of *Book VII* of *The Elements*. Today it is referred to as the *Euclidean algorithm*. This algorithm solves the problem of computing the greatest common divisor of two given natural numbers. The computer scientist Donald Knuth calls it the "granddaddy of all algorithms, because it is the oldest nontrivial algorithm that has survived to the present day." The idea is captivatingly simple. Repeatedly, replace the larger of the two numbers by their difference until both are equal. This last number is then the greatest common divisor. Consider as an example the pair of numbers (3, 8). First, eight minus three equals five, so we obtain the new pair (3, 5); then, five minus three equals two giving (3, 2); next, three minus two equals one which yields (1, 2); and lastly two minus one equals one, which results in (1, 1). Therefore, the greatest common divisor of three and eight is one. This procedure admits a compelling visualization illustrated in Figure 21.1 for the pair of numbers (3, 8). First, construct a 3 × 8 rectangle of unit squares as shown in the upper left. Repeatedly subtract a 3 × 3 square from this rectangle until it is impossible to do so. This leaves a 2 × 3 rectangle remaining (upper right). Now proceed to remove 2 × 2 squares from this 2 × 3 rectangle until it cannot be done. This yields the 2 × 1 rectangle remaining (lower left). Finally, remove a 1 × 1 square from the 2 × 1 rectangle to obtain a 1 × 1 square remaining.

How do we make the Euclidean algorithm generate musical rhythms? The key is to shift our attention from the answer given by the algorithm to the

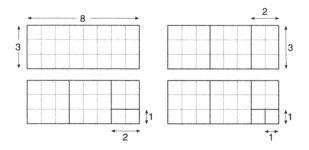

FIGURE 21.1 Visualization of the *Euclidean* algorithm for computing the greatest common divisor of integers 3 and 8 by repeated subtraction.

"history" (or sequence of calculations) of the algorithm, as it progresses towards the answer. This procedure is illustrated with the preceding pair of numbers (3, 8). For this purpose, the smaller number three is associated with the number of onsets that we want the rhythm to have, and the larger number eight, with the total number of pulses (onsets and silent pulses) that determine the rhythmic span or cycle. The algorithm illustrated in Figure 21.2 uses box notation. First write a rhythm of eight pulses and three onsets in which all the onsets are completely on the left at pulses 1, 2, and 3 as in Figure 21.2a. Next take (subtract) a number of silent pulses equal to the number of onsets (in this case three), from the right (pulses 6, 7, and 8) and place them below the others, flush to the left as in Figure 21.2b. Now there is a remainder of two silent pulses. Move these below the rest, also flush to the left, as in Figure 21.2c. Now that we have a single column remaining at pulse position three, the repeated subtraction phase of the algorithm is finished. The concatenation phase of the algorithm follows next. First separate the three columns as in Figure 21.2d. Next, rotate each column to become a row as in Figure 21.2e, and lastly, concatenate the three rows to form the generated rhythm in Figure 21.2f. Note that the rhythm generated by this

procedure is none other than the [3-3-2] pattern found all over the world. In particular it is the Cuban *tresillo* as well as the first half of the *clave son*, which we have encountered repeatedly throughout the book.

This procedure attempts to distribute the three onsets among the eight pulses as evenly as possible. It is this property that, to a large extent, is most responsible for obtaining rhythms that are popular throughout the world. The rhythms generated in this way are called *Euclidean* rhythms, because they are generated using an implementation of the Euclidean algorithm, and will be denoted by $E(k,n)$, where k is the number of onsets and n is the number of pulses in the cycle. Thus, the Cuban tresillo is denoted by $E(3,8)$. Other than k being smaller than n, there are no restrictions on these values. Hence, Euclidean rhythms constitute a broader set of rhythms than the family of *diatonic* rhythms. Jesse Stewart defines a *diatonic* rhythm "as a repeating rhythmic pattern in which an odd number of sounded pulses are spread out as much as possible across the tones of an even-numbered time cycle (generally consisting of 8, 12, or 16 pulses)."[5] Therefore, except for $k = 3$, and $n = 12$, which yield the regular Euclidean rhythm with interval structure [3-3-3-3], all other diatonic rhythms with an odd value of k, and $n = 8$, 12, and 16, are irregular.

When the number of sounded pulses (onsets) is greater than the number of silent pulses, all the silent pulses are moved in the first step of the algorithm. The remainder of the procedure remains the same. This process is illustrated with the numbers (5, 8) in Figure 21.3. First, the three silent pulses (6, 7, 8) are moved, as in Figure 21.3b. Now there are two single-onset columns remaining at pulse positions 4 and 5. These are moved next, as in Figure 21.3c. The concatenation phase in Figure 21.3d–f is the same as before. Note that the resulting Euclidean rhythm $E(5,8)$ is the Cuban cinquillo timeline [2-1-2-1-2].

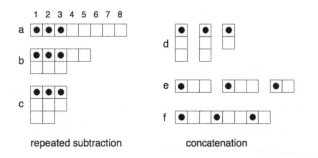

FIGURE 21.2 Generation of the Cuban *tresillo* timeline.

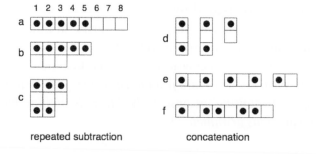

FIGURE 21.3 Generation of the Cuban *cinquillo* timeline.

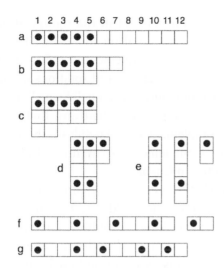

FIGURE 21.4 Generation of an Aka Pygmie timeline.

If we apply this algorithm with the numbers 5 and 12, illustrated in Figure 21.4, we obtain the rhythm E(5,12) = [3-2-3-2-2], which is a timeline played on a metal bell by the Aka Pygmies of Central Africa.[6]

Applying the algorithm to the pair of numbers (7, 12), as shown in Figure 21.5, results in the rhythm E(7,12) = [2-1-2-2-1-2-2], a popular West African bell pattern used in Ghana and Guinea.

Substituting the pair of numbers (5, 16) into the algorithm, illustrated in Figure 21.6, yields the rhythm E(5,16) = [3-3-3-3-4], a popular rhythmic pattern used in modern electronic dance music (EDM), which is a rotation of the bossa-nova timeline.[7]

The reader may have noticed that the Euclidean tresillo rhythm E(3,8) and the Euclidean cinquillo rhythm E(5,8) have a special relation with each other, evident from their circular representations in Figure 21.7. The tresillo [3-3-2] is shown on the left diagram. The *complement* of the tresillo (which consists

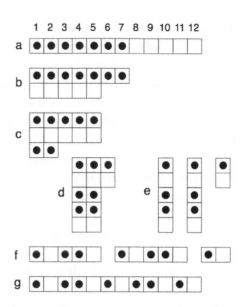

FIGURE 21.5 Generation of a ternary West African timeline.

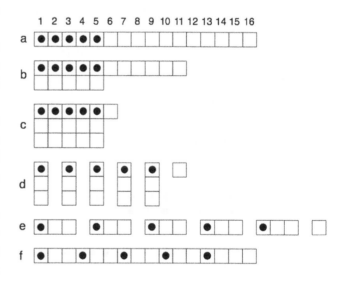

FIGURE 21.6 Generation of a binary West African timeline, which is also the signature rhythm of EDM.

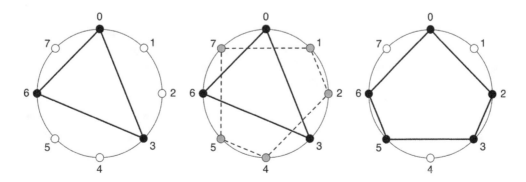

FIGURE 21.7 The tresillo (left), the complement of the tresillo in dashed lines and gray-filled circles (center), and the cinquillo (right).

of the silent pulses) is shown in dashed lines and gray-filled circles, in the center diagram. If the complement of the tresillo is rotated clockwise by a duration of one pulse, one obtains the cinquillo rhythm on the right diagram. You may ask yourself if it is a coincidence that, for the Euclidean tresillo rhythm, its complementary rhythm is also a Euclidean rhythm? Or is this a more general property of Euclidean rhythms? Before we answer this question let us calculate the Euclidean rhythm obtained with 11 onsets and 16 pulses. Note that $11 = 16 - 5$, which corresponds to the number of silent pulses in the West African timeline of Figure 21.6.

The application of the Euclidean algorithm to the sequence of 11 onsets in 16 pulses, depicted in Figure 21.8, yields the Euclidean rhythm E(11,16) = [1-2-1-2-1-2-1-2-1-2-1], which is a rotation of the complementary rhythm of the West African timeline shown in Figure 21.9 (right). The Euclidean rhythm

E(11,16) starts on pulse 14 in the rightmost diagram, where it is superimposed on the West African timeline. Therefore, the complement of E(5,16) is E(11,16). More generally, reflections and rotations of Euclidean rhythms are obviously also Euclidean. Furthermore, it was proved by Jack Douthett and Richard Krantz that the complement of a Euclidean rhythm is also Euclidean.[8] Complementary rhythms will be revisited in more detail in Chapter 25.

Scores of other Euclidean rhythms that are used in music throughout the world may be generated in this way by suitably picking the number n of pulses and the number k of onsets in a cycle. A list of some examples of Euclidean rhythms used in traditional music practice, for values of n and k, such that k does not divide evenly into n, are listed in the following.

E(2,3) = [x x .] = [1-2]

E(2,5) = [x . x ..] = [2-3]

E(2,7) = [x .. x ...] = [3-4]

E(3,4) = [x x x .] = [1-1-2]

E(3,5) = [x . x . x] = [2-2-1]

E(3,7) = [x . x . x ..] = [2-2-3]

E(3,8) = [x .. x .. x .] = [3-3-2]

E(3,10) = [x .. x .. x ...] = [3-3-4]

E(3,11) = [x ... x ... x ..] = [4-4-3]

E(3,14) = [x .... x .... x ...] = [5-5-4]

E(4,5) = [x x x x .] = [1-1-1-2]

E(4,7) = [x . x . x . x] = [2-2-2-1]

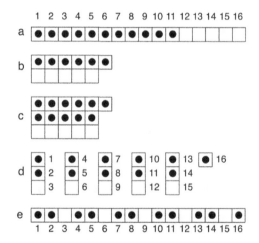

FIGURE 21.8 Generation of the complement of the binary West African timeline.

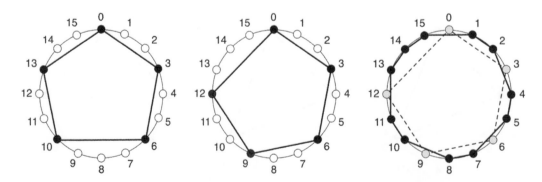

FIGURE 21.9 The bossa-nova (left), the West African timeline (center), and the complement of the West African timeline in black-filled circles (right).

E(4,9) = [x . x . x . x . .] = [2-2-2-3]

E(4,11) = [x . . x . . x . . x .] = [3-3-3-2]

E(4,15) = [x . . . x . . . x . . . x . .] = [4-4-4-3]

E(5,6) = [x x x x x .] = [1-1-1-1-2]

E(5,7) = [x . x x . x x] = [2-1-2-1-1]

E(5,8) = [x . x x . x x .] = [2-1-2-1-2]

E(5,9) = [x . x . x . x . x] = [2-2-2-2-1]

E(5,11) = [x . x . x . x . x . .] = [2-2-2-2-3]

E(5,12) = [x . . x . x . . x . x .] = [3-2-3-2-2]

E(5,13) = [x . . x . x . . x . x . .] = [3-2-3-2-3]

E(5,16) = [x . . x . . . x . . x . . x . . .] = [3-3-3-3-4]

E(6,7) = [x x x x x x .] = [1-1-1-1-1-2]

E(6,13) = [x . x . x . x . x . x . .] = [2-2-2-2-2-3]

E(7,8) = [x x x x x x x .] = [1-1-1-1-1-1-2]

E(7,9) = [x . x x x . x x x] = [2-1-1-2-1-1-1]

E(7,10) = [x . x x . x x . x x] = [2-1-2-1-2-1-1]

E(7,12) = [x . x x . x . x x . x .] = [2-1-2-2-1-2-2]

E(7,15) = [x . x . x . x . x . x . x . .] = [2-2-2-2-2-2-3]

E(7,16) = [x . . x . x . x . . x . x . x .] = [3-2-2-3-2-2-2]

E(7,17) = [x . . x . x . . x . x . . x . x .] = [3-2-3-2-3-2-2]

E(7,18) = [x . . x . x . . x . x . . x . x . .] = [3-2-3-2-3-2-3]

E(8,17) = [x . x . x . x . x . x . x . x . .] = [2-2-2-2-2-2-2-3]

E(8,19) = [x . . x . x . x . . x . x . x . . x .] = [3-2-2-3-2-2-3-2]

E(9,13) = [x . x x . x x . x x . x x] = [2-1-2-1-2-1-2-1-1]

E(9,14) = [x . x x . x x . x x . x x .] = [2-1-2-1-2-1-2-1-2]

E(9,16) = [x . x x . x . x . x x . x . x .] = [2-1-2-2-2-1-2-2-2]

E(9,20) = [x . . x . x . x . x . . x . x . x . x .] = [3-2-2-2-3-2-2-2-2]

E(9,22) = [x . . x . x . . x . x . . x . x . . x . x .] = [3-2-3-2-3-2-3-2-2]

E(9,23) = [x . . x . x . . x . x . . x . x . . x . x . .] = [3-2-3-2-3-2-3-2-3]

E(11,12) = [x x x x x x x x x x x .] = [1-1-1-1-1-1-1-1-1-1-2]

E(11,20) = [x . x x . x . x . x . x x . x . x . x .] = [2-1-2-2-2-2-1-2-2-2-2]

E(11,24) = [x . . x . x . x . x . . x . x . x . x . x .] = [3-2-2-2-2-3-2-2-2-2-2]

E(13,24) = [x . x x . x . x . x . x . x x . x . x . x . x .] = [2-1-2-2-2-2-2-1-2-2-2-2-2]

E(15,34) = [x . . x . x . x . x . . x . x . x . x . . x . x . x . x . x . .] = [3-2-2-2-3-2-2-2-3-2-2-2-3-2-2]

The durational patterns in the earlier list should be viewed more generally as necklaces and bracelets rather than rhythms, since in many instances, rotations and/or reflections of these interonset intervals (IOI) patterns yield other rhythms that are also used in traditional music around the globe. For example, the rhythm E(2,3) = [x x .] when started on the second onset becomes [x . x], which is a rhythm of a *Drum Dance* song of the Slavey Indians of Northern Canada,[9] as well as the hallmark rhythm of the *Lenjengo* recreational dance of the Mandinka people of West Africa.[10] Starting the rhythm E(3,4) = [x x x .] on the third onset yields [x . x x], which is the Arabic rhythm *wahdah sāyirah*.[11] Starting the rhythm E(3,7) = [x . x . x . .] on the third onset yields [x . . x . x .], which is the meter of the *aksak Bešli i čaj tele*.[12] Similarly, starting E(5,16) = [x . . x . . x . . x . . x . . .] = (33334) on the third onset yields the bossa-nova clave [3-3-4-3-3].[13] Furthermore, the patterns in this list are also expressed in their shortest form. Linear stretching of these rhythms will produce other rhythms. For example, when E(2,3) = [x x .] = [1-2] is multiplied by two it becomes [x . x . . .] = [2-4], a variant of the Mexican *son* rhythm.[14]

Euclidean rhythms are closely related to a similar concept studied in theoretical computer science: *Euclidean strings*.[15] Let $V = (v_0, v_1, ..., v_{n-1})$ denote a string of integers such as an interval vector of a rhythm. A string $V = (v_0, v_1, ..., v_{n-1})$ is a *Euclidean string* if increasing the duration v_0 by one, and decreasing the duration v_{n-1} by one, yields a new string that is a rotation of V. For example, if the operation is applied to the Euclidean rhythm E(4,9) = [2-2-2-3], one obtains [3-2-2-2], and since this is a rotation of [2-2-2-3], it follows that E(4,9) is a Euclidean string.[16] On the other hand, the rhythm [3-3-2-2-2] is not a Euclidean string because [4-3-2-2-1] is not a rotation of [3-3-2-2-2].

The reader may have noticed that the Euclidean rhythms are composed of IOIs of two sizes. For example, E(3,8) = [3-3-2], E(3,11) = [4-4-3], E(3,14) = [5-5-4], and E(5,12) = [3-2-3-2-2]. Furthermore, whatever the two IOI values are, they differ by exactly one pulse. However, there

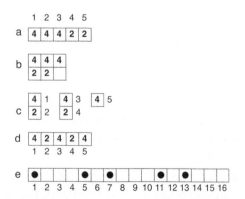

exist rhythm timelines that consist of more than two IOI values, such as the Pygmy BaYaka rhythm [2-2-2-1-2-3], the Pygmy Mbuti rhythm [6-4-2-4], and of course, the ubiquitous *clave son* [3-3-4-2-4]. Furthermore, there exist rhythm timelines that, like Euclidean rhythms, consist of precisely two IOI values, such that these values differ by more than one pulse, such as the Bushmen San timelines [2-4-4-2-4], [6-6-4], and [4-6-4-6].[17] Rhythms that consist of IOIs of precisely two values, such that the IOIs are maximally evenly distributed are called *well-formed* rhythms.[18] The Euclidean algorithm that generates Euclidean rhythms when applied to sequences of two-valued pulse symbols can also be used to generate *well-formed* rhythms by merely substituting the two-valued pulse symbols with two-valued IOI symbols. For example, assume we wish to generate a well-formed rhythm with three IOIs of four pulses and two IOIs of two pulses, and refer to Figure 21.10. We begin in row, Figure 21.10a, by listing the two groups of IOI symbols, as we did with the onsets and silent pulses (rests) for generating Euclidean rhythms. Repeated subtraction until one column is left yields row, Figure 21.10b. Reading, Figure 21.10b, from top to bottom and left to right yields, Figure 21.10c, which when concatenated yields the well-formed IOI pattern [4-2-4-2-4], which in turn yields the well-formed (non-Euclidean) rhythm in box notation in row, Figure 21.10e. Note that this rhythm is a rotation of the 16-pulse Bushmen San timeline [2-4-4-2-4].

To conclude this chapter, it should be noted that the Euclidean algorithm has also been applied to music in a completely different context. Viggo Brun used it in the construction of stringed musical instruments, in which there are constraints on the lengths of the strings as well as on the ratios of these lengths.[19]

FIGURE 21.10 Illustration of how the Euclidean algorithm applied to IOIs can be used to generate the *well-formed* rhythm [4-2-4-2-4].

NOTES

1 Beckman, P., (1971).
2 Toussaint, G. T., (1993), traces the history of the second proposition of Book I of the *Elements*, which in effect states that any problem that can be solved with a straight edge and the *modern* compass can also be calculated with a straight edge and the *collapsing* compass. The distinction between the two compasses is that with the modern compass one is permitted to transfer a distance from one location on the paper to another, whereas this is not allowed with the collapsing compass. Not surprisingly, algorithms that utilize the collapsing compass require more steps than those that employ the modern compass.
3 Bach, E. & Shallit, J., (1996), p. 67.
4 Sacks, O., (1998).
5 Stewart, J., (2010), p. 173.
6 Chemillier, M., (2002), p. 175. The Aka Pygmies of Central Africa use several additional rhythms with a similar pattern consisting of two intervals of duration three followed by two groups of elements of duration two, such that the cardinality of the two groups differs by one. Such rhythms include [3-2-2-3-2-2-2] and [3-2-2-2-2-3-2-2-2-2].
7 Butler, M. J., (2001).
8 Euclidean rhythms are *maximally even*. Douthett, J. & Krantz, R. J., (2008), p. 205 proved that the complement of a maximally even rhythm is maximally even, thereby establishing the result for Euclidean rhythms. See also Amiot, E., (2007), p. 12, for an alternate proof of this property.
9 Asch, M. I., (1975), p. 249.
10 Knight, R., (1974), p. 28.
11 Touma, H. H., (1996), p. 50.
12 Goldberg, D., (2015).
13 Butler, M. J., (2006), p. 147, points out that [3-3-3-3-4] is a rhythm often used in EDM.
14 Stanford, E. T., (1972), p. 79.
15 Ellis, J., Ruskey, F., Sawada, J., & Simpson, J., (2003). Euclidean strings are a topic that falls under the more general *theory of words*. Domínguez, M., Clampitt, D., & Noll, T., (2009), provide a translation bridge between the theory of words and music theory.
16 See Toussaint, G. T., (2005c), for a more detailed comparison of Euclidean rhythms with Euclidean strings in the context of Balkan *aksak* rhythms and sub-Saharan African timelines.
17 Poole, A., (2018).
18 Milne, A. J., Herff, S. A., Bulger, D., Sethares, W. A., & Dean, R., (2016). XronoMorph: Algorithmic generation of perfectly balanced and well-formed rhythms. In *Proceedings of the 2016 International Conference on New Interfaces for Musical Expression*. Brisbane, Australia.
19 Brun, V., (1964), p. 128.

CHAPTER 22

Lunisolar Rhythms

Leap Year Patterns

For many thousands of years, human beings all over the planet have spent countless hours gazing at the stars, the sun, and the moon. If they lived far from the equator, they often experienced vast differences between four seasons: a cold winter with snow and ice; a hot summer; a spring when new leaves appear, bears come out from hibernation, and colorful flowers bloom; and an autumn when the leaves turn to a variety of shades of brown, yellow, and red, before detaching themselves from their branches, and gliding to the ground. Some countries experience only two seasons: the rainy season and dry season. In other countries such as Egypt, these types of seasons are nonexistent. Indeed, in ancient Egypt, the changes that people observed in the desert had to do mainly with the level of water in the river Nile: was it constant, rising, or falling. The ancient Egyptians had three seasons of 4 months each, where each month lasted 30 days, for a total of 360 days. In addition, they concluded from their celestial observations that the number of days spanning the three seasons was 365. Therefore, at the start of each year they added five extra days of festivities, thus making their year last a total of 365 days.[1]

Astronomers also counted how many moons and days were observed in the cycles of their seasons. Such measurements inspired disparate cultures to design calendars in different ways. Let T_s denote the time duration of one revolution of the earth around the sun, more commonly known as a year. Let T_e denote the time duration of one complete rotation of the earth, more commonly known as a day. The exact values of T_s and T_e are of course constantly varying, since the orbits of the stars and planets in the universe are themselves continually changing. Every galaxy exerts gravity on every other galaxy, and all the stars and planets in each galaxy exert gravity on each other. However, from the measurements made of T_s and T_e, we estimate the ratio T_s/T_e to be today approximately 365.24220.[2] It is convenient therefore to make a year last 365 days, as did the ancient Egyptians. The problem that arises, both for keeping track of history, and for making predictions about the future, is that after some time, this seemingly small discrepancy, equal to 0.242199 becomes a large and inconvenient error. Since 0.242199 is almost 6 h, or one fourth of a day, one simple solution is to add one extra day every 4 years. Indeed, in the year 237 BC the ruler of Egypt, Ptolemy III, proposed exactly this modification to the Egyptian calendar. This proposal met some resistance in Egypt, but a few years later, Julius Caesar adopted the practice, and for this reason, this calendar became known as the *Julian* calendar.[3]

A year with one extra day is called a *leap* year. The Julian calendar assumes that a year is exactly 365.25 days long, which is still slightly greater than 365.242199. So now we have an error in the opposite direction, albeit smaller. One solution to this new problem is the *Gregorian* calendar, named after Pope Gregory XIII, who was responsible for its adoption. In the Gregorian calendar, leap years are defined as those divisible by four, except not those divisible by 100, unless these are also divisible by 400. With this rule a year becomes $365 + 1/4 - 1/100 + 1/400 = 365.2425$ days long, a much better approximation.

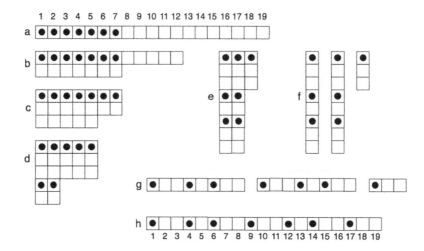

FIGURE 22.1   Generating the Jewish leap year calendar pattern using the Euclidean algorithm.

The methods just described for solving the calendar problem may be aptly described by the computer term *hack*. A hack is an effective, but clumsy, solution to a problem. A more elegant structural approach to designing rules for the introduction of leap years in calendars is based on the idea of rhythmic cycles. One such method is used in the design of the Jewish calendar. Here a regular year has 12 months and a leap year has 13 months. The cycle has a total of 19 years that include 7 leap years.[4] The 7 leap years are distributed as evenly as possible in the 19-year cycle. The cycle is assumed to start with Creation as year one. The remainder obtained by dividing the year number by 19 indicates the resulting position in the cycle. The leap years are 3, 6, 8, 11, 14, 17, and 19. For example, the year $5765 = 303 \times 19 + 8$ and so is a leap year. The year 5766, which begins at sundown on the Gregorian date of October 3, 2205, is $5766 = 303 \times 19 + 9$, and is therefore not a leap year. Applying the Euclidean algorithm to the integers 7 and 19, as shown in Figure 22.1, yields E(7,19) = [x . . x . x . . x . . x . x . . x . .] with a durational pattern [3-2-3-3-2-3-3]. As before, we subtract columns in Figure 22.1a–e until there is a remainder of only one column in (e) consisting of the pattern [x . .]. In Figure 22.1f, the three columns are separated. In Figure 22.1g they are each rotated, and finally in (h), they are concatenated. If the seventh pulse is counted as the first pulse we obtain the pattern [. . x . . x . x . . x . . x . . x . x], which describes precisely the leap year pattern 3, 6, 8, 11, 14, 17, and 19 of the Jewish calendar. Therefore, the leap year pattern of the Jewish calendar is a Euclidean necklace. Figure 22.2 depicts the necklace in an orientation that shows off its vertical

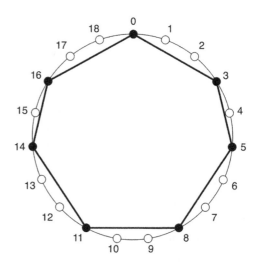

FIGURE 22.2   The Jewish calendar necklace with 7 leap years in a cycle of 19 years.

mirror symmetry. Interestingly enough, this necklace exhibits the same pattern of long–short intervals as the *bembé* rhythm and diatonic scale: [3-3-2-3-3-3-2] for the calendar versus [2-2-1-2-2-2-1] for the rhythm and scale. All the intervals in the diatonic scale are increased by one to obtain the Jewish leap year pattern. In musical terms, they have the same *rhythmic contour*.

Another structural design of a calendar that uses cycles is the Islamic calendar, which is based on the time between two successive new moons (*lunations*), in which 1 year is defined as 12 lunations. This method gives approximately 10,632 days every 30 years, in which 11 leap years are employed. The common approximations used in the Islamic calendar put leap years at positions 2, 5, 7, 10, 13, 16, 18, 21, 24, 26, and 29 in the cycle yielding

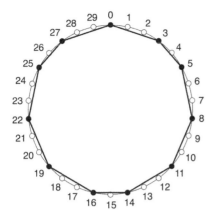

FIGURE 22.3 The Islamic calendar necklace with 11 leap years in a cycle of 30 years.

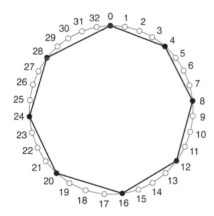

FIGURE 22.4 The Persian calendar necklace with 8 leap years in a cycle of 33 years.

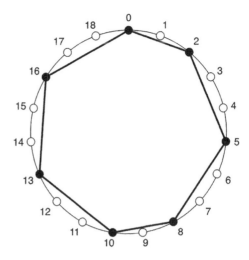

FIGURE 22.5 The Sumerian calendar necklace with 7 leap years in a cycle of 19 years.

Therefore, the Jalali leap year pattern introduced into the Persian calendar is also a Euclidean necklace.

Around the year 3000 BC the Sumerians who settled in Mesopotamia developed a lunar calendar that had 12 months of 29 or 30 days each. This calendar went through several revisions over the years. In the fifth century BC, this Babylonian calendar had seven intercalations in a cycle if 19 years in positions 1, 3, 6, 9, 11, 14, and 17, giving it a durational pattern [2-3-3-2-3-3-3], which is also a Euclidean rhythm shown as a polygon in Figure 22.5.[6] In fact this Sumerian calendar has the same leap year pattern as the Jewish calendar in Figure 22.2, rotated clockwise by five positions.

the duration pattern [3-2-3-3-3-2-3-3-3-2-3], shown in polygon notation in Figure 22.3. Applying Euclid's algorithm to the integers 11 and 30 yields E(11,30) = [x . . x . x . . x . . x . . x . x . . x . . x . x . .]. If we start this rhythm on the 11th pulse, we obtain the pattern [. x . . x . x . . x . . x . . x . x . . x . . x . x .] that describes precisely the leap year pattern 2, 5, 7, 10, 13, 16, 18, 21, 24, 26, and 29 of the Islamic calendar. Therefore, the leap year pattern of the Islamic calendar is also a Euclidean necklace.

The Jalali Persian calendar introduced in the year 1079 is also based on cycles. Omar Khayyám[5] contributed significantly to the design of this calendar. He proposed a cycle of 33 years with 8 leap years with the pattern [4-4-4-4-4-4-4-5]. Euclid's algorithm applied to the integers 8 and 33 yields the rhythm E(8,33) = [x . . . x . . . x . . . x . . . x . . . x . . . x . . . x . . . .], shown in polygon notation in Figure 22.4.

NOTES

1 Winlock, H. E., (1940), p. 458.
2 Assar, G. R. F., (2003), p. 172. This number is derived from the definition of a *tropical year*, which is one revolution of the earth around the sun realizing two consecutive equinoxes. On the other hand, the *sidereal year* consists of 365.25636 days, and is determined by an orbit of the earth around the sun such that the earth returns to the conjunction with the stars that are (or appear to be) fixed in the sky.
3 Reingold, E. M. & Dershowitz, N., (2001). For the relationships between Euclidean rhythms, leap years, and drawing digital straight lines on a computer screen of pixels, see Harris, M. A. & Reingold, E. M., (2004), and Klette, R. & Rosenfeld, A., (2004).
4 Ascher, M., (2002), p. 48.
5 Richards, E. G., (1998), p. 235.
6 Assar, *ibid.*, p. 174.

CHAPTER 23

Almost Maximally Even Rhythms

IN CHAPTER 21 IT WAS DEMONSTRATED with numerous examples taken from music, from around the world, that for a multitude of numerical values of *n* and *k*, the number of pulses and onsets, respectively, there exist many rhythm timelines in cultures all over the world that have the property that they are Euclidean or maximally even. However, for every fixed pair of values of *n* and *k*, the Euclidean algorithm yields only one rhythm necklace. Consider for instance the Euclidean rhythm obtained when *n* = 8 and *k* = 2. Since eight is divisible by two, the rhythm obtained is [x . . . x . . .] = [4-4]. Repeating this pattern yields only a steady pulsation, but not a very interesting rhythm. However, if we displace the second attack by one pulse, say to the left, we obtain a very interesting rhythm [x . . x . . . .] = [3-5]. This rhythm is quite common in Afro-Cuban music, where it is called the *conga*.[1] It has also been incorporated into *rock-n-roll* music, perhaps most notably as the midsong electric guitar solo in the Beach Boy's 1964 best-selling ballad *"Don't Worry Baby,"* ranked by the *Rolling Stone Music* magazine as the 178th greatest song of all time.[2]

Adding one more attack to the conga, so that *n* = 8 and *k* = 3, yields again only one Euclidean rhythm E(3,8) = [x . . x . . x .] = [3-3-2], which when rotated yields additional rhythms [3-2-3] and [2-3-3]. On the other hand, several other rhythms used in practice also consist of three onsets among eight pulses, but they are neither Euclidean rhythms nor rotations thereof. As these two examples illustrate, to generate a larger, more inclusive class of "good" rhythms, the maximally even concept has to be relaxed. One approach is to modify slightly, or mutate, a maximally even rhythm to generate other rhythms that are *almost* maximally even. In a brute-force approach, for a given pair of values of *n* and *k*, we could first generate all possible rhythms, and then calculate according to some chosen distance function, the distance between all these rhythms and the maximally even rhythm, and finally select those rhythms that are close enough to the maximally even rhythm according to a preselected threshold. However, such an approach requires much computation. There are other more direct ways of generating rhythms that are close to maximally even rhythms. As an example, consider again the ubiquitous tresillo rhythm at the top of Figure 23.1, and define a mutation operation as the displacement of its onsets (other than the first) by a duration of one pulse towards the left (anticipation). We can generate three new rhythms this way by moving either the second onset from pulse three to pulse two, the third onset from pulse six to pulse five, or both of these. The three rhythms produced in this way are used in music in many parts of the world. The second rhythm [3-2-3] is used in Beijing Opera, the third [2-4-2] is the *catarete* from Brazil, and the fourth [2-3-3] is a *bossa-nova* rhythm from Brazil, as well as the *nandon bawaa* from Ghana. Two of these are rotations of the tresillo, and therefore they are obviously maximally even. The *catarete*, although not maximally even, is well formed.

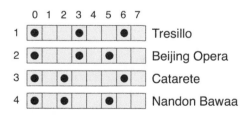

FIGURE 23.1 The four almost maximally even rhythms with three onsets and eight pulses.

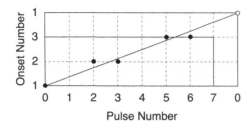

FIGURE 23.2 Generating the almost maximally even rhythms with eight pulses and three onsets.

This description of the algorithm for generating almost maximally even rhythms has another more geometric interpretation that uses a two-dimensional *onset-pulse grid*, as illustrated in Figure 23.2. First construct a 3×8 rectangular grid with height three units and width eight units. The height marks the onset number, and the width the pulse number. Note that the width also indicates time, since there are eight pulses in the cycle. Next draw a straight line connecting the corner point (0,1) on the lower left to the corner point (0,1) on the upper right.

Corresponding to each onset number, there is a horizontal dashed grid line intersected by a diagonal line (solid) that connects the two opposite corners of the rectangle. These intersection points have the time coordinates: two and two thirds for the first, and five and one third for the second. They divide the time span of eight pulses into three equal intervals. Onsets played at these positions would generate an isochronous perfectly even rhythm. In some forms of music such as electronic music, a composer may be perfectly happy to insert onsets such as these anywhere in the cycle, without respecting the integer locations of the pulses, in effect treating the time axis as a continuum of real numbers rather than integers. However, in the present context, and for our purposes, the onsets must be played only on the integer values of time, i.e., the

pulse numbers. The Euclidean rhythm, in this case, is obtained by snapping every intersection point to the pulse that is the nearest neighbor to the right. On the other hand, the family of *almost* maximally even rhythms is made up of rhythms obtained by all combinations of snapping intersection points to both, the nearest right and the nearest left pulses.[3]

Figure 23.3 shows the onset–pulse grid diagram for generating the almost maximally even rhythms with 16 pulses and five onsets, and the resulting 16 almost maximally even rhythms are pictured in Figure 23.4. About half of these 16 rhythms are used in musical practice, and they are highlighted and labeled with one of their more common identifying names.[4]

It is worth noting that the rhythms generated in this fashion are not equally almost maximally even. All the rhythms do "live" in a thin strip near the diagonal line in Figure 23.3, and thus their unevenness is bounded from above, but if for each rhythm we sum the horizontal deviations of their attacks from this diagonal line, some are more even than others. This is not surprising since a generated rhythm may be obtained from the maximally even rhythm by a number of mutations that can vary between one and four.

Timelines with 12 pulses and five onsets also figure prominently in much music, and Figure 23.5 shows the onset–pulse grid diagram for generating them. There are again 16 almost maximally even rhythms in this case, and they are pictured in Figure 23.6. Again, about half of these rhythms are used in musical practice, and they are highlighted and labeled with some commonly assigned names.

In the preceding examples, all the almost maximally even rhythms generated from a seed rhythm (specified number of onsets) had the same number of onsets as the seed rhythm. This resulted from the fact that the number of pulses was large relative to the number of

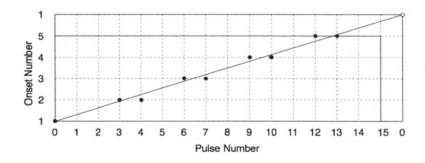

FIGURE 23.3 Generating the almost maximally even rhythms with 16 pulses and five onsets.

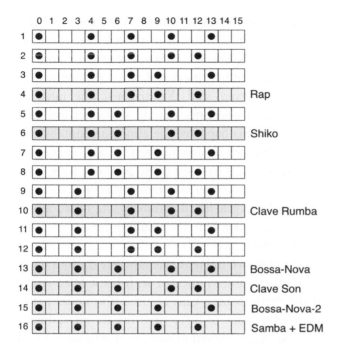

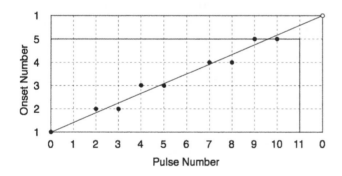

FIGURE 23.4 The 16 almost maximally even rhythms with 16 pulses and five onsets. (Modified with permission from Toussaint, G. T., *Percussive Notes*, 2011, November issue, pp. 52–59.)

FIGURE 23.5 Generation of the almost maximally even rhythms with 12 pulses and five onsets.

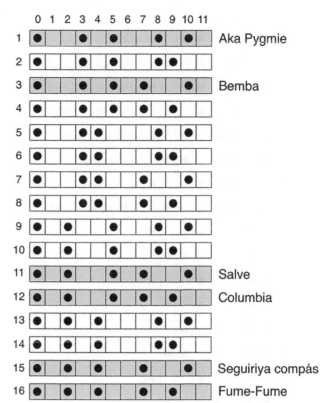

FIGURE 23.6 The 16 almost maximally even rhythms with 12 pulses and five onsets.

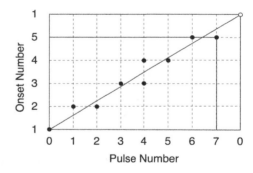

FIGURE 23.7 Generating the almost maximally even rhythms with eight pulses and five onsets.

onsets, thus giving the onsets enough space to snap to their nearest pulses in either direction without "colliding" with other already assigned pulses. However, when the number of onsets is larger relative to the number of pulses, the resulting overcrowding causes such collisions to happen, as illustrated with the timelines that have five onsets and 12 pulses that yield the onset–pulse grid shown in Figure 23.7.

The 16 almost maximally even rhythms obtained from this onset–pulse grid are shown in Figure 23.8. Again, about half of these are used in musical practice,

and are identified by a name or genre in which they are used.[5] Note that this time, although the seed rhythm has five onsets, there are four rhythms numbered 3, 4, 11, and 12 that have only four onsets due to collisions resulting from snap operations. For example, in Figure 23.7, if the intersection point on onset line three is snapped to the right and the intersection point on onset line four is snapped to the left, then both intersection points are snapped to pulse number four.

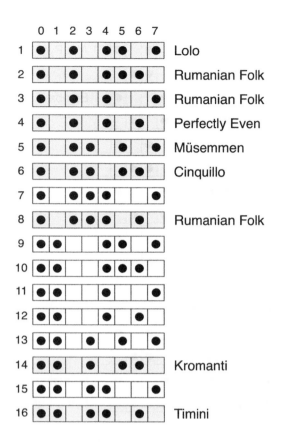

FIGURE 23.8 The 16 almost maximally even rhythms with eight pulses and five onsets.

In all the examples examined so far, the number of onsets is evenly divisible into the number of pulses. However, in the special cases in which this does not happen, the resulting maximally even rhythm is actually *perfectly* even, and the diagonal line intersects the horizontal grid lines exactly at the vertical pulse lines. When this happens, the almost maximally even rhythms obtained by moving the intersection points to their nearest right and left pulse positions have a tendency to acquire larger interonset intervals.[6] Consider for example the grid line diagram for four onsets among 16 pulses in Figure 23.9. The resulting family consisting of eight almost maximally even rhythms generated with this grid is shown in Figure 23.10, where the four onsets of each rhythm are indicated with black filled circles. All the rhythms have gaps of either five or six pulses. Nevertheless, four of these rhythms are interesting in that their onsets consist of subsets of rotations of well-known rhythms used in practice.

In Figure 23.10 the onsets marked with white-filled circles indicate those onsets that would have to be inserted for the rhythm to become the one labeled on the right. So, if an onset is inserted to rhythm No. 1 at pulse 13, the resulting five-onset rhythm would be the gahu if started at pulse 13. Similarly, if an onset is

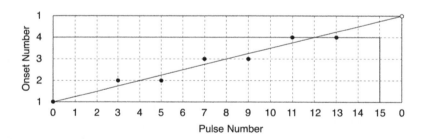

FIGURE 23.9 Generation of the almost maximally even rhythms with 16 pulses and four onsets.

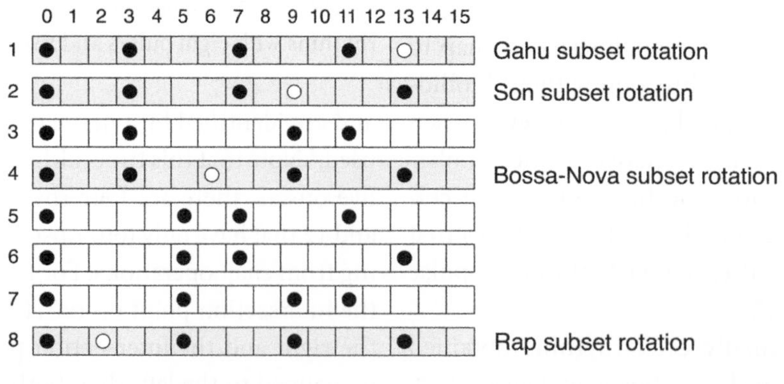

FIGURE 23.10 The eight almost maximally even rhythms with 16 pulses and four onsets. (With permission from Toussaint, G. T., *Percussive Notes*, 2011, November issue, pp. 52–59.)

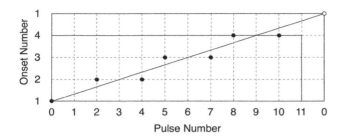

FIGURE 23.11 Generation of the almost maximally even rhythms with 12 pulses and four onsets.

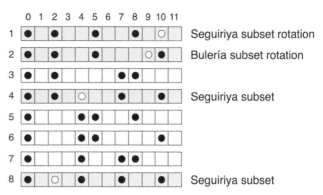

FIGURE 23.12 The eight almost maximally even rhythms with 12 pulses and four onsets.

added to rhythm No. 2 at pulse nine, the resulting five-onset rhythm would be the *clave son* if started at pulse 13. Also, if an onset is added to rhythm No. 4 at pulse six, the resulting rhythm would be the clave bossa-nova if started at pulse three. Finally, if an onset is added to rhythm No. 8 at pulse two, the resulting rhythm would be the rap timeline if started at pulse nine.

For a second example take the case of four onsets among 12 pulses, for which the grid line diagram is shown in Figure 23.11, and the rhythm it generates in Figure 23.12.

Four of the rhythms in Figure 23.12 are derived from the rhythms labeled on the right. So, if an onset is inserted to rhythm No. 1 at pulse ten, the resulting five-onset rhythm would be the seguiriya meter if started at pulse ten. Similarly, if an onset is added to rhythm No. 2 at pulse nine, the resulting five-onset rhythm would be the buleria meter if started at pulse three. Also, if an onset is added to rhythm No. 4 at pulse four, the resulting rhythm would be the seguiriya meter. Finally, if an onset is added to rhythm No. 8 at pulse two, the resulting rhythm would again be the seguiriya meter.

As the examples examined in this chapter show, for small values of the number of pulses and onsets, nearly half of all members of the families of almost maximally even rhythms are used in practice as timelines. Furthermore, for the important case of five onsets among 16 pulses the almost maximally even set includes the *clave son*. However, for these and other similar values, as well as for larger values of onsets and pulses, although the almost maximally even sets include a significant fraction of rhythms that are used in practice, they also include rhythms that appear to have not been adopted. Therefore, although the property of *almost maximal evenness* is a salient condition for a rhythm to be good, it seems to be insufficient to characterize all rhythms that have been adopted as timelines by cultures

in the past. To be sure, this property alone does not characterize the *clave son* uniquely, nor explain how it managed to become so successful throughout the world.

NOTES

1 Rey, M., (2006), p. 192.

2 www.rollingstone.com/music/lists/the-500-greatest-songs-of-all-time-20110407.

3 The field of computer science that deals with the automatic generation of symbol sequences provides many similar methods of generating binary sequences usually called *words* (rhythms in our context). See Lothaire, M., (2002) and Allouche, J.-P. & Shallit, O., (2002).

4 Entry number 16 in Figure 23.4 is a bossa-nova variant frequently encountered in electronic dance music, see Butler, M. J., (2006), p. 83. It is used for example as a snare-drum ostinato in Mario Più's *Communication*. See Butler, M. J., (2006), p. 147.

5 The names used here in no way suggest a complete list of either the countries in which the rhythms are played or what they are called in different cultures. For example, the Rumanian folk rhythm (third on the list) with durational pattern [2-2-3-1], when started on the third onset becomes [3-1-2-2], the time-keeping rhythm of the *sabar* of Senegal, played on the side of the drum with a stick. See Tang, P., (2007), p. 98.

6 There exist other methods for generating approximately even rhythms that handle the intersections of horizontal and vertical lines in a different manner. For example, if vertical and horizontal lines generate onsets and rests, respectively, when intersected by the diagonal line, and when all three lines meet at a point, an onset–rest pair is generated, and then the grid in Figure 23.11 yields the rhythm [x . x x x . x x x . x x x . x x]. Series, C., (1985), p. 21, describes a characterization of the class of rhythms obtained in this way. See also the *cutting sequences* in: Domínguez, M., Clampitt, D., & Noll, T., (2009), p. 480, and the book by Lothaire, M., (2002), p. 109, for several related methods and references.

Homometric Rhythms and Crystallography

Have you ever had the experience of viewing two different objects from one vantage point, concluding that they are identical, only to discover after obtaining more information that they are in fact vastly different? The example in Figure 24.1 illustrates this point succinctly. The object on the left is a cube, and the one in the center a cylinder. These 2 three-dimensional objects are poles apart from each other, and both objects look very different from the top (a square vs. a circle), but if projected from the front they look like the same square on the right.

In the 1920s and 1930s crystallographers were studying the atomic structure of crystals with a revolutionary and powerful new technological tool that had just become available: *X-rays*. These scientists were interested in reconstructing the positions of the crystal atoms relative to each other, from only the distances between the pairs of atoms. However, X-rays are not as powerful as the computed tomography scanners of today that yield three-dimensional pictures of the objects under scrutiny. In those days X-rays were bombarded through the crystals under investigation, providing only projections, or two-dimensional pictures that did not contain

the exact coordinates of the atoms, but only evidence of their location. These pictures, called X-ray *diffraction patterns*, were then used to infer the distances between the atoms. The lofty final goal of this research project was nothing less than the construction of the exact three-dimensional structural models of the molecules under investigation. Unfortunately, the scientists encountered a stumbling block. To their chagrin they discovered that, just as in the example of Figure 24.1, there exist pairs of molecules that have different three-dimensional atomic structures, but their two-dimensional X-ray diffraction patterns yield exactly the same collection of interatomic distances.

In 1944 the Massachussetts Institute of Technology physicist and crystallographer A. Lindo Patterson published a landmark paper in the *Physical Review* titled *"Ambiguities in the X-ray Analysis of Crystal Structures."*[1] In this paper Patterson proposed to view crystallography in an original and unconventional manner. In order to understand clearly the anomaly evident in earlier experiments, instead of producing X-ray diffraction patterns from three-dimensional crystals in the laboratory, he analyzed, one-dimensional idealized "crystals" with paper and pencil using the tools of distance geometry. Since one-dimensional "crystals" are periodic patterns much like the keyboard on a piano, which repeat octave after octave, Patterson took one period of the "crystal" and wrapped it around a circle for convenience. Thus, he obtained a circle with black and white equally spaced points. He called these sets *cyclotomic sets*. In mathematical language, they are

FIGURE 24.1 Two different objects that look the same from the front.

called *circular lattices*, and in this book they are called musical rhythms. In that same paper Patterson published an exhaustive combinatorial analysis of all possible different necklace patterns that could be obtained in a circle of *n* points by coloring *k* of them black and *n* − *k* of them white, for *n* = 8, 9, …, 16 and *k* = 1, 2, …, 8. Recall that if one configuration can be brought into correspondence with another by rotations, reflections, or a combination of these operations, then the two configurations are considered to belong to the same bracelet. If reflections are left out of this definition, then we are left with necklaces. Clearly, if two cyclotomic sets are instances of the same bracelet, then their histograms of distances are also the same. For *n* = 8 and *k* = 4, there are eight different necklaces shown in Figure 24.2.

Note that although all eight patterns constitute eight different necklaces, i.e., no two are congruent to each other, the third and fourth patterns have the same set

of distance histograms as shown in Figure 24.3, where the two patterns are redrawn in such a way so as to emphasize that they are complements of each other. Thus, Patterson provided, for the first time, a precise geometric equivalent of the experimental anomaly discovered in the X-ray diffraction patterns of molecules. The black points represent atoms, and the numbers along the chords connecting the black points represent the distances between atoms. That one of them cannot be rotated or reflected to obtain the other is evident from the simple observation that in the pattern on the left the two intervals of length one are separated from each other by other intervals, whereas in the pattern on the right, they are adjacent. Distances one and three occur twice each, and distances two and four occur once each. As the reader may have observed, cyclotomic sets are identical to the clock diagrams of rhythms used throughout this book. As an interesting

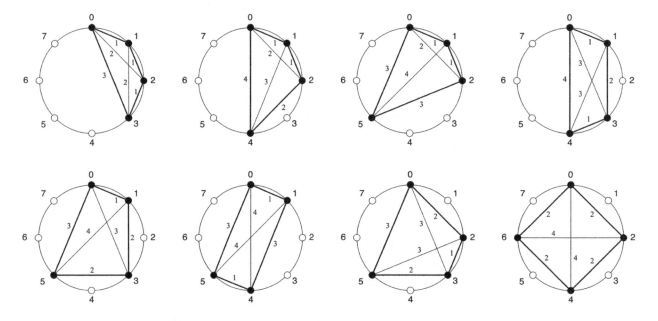

FIGURE 24.2   The eight cyclotomic (rhythm) necklaces obtained with *n* = 8 and *k* = 4.

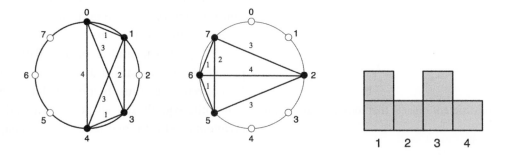

FIGURE 24.3   Two different rhythms that have the same interval-content histogram.

side comment, it is worth pointing out that these two rhythms are used together frequently in best-selling popular music. The rhythm on the left with interonset-interval sequence [1-2-1-4] is a common bass pattern, whereas the complementary rhythm on the right is usually played on the high hat or the ride cymbal. The fifth, sixth, and seventh patterns in Figure 24.2 also represent rhythms that are heard frequently in sub-Saharan African music. It is doubtful, however, that Patterson had made this connection between crystal molecules and musical rhythms.

Patterson's discovery was a shocking revelation to crystallographers. If this example was a unique monster, an exception rather than the rule, then perhaps there was not much to worry about, since the molecules in the real world, of most interest to humanity, are more complex than this simple example. However, if there existed sufficiently many other such monsters, then the reconstruction of molecular structures from their interatomic distances would face a formidable obstacle. Patterson sent his example to the young prodigious geometer Paul Erdős, one of the most prolific mathematicians of the twentieth century, and asked him if it might be possible to construct other similar examples.[2] Erdős, unfortunately, did not have good news for Patterson. He wrote back that there existed an infinite number of such monsters and included the four-point example shown in Figure 24.4. These types of problems now form part of the discipline called *distance geometry*, which is concerned mainly with the reconstruction of configurations of points in space from their interpoint distances.[3]

This family of patterns is determined by a parameter a that measures the arclength (equivalent to time) along a circle of unit circumference, and may take any positive value less than one half. As the reader may verify, it is impossible to either rotate or reflect one of the two patterns to obtain the other, because in the pattern on the left the intervals of length a and $(1/4) - a$ are not adjacent to each other, whereas in the pattern on the right they are. Nevertheless, both patterns contain the same set of geodesic distances, namely: a, ½, $(1/2) - a$, ¼, $(1/4) - a$, and $(1/4) + a$.

The infinite family of homometric cyclotomic sets that Erdős sent to Patterson is defined on the continuous circle, since the parameter a may take on any positive real value between zero and one half. However, musical rhythms most often live in a discrete universe of an integer number of pulses. By choosing the value of a to be an integer, and setting the number of pulses in the cycle appropriately, we may create other examples of discrete homometric rhythms to add to Patterson's four-onset, eight-pulse example. For instance, the resulting homometric pair for 16 pulses with a equal to 1 is shown in Figure 24.5.

Patterson generalized the continuous example of Erdős shown in Figure 24.4 to the case of five points arranged on a unit-circumference circle, as pictured in Figure 24.6. Both rhythms in this family of homometric pairs have in common the equilateral triangle of sides $1/m$, where $m = k - 2$, and k is the number of points, in this case five. Furthermore, a and b are chosen so that $a + b = 1/(2m)$.

By selecting the number of pulses n, as well as the integer values for a and b, appropriately, it is possible to convert this continuous model to discrete version as well, as shown in Figure 24.7. Here $n = 24$ so that $1/m$ becomes eight, and $a = 1$ and $b = 3$, so that the equation $a + b = 24/(2m)$ is satisfied.

By searching for cyclotomic sets with a higher number of black points, Patterson also found examples where

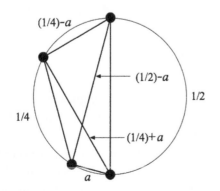
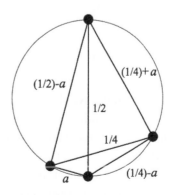

FIGURE 24.4   The infinite family of four-onset homometric pairs constructed by Paul Erdős.

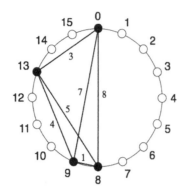
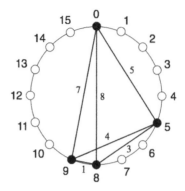

FIGURE 24.5 The Erdős-type construction for rhythms with 16 pulses and $a = 1$.

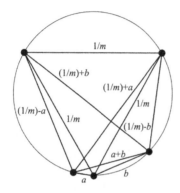
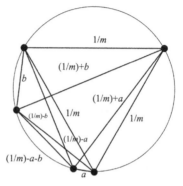

FIGURE 24.6 Patterson's continuous example for five points.

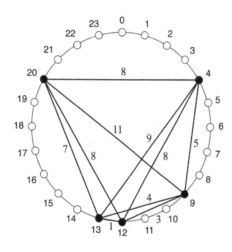
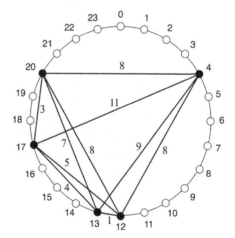

FIGURE 24.7 A discrete version of Patterson's example of Figure 24.6.

more than two noncongruent sets have the same set of distances. One such triplet that contains six black points among 16 pulses (see Figure 24.8) is most relevant to the rhythmic analysis considered here. Note that the three rhythms contain all the possible interonset interval values ranging from one to eight, and have the interval vector (histogram) given by [2,1,2,2,2,3,2,1]. The intervals

of length one and eight occur once each, the interval of length six occurs three times, and all the other intervals occur twice.[4]

It is truly amazing that among the rare instances of homometric triplets consisting of six onsets in a cycle of 16 pulses, there should occur the rhythm on the left in Figure 24.8, which sounds and is almost identical

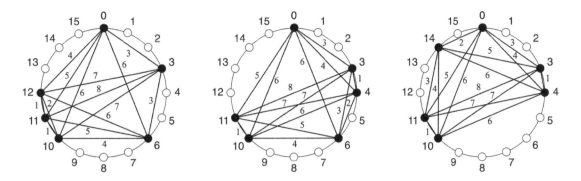

FIGURE 24.8 A homometric *triplet* discovered by Patterson.

to the five-onset *clave son*, the only difference being that this rhythm has one additional decorative onset at pulse number 11. Furthermore, the rhythm in the middle of Figure 24.8 sounds and is almost identical to the soukous timeline. The only difference here is that this rhythm has an additional decorative onset at pulse number four. To top it all off, these four rhythms (the soukous, the son, and the two leftmost rhythms in Figure 24.8) sound quite similar to each other. There are people in some parts of Africa that upon hearing these four rhythms would say that they were all the same rhythm. Musicologists may make the milder claim that they are *variants* of each other. It might be tempting to entertain the idea that if two rhythms have the same interval-content histograms then they would necessarily sound similar. Indeed, some researchers in the field of music information retrieval compute a variety of features of these histograms to characterize different rhythms. Unfortunately, this is not the case, as the third example in Figure 24.8 illustrates. This rhythm sounds completely different from the other

two, and yet has the same interval-content histogram. It turns out that finding features of rhythms that characterize their uniqueness and are good for measuring their similarity in a way that correlates well with human judgments is a difficult problem.

In closing this chapter, it should be noted that homometric sets have been explored in music for some time in the pitch domain, where they are called Z-related sets.[5]

NOTES

1 Patterson, A. L., (1944). See also Franklin, J. N., (1974), for other examples of homometric sets, and Erdős, P. & Turán, P., (1941) for a mathematical perspective.

2 Hoffman, P., (1999).

3 Skiena, S. S., Smith, W. D., & Lemke, P., (1990).

4 In the pitch domain Lewin, D., (1982) also discovered triples of homometric sets in the analogous 16-tone system. See also Lewin, D., (1976) regarding the interval content of invertible hexachords.

5 Mandereau, J., Ghisi, D., Amiot, E., Andreatta, M., & Agon, C., (2011) and Soderberg, S., (1995).

CHAPTER 25

Complementary Rhythms

AT FIRST GLANCE THE OBJECT DEPICTED IN FIGURE 25.1 APPEARS to be a black candlestick holder, or vase, on a white background. However, if the viewer focuses attention on the contour boundary between the white and black regions, two white faces (in profile view) staring at each other over a black background may be perceived instead. This visual perception phenomenon is known as the figure-ground effect: "the ground is perceived as extending continuously behind the figure."[1] This process is also known as *figure-ground reversal*, and patterns that exhibit this reversal are referred to as *bistable*.[2] As we gaze at visual stimuli such as these, the *figure* (or foreground) and the *ground* (or background) spontaneously switch their roles. This perceptual phenomenon has been exploited in the work of several artists, including Salvador Dali, who used it

quite successfully in several paintings such as the *Slave Market with Disappearing Bust of Voltaire*.[3]

A similar phenomenon occurs in the domain of aural perception of rhythms. Consider the two rhythms shown in Figure 25.2. Let the rhythm on the left be the lead rhythm (figure) played on a low-pitched conga, and the one in the middle be the accompaniment (background) played on a high-pitched conga. Here the space in which the rhythms are embedded consists of the 12-point circular lattice. The rhythm on the left occupies positions 0, 2, 4, 5, 7, and 9, and the rhythm in the middle occupies the empty space (background) at positions 1, 3, 6, 8, 10, and 11. If both rhythms are played simultaneously (rightmost diagram) at the same volume, then the listener's attention may shift back and forth spontaneously, or at will, to perceive either the rhythm on the left or the one on the right.[4]

In Figure 25.2 since the rhythm in the middle consists of onsets at the positions of silent pulses contained in the rhythm on the left, and vice versa, these rhythms are called *complementary*.[5] Together, they fill the entire set of pulse locations in the cycle and are thus referred to as *interlocking* rhythms.[6] Interlocking rhythms constitute one of the main principles of rhythm integration in African drumming.[7] Similar interlocking rhythms are found in the African Diaspora. An example from the Afro-Peruvian Black Pacific *landó* music is the *zamacueca* rhythm (Figure 25.3) performed on the *cajón*, a box-shaped percussion instrument that affords both low-pitched and high-pitched sounds.[8] The dominant rhythm is the high-pitched pattern on the leftmost diagram with interonset-interval (IOI) structure [3-2-1-3-2-1]. The softer low-pitched rhythm is shown in the middle diagram, and the resultant of the two combined complementary

FIGURE 25.1 Visual figure-ground reversal.

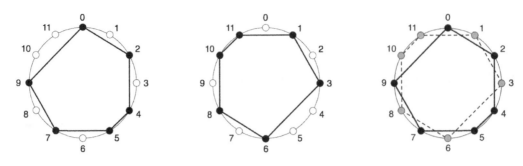

FIGURE 25.2   An example of aural rhythmic figure-ground reversal.

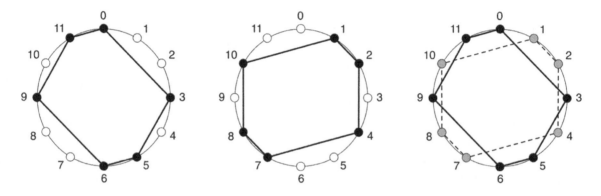

FIGURE 25.3   The *zamacueca* complementary (interlocking) rhythms in the Afro-Peruvian Black-Pacific *landó* music.

rhythms appears on the rightmost diagram. Note again that the low-pitched onsets of the rhythm in the middle diagram consist of the pulses (rests) of the high-pitched rhythm in the leftmost diagram.

In the pair of complementary rhythms in Figure 25.2, both rhythms are instances of the same necklace: it is sufficient to rotate one rhythm to bring it into correspondence with the other. In this case it is also possible to achieve correspondence by reflecting one rhythm about an axis of reflection determined by the bisector of pulse pairs (1, 2) and (7, 8). The pairs of complementary rhythms in Figure 25.3 are not instances of the same necklace, since one cannot be brought into correspondence with the other by a rotation. This follows from the fact that the IOI structure of the rhythm on the leftmost diagram has IOI structure [3-2-1-3-2-1] in a clockwise orientation, whereas the rhythm in the center diagram has this same IOI structure in a counterclockwise orientation. Neither can correspondence be achieved by a mirror reflection. However, correspondence can be achieved if both rotation and mirror reflection are carried out. If the rhythm in the center diagram is rotated counterclockwise by a duration of one pulse and then reflected about a vertical line through pulses 0 and 6, it

is transformed into the rhythm in the leftmost diagram. Therefore, these two rhythms are instances of the same bracelet. Thus, the aural examples in Figures 25.2 and 25.3 are not entirely analogous to the visual example in Figure 25.1, where the figure (candle holder) and the ground (two faces) are not rotations and/or reflections of each other, but rather completely different patterns.

Regarding the more general figure-ground problem for the case of sound, in 1987 the Stanford University information theorist, Thomas M. Cover asked the question: "Is it possible that the silence that lies between these bursts of sound also qualifies as a rhythm? Not the same rhythm but one of equally compelling artistic merit?" and suggested an open problem: "It remains to discover a rhythm the complement of which is also a rhythm and to choose the sounds appropriately."[9] Indeed, examples of pairs of complementary rhythms that have unequal cardinalities of onsets (and are thus not the same), such that both are of equally compelling artistic merit, do exist, as Figure 25.4 illustrates. The rhythm on the left is the illustrious *bembé* rhythm (also diatonic scale in the pitch domain), and the one on the right is a rotation of the equally prominent *fume–fume* rhythm (also pentatonic scale in the pitch domain). The "audible" beats

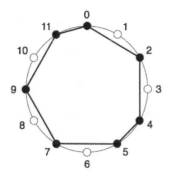 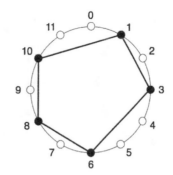

FIGURE 25.4 The *bembé* rhythm (left) and its complement (right).

of one rhythm are the "silent" beats of the other.[10] It is interesting that both the seven-onset *bembé* timeline and its five-onset complement are both highly prominent in sub-Saharan African traditional music. Playing both together on different drums (with different timbres) at the same intensity (or on the same drum at different intensities) can also produce in the listener spontaneous figure-ground reversals between the two rhythms.

Complementary rhythms that have the same number of onsets enjoy the very special and rather surprising property that their IOI histograms are always identical to each other. In other words, such rhythms and their complements are *homometric*. That the complementary six-onset rhythm pairs in Figures 25.2 and 25.3 are homometric is not unexpected since one rhythm is a rotation of the other, and this property is invariant under rotations and reflections. What is surprising is that even if one rhythm is not a rotation or mirror image of the other, as long as both are complementary and have the same number of onsets, they are homometric. One such example is the rhythm pictured in Figure 25.5 (left). This rhythm consists of the union of two regular rhythms: the three-IOI regular rhythm [4-4-4] at pulses (0,4,8) and the four-IOI regular rhythm

[3-3-3-3] at pulses (0,3,6,9).[11] Clearly the rhythm on the left is neither a rotation of the rhythm on the right nor a combination of rotation and mirror image reflection, since the two adjacent interonset durations of size two lie next to each other in the first rhythm and opposite to each other in the complementary rhythm. Nevertheless, as the reader may readily verify, both rhythms have the same interval histogram (center) and are therefore homometric. This is sometimes called the *hexachordal* theorem in the music literature, because it was originally proved there in the context of pitch and chords in a 12-tone system.

There is a simple and straightforward argument to show why the hexachordal theorem must be true. First, consider a rhythm of 12 pulses with an onset on every pulse and refer to Figure 25.6. The first diagram shows that distance one determines a 12-sided polygon. The second diagram contains two hexagons consisting of edges of distance two for a total of 12 distances. The third figure contains three squares with a side length of three, which makes 12 distances. The fourth figure contains four triangles that determine 12 edges of length four each. The fifth figure is a 12-sided star polygon with sides of length five, as a result of the fact that the numbers 5 and 12 are

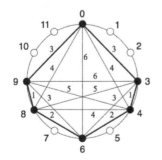 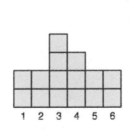 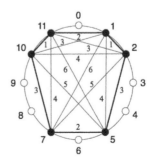

FIGURE 25.5 The three-over-four rhythm (left), its complement (right), and their common interval histogram (center). (With permission from Liu, Y. and Toussaint, G. T. 2012. *International Journal of Machine Learning and Computing*, 2(3):261–265.)

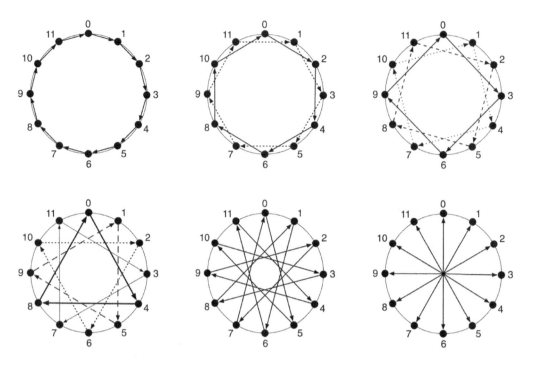

FIGURE 25.6   In a 12-pulse cycle every distance occurs 12 times.

relatively prime. Recall that two numbers are relatively prime if they have no factors in common other than one. In this case, 5 and 12 are relatively prime. Finally, the sixth figure contains six bigons (two-sided polygons) or 12 distances of length six, one from each onset to its diametrically opposite onset.[12] It follows from these observations that when there is an onset at every pulse of a 12-pulse cycle, every distinct distance value occurs 12 times, and this generalizes to any number of pulses.

The second step in the argument involves the analysis of how these distances change when not all the pulses are onsets. Accordingly, let there be n pulses in total, where n is an even number, and let there be p onsets (called black points) and $q = n - p$ silent pulses (called white points). For any fixed distance $d = 1, 2, \ldots, n/2$, let N_{ww}, N_{bb}, and $N_{bw} = N_{wb}$ be the number of edges of distance d connecting two white points, two black points, and a black–white pair, respectively. For example, in Figure 25.7, for the case of $n = 6$, $k = 4$, and $d = 2$, the number of white–white, black–black, and black–white pairs is 1, 3, and 2, respectively.

In particular, the second step in the argument establishes the following two simple relationships between these five quantities:

$$p = N_{bb} + (1/2)N_{wb}$$

$$q = N_{ww} + (1/2)N_{wb}$$

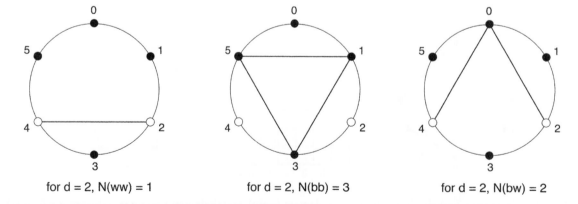

for d = 2, N(ww) = 1   for d = 2, N(bb) = 3   for d = 2, N(bw) = 2

FIGURE 25.7   The number of white–white, black–black, and black–white pairs at distance 2.

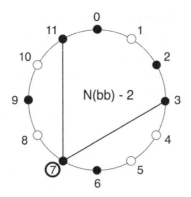 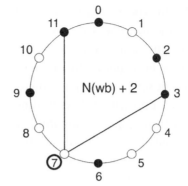

FIGURE 25.8   Both *d*-neighbors of a black point are black.

When all *n* points are black, the quantities N_{ww} and N_{wb} are equal to zero, and $N_{bb} = n$. Therefore, $q = 0$ and $p = n$, which confirms that $p + q = n$. It must be shown that these relationships continue to hold when some black points are changed to white. Initially all points are black and one of them will be changed to white. This case is illustrated in Figure 25.8 for the more general situation in which some black points have already been changed to white. For concreteness and without loss of generality, the black onset at pulse seven is changed to white. For any specified distance value *d,* this case assumes that there are two black onsets that realize this distance; call them *d*-neighbors. In changing this onset seven from black to white, the number of black–black *d*-neighbors N_{bb} goes down by 2, the number of white–black *d*-neighbors N_{wb} goes up by 2, and the number of white–white *d*-neighbors N_{ww} remains unchanged. It follows from the two preceding equations that *p* goes down by one and *q* goes up by one, confirming that $p + q = n$.

The case when both *d*-neighbors are white is illustrated in Figure 25.9. This change causes N_{wb} to decrease by 2, N_{ww} to increase by 2, and leaves N_{bb} unchanged. It follows again from the two preceding equations that

p goes down by one and *q* goes up by one, again validating that $p + q = n$.

Finally, the case where one neighbor is black and the other white is pictured in Figure 25.10. Now N_{bb} goes down by 1, N_{ww} goes up by 1, and N_{bw} remains unchanged. Substituting these changes into the two equations makes *p* go down by one and *q* to go up by one, once more supporting the claim that $p + q = n$.

This case analysis establishes that no matter how many, or which, black points are changed from black to white, the preceding equations that relate *p* and *q* to N_{ww}, N_{bw}, and N_{bb} remain true.

The final step in the argument for proving the hexachordal theorem involves setting *p* equal to *q* in the earlier equations, which yields the equation:

$$N_{bb} + (1/2)N_{wb} = N_{ww} + (1/2)N_{wb}$$

Since the two terms on each side of the equality that contain N_{wb} cancel each other out, we conclude $N_{bb} = N_{ww}$, which proves the result that for any distance value *d*, the number of times it occurs in the rhythm (N_{bb}) is the same as in its complementary rhythm (N_{ww}).

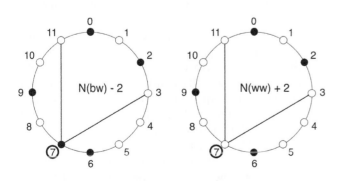

FIGURE 25.9   Both *d*-neighbors of a black point are white.

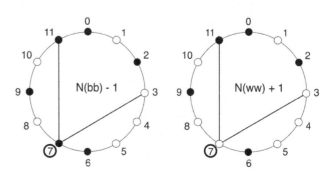

FIGURE 25.10   One *d*-neighbor of a black point is black and the other is white.

In 1995, thanks to the Internet's ability to provide wide and instant search of knowledge, I discovered to my surprise a large scientific literature in the crystallography journals regarding the hexachordal theorem and related results on homometric sets that the music theory community was unaware of. I presented these findings at the *Fourth International Workshop on Computational Music Theory* at the Universidad Politécnica de Madrid, Escuela Universitaria de Informática, in July 24–28, 2006.[13] Interestingly, the crystallography literature contained a variety of different proofs of the hexachordal theorem, which I presented at a *Special Session on Mathematical Techniques in Music Analysis*, at the 113th Annual Meeting of the American Mathematical Society, in New Orleans on January 6, 2007.[14] It appears that the music and crystallography research communities had not been aware of their common interests.

The term homometric was introduced in 1939 by Lindo Patterson.[15] The earliest proof of the hexachordal theorem in the music literature is due to Milton Babbitt[16] and David Lewin.[17] It used some powerful machinery from topology. Later Lewin obtained new proofs using group theory. Later still, Eric Regener, my colleague in the School of Computer Science at McGill University, found an elementary simple proof of this theorem.[18] This theorem was known to crystallographers at least 20 years earlier.[19] It was claimed to have been proved by Lindo Patterson around 1940, but he did not publish a proof. In the crystallography literature, the hexachordal theorem is called Patterson's theorem.[20] The first published proof in the crystallography literature is due to Buerger; it is based on image algebra and is nonintuitive. A much simpler and elegant elementary induction proof was later found by Iglesias, which has been simplified in the description presented earlier.[21] More recently, another simple and elementary induction proof was discovered by Steven Blau.[22] Several other proofs of this theorem have been published over the years in the music theory, crystallography, and mathematical literatures. Some are shorter than the proof described earlier, a few are more general, and others require advanced mathematical knowledge.[23]

The simplified version of Iglesias' proof presented earlier establishes a more general result, from which the hexachordal theorem, among other results discussed by Iglesias, follows as a corollary. The discrete hexachordal theorem considered in this chapter has also been generalized to the continuous case.[24] However, if we are interested in proving just the hexachordal theorem, then the simplest proof I have heard comes from the Harvard University mathematician Noam Elkies, via Dmitri Tymocsko. It is fitting to close this chapter with this gem.

Using the same notation as earlier, for any fixed distance $d = 1, 2, \ldots, n/2$, let B and W denote the black and white sets of points, respectively, each of cardinality $n/2$. For each value of d each point in B is d units away from two other points (that may be either black or white), one advancing in a clockwise direction around the circle, and the other in a counterclockwise direction. Therefore, in total, B has n distances of size d. Now let N_{ww}, N_{bb}, and $N_{bw} = N_{wb}$, be the number of distances d between two white points, two black points, and a black–white pair, respectively. It follows that for all the points in B we have that $N_{bb} + N_{bw} = n$. By the same argument, A has n distances of size d, and it follows that $N_{ww} + N_{wb} = n$. Therefore, $N_{bb} + N_{bw} = N_{ww} + N_{wb}$, and since $N_{bw} = N_{wb}$, the two terms cancel out leaving $N_{bb} = N_{ww}$. Therefore, each distance d occurs the same number of times between pairs of black points as pairs of white points.

NOTES

1 Thurlow, W., (1957), p. 653, gives an example of a figure-ground effect in the auditory domain.

2 Hasson, U., Hendler, T., Ben Bashat, D., & Malach, R., (2001), p. 744. This illusion is sometimes called the "vase or face" illusion. These authors uncover a neural correlate of shape-selective grouping processes in the human brain.

3 Fisher, G. H., (1967), p. 555.

4 See Anku, W., (2007), for a discussion of psychoacoustic considerations of rhythm in a cultural context.

5 Morris, R. D., (1990).

6 Kubik, G., (2010a), p. 42. The term *interlocking* rhythms sometimes also applies to situations where some attacks of the two rhythms coincide. Of course, with complementary rhythms, this generalization is automatically excluded.

7 Anku, W., (1997), p. 212, analyses interlocking rhythms as well as overlapping, and adjacency and alternation rhythms. For further details, see Anku, W., (1995, 2002a, 2002b), as well as Cuthbert, M. C., (2006), for an evaluation and expansion of Willie Anku's theories.

8 Miranda-Medina, J. F. & Tro, J., (2014), p. 217 and Feldman, H. C., (2005), p. 215.

9 Cover, T. M., (1987), p. 171.

10 Sachs, C., (1943), p. 190, uses the terms "audible" and "empty" beats in discussions of complementary rhythms. He notes that Indian *tabla* drummers often

play the "audible" beats with the right hand and the "empty" beats with the left. Kubik, G., (1999), p. 54, notes that: "One of the two phenotypes is usually dominant within a culture, while the other is implied or may sometimes be silently tapped with a finger." In other cultures, both rhythms are struck with equal force, one with each hand, and the discrimination is based on differences in timbre.

11 This is a common clapping pattern in West African music. See Jones, A. M., (1954a), p. 35.

12 Note that the distances are directed distances. Thus, between pulses zero and six there are two distances, one from zero to six and one from six to zero.

13 Toussaint, G. T., (2006a).

14 Toussaint, G. T., (2007).

15 Mardix, S., (1990).

16 Dembski, S. & Straus, J. N., Eds., (1986).

17 Lewin, D., (1959, 1960).

18 Regener, R., (1974).

19 Patterson, A. L., (1944).

20 Buerger, M. J., (1976).

21 Iglesias, J. E., (1981).

22 Blau, S. K., (1999).

23 For additional proofs of the hexachordal theorem, see Rahn, J., (1980), Chapter 5, Senechal, M., (2008). Amiot, E., (2009), gives a concise proof of the hexachordal theorem using a mathematical tool called the fast Fourier transform. See also Hosemann, R. & Bagchi, S. N., (1954) for further information on homometric structures. For a nice discussion of homometry in musical distributions, see Mandereau, J., Ghisi, D., Amiot, E., Andreatta, M., & Agon, C., (2011). For elegant proofs of this and related theorems see McCartin, B. J., (2012), McCartin, B. J., (2014), McCartin, B. J., (2015), McCartin, B. J., and McCartin, B. J., (2016).

24 Mandereau, J., Ghisi, D., Amiot, E., Andreatta, M., & Agon, C., (2011), p. 14.

Flat Rhythms and Radio Astronomy

CHAPTER 24 EXPLORED THE UNLIKELY CONNECTION between musical rhythms and crystallography. This chapter considers another surprising connection, this time between musical rhythm and radio astronomy. When I say "surprising," I mean surprising at first thought. Once it is realized that two areas are concerned with distances between pairs of elements, there is bound to be some connection. Radio astronomers are interested in receiving signals from outer space to discover new planets in other solar systems, perhaps stumble on alien intelligent life, as well as answer a variety of questions related to the structure of matter in the universe. For this purpose, they construct colossal and expensive radio telescopes or dish antennas: larger dishes provide better signals. However, very large dishes become prohibitively expensive. One approach to deal with this problem is to use several small dishes instead of one large one, and to arrange them some distance apart, thus ensuring that the waves of the signals arriving at each dish are out of phase with each other, resulting in interference. This interference causes the amplitude of the resultant wave obtained by superimposing the two waves to exhibit greater variation. For example, Figure 26.1 shows a two-element radio telescope at a unit distance apart. By analyzing this interference between the signals received in the two receivers, astronomers can improve signal detection.

By using more than two elements, much greater interference information is obtained, and the increased signal detection capability makes more efficient use of the individual elements. The two-element radio telescope provides only one separation distance between them that can be used to analyze the interference pattern between the two signals. However, using several elements allows the possibility of realizing distances between all pairs of elements. For example, the four-element radio telescope shown in Figure 26.2 realizes six pairwise distances [1, 1, 1, 2, 2, 3]. However, this arrangement is far from optimal since among the six distances realized, this set contains only three distinct distances: distance one is repeated three times and distance two is repeated twice, thus introducing considerable redundancy. In this application, redundancy is wasteful. What is required is a placement of the elements in such a way that as many distinct distances as possible are realized.

It is not too difficult to spread out the elements of a radio telescope so that all pairs of distances between them are different, especially if the number of elements used is not large, and one has unlimited space available. The example with four elements shown in

FIGURE 26.1 A two-element radio telescope separated by a unit distance.

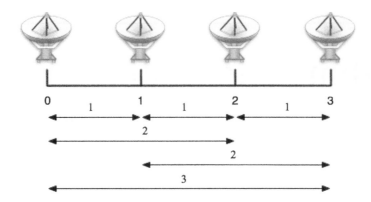

FIGURE 26.2 A linear four-element radio telescope with uniform spacing that realizes three distinct distances.

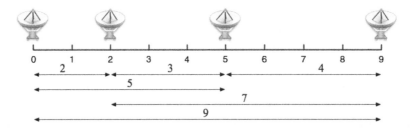

FIGURE 26.3 A linear four-element radio telescope array of length nine that realizes six distinct distances.

Figure 26.3 realizes six different distances [2, 3, 4, 5, 7, 9]. If each unit in this example represents 1 km in the real world, this telescope would need a space 9 km long. In practice it is desirable to use as little space as possible and still realize the maximum number of different distances. Figure 26.4 shows the optimal placement of four elements. The distances realized are [1, 2, 3, 4, 5, 6], and the required total length is only six units.

The problem of distributing radio telescope elements so that all the distances between pairs of elements are different is structurally identical to the problem of designing *Golomb rulers*. Ordinary rulers that are filled with *many* equally spaced marks on the sides are used for measuring arbitrary distances and drawing straight lines between pairs of points. A Golomb ruler, on the other hand, is a ruler with very *few* nonequally spaced marks.[1] Furthermore, since one can only measure distances between pairs of marks, the marks in a Golomb ruler are arranged so that as many *different* distances as possible can be measured. Although the design of such rulers was first investigated by Simon Sidon[2] in 1932, they are named after the mathematician Solomon Golomb, who was one of several independent rediscoverers of the problem and played a singular role in making them popular. A Golomb ruler with k marks is called *optimal* if no other shorter Golomb ruler with k marks exists. Furthermore, if the Golomb ruler measures *all* the distances ranging from one to the length of the ruler, it is called *perfect*. Figure 26.5 shows a perfect Golomb ruler with four marks.

It is not easy to characterize optimal Golomb rulers and thus obtain an efficient algorithm for generating all

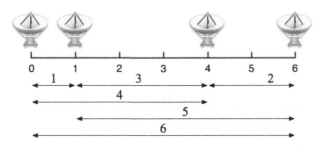

FIGURE 26.4 A linear four-element radio telescope array that realizes six distinct distances, and has length six.

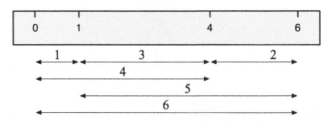

FIGURE 26.5 A *perfect* Golomb ruler with four marks.

of them. So far, optimal Golomb rulers have been found with up to 26 marks only. The search for an optimal ruler with 27 marks continues to this day. However, sufficient conditions have been found for generating special cases of Golomb rulers. For example, Paul Erdős and Pál Turán showed that for every odd prime number p a Golomb ruler with p marks may be constructed with the following algorithm: $2pk + (k^2 \bmod p)$, where k varies between 0 and $p - 1$, and *mod* stands for the modulo function or what remains when k^2 is divided by p. If we let $p = 5$, this formula yields marks with integer values 0, 11, 26, 34, and 41. These five marks determine the ten distinct distances given by 7, 10, 11, 13, 17, 23, 24, 30, 31, and 41. Although much shorter Golomb rulers with five marks exist, the power of this algorithm lies in its generality. A five-mark Golomb ruler with length 11 is shown in Figure 26.6.

The problem of whether there exist rulers with a different set of marks that measure the same set of distances has also been investigated. This is the problem of the existence of homometric sets on the line, also called the *turnpike* problem in computer science. Figure 26.7 shows two six-mark homometric rulers discovered by R. Hosemann and S. N. Bagchi in 1954. These are not Golomb rulers, because some distances occur more than once: distances two and six each occur twice. However, these two rulers have another redeeming property, and that is that they are able to measure *all* the integer distances between 0 and 11. Such rulers have been called *spanning rulers* by Roger Alperin and Vladimir Drobot, who studied some of their properties.[3] They also defined a *minimal spanning ruler* as one that stops being a spanning ruler if one of its marks is erased.

In 1974 J. Franklin provided a variety of constructions and examples of pairs of homometric rulers.

However, like the example in Figure 26.7, none of his examples were Golomb rulers. Franklin's rulers were consistent with a published theorem dating back to 1939 due to S. Piccard,[4] which stated that if the sets of distances determined by the marks of two rulers had no repetitions (i.e., the rulers were Golomb), then a necessary and sufficient condition for the two distance sets to be equal is that the two rulers should be congruent, meaning that one ruler may be translated or rotated to bring it into perfect correspondence with the other. However, in 1975 Gary Bloom discovered a counterexample to Piccard's theorem.[5] He constructed two Golomb rulers shown in Figures 26.8 and 26.9 that measure the same set of distances between all pairs of their marks, and yet are not congruent.

By connecting the ends of a Golomb ruler together to form a circle, and considering the marks on the ruler as possible positions for the locations of the onsets of musical notes, the ruler determines a rhythm. If we replace the straight-line distances between the marks on a ruler by geodesic distances on the circle that represent time durations, then the Golomb ruler problem becomes that of constructing rhythms in which all interonset durations are distinct. This means that the interonset duration histograms of such rhythms are flat in the sense that all histogram bins have height either one or zero. Accordingly, we will use Jon Wild's terminology and call such rhythms *flat* rhythms.[6] It is very easy to create flat rhythms if the number of onsets is 3, because then there are only three interonset intervals (IOIs). One merely has to ensure that the three distances are distinct. From the geometrical point of view, this means that the rhythm triangle should not be *isosceles*. Figure 26.10 shows two flat rhythms, left and center, with IOI sequences [2-4-6]

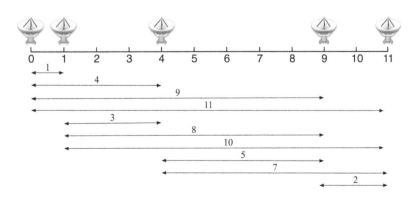

FIGURE 26.6   A Golomb ruler of length 11 with five marks.

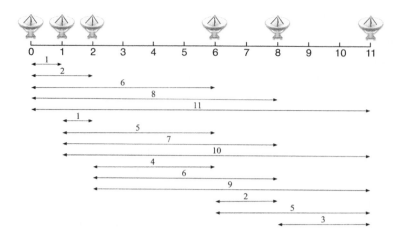

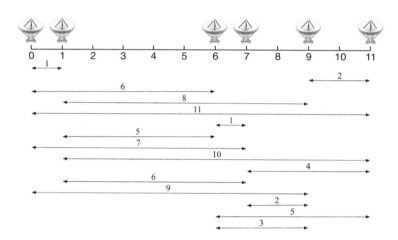

FIGURE 26.7 Two 6-mark homometric *spanning* rulers that are not Golomb rulers.

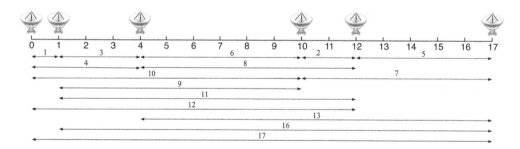

FIGURE 26.8 First Golomb ruler of Bloom's counterexample.

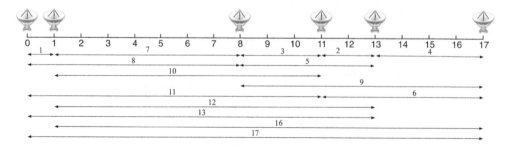

FIGURE 26.9 Second Golomb ruler of Bloom's counterexample.

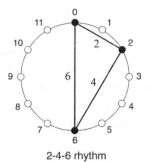
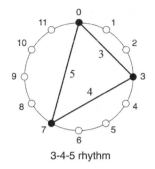
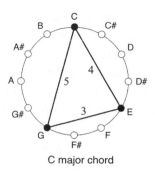

2-4-6 rhythm   3-4-5 rhythm   C major chord

FIGURE 26.10   Rhythms and chords with unique distinct intervals.

and [3-4-5], respectively. On the right is the C major chord visualized as a triangle connecting the C, E, and G notes. It also has three distinct and unique intervals [4-3-5].

Although in the pitch domain the IOIs {3, 4, 5} are common (in the form of the C major chord), this pattern of intervals is not commonly employed in the rhythm domain. However, one example is found in the Afro-Peruvian *landó*.[7] This rhythm is played on a wooden box with a hole in the back, called the cajón, and is illustrated in Figure 26.11. The cajón is played by sitting on it and striking the high tones with the fingers on the upper edge of the cajón with one hand, and the low tones with

the palm of the hand near the center of the box. The rhythm of the sharp dominating high tones is shown in the left diagram with IOIs {3, 4, 5}, at pulses 2, 7, and 10. The softer low tones with IOI pattern [4-1-3-3-1], and the combination of the two rhythms, are shown in center and right diagrams, respectively.

More difficult is the creation of rhythms with distinct IOIs when there are more than three onsets, although such rhythms also exist. Two examples of flat rhythms with four onsets and 16 pulses are shown in Figure 26.12. Both rhythms have exactly the same set of distances (all except two and six), and since they are different bracelets, they are also homometric rhythms.

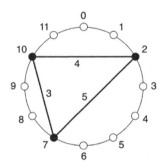
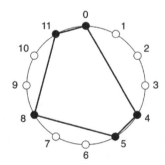
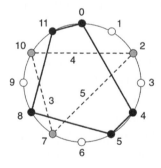

FIGURE 26.11   The Afro-Peruvian Landó rhythm played on the *cajón*. The high tone (left), the low tone (center), and their combination (right).

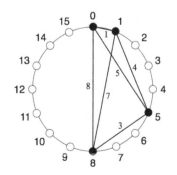
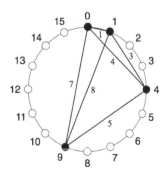

FIGURE 26.12   Two 4-onset, 16-pulse homometric rhythms with flat interval-content histograms.

Just as there are perfect Golomb rulers that measure all the integer distances up to their maximum value, so there are rhythms that contain *all* durations exactly once, thus producing flat histograms without gaps. Inspired by the notion of perfect Golomb rulers, we will call such rhythms *perfect flat* rhythms. Two examples of perfect flat (and homometric) rhythms with four onsets and 12 pulses are given in Figure 26.13. Note that they are also *minimal spanning* rhythms, since deleting any onset from either of them prevents them from being able to produce more than three durations. In the pitch domain, these two polygons are the well-known *all-interval tetrachords*. These two chords have been explored extensively, by both music theorists and composers, for several decades.[8]

One may wonder how many more pairs of noncongruent perfect homometric rhythms exist in a 12-pulse time span. Recall that rhythms that are congruent may be transformed into each other by rotations and/or reflections, and thus are instances of bracelets. It turns out that the bracelets pictured in Figure 26.13 are the only two possibilities. An elementary case analysis will prove the point.[9] Since the time span has 12 pulses, there are a total of six distinct durations to be realized by all pairs of onsets. Therefore, the number of pairs of onsets must equal six, implying that the number of allowable onsets must satisfy the equation $k(k-1)/2 = 6$. The only solution of this equation is $k = 4$. It follows that only quadrilaterals need be considered in our search. Since the distance six must occur once in this quadrilateral it must be either a diagonal or an edge of the quadrilateral. Consider first the case where it is a diagonal. Without loss of generality, let this diagonal connect pulses zero and six. Then there are eight possibilities for constructing the quadrilateral such that one onset lies

in each half circle determined by the diagonal, and the sum of the two edges on each side equals six, as shown in Figure 26.14. The first and second quadrilaterals (also the third and fourth) are reflections of each other about the vertical axis through pulses zero and six. In addition, the first and fourth (also the second and third) are reflections of each other about the horizontal axis through pulses nine and three. Similar relations hold for the bottom four quadrilaterals.

The four quadrilaterals at the bottom of Figure 26.14 are very different from the four at the top, in that they do not contain the interval of length three. Instead, they have two instances of the interval of length five. Therefore, they may be discarded from consideration. The four at the top do contain all six intervals, but they are all instances of the same bracelet. Thus, we have found one of the candidates in this set.

The second situation arises when the duration of length six is an edge of the quadrilateral. Six arrangements of the other three edges are possible as shown in Figure 26.15. The middle edges may have lengths one, two, or three, and for each of these three choices, each of the two remaining edge lengths may be inserted at the top or the bottom. Notice that in the four arrangements on the left the diagonal incident to pulse three has duration three, which in all four cases is already present as an edge. Hence, these four candidates may be discarded. The two candidates on the right do contain all six intervals, but they are reflections of each other about the axis through pulses nine and three, and are therefore instances of the same necklace. In conclusion, there are only two rhythm bracelets that have perfect flat interval histograms in a 12-pulse time span.

We close this chapter by including a few of Jon Wild's examples of flat rhythms in Figures 26.16–26.18.

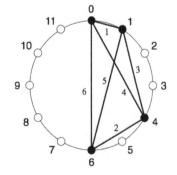
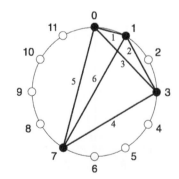

FIGURE 26.13 Two 4-onset 12-pulse minimal-spanning, homometric rhythms with *perfect* flat interval-content histograms.

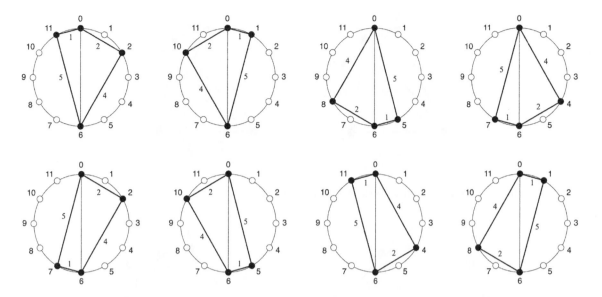

FIGURE 26.14 The eight cases in which the interval of length six is a diagonal.

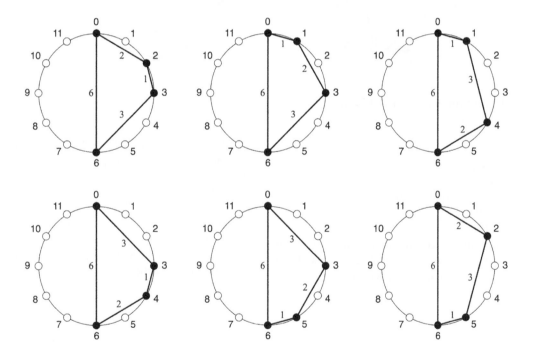

FIGURE 26.15 The six cases in which the interval of length six is an edge of the polygon.

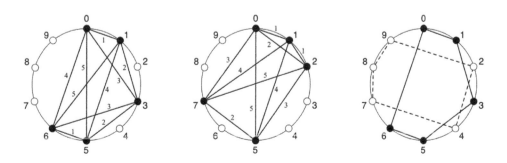

FIGURE 26.16 Flat rhythms for $n = 10$ and $k = 5$.

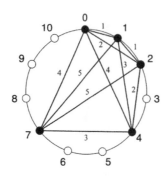
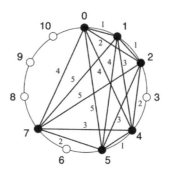

FIGURE 26.17 Flat rhythms for $n = 11$ and $k = 5$ (left), and $k = 6$ (right).

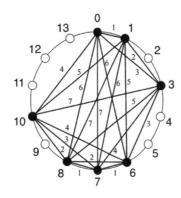
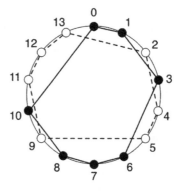

FIGURE 26.18 Flat rhythms for $n = 14$ and $k = 7$.

NOTES

1 Alperin, R. C. & Drobot, V., (2011).
2 Erdős, P. & Turán, P., (1941).
3 Alperin, R. C. & Drobot, V., (2011).
4 Piccard, S., (1939).
5 Bloom, G. S., (1977).
6 Wild, J., (2007), introduced the notion of *flat interval distributions* in the context of pitch class sets. Here Wild's terminology is adapted to rhythms.

7 Miranda-Medina, J. F. & Tro, J., (2014), p. 217 and Feldman, H. C., (2005), p. 215.
8 Block, S. & Douthett, J., (1994), p. 35. See Goyette, J., (2011), Childs, A. P., (2006), and Peck, R. W., (2015), for discussions on the structural and transformational properties of the all-interval structures as well as additional references.
9 The proof by case analysis given here is similar to that found in McCartin, B. J., (1998), p. 360.

CHAPTER 27

Deep Rhythms

IN THE LAND OF THE PHARAOHS in ancient Egypt, the builders of the majestic pyramids and temples that adorn the surroundings of the Nile River made use of an amazing low-tech, biodegradable, zero-radiation, inexpensive, and low-maintenance computer to find solutions to a variety of geometric problems that they encountered in their daily lives: the *knotted rope*. As the name suggests, this computer consists of a rope of suitable length tied together at the ends and interspersed with a preselected fixed number of equally spaced knots. There is evidence that one popular model of this computer comprised 12 knots as shown in Figure 27.1, in two configurations: loose (left) and taught (center). The users of this computer knew from experience in the field that if three people holding the rope at the knots numbered one, five, and eight walked away from each other until all three strands of rope between them were taught, the final shape of the rope would be a triangle.[1] Furthermore, and this being the crucial point, they were confident that the triangle would be a *right triangle*: the angle made between the two short sides of lengths three and four is a 90° angle. One of the most useful applications of this computer was therefore the construction of 90° angles in architecture. To illustrate this application,

assume for example that the workers had to build a new wall that made a 90° angle with another wall already built, and refer to Figure 27.1 (right), where the old wall is shown in the horizontal position. Assume further that this new wall is required to meet the old wall at the point marked A. First one worker holds the rope at the fifth knot and stands at position A. The second worker then takes knot number one and walks away from the first worker along the old wall until the rope between them is taught, ending at position B. Finally, the third worker takes knot number eight and walks away from the other two workers, adjusting its position until both strands are taut. The engineers were convinced that if the new wall was built in a straight line from A to C, the angle BAC would be a 90° angle.

You may ask yourself how the workers came to be so assured that in a triangle with side lengths consisting of 3, 4, and 5 units (called a 3-4-5 triangle), the angle between the two short sides was a right angle. In those days, the workers followed algorithms outlined in manuals issued by the pharaohs. Very likely, in one such manual, the procedure outlined earlier for constructing walls at 90° angles with the knotted rope was followed by an official statement of the form: *If this algorithm is*

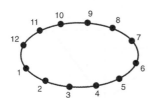 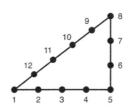 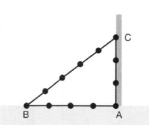

FIGURE 27.1   The knotted string computer for constructing right angles.

followed, then the correct solution is guaranteed to be found. Next to this statement would usually be stamped the seal of approval of the Pharaoh. This process might be called a proof by governmental decree. It would be several thousand years later before the correctness of this algorithm was formally established using a more democratic method: a mathematical proof.

In the sixth century BC, the great mathematician Pythagoras of Samos in ancient Greece proved a wonderful theorem about right-angled triangles, of which the 3-4-5 triangle used in the knotted rope computer is a special case.[2] This theorem, now called Pythagoras' Theorem, is taught to every child in school. The theorem states that in a right-angled triangle, the area of the square with side equal to the longest side of the triangle (called the hypotenuse) is equal to the sum of the areas of the squares with sides equal to the other two sides of the triangle. The proof of Pythagoras' Theorem that appears in Euclid's *Elements* is rather lengthy and detailed. Since then hundreds of different proofs have been discovered, some of them quite elegant, short, and simple. One example that falls in the latter category is illustrated in Figure 27.2. The right-angled triangle in question is the dark gray triangle with sides a, b, and c, that occurs four times in the left diagram. It is required to be proved that the area of the square with side a plus the area of the square with side b (both drawn in light gray on the right diagram) equals the area of the large white square with side c (on the left diagram). The two complete squares, on the left and right diagrams, both have sides of length $a + b$, and are therefore equal. Also, each complete square contains four copies of the dark gray right-angled triangle, arranged in different ways. Therefore, removing these four triangles from both complete squares leaves equal areas. The remaining area on the left is the white

square of side c, and the remaining area on the right consists of two light-gray squares of sides a and b, thus proving the theorem.

This proof of Pythagoras' Theorem holds for any positive values of a, b, and c, as long as the given triangle has a right angle. For the special case $a = 3$, $b = 4$, and $c = 5$, a simple and original proof illustrated in Figure 27.3 may be found in the ancient Chinese classic book *Chou Pei Suan Ching*, probably written during the Han dynasty period dated between 500 BC and AD 200. Given a triangle with a 90° angle flanked by sides of lengths three and four, it must be proved that the area of the square with side length eh is $3^2 + 4^2 = 9 + 16 = 27$. First four copies of the triangle (shaded in light gray) are embedded in a 7×7 square array that consists of 49 unit squares, as shown in the figure. This leaves out four identical triangles at the corners of the array as well as one small dark gray square in the center. The area of the entire array is 49. Therefore, the area of the gray and white triangles is 48. But the area of the white triangles equals the area of the gray triangles. Therefore, the area of the gray triangle is 24. The area of the square with side eh is equal to the area of the four gray triangles plus the small unit square in the middle, and is thus 27, as was required to prove.

When students are asked how they would prove that the knotted-rope algorithm used by the ancient Egyptians to construct a right angle is correct, some are quick to answer: *apply the theorem of Pythagoras.* They are then surprised to discover that Pythagoras' Theorem does not do the job. The Theorem of Pythagoras proves that if we are given a right-angled triangle, then the square of the hypotenuse equals the sum of squares of the other two sides. The theorem required to prove the correctness of the knotted-rope algorithm would have to state that if we are given a triangle in which the square

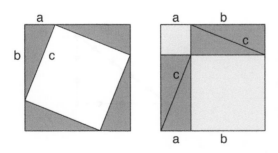

FIGURE 27.2 An elegant, simple, and short proof of Pythagoras' Theorem.

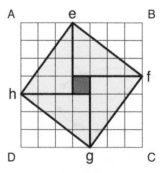

FIGURE 27.3 Proof of the Pythagoras theorem in the Chinese classic *Chou Pei Suan Ching*.

of the hypotenuse equals the sum of the squares of the other two sides, then the angle opposite the hypotenuse must be a right angle. In other words, we require the *converse* of Pythagoras' Theorem. It is the difference between the logical statements *if* and *only if*. The good news is that a proof of this converse theorem also appears in the *Elements* of Euclid. In fact, it appears just after the proof of Pythagoras' Theorem.

Pythagoras was also a great music theorist, and laid some of the foundations of scales, chords, and tuning in music. The quotation "*There is geometry in the humming of strings*," is attributed to him. Pythagoras was referring to the pitch aspects of music in relation to the plucking of strings of different lengths. However, he could just as well have added that there is geometry in the drumming of drums. Pythagoras experimented with strings of different lengths that had the same tension and discovered that they sounded well together when the ratios of their lengths were related by small integers such as 1:2 and 2:3. In the ubiquitous diatonic scale used today the major and minor three-note chords (triads) are considered to be the most stable and resonant because they are made up of the three most stable and resonant intervals: three, four, and five, the same integers that make up the smallest integers that satisfy the Pythagorean Theorem. The C major and minor chords are shown as triangles in Figure 27.4. Interestingly, even though the major and minor chords have exactly the same intervals, the fact that the order of the intervals is reversed in going from one chord to the other makes the chords sound quite different from each other.

There are many ways in which 12 may be divided into three intervals, and the partition used by the major and minor chords, is very special. The most even division would yield a triangle with three edges of the same length of four. Then there are many partitions that would use two intervals of the same length, such as [3-3-6]. In addition, there are partitions that have very small and very large intervals such as [1-2-9]. The major chord on the other hand is the partition closest to [4-4-4] that has three distinct distances.

More than 2,500 years after Pythagoras, in the twentieth century, another great and prolific mathematician, Paul Erdős, asked a very simple geometry question that also concerns distinct distances, musical rhythms, and scales, and that still has mathematicians baffled. Erdős asked if one can arrange n points in the plane with the restrictions that no three of the points should lie on a line, and no four of them may lie on a circle, so that for every value of $i = 1, 2, ..., n - 1$ there is a distance determined by pairs of these points that occurs exactly i times. In other words, each distance realized by pairs of points should occur a *unique* (distinct) number of times. We call such a set a *deep* set. The corner points of a rectangle of width a and height b does not yield unique distances because all three distances, a, b, and c, where c is the diagonal of the rectangle, occur the same number of times, namely twice. Furthermore, all four points lie on a circle, and thus violate one of the constraints. Therefore, the corners of a rectangle do not constitute a deep set.

You may ask yourself why Erdős disallowed the points to lie on a line or a circle. The answer is that the problem is too easy and uninteresting when these allowances are made. Consider for example the seven points on a line separated by unit distances, pictured in Figure 27.5. As the picture makes clear, distance six occurs once, distance five occurs twice, distance four thrice, distance three four times, distance two five times, and distance one six times. In other words,

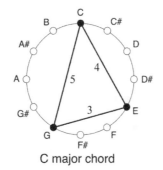
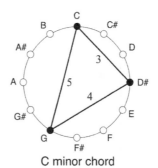

FIGURE 27.4 The C major and C minor triangles with arclengths 3, 4, and 5.

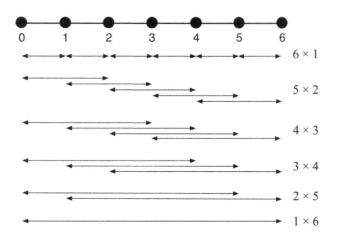

FIGURE 27.5 For evenly spaced out points on a line a unit distance apart, every distance occurs a unique number of times.

no distance occurs the same number of times as any other distance. In other words, the multiplicity of each distance is unique.

A similar situation arises with points evenly spaced out on a circle. Consider the circle with 12 pulses and seven onsets at pulses 0 to 6, and refer to Figure 27.6. Again, each circular arc length (geodesic distance) occurs a unique number of times. This remains true for any number of pulses and onsets as long as the onsets span at most one semicircle.

However, let us return for a moment to the original problem that Erdős posed for the two-dimensional plane. An example of a valid solution for $n = 4$ is illustrated in Figure 27.7. Two of the four points are located on the horizontal axis at coordinates 1 and −1, and the other two points are located on the vertical axis at coordinates −1 and $\sqrt{3}$. The histogram of the three distinct distances that occur between the points is shown on the right. The smallest distance $d_1 = \sqrt{2}$ occurs twice, the next largest distance $d_2 = 2$ occurs thrice forming an equilateral triangle, and the largest distance $d_3 = 1 + \sqrt{3}$ occurs once between the two points on the vertical axis.

So far, the only solutions that have been found for this problem are for $n = 2, 3, 4, 5, 6, 7,$ and 8. The solution for $n = 8$ illustrated in Figure 27.8 is quite elegant. The first diagram shows the arrangement of the eight points, along with all their pairwise distances. Since the eight points are located at the vertices of a triangular

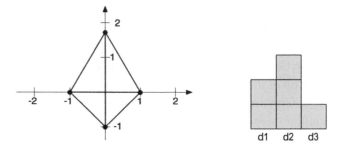

FIGURE 27.7 A set of four points with distinct distance multiplicities (left) and their histogram (right).

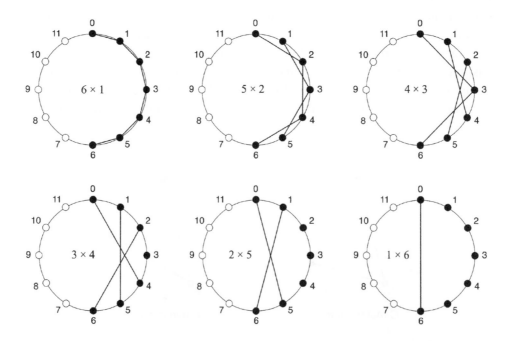

FIGURE 27.6 For evenly spaced out points in a semicircle, every distance occurs a unique number of times.

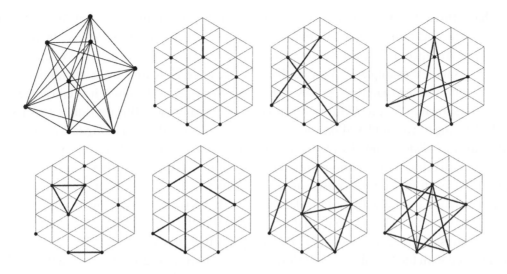

FIGURE 27.8   A *deep* set of eight points in the plane.

lattice, it helps to see the distances embedded in the triangulation. The remaining seven diagrams show each of the seven distinct distances by themselves in order of increasing multiplicity. The distance equal to one side of a triangle occurs once. The distance equal to two sides of a triangle occurs five times. The distance equal to two heights of a triangle occurs four times. For the remaining distances, it helps to view every pair of adjacent triangles as a lozenge, and combine adjacent lozenges into parallelograms. Then the distance corresponding to the long diagonal of a 1 × 2 parallelogram occurs six times. The distance equal to the long diagonal of a 1 × 3 parallelogram occurs seven times. The distance equal to the long diagonal of a 1 × 4 parallelogram occurs three times. Finally, the distance equal to the long diagonal of a 2 × 3 parallelogram occurs twice.

As already pointed out, one reason why the problem that Erdős posed is so difficult is that the points must lie in the plane: a two-dimensional universe. However, if the problem is considered in a one-dimensional universe, the circle, then the problem is more tractable, even for the case when the points are not restricted to lie in a semicircle. Furthermore, it is precisely in this form that the problem is most relevant to musical rhythms and scales.

The simplest deep rhythms are those with three onsets—triangles. Geometrically, they are represented by *isosceles* triangles, *i.e.*, those that have two sides equal and one side different from the other two. For example, for three onsets and 12 pulses there are four possible such rhythms with the apex of the isosceles triangle positioned at pulse zero, as illustrated in Figure 27.9. From left to right, these rhythms have durations [1-10-1], [2-8-2], [3-6-3], and [5-2-5]. Since this vertex may be anchored on any of the 12 pulses, the total number of such rhythms is 48. For more than three onsets, deep rhythms are less straightforward to generate.

Deep rhythms with four onsets contain six distances. One method with which to realize a multiset of six

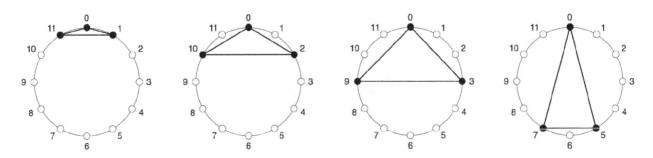

FIGURE 27.9   The four deep rhythms with three onsets and 12 pulses anchored at pulse zero.

distances is to have three distinct distances: one that occurs once, one twice, and a third thrice. An easy way to obtain three equal distances is by means of an equilateral triangle. The smallest example with an equilateral triangle that contains more than three pulses is the three-in-six rhythm shown in Figure 27.10 (right). It is the rhythm called *nyunga-nyunga* played on the *mbira* (also called the *sanza*, *kalimba*, and *thumb-piano*) by the Shona people of East Africa. The thumb-piano illustrated in Figure 27.11 is an instrument consisting of a resonating chamber made from one half of a calabash (or gourd) shell covered with a slab of wood that contains a hole, on which is mounted a series of long metal blades that are plucked with the thumbs. The rhythm in Figure 27.10 is the composition of two regular rhythms, [2-2-2] played with the left thumb (left) and [3-3] played with the right thumb (center). When the two are combined (right), a deep rhythm is obtained, with distance two occurring thrice, distance one twice, and distance three once. Note that, by convention, if a distance spans the diameter of the circle it is considered to occur once rather than twice. This rhythm is also a Korean shaman timeline played on a *tsching* (a Korean gong), a rhythm of the Cuban "canto de clave,"[3] the rhythm of the *zarabanda*,[4] and the simple hemiola, "common to the point of cliché in the cadences of the Baroque era."[5] When the rhythm is started on the fourth onset to obtain the durational pattern [2-2-1-1], it is the fundamental rhythmic pattern of the Colombian *bambuco*.[6]

Another unforgettable deep rhythm with four onsets is George Gershwin's rhythmic riff in *I Got Rhythm*, given by [. . x . . x . . x . . x . . . .].[7] Here distance seven occurs

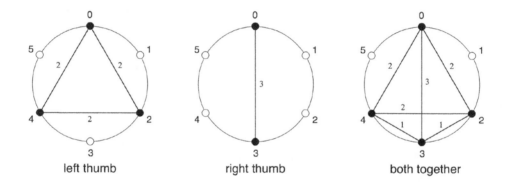

left thumb   right thumb   both together

FIGURE 27.10 The *nyunga-nyunga mbira* rhythm of the Shona people of East Africa.

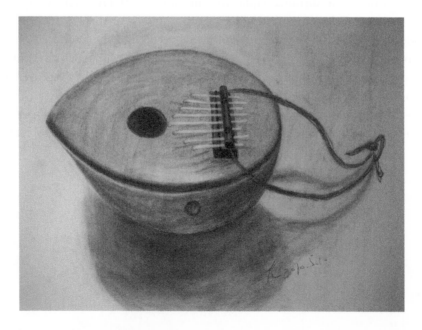

FIGURE 27.11 The *thumb-piano*. (Courtesy of Yang Liu.)

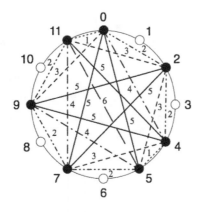

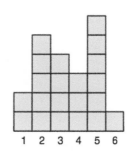

FIGURE 27.12   The *bembé* deep rhythm and its interval-content histogram.

once, distance six twice, and distance three thrice. But perhaps the most illustrious rhythm that contains the *deepness* property is the pattern of seven onsets among 12 pulses that defines the *bembé* rhythm shown in Figure 27.12 (left), along with all the distances between every pair of onsets indicated. As the histogram on the right clearly shows, every distance ranging from one to six occurs a distinct number of times. As already pointed out, this pattern is identical to the diatonic scale pattern in the pitch domain. Indeed, it was in this domain that in 1966 Terry Winograd and in 1967 Carlton Gamer studied scales with this property and christened them *deep scales*.[8] It is for this reason that rhythms possessing this property are here called *deep rhythms*.

Another well-known example of a deep rhythm is the five-onset, 12-pulse *fume–fume* timeline (also the pentatonic scale) shown alongside its interval-content histogram in Figure 27.13.

There is a simple algorithm that will generate deep rhythms that are more interesting than those that have all their onsets in one semicircle. This algorithm will generate different families of deep rhythms that depend on a parameter d, and are illustrated in Figure 27.14 for a cycle with 16 pulses.[9] In this case, the algorithm will generate deep rhythms with any number of onsets between two and nine. The algorithm places the first onset at pulse zero. To place the remaining onsets, the pulses in the cycle are scanned in a clockwise order, and a visited pulse is selected as an onset every time a distance d has been advanced. The value of d can be any integer that is relatively prime to the number n of pulses in the cycle, *i.e.*, n and d have no common divisors other than one. Therefore, for $n = 16$, d may take on the values three, five, and seven. The example in Figure 27.14 uses $d = 7$. Therefore, the remaining eight onsets are selected in order at pulses 7, 14, 5, 12, 3, 10, 1, and 8. Below each rhythm is shown its histogram of the interonset distances, and it can be seen that after every insertion of a new onset the new rhythm remains *deep*. Also, worth

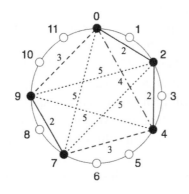

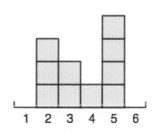

FIGURE 27.13   The *fume–fume* deep rhythm and its interval-content histogram.

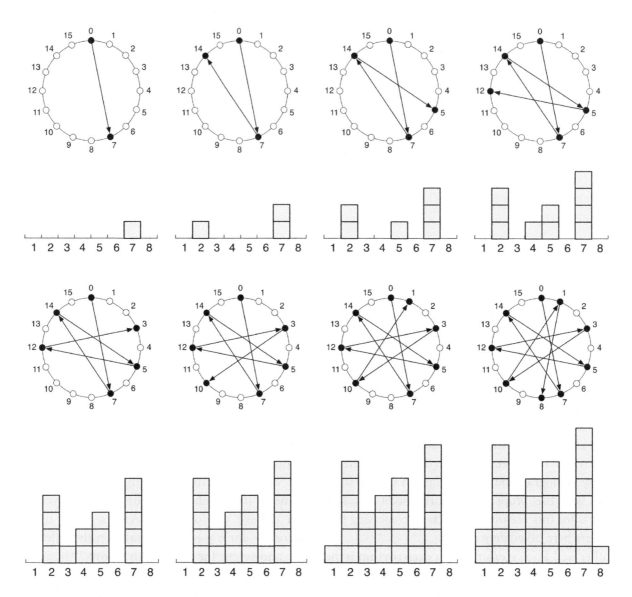

FIGURE 27.14 Illustrating the algorithm for generating *deep* rhythms with $d = 7$.

noting is that when d is relatively prime to n, and d is small enough, a new attack is never inserted in a location diametrically opposite to an already existing attack, thus preserving the property of rhythmic oddity. If the *Hop-and-Jump* algorithm of Chapter 15 is stopped when seven attacks are obtained, then the rhythm necklace generated is the same as the seven-attack rhythm necklace generated with the procedure described here.

When $d = 5$ the algorithm generates deep rhythms and histograms shown in Figure 27.15. Note that the two families of rhythms are quite different from each other. When $d = 7$ the rhythms generated quickly become maximally even, whereas when $d = 5$ the onsets tend to group together into three separate clusters, making them *minimally* even. This shows that deepness and maximal evenness are properties of a very different nature.

Figure 27.16 shows the deep rhythms and histograms generated by the algorithm when $d = 3$. This family of rhythms is also quite different from the other two families. Whereas when $d = 5$ the onsets group together into three separate clusters of three rhythms each, making them sort of minimally even, when $d = 3$ the onsets cluster into four smaller groups of two rhythms each.

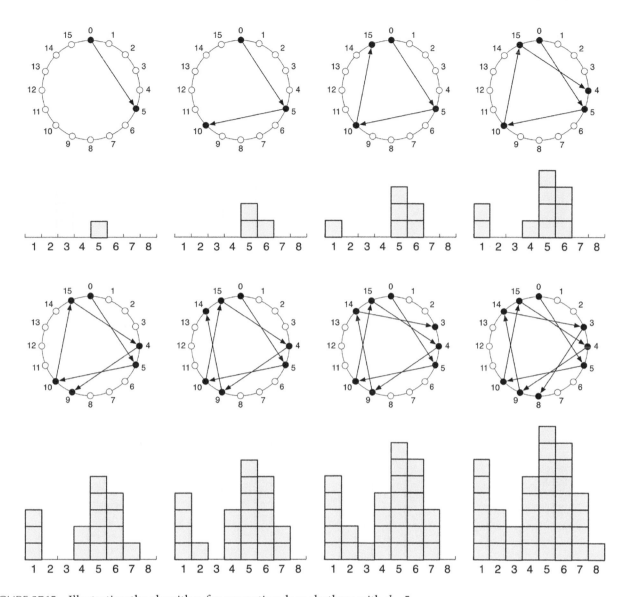

FIGURE 27.15   Illustrating the algorithm for generating deep rhythms with $d = 5$.

NOTES

1 Ross, C., (2007), p. 157. An alternate view argues that the use of the knotted rope for laying out right angles in ancient Egypt is speculation, and that there is little reliable evidence that the builders were aware of the Pythagorean Theorem. Imhausen, A., (2006), discusses additional ancient Egyptian mathematics.

2 An ancient Babylonian clay tablet, dating back to 1800 BC, known as Plimpton 322 at Columbia University, contains other such Pythagorean triples of numbers, leading some authors to suggest that the Pythagorean Theorem may have been known well before Pythagoras. See Robson, E., (2001, 2002) and Polster, B., (2004), p. 4. However, knowledge of examples that satisfy a theorem does not by itself imply knowledge of the theorem; for that a proof is required.

3 Orovio, H., (1992), p. 111.

4 Van der Lee, P., (1995), p. 202. In this *zarabanda* rhythm, the second and fourth onsets are played with a lower pitch to obtain [x . x x x .].

5 Gotham, M., (2013). See also Ritchie, S., (2012).

6 Varney, J., (2001), p. 140. In this *bambuco* rhythm, the first and third onsets are played with a lower pitch to obtain [x . x . x x]. See also the PhD thesis: Varney, J., (1999).

7 Crawford, R., (2004), p. 163. See also its relation to Acid Jazz discussed in Chapter 15.

8 Johnson, T. A., (2003), p. 41.

9 The algorithm described here is used in the pitch domain to generate scales. There exists a variety of similar algorithms that generate different families of scales that may be applied to the time domain as well. See Clough, J., Engebretsen, N., & Kochavi, J., (1999) and Carey, N. & Clampitt, D., (1996, 1989), for further details and references.

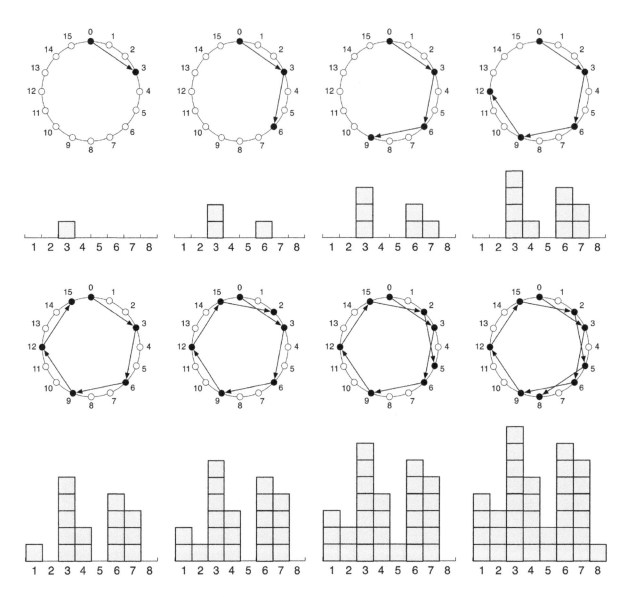

FIGURE 27.16 Illustrating the algorithm for generating deep rhythms with $d = 3$.

CHAPTER 28

Shelling Rhythms

IN 1956 ELVIS PRESLEY released his own faster and more syncopated version of a slow blues song recorded 4 years earlier by Willie Mae "Big Mama" Thornton. For both Thornton and Presley this song, titled "*Hound Dog*," rose to the top of the charts. In Elvis Presley's rendition, the rhythmic patterns produced by the snare drum, the bass, and the handclaps are typical traditional African rhythms, a common thread that meanders through the rockabilly music of the 1950s.[1] Less well known is the fact that these rhythms have two interesting mathematical properties: they are all *deep*, and three of them exhibit what is called the *shelling* property with respect to the *deepness* attribute. The four rhythms that permeate the song are shown in Figure 28.1 along with their interonset-interval histograms directly underneath each of them. From these histograms, it is clear that they are all deep: they have

the property that each interval in the rhythm appears a unique number of times.[2] The bass pattern on the left is the well-known Cuban *tresillo* pattern. In "Big Mama" Thornton's version the less syncopated duration pattern of the bass rhythm is [x . . . x . x .]. The clap pattern in Elvis' version (second from left), known as the *habanera* rhythm,[3] is obtained by inserting an onset at pulse position five (here, inserting refers to replacing a silent pulse with a sounded one). In "Big Mama" Thornton's version of *Hound Dog* the unsyncopated clap pattern is a straight pulsation given by [. . x . . . x .]. Elvis' snare-drum pattern, on the right is the signature "backbeat"[4] pattern of "rock-n-roll" music, and plays the role of the clap in "Big Mama" Thornton's version. If the snare-drum onset at pulse two is added to the clapping pattern, then the well-known *cinquillo* pattern (second from right) is obtained.[5]

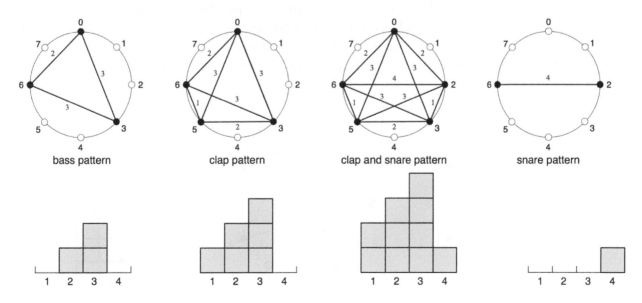

FIGURE 28.1 The rhythms and interval histograms used in Elvis Presley's "*Hound Dog*."

A rhythm admits a *shelling*, with respect to some property *P* if there exists a sequence of insertions or deletions of onsets so that, after each insertion or deletion, the rhythm thus obtained continues to have property *P*.[6] Again, here and throughout this chapter, the terms insertions (or deletions) refer to replacements of a silent (or sounded) pulse with a sounded (or silent) pulse. In the present case, starting with the bass pattern, onsets may be inserted in positions 5 and 2, in this order, while maintaining the deepness property of the resulting rhythms. Similarly, starting with the cinquillo rhythm and deleting the onsets in the reverse order (2 before 5) preserves the deepness property. In fact, since, by convention, a pair of onsets defines just one duration interval corresponding to the shortest circular arc determined by the pair, another onset may be deleted from the bass pattern on the left, while still maintaining the deepness property, albeit at its most elemental level. Deleting the third onset at pulse six yields the Cuban timeline [x . . x . . . .] called the *conga*,[7] which at a faster pace becomes the *charleston* rhythm.[8] This rhythm, probably the simplest, minimalist, yet effective, electric guitar solo ever used in a rock-n-roll megahit, is played in the middle section of the 1964 song *Don't Worry Baby*, by the Beach Boys. Their solo ends by inserting the third onset, thus converting it to the tresillo. Deleting the first onset of the tresillo yields [. . . x . . x .], also a Cuban rhythm called the conga.[9]

The 12-pulse ternary timelines used in much sub-Saharan African music provide another example of a family of rhythms that admit shelling with respect to the deepness property. Consider the *bembé* rhythm and its interval-content histogram shown in Figure 28.2.

There are seven different onsets each of which may be deleted to obtain a six-onset rhythm. Each of the seven onsets has a spectrum of six distances determined by the other six onsets. To ensure that the rhythm remains deep when an onset is deleted, the onset to be deleted must contain a spectrum with exactly one instance of every distance present in the histogram of the rhythm before the onset is deleted. In that way, the height of each column of the histogram will be reduced by one unit, thus preserving the deepness property. The spectrum for the onset at pulse 11 is pictured in Figure 28.2 (left). Note that only the onsets at pulses 5 and 11 have spectra that contain all the distances, and therefore, only one of these two onsets may be deleted. If we remove the onset at pulse 11 we obtain the rhythm shown in the middle with its resulting histogram on the right. This rhythm is also a bell timeline used by the Yoruba people of Nigeria. Of the six onsets, only the onset at pulse five has all five distances in its spectrum. Therefore, only this onset may be deleted next if deepness is to be preserved, and doing so yields the fume–fume rhythm.

Another characteristic property of rhythms is the *all-interval property*. A rhythm with this property contains all the possible duration intervals, and we may ask whether a shelling exists that maintains this property. In addition, if a rhythm admits the deletion of some onsets that preserves the all-interval property, does it admit a sequence of single deletions that will yield a *minimum all-interval rhythm*? A rhythm is a minimum all-interval rhythm if it has the minimum number of onsets such that deleting any of its onsets results in a rhythm without the all-interval property. Figure 28.4 contains two minimum all-interval rhythms in a 12-pulse time span.

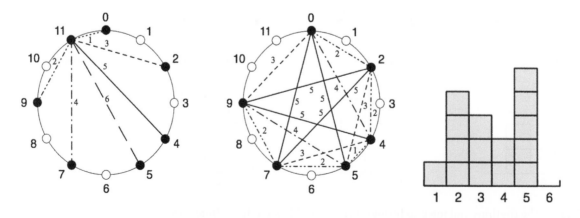

FIGURE 28.2 Deleting the onset at pulse 11 (left) from the *bembé* rhythm yields a Yoruba bell timeline (center) with a deep histogram (right).

To illustrate the process of all-interval shelling consider the *bembé* rhythm (isomorphic to the diatonic scale in the pitch domain). Since it is a deep rhythm, it obviously also has the all-interval property. Can this rhythm be shelled to the two minimum all-interval bracelets given by durational patterns [1-3-2-6] and [1-2-4-5]? That it can for one and not for the other is illustrated in Figure 28.3, which shows the seven initial onsets of the *bembé* with black-filled pulses. Since we must retain the interval of duration one, we must keep either the interval between pulses 0 and 11 or pulses 4 and 5. Because the rhythm is symmetric about the line through pulses 2 and 8, we need to be only concerned with one of these choices. Therefore, let us keep the onsets at pulses 0 and 11 and refer to the left diagram in Figure 28.3. Since we need an interval of duration six, the onset at pulse five must be kept. Thus, the only remaining choice for the fourth onset is at pulse three, which is invalid since it is not contained in the *bembé*.

The only other possible choice for this bracelet, which keeps pulses 0 and 11, is shown in the middle diagram, but here onsets are needed at pulses 6 and 8, neither of which is contained in the *bembé*. Therefore, this minimum all-interval bracelet is not reachable from the *bembé* via single-onset deletion shelling operations.[10]

Consider then the second all-interval bracelet contained in the *bembé* shown in the right diagram of Figure 28.4. This rhythm is a shelling descendant of the *bembé*, and it may be reached via several shelling sequences. One sequence is as follows. First the onsets at pulses 0 or 4 may be deleted. Then the onset at pulse 2 may be dropped, followed by the onset at pulse 9. Another sequence first deletes the onset at pulse 2, and then either 0 followed by 9, or 4 followed by 7.

The two minimum all-interval subsets of the *bembé* rhythm shown in Figure 28.3 are *perfect* in the sense that each interonset interval is used exactly once. However, there may exist shelling sequences that lead to minimum all-interval descendants that are not perfect, such as the two shown in Figure 28.4, where the onsets that may be deleted are shaded gray. A symmetric pentagon (left) is obtained by deleting the onsets at pulses 7 and 9. That this is a minimum may be ascertained by inspection. If either onset 11 or 5 is dropped, then the interval of duration six is lost. If either onset 0 or 4 is deleted, the interval of duration four is lost, and if the onset

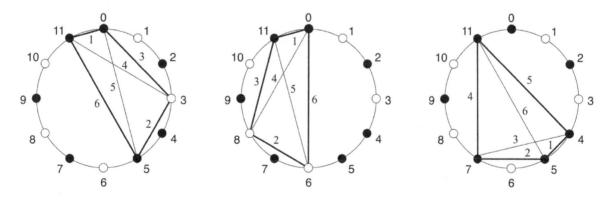

FIGURE 28.3   The two *minimum* all-interval rhythm bracelets.

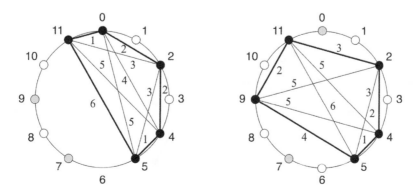

FIGURE 28.4   Two minimum all-interval shelling descendants of the *bembé*.

at pulse 2 is removed, the interval of duration three is no longer present. An asymmetric pentagon (right) is obtained when the onsets at pulses 0 and 7 are dropped. It may also be verified by inspection that this rhythm is all-interval minimal.

The process of shelling rhythms is a natural technique that is routinely applied by drummers during their solo improvisations, and by drummers that support the soloist, to add spice by introducing rhythmic variation. A characteristic device used in sub-Saharan African music is the shelling of rhythms by the deletion of one or more onsets, to challenge and increase the attention of dancers and listeners. As an example, consider the *bembé* timeline (or "standard pattern") in Figure 28.5 (upper left). The ethnomusicologist and percussionist J. M. Chernoff documented the seven shellings of *bembé*, shown in the figure, that he heard while traveling in Africa.[11] He characterizes this type of shelling as "omission" of onsets that increase syncopation. The seven shellings are shown in order of decreasing cardinalities of onsets. These shellings have several interesting properties worthy of note. Recall that one measure of syncopation (or complexity) is the rhythmic oddity property discussed in Chapter 15. The *bembé* rhythm does not have this property since it contains two antipodal onsets at pulses 5 and 11, that partition the cycle into two equal half cycles. By contrast, all seven

shellings do have the rhythmic oddity property, and thus these shellings create greater syncopation than the *bembé*, supporting Chernoff's observation. We may thus characterize these shellings as *rhythmic-oddity-inducing* shellings. Note also that all seven shellings preserve the last of the four regular downbeats at pulse nine, and the first five shellings preserve both the first and last downbeats at pulses 0 and 9, respectively (the only two downbeats of the *bembé*). Therefore, these shellings are also *downbeat-preserving* shellings. Finally, observe that the four shellings that contain four and five onsets are not deep, whereas the remaining three rhythms are deep. The two-onset rhythm and the rhythm with intervals [2-7-3] are deep by default. Therefore, although good timelines have a tendency to be deep (as is the *bembé* in this example), the deepness property need not be preserved when salient shellings of timelines are introduced during a performance.

Shelling is also used as a composition tool in contemporary music. Indeed, the minimalist composer Steve Reich applied this technique, which he called *rhythmic construction*, for the first time, in his 1973 composition titled *Drumming*, and described it as "the process of gradually substituting beats for rests (or rests for beats)."[12] From the empirical examination of preferred rhythms used in several cultural traditions,[13] it follows that deepness is a mathematical property that appears to

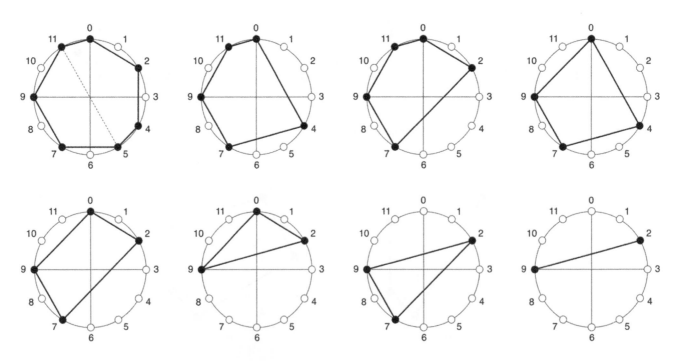

FIGURE 28.5 The seven shellings of the *bembé* documented by J. M. Chernoff (1991).

reflect this cultural preference. Therefore, the process of shelling rhythms while maintaining this property provides an algorithm that may find application to the automatic generation of "good" rhythms. Whether deepness by itself is a property that has a psychologically robust reality is yet to be determined. However, Reich did not preserve the deepness property in the shelling employed in *Drumming*. The first and last eight measures of this 75-min composition that consists of 100 measures are shown using polygon notation in Figure 28.6. The first eight are labeled (1)–(8), and the last eight (93)–(100). The polygon notation makes it easy to discover the structural rhythmic properties utilized in both the *construction* and *reduction* phases at the beginning and ending of the piece. The first observation is that, apart from the obvious fact that the ending is not the same as the beginning

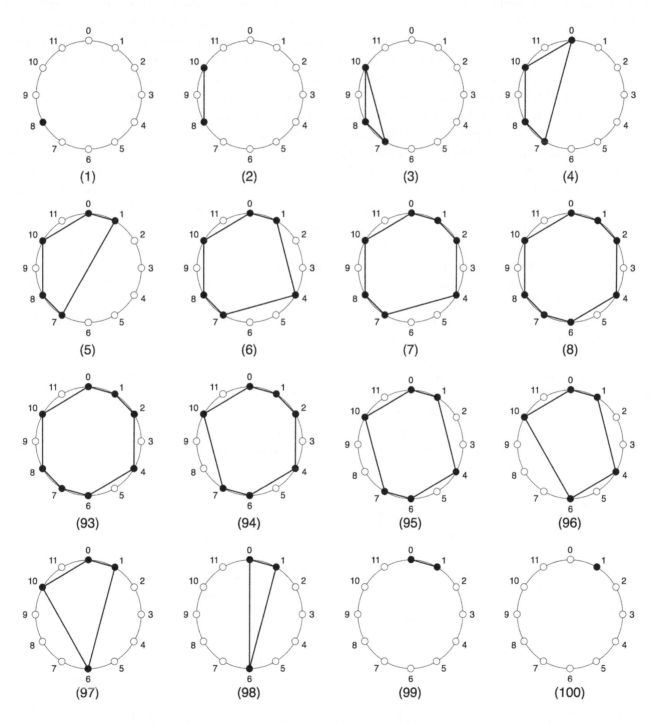

FIGURE 28.6 The *construction* and *reduction* method employed in Steve Reich's *Drumming*.

played backwards, the rhythms with 3, 4, and 5 onsets (triangles, quadrilaterals, and pentagons) all lie in a half cycle (semicircle) in the construction phase, but do not in the reduction phase. The rhythms with 5 and 6 onsets exhibit mirror symmetry (about the line through pulses 10 and 4) in the construction phase (measures five and six), but not in the reduction phase (measures 95 and 96). Nevertheless, measure 95 does have 180° rotational symmetry about the same line through pulses 4 and 10. Furthermore, the rhythms in measures 6 and 95 have an interesting symmetry relationship to each other. If the rhythm in measure 6 is decomposed into two half cycles by cutting it along the line through pulses 4 and 10, then reversing the order of the lower chain of inter-onset intervals converts the rhythm of measure 6 to that of measure 95. Referring to Figure 28.7, this is equivalent to performing a mirror symmetry operation of the shaded half cycle about the line through pulses 1 and 7.[14]

Some musicologists may find the mathematical mirror symmetry operation, defined by reflecting part of a rhythm, as being musically dubious. In the detailed and discerning review of the first edition of this book, Mark Gotham[15] considers such an operation to be particularly questionable, suggesting that "it would be easy to follow and credit an 'edit distance' approach concerning the number of changes necessary to transform one rhythm into the other. Here, that would be a question of moving either the two-onset pair by a distance of one step together, or one onset by a distance of two steps (between positions 6 and 8)." These operations posit two interesting ways of viewing the end result obtained with the standard edit distance transformation. Moving one onset by two positions is equivalent to applying a

swap operation twice.[16] On the other hand, moving the two-onset pair together is a novel extension of the set of allowable operations in the standard edit distance, which are (1) insertions, (2) deletions, and (3) substitutions of single onsets. Therefore, the edit distance transformation operating on the left rhythm substitutes the onset at position 8 with a silent pulse, and substitutes the silent pulse at position 6 with an onset. But this operation results in precisely a mirror symmetry reflection of *part* of the rhythm, namely the three-pulse portion that extends between positions 6 and 8, with respect to the axis of reflection determined by onsets 1 and 7. The mirror-symmetric reflection operating on part of a rhythm is not intended to capture some specific property of rhythm similarity or to model the way a composer might modulate rhythms, but rather to suggest a method to generate new rhythms from old, just as Steve Reich converted measure 6 to measure 95. It is illustrated here as an alternate manner of viewing the edit distance transformation, and to make yet another connection with the field of mathematics.

Reflecting a portion of a rhythm also serves to connect several well-known rhythms, and thus to understand them better. Two examples to illustrate the point are given later. Figure 28.8 (left) shows the 12-pulse ternary *fume–fume* timeline with interval structure [2-2-3-2-3]. A mirror reflection of the portion of this rhythm from positions 2 through 7, about the axis of symmetry that bisects intervals (4,5) and (10,11), yields the *rumba columbia* timeline [2-3-2-2-3] popular in rural Cuba, on the right.

For the second example consider the signature rhythm of electronic dance music (EDM) with interval structure

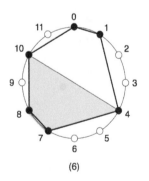
(6)

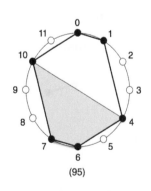
(95)

FIGURE 28.7 Transformation of the rhythm of measure 6 of *Drumming* to that of measure 95 by flipping over the shaded half cycle using mirror symmetry, about the reflection axis determined by pulses 1 and 7.

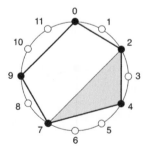

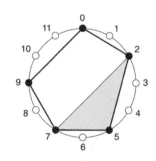

FIGURE 28.8 Transformation of the *fume–fume* timeline to the *rumba columbia* timeline by the mirror symmetry operation of the part between positions 2 and 7 about the axis of reflection determined by the midpoints of onset pair (4,5) and onset pair (10,11).

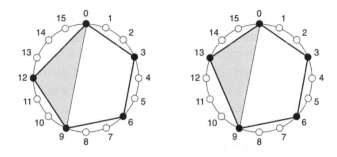

FIGURE 28.9 Transformation of the ubiquitous EDM time-line to a variant of the *bossa-nova* timeline by the mirror symmetry operation of the part between positions 9 and 0 about the axis of reflection defined by the midpoints of onset pair (12,13) and onset pair (4,5).

[3-3-3-3-4] shown in Figure 28.9 (left).[17] Performing a reflection operation of the part of the rhythm from positions 9 to 0, about an axis of symmetry that bisects onset pairs (12,13) and (4,5), yields a variant of the *bossa-nova* [3-3-3-4-3] (right).

The rhythm in measure 95 of *Drumming*, in Figure 28.7 (right), belongs to a special class of polygons known as *zonogons*. Zonogons have the property that for every edge in the polygon there is an opposite edge of

equal length parallel to it. In the polygon of measure 95, edges (0,1), (1,4), and (4,6) are parallel to, and of equal length as, edges (6,7), (7,10), and (10,0), respectively. It is therefore appropriate to designate such rhythms as zonogon rhythms. Regular rhythms with an even number of pulses greater than or equal to four are obvious members of the zonogon family. However, there are many zonogon rhythms that are irregular (not regular), and Figure 28.10 shows six examples. In the top left is the 16-pulse Persian rhythm *al-thaqil al-thani* with interval structure [3-3-2-3-3-2] equivalent to the tresillo concatenated with itself to span a 16-pulse cycle.[18] Clearly, any rhythm with *n* pulses concatenated with itself will produce a zonogon rhythm with 2*n* pulses. In the top center is the 12-pulse rhythmic ostinato used by Steve Reich in measure No. 8 of *Drumming*, with interval structure [1-1-2-2-1-1-2-2]. In the top right we have the 12-pulse high-pitched *zamacueca* rhythm from the Afro-Peruvian *landó* music, with interval structure [3-2-1-3-2-1].[19] Note that this rhythm is a rotation of the rhythm in measure 95 of Steve Reich's *Drumming*, by one pulse duration, in a counterclockwise direction. In the bottom left is the 12-pulse *kéné foli* rhythm of Guinea with interval structure [2-1-2-1-2-1-2-1].[20] In the

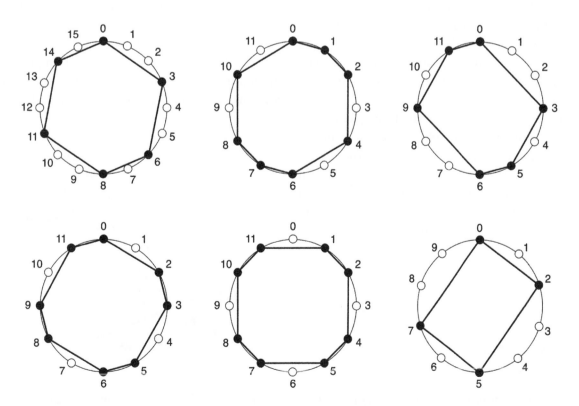

FIGURE 28.10 Further examples of irregular zonogon rhythms.

bottom center is the 12-pulse rhythmic ostinato played by bell No. 4 in the *Gamamla*, which is a rotation of the *kéné foli* rhythm.[21] Of course any rotation or mirror image reflection of a zonogon rhythm is trivially also a zonogon rhythm.

Finally, in the bottom right we have the ten-pulse rectangular *khalif-al-ramal* rhythm from the thirteenth-century manuscript by Safi al-Din, with interval structure [2-3-2-3].[22] All rectangular rhythms are obviously also zonogon rhythms.

Zonogon rhythms have several noteworthy properties. One property is that they are periodic, meaning that they consist of repetitions of portions of the rhythm. The top-right rhythm in Figure 28.10 repeats the phrase [3-2-1] twice, and thus has period 2, or twofold rotational symmetry. On the other hand, the bottom-left rhythm repeats the phrase [2-1] four times, and has period 4, or fourfold rotational symmetry.

A second property of zonogons is that they have *central symmetry*.[23] This means that a zonogon Z contains a point c (the center) such that for every point x in Z there exists another point x' in Z such that c is the midpoint of the line segment connecting x and x', as illustrated in Figure 28.11. For the case of a zonogon rhythm in circular notation, its center is the center of the circle, implying that for every onset of the rhythm there exists an antipodal onset diametrically apart. Therefore, zonogon rhythms have no rhythmic oddity at all, since they contain the maximum possible number of equal bipartitions. A zonogon rhythm with $2n$ onsets has n bipartitions.

We have seen in Chapter 25 that complementary rhythms play an important role in the music of the African Diaspora. Another interesting property of zonogon rhythms is that the complement of a zonogon rhythm, if it exists, is also a zonogon rhythm. A rhythm with n pulses that contains n onsets has no possibility of having a complementary rhythm. Therefore, assume

that the number of onsets is at most $(n-4)$. To see this, refer to Figure 28.12, and assume Z is a zonogon rhythm in a span on n pulses with $n \geq 8$. Consider a sequence of adjacent onsets of Z, say, s, $s+1$, and $s+2$, such that both pulses $s-1$ and $s+3$ are silent pulses. The number three here is fixed just for clarity, but the argument holds for a string of any length flanked by silent pulses on both sides. This means that the complementary rhythm must have onsets at pulses $s-1$ and $s+3$, and its polygon must have an edge connecting $s-1$ and $s+3$. Since Z is a zonogon it has central symmetry with respect to the center C. Therefore, Z must have onsets at the antipodal positions t, $t+1$, and $t+2$, such that both pulses $t-1$ and $t+3$ are silent pulses. This means that the complementary rhythm of Z must have onsets at pulses $t-1$ and $t+3$, and its polygon must have an edge connecting $t-1$ and $t+3$. But pulse $t-1$ is the antipode of pulse $s-1$, and pulse $t+3$ is the antipode of $s+3$. Therefore, the edge connecting $t-1$ and $t+3$ has the same length as the edge connecting $s-1$ and $s+3$, and is parallel to it. Since this argument holds for all contiguous sequences of onsets flanked by silent pulses in Z, it follows that the complement of Z is also a zonogon.

Returning to Steve Reich's *Drumming*, also noteworthy is the rhythm used in measure 97, which has a perfectly flat interonset interval histogram: it contains exactly one instance of every interval ranging from one to six (this is the infamous all-interval tetrachord in pitch theory). It is also worth pointing out that the rhythm used in measure 6 is the union of two regular rhythms: one with three onsets at pulses 0, 4, and 8, and

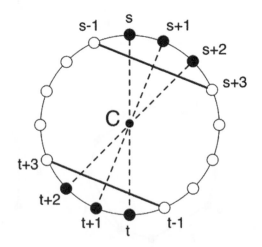

FIGURE 28.12 The complement of a zonogon rhythm is also a zonogon rhythm.

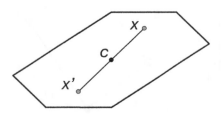

FIGURE 28.11 Illustrating the central symmetry property of zonogons.

one with four onsets at pulses 1, 4, 7, and 10. This combination of rhythms is used frequently in the music of sub-Saharan Africa.

The geometric analysis of the shelling technique that Steve Reich used in the construction and reduction phases of the beginning and ending of his piece *Drumming* reveals that the main property that is preserved during this process may be characterized by the notion of symmetry. This symmetry is sometimes perfect mirror symmetry as in measures 5, 6, 8, and 93, sometimes rotational symmetry as in measure 95, and at other times near-mirror symmetry.[24] Thus we may characterize Reich's construction and reduction techniques as being dominated by *symmetry-preserving rhythmic shelling*.

A convenient and useful representation of shelling rhythms is via graphs, in which the nodes (vertices) of the graphs represent rhythms, and pairs of nodes are connected with edges provided that their corresponding rhythms share some specified property. As a simple example, consider all the 16 rhythms consisting of four pulses.[25] If an edge is connected between every pair of nodes corresponding to pairs of rhythms that have the property that they differ from each other by only one attack then the graph obtained is the hypercube shown in Figure 28.13. In this figure, all the rhythms that have the property of deepness are shaded black for the three-attack deep rhythms, and gray for the elementary (trivially deep) two-attack rhythms. Furthermore, the edges connecting pairs of deep rhythms are highlighted in bold lines. Thus, the shaded vertices and bold edges determine the *deep-rhythm-subgraph* of the hypercube.

Note that this subgraph is *connected*. This means that with shelling we can traverse from any one deep rhythm to any other deep rhythm by a sequence of insertions or deletions of onsets, while preserving the deepness property of all rhythms traversed along the way.

Just as we constructed a graph for the deepness property shown in Figure 28.12, a similar subgraph of the hypercube may be constructed for any other property of a rhythm, by connecting the two vertices of the corresponding rhythms, provided they share that property. The representation of rhythms using graphs, as described here, was inspired by similar ideas that were previously explored by music theorists working in the pitch domain.[26] In 1964 Allen Forte defined what he called a *set-complex* $K(T)$ for chords. For any chord T, another chord S is a member of the set-complex $K(T)$, provided that S is a subset or superset of either T or its complement. Furthermore, S must have a number of elements different from T or its complement, and the number k of elements of both S and T is bounded by $3 \leq k \leq 9$.[27] With such a loose definition, Forte discovered that these set complexes were too large to be useful, and subsequently refined his definition to a subcomplex that he called $Kh(T)$. To belong to $Kh(T)$, a chord must be included in *both* T and its complement. In 1974 Eric Regener went further in the direction outlined earlier. First, he defined the general inclusion graph (lattice) of all pitch class sets. Every possible chord corresponds to a node in the graph. Two nodes, corresponding to chords T and S in this graph, are connected with an edge, if T contains S, and T has exactly one more element than S.[28]

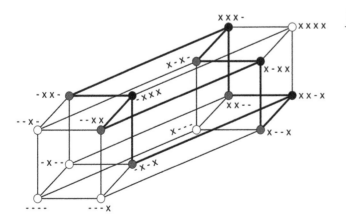

FIGURE 28.13 The hypercube representation of all four-pulse rhythms that differ by at most one attack. Deep rhythms are shaded and connected with bold edges.

NOTES

1 It is ironic that a non-African-American such as Elvis Presley should modify the music of the African-American blues singers by incorporating more syncopated African rhythms into their blues songs. Elvis appropriated several other songs from black Americans, that went on to become hits, such as *Mystery Train*, by Little Junior Parker, and *That's All Right*, by Arthur "Big Boy" Crudup. See Fryer, P., (1998), p. 2.

2 In the case of a two-onset rhythm, such as the snare drum backbeat in the rightmost diagram of Figure 28.1, with onsets on pulses two and six, the duration interval of four units may be counted once or twice depending on the context. In this case, given two onsets a and b that determine two durations $[ab]$ and $[ba]$, we take the smallest of the two as the single duration determined by the pair of onsets. Thus if $[ab] = [ba]$, as in the snare

drum example, we count just one of the intervals. Therefore, a rhythm with only two onsets is deep by default.

3 Rey, M., (2006), p. 192.

4 Wade, B. C., (2004), p. 61.

5 According to Brewer, R., (1999), p. 307 (Example 10), the clapping pattern in Elvis Presley's *"Hound Dog,"* is the five-onset *cinquillo*. However, in the recording the clap on pulse two cannot be heard as it is drowned out by the attack on the snare drum.

6 Regarding the operation of *shelling*, it is useful to distinguish between restricting the deletions or insertions to single onsets (one at a time), and multiple onsets deleted or inserted simultaneously.

7 Rey, M., (2006), p. 192.

8 Middleton, R., (1983), p. 252.

9 Mauleón, R., (1997), p. 17.

10 Music theorists may find that this example follows trivially from the generative theory of diatonic scales. A generated scale may be constructed by repeatedly adding a constant integer around the pitch circle. For further details, see Gamer, C., (1967) and Johnson, T. A., (2003). p. 83. However, this book is not aimed primarily at music theorists, and it is hoped that the general idea expounded in this chapter, of shelling with respect to *any* property (either in the pitch or time domain), will suggest new avenues for research in music theory.

11 Chernoff, J. M., (1991), p. 1100.

12 Reich, S., (2002), p. 64.

13 See the list of Euclidean rhythms in Toussaint, G. T., (2005c).

14 In the computational geometry literature, such a restructuring operation performed on a portion of a polygon is called a *flip-turn*. See Toussaint, G. T., (2005e).

15 Gotham, M., (2013).

16 Toussaint, G. T., (2006b).

17 Butler, M. J., (2001).

18 Wright, O., (1978).

19 Miranda-Medina, J. F. & Tro, J., (2014), p. 217 and Feldman, H. C., (2005), p. 215.

20 Konaté, F. & Ott, T., (2000), p. 89.

21 Klöwer, T., (1997), p. 175.

22 Wright, O., (1978).

23 Steinhardt, P. J. & Ostlund, S., (1987), p. 509. See also Craizer, M., Teixeira, R. C., & da Silva, M. A. H. B., (2013).

24 Traditionally mirror-symmetry, or symmetry in general, is viewed in a binary manner: an object either has or does not have symmetry. However, measures exist for determining the degree of symmetry possessed by an object such as rhythm or a visual pattern. For further details, see Toussaint, G. T., Onea, N., & Vuong, Q., (2015), Toussaint, G. T. & Beltran, J. F., (2013), Zabrodsky, H., Peleg, S., & Avnir, D., (1992), and Eades, P., (1988). The concept of near-symmetry is also discussed in the context of voice-leading in Tymoczko, D., (2011). See Schutz, M., (2008) regarding what musicians need to know about vision.

25 The ideas described apply to rhythms with any number of pulses but the drawings for more than four pulses become rather cumbersome.

26 I am indebted to Dmitri Tymoczko for bringing this connection to my attention.

27 Forte, A., (1964).

28 Regener, R., (1974), pp. 208–209.

CHAPTER 29

Phase Rhythms

The "Good," the "Bad," and the "Ugly"

IMAGINE TWO WHEELS ROTATING independently but simultaneously in a clockwise direction at the same speed, each with four spokes evenly spaced apart along the rims, such that the spokes of the two wheels are perfectly aligned with each other, as illustrated in Figure 29.1 (left), where the locations of the four spokes in each wheel are indicated by four black dots. Imagine further that a sound is made by an instrument every time one of the spokes of either wheel crosses the "north pole" location of the wheel. Such a mechanism will generate the simplest kind of rhythm, *i.e.,* an isochronous regular pulsation of sound onsets, since the spokes of both wheels will generate sounds in unison. Now imagine that at some time t_0, the inner (smaller) wheel starts rotating at a slightly faster speed than the outer (larger) wheel. After some interval of time at say t_1, the sounds made by the inner wheel will occur slightly after the sounds made by the larger wheel, resulting in a slight echo, or what drummers call a *flam* (illustrated in the middle diagram of Figure 29.1). Still later, there will come a time, at say t_2, at which the sounds of the smaller wheel will occur at the midpoints of the sounds made by

the larger wheel (illustrated in the rightmost diagram of Figure 29.1), generating a momentary isochronous pulsation with twice as many beats per cycle as the regular pulsation at the start time t_0. Eventually, the two wheels will sound in unison again and generate the same regular isochronous pulsation that occurred at time t_0. From the start to the end of this cycle, this mathematical mechanism (or algorithm) will have generated an automated musical (rhythmic) composition belonging to the genre referred to as *phase music*.

Phase music was made popular by the composer Steve Reich, who created a series of phase music pieces for voice, piano and violin.[1] Needless to say, the performance of such phase pieces is much easier when implemented by two tape recorders (a method Reich favored in his early pieces) than carried out by live musicians. This is due in large part to the highly demanding *continuous* increasing and decreasing separation between the sounds that have to be made by the performers. In 1972, Reich composed a strictly rhythmic phase piece, which has become a classic, titled *Clapping Music*,[2] which consists of the 12-pulse rhythmic pattern [x x x . x x . x . x x .] played by two performers, each clapping the rhythm with their hands. One performer repeats this pattern unchangingly throughout, while the other shifts the pattern by one unit of time after a fixed number of repetitions. This shifting continues until the performers are once again playing in unison, signaling the end of the piece. Russell Hartenberger, who performed the piece with Reich, recounts that originally *Clapping Music* was meant to be performed in the style of

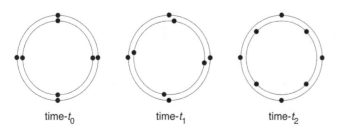

FIGURE 29.1   A *continuous phase* rhythm.

continuous phasing.[3] When rehearsing the piece in the continuous phasing mode proved difficult, Reich asked Hartenberger to just repeat the rhythmic pattern a fixed number of times and then jump in a *discrete* manner to the pattern rotated by a specified interval. They found the result worked well, and this is how *Clapping Music* has been performed ever since.

One of the most common activities of human beings of all ages evident all over the world consists of striking the hands together to produce a sound—*hand clapping*. In both traditional and modern music, hand clapping is used predominantly as a percussion instrument, with primary focus on the rhythmic patterns played, rather than the properties of the sound of the clap. For example, A. M. Jones[4] and, more recently, Kofi Agawu[5] provide detailed accounts of the roles of hand-clapping rhythms used in the traditional music of sub-Saharan Africa.

Scientists have also studied various aspects of hand clapping. In the earliest studies of hand clapping, psychologists explored the perception of hand-clapping sounds produced by different hand configurations that generated a variety of distinct timbres.[6] In perception experiments it was found that listeners were able to accurately identify distinct hand configurations from the resulting sounds. Bayesian and hidden Markov models have been used for person identification based on the timbre properties of their hand-clapping sounds.[7] Some work has also been done on electronically synthesizing realistic hand-clapping sounds.[8]

The pattern and the structure of *Clapping Music* are illustrated using 12-pulse circular notation in Figure 29.2. Each clap (sounded pulse) is denoted by a black-filled circle, and each rest (silent pulse) by a white-filled circle. The leftmost diagram in the figure shows Reich's pattern, which is played continuously throughout the piece by one of the performers (the outer circle). At the start, both players clap this pattern a predetermined fixed number of times (Version-0).[9] After the first rotation, the second player starts the pattern one pulse ahead as in the middle diagram (Version-1). Figure 29.2 shows only the first two rotations (modules) of the pattern, but this process continues until the inner circle pattern is again perfectly aligned with the outer circle (Version-13 equal to Version-0).

On the published sheet music, Steve Reich writes the following notable performance instruction: "The choice of a particular clapping sound, *i.e.*, with cupped or flat hands, is left up to the performers. Whichever timbre is chosen, both performers should try and get the same one so that their two parts will blend to produce one overall resulting pattern."[10] These instructions raise two questions: (1) is it possible for two performers to produce claps that have the exactly the same timbre? and (2) if they could, how would the resulting piece sound? That it is difficult to produce hand claps with exactly the same timbre is suggested by Reich's phrase "should try and get the same one." On the scientific side, it has been observed, by means of examples that produce the psychological phenomenon called *streaming*, that the human auditory system is extremely sensitive to the minutest variations in timbre.[11] Streaming is the phenomenon of perceiving two separate rhythms as opposed to one, a topic that will be covered in detail in Chapter 30. Thus, it may well be practically impossible for humans to perform a sequence of hand claps that have perceptibly the same timbre. However, for computers this is not a problem. Therefore, computers permit us to perform Steve Reich's piece by following his instructions to the letter.

Upon my relocation from Harvard University to New York University Abu Dhabi (United Arab Emirates) in August 2011, I was eager to test Reich's instructions

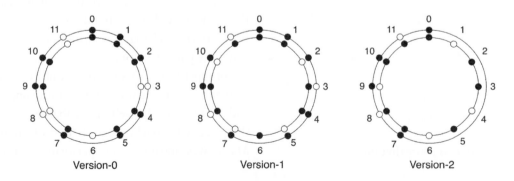

FIGURE 29.2 The first three modules of the *discrete* phase rhythm of *Clapping Music*.

with precision using the aid of a computer. I therefore designed a software program called *Clapping Music Juke Box*, to easily change the various parameters of Reich's composition of *Clapping Music*, for the purpose of exploring how different rhythmic patterns would sound in the context of discrete phasing, compared to Reich's original pattern, following his specific performance instructions.[12] I was also curious to compare discrete phase pieces that used Reich's pattern [x x x . x x . x . x x .] with phase pieces that used the *Yoruba* bell pattern (a rotation of the *bembé* bell pattern[13]) [x . x . x x . x . x x .], as well as the pattern obtained by Joel Haack following the application of a series of mathematical combinatorial constraints [x x x x . x . x x . x .].[14] I hoped that with this software, I would be able to listen to numerous such compositions without having to find a partner and without the need for the time-consuming training sessions that such pieces demand, with endurance being one of the main difficulties. In addition, the software would allow me to substitute hand-clapping sounds with other musical tones, to explore a range of further interesting musical possibilities.

The opening page of the graphical user interface (GUI) of *Clapping Music Juke Box* is shown in Figure 29.3. The user can specify the number of segments in the cycle (pulses that are sounded or silent) that comprises the rhythmic pattern, the radius of the circle for the circular box notation of the cyclic rhythmic pattern, the tempo (in milliseconds per beat), and the number of revolutions that each module is performed before advancing to the subsequent module. The default values of 12 pulses, 125 mm, 210 ms, and 4 revolutions, can be selected by clicking the "default" button. The user can also select among ten claps of differing timbres for each of the two

"performers" labeled "Clap 1" and "Clap 2." The 20 clap sounds were recorded from real human hand claps rendered with a variety of different hand positions ranging from flat palms to cupped hands.

Once the "Go" button is clicked, the second GUI page appears. Figure 29.4 shows what the second page looks like for the default settings and the selected claps for each "performer" (claps 4 and 9 in this case), once the 12 circular boxes, which are initially empty, are selected. Clicking the "Play" button starts the execution of the performance. In the actual GUI, the dots of the outer and inner rhythms are colored red and green so as to more clearly follow them visually if desired. In addition, the vertical line originating at the center of the circle rotates clockwise at the chosen tempo (much like the *seconds* hand in a clock) to help the listener–viewer follow the rendition of the score.

The first discovery I made using this software was that the human ear is much more sensitive to the slightest variation in timbre than I anticipated. This observation became evident when I selected two claps (one for each "performer") that sounded quite similar to each other when heard in isolation. The resulting performance of *Clapping Music*, however, exhibited a pronounced streaming effect, yielding a clear perception of two distinct rhythmic patterns being played simultaneously, with no perception of the *resultant* pattern, one of the intentions spelled out in Reich's instructions. The resulting performances were not satisfactory from a musical aesthetic point of view. This effect was probably exacerbated by the fact that the claps used for each "performer" consisted of one and the same sound, thus rendering the two performers as perfect duplicators of

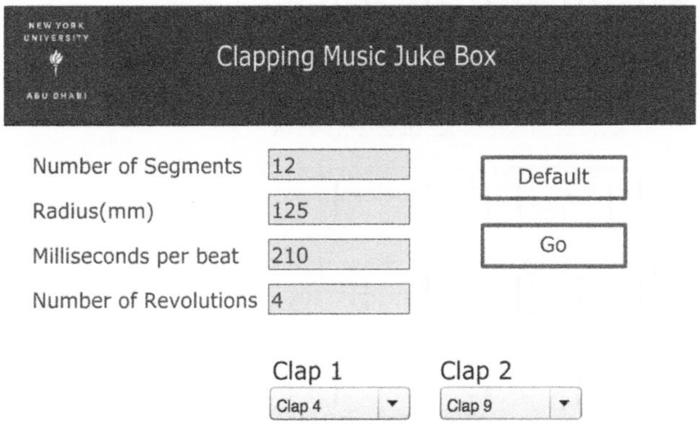

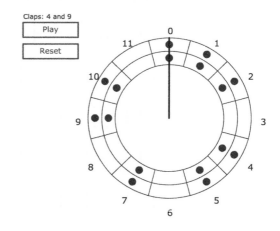

FIGURE 29.3 The GUI of the *Clapping Music Juke Box* software: The parameter and clap sound selection page.

FIGURE 29.4 The GUI of the *Clapping Music Juke Box* software: The performance animation page.

their respective claps. These parameters produced renditions of *Clapping Music* with pronounced *streaming* effects that I call the "Bad" performance.

To test the extreme situation suggested by Reich's instructions, to make the claps of both performers as close as possible to each other, I selected the same clap for both performers, whether they clapped in unison (double claps) or not (single claps). In other words, every time one or the other performer clapped in isolation, or both clapped together, the same clap sound was produced. This scenario does not model exactly what would be expected in practice, since two isotonic claps played in unison should sound somewhat louder than one clap alone (more on this feature later). Figure 29.5 shows the resultant pattern after the first rotation. As the reader may easily imagine, hearing a regular pulsation of 12 isotonic and isochronous claps repeated 12 times sounds downright boring, to the point of being unbearably ridiculous. I consider as "Ugly" those performances that

used parameters that tend to yield isotonic isochronous renditions of *Clapping Music*.

The experiments described in the preceding lead to the intriguing and much more difficult question of which parameters are necessary and sufficient to yield the "Good" performances of *Clapping Music*. To answer this question, I was lucky that in 2013 William (Bill) Sethares of the University of Wisconsin in Madison (USA) visited me in Abu Dhabi to collaborate on this project for a period of 2 months. Recall that the *Clapping Music Juke Box* permits the exploration of only isotonic rhythms, and a very limited version of them at that, since the sounds of each individual performer were held constant throughout the performance. In real life, however, humans do not perform perfect isochronous pulsations, even if they try. Therefore, the two global parameters that Bill and I explored were isotonality (or expressive timbre for the case of claps) and isochronicity (or expressive timing[15]). The 12 modules in the performance of *Clapping Music* are depicted graphically in Figure 29.6, where bold vertical line segments indicate double claps. Empty boxes indicate rests or silent pulses, and the remaining boxes indicate the single claps of each performer.

All the single and double claps were extracted from seven performances of *Clapping Music*, which provided a large database of audio sounds. Analysis of the single claps revealed that performers were not constant with respect to expressive timbre. Furthermore, it was possible for human subjects to rank the timbre of the

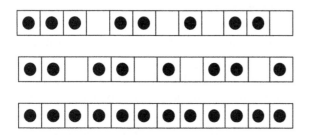

FIGURE 29.5 Reich's pattern (top), its first rotation (center), and the resultant pattern (bottom).

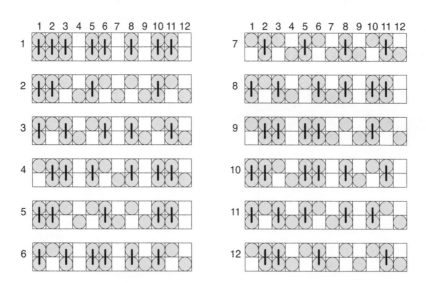

FIGURE 29.6 Graphical score of the 12 modules of Clapping Music. (With permission from Sethares, W. A. & Toussaint, G. T. 2014. *Journal of New Music Research*, 44(1):11–24.)

claps by "tone" (high or low) so as to correlate highly with objective features of acoustic signals. The interclap intervals were also not held constant to varying degrees by the seven different pairs of performers. This feature of the performance for the double claps had an effect on the resulting timbre of the double claps. Since the individual claps of the two performers producing double claps hardly ever sound *exactly* at the same time, the resulting timbre of the double claps depends on their microtiming delay between the two individual claps. As the delay between the claps increases, the perceived pitch of the timbre of the double claps decreases.[16]

Clapping Music involves a simple cooperative "sensorimotor syncopation" timing task that involves a great deal of clapping by one performer at midpoints between claps of another performer. This is an instance of producing a rhythm that stays off the beat.[17] A standard exercise for hand-clapping training in the flamenco music of southern Spain involves two persons clapping an isochronous sequence of claps, such that one person claps in the middle of the other person's interclap intervals. The clapping tempo starts at a slow rate, and steadily increases until one person suddenly shifts to clap in unison with the other, signaling failure. The purpose of the exercise is to delay this shift as long as possible and at a tempo as fast as possible. At a fast tempo, before the shift occurs, the isochrony of the clapping is unstable, and it becomes difficult for one person to clap exactly at the midpoints of the interclap intervals of the other person. This observation suggests that, in a performance of *Clapping Music*, the modules in which this phenomenon is present would be more difficult to play isochronously and would exhibit a higher deviation in interclap intervals. To test this hypothesis, we computed for each performer and each module the average deviation between the positions of the claps as performed, and as notated in the score. Figure 29.6 shows the 12 modules in *Clapping Music*, with vertical bars indicating the location of double claps. Figure 29.7 shows the histogram of the number of double claps in each module. A smaller number of double claps in a module indicates that the performers will play more claps in each other's empty spaces, and thus suggests the hypothesis that the module will be more difficult to play and will exhibit a greater deviation. The first module with eight doubles should be the easiest to perform, whereas the second, seventh, and twelfth modules, each with only four doubles should be the most difficult.[18]

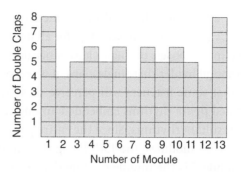

FIGURE 29.7   Histogram of the number of double claps in *Clapping Music*. (With permission from Sethares, W. A. & Toussaint, G. T. 2014. *Journal of New Music Research*, 44(1):11–24.)

Each module *i* contains a combination of single claps s_i, doubles d_i, and rests r_i (where neither performer claps). By design, the number of rests plus the number of single claps is equal to 12 minus the number of doubles within each module. Call this count $c_i = 12 - d_i = s_i + r_i$, the number of time slots where doubles do not occur. This is shown in Figure 29.8 for all 13 modules. The heights of the histogram columns provide an ordered sequence of numbers representing the expected difficulty of performing each module. Let us denote this sequence as ED = (4, 8, 7, 6, 7, 6, 8, 6, 7, 6, 7, 8, 4). For each module and each of the seven performances, the average deviation from the score was computed. This average curve also provides us with a sequence of numbers (in milliseconds) denoted by AD = (4.76, 9.18, 7.62, 7.05, 7.69, 5.89, 8.53, 7.69, 7.98, 6.80, 8.39, 10.00, 3.56). The Spearman rank correlation test for the sequences ED and AD is $r = 0.93$ with $p < 0.000002$, supporting the hypothesis.

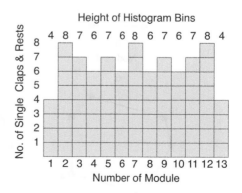

FIGURE 29.8   Histogram of the number of single claps and rests in *Clapping Music*. (With permission from Sethares, W. A. & Toussaint, G. T. 2014. *Journal of New Music Research*, 44(1):11–24.)

Figure 29.9 shows the rhythms extracted from each module, by including an onset (clap) for all double claps, and deleting all onsets of single claps. Since double claps tend to sound louder that single claps, it is reasonable to expect that some listeners will perceive these rhythms as accented *figure* rhythms embedded among the *background* resultant rhythms. Since the entire performance consists of 13 modules, the 13th being the same as the first, module 7 is the middle module that separates the first six modules form the last six. Note that module 7 is a regular pulsation of four beats per cycle. Also, modules 2–6 are rotations of modules 8–12. Also noteworthy is that some of these rhythms are well-known timelines (bell patterns). Module 11 is the *fume–fume* [2-2-3-2-3] (see Chapter 12). Module 3 is a rhythm found

in Northern Zimbabwe called the *bemba* [2-3-2-3-2] (see Chapter 12). Module 6 is the *double paradiddle* [2-2-1-2-2-3] drum rhythm, and is a popular *Yoruba* bell timeline from West Africa. It is also a *Kpelle* rhythm of Liberia (see Chapters 28 and 31).

All the double-clap rhythms exhibit mirror-reflection symmetry. This is clearly visible from Figure 29.10 which depicts the first six modules in circular polygon notation, and the dashed lines indicate the axes of reflection mirror symmetry. Since the double-clap rhythms in modules 2–6 are rotations of the double-clap rhythms in modules 8–12, the rhythms in all modules possess reflection mirror symmetry. Note also that as the piece advances from module to module, the axes of mirror symmetry of each rhythm rotate

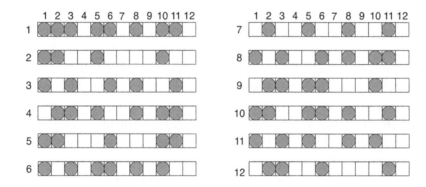

FIGURE 29.9   Graphical score of the double claps in the 12 modules of *Clapping Music*.

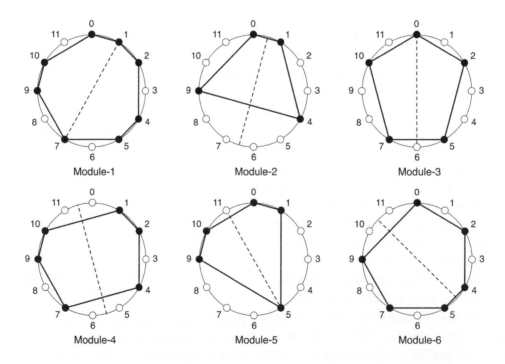

FIGURE 29.10   The double-clap rhythms in circular polygon notation and their axes of mirror reflection symmetry.

counterclockwise by half a pulse each time a module is advanced.

We are now equipped with enough material to provide some partial answers to the question of what makes a "Good" performance of *Clapping Music*. There are several timbre and timing parameters that contribute to making a performance of *Clapping Music* be a member of the "Good." Interestingly, if these parameters are implemented perfectly by a computer, the resulting performances will fall in the "Bad" or "Ugly" groups. It is rather the inability of human performers to implement the parameters perfectly that results in "Good" performances. The following four parameters are deemed important.

1. The timbre and volume of the claps of both performers should be as similar to each other as *humanly* possible. Differences in timbre, volume, or any other type of accent for that matter will tend to generate auditory streaming, especially if the accents are unvarying, resulting in "Bad" performances. Since attaining this goal is impossible, its approximate attainment results in facilitating the perception of resultant rhythm, tending to yield "Good" performances.

2. The timbre and volume of the claps of the individual performers should be as unvarying as *humanly* possible. Again, since attaining this goal will tend to facilitate auditory streaming if the timbre of the claps of two performers differ, and since its attainment is impossible, slight variations of an individual performer's claps will tend to prevent streaming and contribute to obtaining "Good" performances.

3. The performers should try to make their double claps to sound in unison as much as *humanly* possible. Accenting the double claps will facilitate the perception of double-clap rhythms shown in Figure 29.9, which will increase variability in the perceived resultant rhythms across the various modules and tend to make the performance "Good." Since this is humanly impossible, the resulting mismatch between the two claps creates a variety of different pitch effects. Even if humans were able to clap perfectly together, this would increase the volume of double claps. In either case the resulting accents would contribute to making the performances "Good" by adding expressive timing to the resultant rhythms.

4. Each performer should try to execute single claps that occur at the midpoints of the interclap intervals of the other performer, as accurately as *humanly* possible. Attaining this goal perfectly, however, tends to create "mechanically" produced rhythms which do not sound good. The unavoidable variability of this parameter tends to produce a desirable expressive timing effect. Indeed, expressive timing is a common phenomenon in rhythmic performance. A study by Goldberg (2015) provides evidence that expressive timing (or "timing variation") is concomitant with melodic grouping as well as with interactions between performers in a group and between the performers and audience.

Although the four parameters listed earlier provide some partial answers to the question of what makes a "Good" performance of *Clapping Music*, they do not provide an airtight characterization in terms of properties that are both necessary and sufficient for such a performance to be "good." This remains a tantalizing open problem and suggests further experiments to solve it.

The results described in this chapter suggest new avenues for research in ethnomusicology. For example, many of the clapping patterns in the African music described by Jones (1954a) contain two or three patterns played by different performers that create an overall resultant pattern. It would be interesting to study not only the rhythm of the clapping patterns themselves, as has been done in the past, but also the effects of timbre variations among the performers of clapping patterns, as well as the degree of concomitant auditory streaming evident in the performances of such traditional music. This research also illustrates the serendipitous discoveries that can occur as a result of using computers to assess the instructions of the composers. Since human beings are unable to make hand-clapping sounds that are uniformly identical in terms of timbre, and much less, identical to hand claps of another person, the computer inadvertently provides us with a rich new field of interdisciplinary research.

NOTES

1 Cohn, R., (1992a). See also the interview of Steve Reich by Hartenberger, R., (2016). Saltini, R. A., (1983) analyzes structural levels and the choice of beat-class sets in Steve Reich's phase shifting music.

2 Reich, S., (2002).
3 Hartenberger, R., (2016).
4 Jones, A. M., (1954ab).
5 Agawu, K., (2016).
6 Repp, B. H., (1987). See also Tousman, S. A., Pastore, R. E., & Schwartz, S., (1989).
7 Jylhä, A., Erkut, C., Şimşekli, U., & Cemgil, A. T., (2012). See also Jylhä, A. & Erkut, C., (2008).
8 Ahmad, W. & Kondoz, A. M., (2011). See also Peltola, L., Erkut, C., Cook, P. R., & Välimäki, V., (2007).
9 The original score instructs the performers to repeat each module 12 times, but more recently, performers reduce this number to eight repetitions or less.
10 Reich, S., (1980).
11 McAdams, S. & Bregman, A., (1979).
12 Toussaint, G. T., (2013b). The programming and implementation of the *Clapping Music Juke Box* design was carried out by New York University Abu Dhabi student Abishek Ramdas.
13 This pattern is also called the *atsiagbekor* African bell pattern (see Hartenberger, R., (2016), p. 158).
14 Refer to Chapter 17 for a detailed explanation of Joel Haack's combinatorial analysis of Reich's pattern by means of musically inspired mathematical constraints.
15 Goldberg, D., (2015).
16 Sethares, W. A. & Toussaint, G. T., (2014). Some of the text in this chapter is taken from this joint paper with permission from William Sethares and Taylor & Francis. doi: 10.1080/09298215.2014.935736.
17 Keller, P. E. & Repp, B. H., (2005), p. 292.
18 Other measures of difficulties encountered in performances of *Clapping Music* have also been studied. In particular, the difficulty of the transitions between modules has been investigated by Duffy, S. & Pearce, M., (2018). Transitions between modules depends on the dissimilarity of adjacent modules, and the number of double claps is one measure of module dissimilarity. The modules in *Clapping Music* have been compared also with other rhythm dissimilarity measures by Cameron, D., Potter, K., Wiggins, G., & Pearce, M. T., (2012).

CHAPTER 30

Phantom Rhythms

PICTURE YOURSELF STANDING IN FRONT of a large kettle drum holding a mallet, playing a three-onset, four-pulse rhythm with adjacent interonset durational pattern [2-2-4]. While playing this rhythm observe the motion of your mallet as it travels through space. In particular, focus on the distance between your mallet and the drum, as a function of time. Chances are that this distance function looks something like the curve shown in Figure 30.1, which illustrates two cycles of the rhythm, and shows the time points (pulses) at which your mallet strikes the instrument, namely zero, two, four, zero, two, four, and zero. The numbering here repeats every eight pulses because the period of this cyclic rhythm consists of eight pulses.

We have all heard the expression "what goes up must come down." A natural corollary to this expression is "what goes up and comes down must reach a point of maximum height somewhere in between the going up and the coming down."[1] The points in time at which your arm achieves this maximum height are likely to be the *midpoints* of the interonset intervals (IOIs), indicated in Figure 30.1 by vertical lines at the pulses numbered one, three, six, one, three, and six. These midpoints themselves may be interpreted as determining another (silent) rhythm lurking in the motor areas of the brain and the subconscious mind, a phantom of

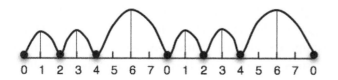

FIGURE 30.1 The vertical motion of a hand, arm, or mallet as a function of time.

the rhythm actually sounded. If we represent the original rhythm [2-2-4] on a circle, as in Figure 30.2 (upper left), then the resulting rhythm of silent beats becomes the rhythm pictured in Figure 30.2 (upper right), and is sometimes called the *shadow* of the rhythm [2-2-4].[2] Although the motor image here is not aligned with the sounding image,[3] the muscles of the arm change their function in a significant way at these midpoints in time, and thus it is reasonable to hypothesize that the motor system of the brain must register these distinctive moments, either consciously or subliminally.[4] For this reason, some musicologists such as Gerhard Kubik believe that these shadow rhythms are physiologically and psychologically relevant to the proper study and understanding of rhythm cognition and performance, going as far as to claim that motion and motor action are essential to the explanation of rhythm.[5]

Shadow rhythms should not be confused with *subjective* rhythms. Subjective rhythms are those that are *perceived* by the listener, but not actually produced acoustically. That is, a subjective rhythm is an aural illusion that exists in the mind of the perceiver, but cannot be measured with scientific equipment in the outside world, since it lacks a concomitant acoustic signal. In other words, a subjective rhythm cannot be detected with electronic measuring devices in a recording of the sounded rhythm. A shadow rhythm, on the other hand, although consisting of silent beats, is embedded in the acoustic signal of its progenitor, and can be detected in a sound recording of its progenitor by calculating the midpoints of the progenitor's adjacent IOIs. This is not to suggest that a subjective rhythm cannot be measured with scientific equipment that is able to read brain activity, such as functional magnetic

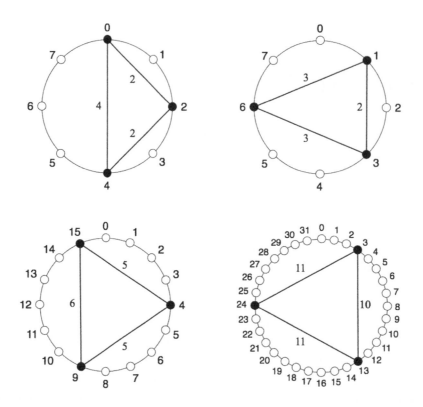

FIGURE 30.2 The rhythm [2-2-4] (upper left) and its first three symmetric *shadows*: upper right, lower left, and lower right.

resonance imagery (fMRI) or magnetoencephalography (MEG). Indeed, such techniques have revealed that a silent beat, *i.e.*, a nonexistent acoustic input, can trigger a brain response if a beat was expected to occur at a specified point in time, but failed to materialize.[6] Of course, the motor action rhythm produced by the points of maximum height of the arm also lacks an acoustic signal. However, it has a physical reality that may also be measured by a scientific instrument. Shadow rhythms may or may not be perceived depending on the particular context. Subjective rhythms are auditory analogs of *subjective contours* (also called illusory contours) encountered in the domain of visual pattern perception illustrated in Figure 30.3, where, on the left, we perceive the contour of a white triangle that is not present and,

on the right, a circle that is absent.[7] It should be emphasized that the subjective contours referred to here are not rational conclusions resulting from cognitive deliberations about what must be out there on the paper, but rather actual perceptions in the mind. In other words, with respect to the subjective triangle on the left, what is relevant is not the deduction from some occlusion properties of objects in space, that there must be a white triangle superimposed on three black disks, but rather, that the materially nonexistent lines (contours defining the boundary of an imaginary triangle) joining the pairs of disks are perceived by the viewer. Similarly, with the figure on the right, the subjective contour does not refer to the fact that the viewer may conclude there is a white disk covering a set of rays that meet at a central common point, but rather that a nonexistent contour in the shape of a circle is perceived, a circle that has no visual signal on the paper that can be measured with scientific equipment such as a camera. As with the aural example of subjective rhythms, this is not to imply that subjective contours do not have neural correlates that can be measured with fMRI or MEG. It is worth noting that in both examples the "whites" of the triangle and the disk appear to be whiter than the "whites" outside the triangle and disk. This is not surprising and, indeed, must be

FIGURE 30.3 Two examples of *subjective* contours.

so if subjective contours are to be perceived. Subjective contours in visual images are analogous to meter in the musical rhythm domain, which emerges from features in the stimulus, but extends beyond them to engender the perception of silent beats (also called "loud rests"[8]) at locations where they are expected.[9]

Shadow rhythms and subjective rhythms should also be distinguished from what are called *inherent* rhythms, a term coined by the ethnomusicologist Gerhard Kubik, who originally defined inherent rhythms are those that are heard by the listener but not played by any single individual musician or instrument. Such rhythms were said to emerge from the interaction of different rhythms played on different instruments or on the same instrument but with different tones.[10] John Collins calls this phenomenon a *polyvalent acoustic gestalt* and recounts that Ghanaian musicians call it *inside rhythm*.[11]

In the psychology literature, the phenomenon responsible for the existence of inherent rhythms is called *auditory stream segregation* or simply *streaming* for short.[12] It appears to have been discovered independently in 1950 by Miller and Heise (1950) in the laboratory at Harvard University, and 10 years later by G. Kubik, while doing field work in Kampala, Uganda, where he was learning to play the *amadinda* xylophone.[13] Miller and Heise originally called the phenomenon the "melodic fission effect," Robert Erickson used the term "channeling,"[14] but more recently, the term "inherent patterns" is being used. The perception of inherent patterns (or streaming) can be explained with the grouping principles of Gestalt psychology and, predominantly, with the principle of *proximity*.[15] Several rhythmic grouping principles based on gestalt principles were proposed in the celebrated theoretical work of Lerdahl and Jackendoff (1983).[16] Kubik is clearly aware of the gestalt processes that are involved in streaming and, yet oddly enough, proposes that inherent patterns are analogous to the optical illusions illustrated in Figure 30.3, which give rise to subjective contours.[17] However, subjective contours do not have material existence (other than in the human brain), whereas the inherent patterns that listeners perceive do. Thus, inherent rhythms are not like subjective rhythms, and not analogous to the subjective contours of visual perception. Inherent rhythms emerge from clustering due to proximity relations. Perhaps the easiest way to describe the difference between subjective contours and inherent patterns, at the risk of oversimplification, is as follows. With subjective contours, we perceive there

is something there when objectively there is nothing there, whereas with inherent patterns, what we perceive is objectively there, and can be detected with a measuring device. This conclusion is at odds with Kubik's analysis. Concerning inherent patterns, he states: "they cannot be detected in a recording by electronic measuring devices. So far, no "thinking robot" has been able to recognize these phantom images."[18] In computational terms, Kubik implies that no algorithm exists that can extract from the complete input rhythm, the "gestalt" inherent patterns formed in the minds of the listeners. On the contrary, it will be shown that even a rather "stupid robot" can accomplish this task. McAdams and Bregman (1979) explored in depth some of the parameters that cause streaming.[19] In particular, they determined, experimentally, the tradeoff that exists between the sizes of the durational IOIs and the pitch intervals of the tones used, as well as the thresholds necessary for streaming to emerge. Thus, the problem of designing an algorithm to exhibit "gestalt" inherent patterns may be converted to a two-dimensional problem of clustering points in the plane in a perceptually meaningful way, a classical problem in the field of pattern recognition and computer vision. In one version of this problem (connect the dots) a set of points are arranged in the plane, and the robot must select which pairs of points should be joined with edges so as to create a perceptually meaningful line drawing. As an example, consider the set of dots in Figure 30.4.

The human brain can perfectly easily perceive four distinct clusters of points, or separate "gestalt" inherent patterns that together suggest a face: namely two eyes, a nose, and a smiling mouth. Endowing a robot

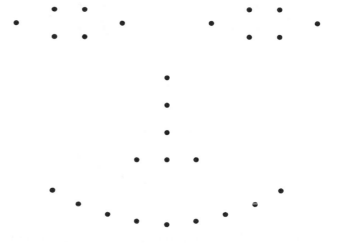

FIGURE 30.4 A set of dots in the plane.

with artificial intelligence to accomplish such a feat was considered a difficult problem, and has received a great deal of attention from psychologists and computer scientists.[20] Most methods are rather complicated, use heuristics, and contain parameters that must be tuned to give good results on the application at hand. However, it is possible to solve the problem with a simple geometric algorithm that does not require any parameter to be tuned. Neither does the algorithm need heuristics from artificial intelligence in order to produce gestalt clusters that agree with human perception. Indeed, the algorithm may be described as part of the vision system of a "stupid" or "nonthinking" robot. The algorithm consists of two steps. The first step involves the construction of a circle around each dot such that the center of the circle coincides with the dot, and the radius of the circle is equal to the distance to its closest neighboring dot. If this is done for the dot pattern of Figure 30.4, then the diagram in Figure 30.5 is obtained.

The second step connects two dots with an edge, provided that their corresponding circles overlap. Applying this step to the dot-circle arrangement of Figure 30.5 yields the line drawing of Figure 30.6. This line drawing of the points is called the *sphere-of-influence* graph (SIG)[21] and is a computational model of the concept of the *primal sketch* in the theory of visual perception.[22]

Armed with this computational tool, let us return to the problem of computing the gestalt inherent patterns from a rhythm. Consider a rhythm consisting of eight isochronous and isotonic pulses. If we represent time

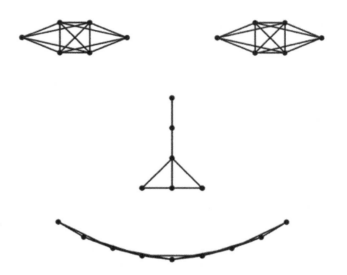

FIGURE 30.6 The SIG graph determined by joining two dots if their corresponding circles in Figure 30.4 intersect with each other.

along the horizontal axis and pitch along the vertical axis (as is done in Western music notation), then we can represent this rhythm by the dot pattern shown at the top of Figure 30.7. The center of the figure shows the dots with their nearest-neighbor circles, and the resulting SIG is pictured at the bottom. The SIG consists of a single connected component, and thus considers the rhythm as one gestalt without the emergence of any inherent pattern. Now let the eight isochronous pulses be made up of two tones instead of one. In particular, let the pulses numbered one, two, four, and seven have an isotonic high tone, and let those numbered three, five, six, and eight have an isotonic low tone. The new

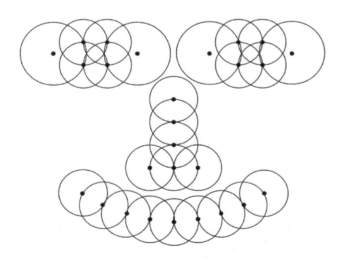

FIGURE 30.5 The dot pattern of Figure 30.4 with the nearest neighbor circles.

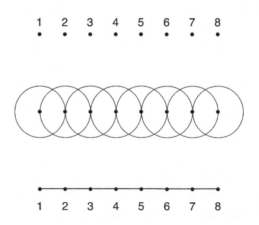

FIGURE 30.7 A sequence of points representing eight isochronous and isotonic tones (top), the circles centered at each point with radius equal to the distance to its nearest neighbor (middle), and the *SIG* graph of the eight points, consisting of a single connected component (bottom).

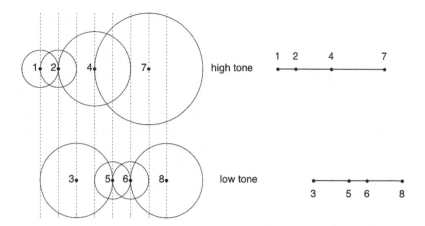

FIGURE 30.8   The same set of eight isochronous points of Figure 30.7, but this time with two tones. The left side of the figure shows the circles centered at each point with radius equal to the distance to its nearest neighbor, and the right side shows the *SIG* graph of the eight points, yielding two disconnected components.

configuration of points, along with the nearest neighbor circles, is shown on the left of Figure 30.8. The resulting SIG, shown on the right, now consists of two disconnected components indicating the emergence of two gestalt-inherent patterns. This example illustrates the power of a variety of new computational tools, such as proximity graphs, that are becoming increasingly available to researchers in comparative musicology.[23] That the concept of streaming provides a useful tool for music analysis has already been established by John Roeder, for example, who used *pulse streaming analysis* to study Schoenberg's middle-period music. Pulse streaming "represents polyphony as concurrent pulse streams created by regular recurring accents."[24] Hopefully, proximity graphs such as the SIG will, someday, find application to the automatic computation of pulse streams.

Before closing this chapter, let's return to the topic of shadow rhythms and a beautiful geometric question that they bring up, which is not only relevant to the theory of musical rhythm but also an interesting mathematical puzzle in its own right. This problem becomes evident by comparing the polygonal representations of the rhythm [2-2-4] with its shadow at the top of Figure 30.2, which reveals that the shadow of the rhythm (top right) has IOIs with durations that have less variability (or are more regular) than its parent rhythm (top left). The variability of a rhythm's intervals may be conveniently measured by the ratio of its longest to its shortest duration interval. For the rhythm [2-2-4] this ratio is $4/2 = 2$, whereas for its shadow, it is $3/2 = 1.5$. In other words, the shadow of rhythm [2-2-4] is more *regular* than [2-2-4].

A natural geometric question that arises is whether this property is true for all rhythms. Consider the second shadow of rhythm [2-2-4], that is, the shadow of the rhythm in the top right of Figure 30.2. The midpoint of the interval between pulses three and six is the midpoint between pulses four and five. Therefore, to represent this shadow rhythm as points on the circle, a cycle of 16 pulses is required. The resulting shadow rhythm is show at the bottom left and has a longest/shortest duration ratio of $6/5 = 1.2$. Continuing this process, the shadow of this rhythm in turn requires a 32-pulse cycle, shown at the bottom right, which has a longest/shortest ratio of $11/10 = 1.1$. In this example, the sequence of ratios of longest to shortest intervals, as we continue to apply the shadow operation, is the decreasing sequence 2, 1.5, 1.2, and 1.1. It turns out that the shadow operation makes *all* nonregular rhythms more regular, and furthermore, if this operation is continued, eventually, the rhythm becomes perfectly regular, that is, all adjacent IOI durations become equal.[25]

With the aid of slow motion film, the ethnomusicologist Gerhard Kubik made some interesting observations from the analysis of upward and downward motions of the arms of xylophone players. In one study, he compared European players with Ugandan players, and found that the upward and downward motions between two consecutive attacks of the Europeans tended to have equal durations, but for the Ugandan players, the upswing lasted twice as long as the downswing. Therefore, it is natural to call the shadow rhythms resulting from the former process *symmetric* shadows, and from the latter process *asymmetric* shadows. Figure 30.9 shows an asymmetric

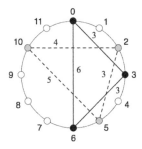

FIGURE 30.9   The *asymmetric shadow* of the rhythm [2-2-6].

shadow rhythm in which the upward swing lasts twice as long as the downward swing. The rhythm actually played is [x . . x . . x . . . . .]. Using this rule, and considering that the duration between the first and second onsets is three pulses long, the maximum height of the arm swing between these two onsets occurs at pulse 2. Thus, the resulting two-third–one-third shadow rhythm has interval structure [. . x . . x . . . . x .]. Comparing this rhythm to its asymmetric shadow rhythm reveals that the ratio of the longest to the shortest duration interval has changed from 6/3 to 5/3, that is, the asymmetric shadow is still more regular than the parent rhythm. Even in the case of asymmetric shadow rhythms, if the shadow operation is repeatedly applied, the resulting rhythm will also eventually become perfectly regular.

"Perceiving temporal regularity in an auditory stimulus is considered one of the basic features of musicality."[26] Since regular rhythms make up the fundamental isochronous pulsations and meters in much of the world's music, and are easy with which to entrain (by clapping, foot tapping, or dancing), compared to nonisochronous meters, an interesting problem regarding shadow rhythms is the characterization of those rhythms that have regular rhythms as their shadows. It is possible, but this is mere speculation, that regular shadow rhythms facilitate the perception and production of rhythms that

engendered them. Obviously, the shadows of regular rhythms are also regular rhythms. Therefore, the more interesting question is to consider the case of irregular rhythms. For the simplest special case, in which irregular rhythms consist of just two onsets, their shadows converge to regular rhythms in a single step. In other words, the shadow of any two-onset irregular rhythm is a regular rhythm. This must be true because the bisector of any chord of a circle must intersect the center of the circle and determine a diameter. To see this, examine the sample of rhythms pictured in Figure 30.10. In the first three cases the rhythms have IOIs [2-14], [4-12], and [6-10], respectively, indicated with bold lines connecting the black-filled pulses. Their shadows are shown with thin solid lines connecting the gray-filled pulses, and the shadows are shown in dotted lines connecting white-filled pulses. After the first step, the subsequent shadows continually alternate between two orthogonal diametrically opposite pulse pairs, forever. The rightmost rhythm with IOIs [8-8] is already regular.

Slightly more complex than the two-onset rhythms are the four-onset irregular rectangle rhythms, such as the [2-6-2-6] rhythm shown in Figure 30.11, with its square shadow at positions 1, 5, 9, and 13. All rectangle rhythms also have regular shadow rhythms. To see this, consider any rectangle R = ABCD inscribed in a circle, as illustrated in Figure 30.11 (right). An edge of R, such as BC, is a chord of the circle, and as such, its perpendicular bisector qs passes through the center O of the circle, and bisects the arc BC of the circle at point q. Since the edge AD of R is parallel to BC, its perpendicular bisector sq coincides with qs, and also bisects the arc AD at point s. Similarly, the bisector of edges AB and CD is a diameter of the circle (passing through the center O), which intersects the circle at points p and r, such that p bisects the arc AB and r bisects the arc CD. Since the interior angles of the rectangle R are right angles (equal to

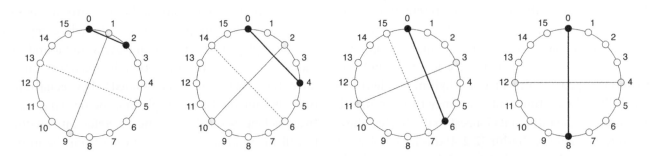

FIGURE 30.10   Several two-onset rhythms (bold lines) and their shadows (thin solid lines).

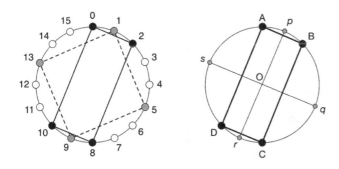

FIGURE 30.11 An irregular (rectangle) rhythm [2-6-2-6] in solid lines and its shadow in dashed lines (left) illustrate the proof that the shadow of any rectangular rhythm is a regular rhythm.

90°), the bisectors *pr* and *qs* are orthogonal to each other forming four right angles. Therefore, the four straight line segments *pq*, *qr*, *rs*, and *sp* (not shown) all have the same length. The theorem of Thales (the oldest theorem in the world[27]) states that any point on a circle subtends a 90° angle with the endpoints of any diameter of the circle. Therefore, the angles *pqr*, *qrs*, *rsp*, and *spq* are all right angles. It follows that *pqrs* forms a square corresponding to the regular shadow of the rectangle *ABCD*.

There also exist irregular rhythms with more than four edges that have regular shadow rhythms. One example is the ternary 12-pulse kéné foli rhythms from Guinea, with interval structure [2-1-2-1-2-1-2-1], shown in polygon notation in Figure 30.12 (left). The shadow rhythm of this rhythm is the regular rhythm in the center figure with eight IOIs of duration 1.5 each (with onsets shows as gray-filled circles). Therefore, the shadow rhythm "lives" in a lattice of 24 pulses. The rhythm and its shadow are superimposed on each other in the diagram on the right. The kéné foli rhythm polygon belongs to the class of polygons knows as *zonogons*, which have the property that for every edge there is an

opposite edge of equal length parallel to it.[28] Squares and rectangles (and of course all regular polygons with an even number of sides) are special classes of zonogons. The musical and psychological implications of rhythm shadows remain a little explored area of research. It would be interesting to determine, for example, if rhythms that have regular shadows are perceived as more regular than rhythms that have irregular shadows.

The following figures exemplify some additional rhythms used in musical practice alongside their shadows. Figure 30.13 shows that one of the six distinguished timelines, the shiko on the left, has as its shadow a rotation of another distinguished timeline, the bossa-nova on the right. The bossa-nova starts on pulse eight. The fact that when a performer plays the shiko, his or her motions implicitly also perform the bossa-nova as a shadow rhythm suggests that the relationship of these two rhythms should be explored further from the psychological or neurological points of view to determine whether shadow rhythms exhibit any neural correlates. Note however that the shadow operation is of course not necessarily a symmetric relation. The shadow of the bossa-nova is not the shiko, but rather an even more regular rhythm that contains three intervals of duration three and two intervals of duration 3.5.

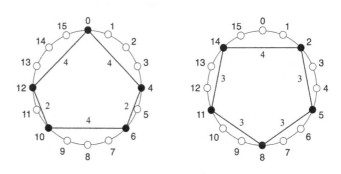

FIGURE 30.13 The *shiko* timeline (left) and its shadow (right).

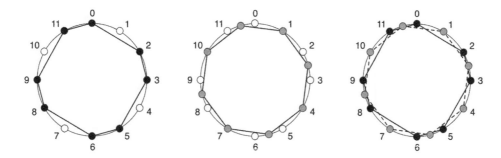

FIGURE 30.12 The *kéné foli* irregular octagonal rhythm (left), its shadow (center), and the two superimposed over each other (right).

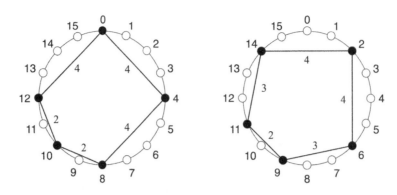

FIGURE 30.14 The classical-music rhythmic ostinato in Rossini's *William Tell Overture* (left) and its shadow (right).

The classical-music rhythmic ostinato employed by Rossini in the *Finale* of the *William Tell Overture*,[29] in Figure 30.14 (left) with durational pattern [4-4-2-2-4] or [2-2-1-1-2], has as its shadow a rotation of one of the most popular timelines used in rap music (right).[30] The rap timeline starts on pulse number two. A rotation of the pattern on the left, given by [4-2-2-4-4] or [2-1-1-2-2], is also the beloved metric pattern of the *Allegretto* in Beethoven's *Seventh Symphony*. In the words of Max Kenyon,[31] "this meter does more than appeal to the soul: it appeals to the body as well." Did Beethoven "hear" the shadow rap rhythm of this metric pattern when he chose it for his composition? Do rappers rapping to the pattern [4-3-2-3-4] feel Beethoven's umbrella rhythm hovering over it?[32]

The shadow rhythms explored in this chapter are closely related to the complementary rhythms covered in Chapter 25. For some rhythms, their shadows are identical to their complements modulo rotation, as is the case with regular rhythms and the interlocking mirror-symmetric rhythms in Figure 25.2. But there are some examples where the shadow of a timeline is trimmed so that it becomes the timeline's complementary rhythm.

A case in point arises in the way the shekere (a gourd enveloped in a net of beads as illustrated in Figure 17.4)

is played in the *agbadza* drum dance of the Ewe people of Ghana.[33] The shekere is held in the right hand and struck downwards on the thigh to the rhythm of the *bembé* pattern [2-2-1-2-2-2-1] shown in Figure 30.15 (left) (in unison with the bell). However, the shekere's upward motion is blocked with the left hand. This upward strike of the shekere on the hand makes a different sound from the downward strike on the thigh. Since the upward strikes on the hand fall in between the downward strikes on the thigh, the IOI pattern of the former rhythm is the shadow of the latter rhythm, and is a rotation of [2-1.5-1.5-2-2-1.5-1.5], as shown in Figure 30.15 (middle) with gray-filled circles. However, the upward strike on the hand is skipped between the pair of onsets in the *bembé* that are one pulse apart (between pulses 4 and 5 as well as pulses 11 and 0). The overall result is that the rhythm produced by the hand is a trimmed partial shadow of the rhythm produced by the thigh, shown in Figure 30.15 (right) with gray-filled circles. This rhythm is the complementary rhythm of the *bembé*, and is a rotation of the fume–fume [2-2-3-2-3] starting on pulse number six.

We close this chapter with a note that, in the pitch domain, some chords also have shadow chords.

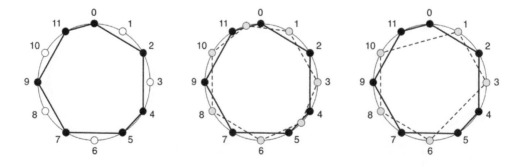

FIGURE 30.15 The *bembé* timeline (left), the *bembé* and its shadow (middle), and the *bembé* and its complement, a rotation of the fume–fume (right).

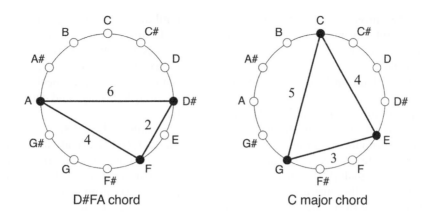

FIGURE 30.16　The C-major chord is the shadow of the D#FA-chord.

In Figure 30.16 the famous C-major chord consisting of the notes CEG (right) with internote intervals [4-3-5] is the shadow of the D#FA-chord (left) with intervals [4-6-2] with root F, because E is the midpoint of the dyad D#F, G is the midpoint of the dyad FA, and C is the midpoint of dyad AD#. Whether the shadow concept has interesting musicological and psychological implications in the pitch domain of chords and scales remains an open question yet to be investigated.

NOTES

1 In the real Newtonian world, this happens with footballs and baseballs if they do not go into outer space. In the mathematical world where almost anything can happen, we have to add of course the assumption of function continuity of some sort.

2 The term "shadow rhythm" as utilized here is adopted from the work of Rahn, J., (1996), p. 82, and should be distinguished from the term "shadow meter" employed by Eytan Agmon, who uses the term in the context of a conflict between two metrical divisions of equal duration. Agmon calls the primary and secondary divisions, the meter and shadow meter, respectively, and no midpoint relation between the two meters is implied. Agmon, E., (1997), p. 64. According to Agmon, the term "shadow meter" was coined by Frank Samarotto. *ibid* p. 74. See also London, J., (2004), p. 81. It is also at variance with the term "shadow" of a rhythm as used by Flatischler, R., (1992), p. 118, who uses the term to mean the *complement* of a rhythm, composed of the unused pulses of the original rhythm (also called complementary rhythms).

3 Scherzinger, M., (2010), refers to those rhythms where the sounding image is aligned with the motor image as *kinesthetic patterns.*

4 It would be interesting to try to determine using a MEG whether there exist any neural correlates of these points in time. Grahn, J. A. & Brett, M., (2007) report on experiments that suggest that rhythms with regular accents produce increased activity in the motor areas of the brain.

5 For example, in his comparative analysis of sub-Saharan African rhythm with European rhythm, von Hornbostel, E. M., (1928), p. 52, particularly in the context of striking drums or balafons, with sticks or mallets, proposes that the European analyzes rhythm via hearing, whereas the African generates rhythm by means of motion. This is probably a caricature with little scientific basis (a European also has to raise the arm with a mallet before striking a drum). In any case, raising the mallet requires muscular contractions and strains the arm muscles before relaxing them, to allow the mallet, with the aid of gravity, to meet its target. Shadow rhythms also have implications for understanding the notion of syncopation. In Western cultures, the downbeat is generally associated with striking the drum downwards, thus defining the accents, whereas in other cultures the tense lifting of the stick is considered to be the accent, thus swapping downbeats with upbeats, and turning syncopation upside down. For further discussion, see Blacking, J., (1955), p. 20, Blacking, J., (1973), and Stone, R. M., (2007). In short, a rhythm that is considered syncopated in one culture may sound perfectly normal in another and vice versa. However, as pointed out by Sachs, C., (1943), p. 47, this distinction may lose much of its impact when light sticks or fingers are used at a high tempo. For further discussion on the topic of motor accents on upstrokes, see Kubik, G., (2010a), p. 87. See also Fitch, W. T., & Rosenfeld, A. J., (2016) for the intimate relation between rhythm and dance.

6 See Jones, M. R. & Boltz, M., (1989) and Grahn, J. A., (2009), p. 262. A delightful and in-depth coverage of the psychology of expectation in time is the book by Huron, D., (2007).

7 The triangle in Figure 30.3 (left) is known as the Kanizsa triangle named after the Italian psychologist Gaetano Kanizsa, whose groundbreaking research into

illusory contours at the University of Trieste during the 1950s stimulated much subsequent research in this area. Winckelgren, I., (1992), p. 1520.

8 Honing, H., (2012), p. 86.

9 Cao, E., Lotstein, M., & Johnson-Laird, P. N., (2014), p. 445.

10 Kubik, G., (1962), p. 33. Kubik illustrates how African instrumental music exploits the many illusions that are present in aural perception, and in his 1962 paper describes four conditions for inherent rhythms to be generated by the acoustic signal: (1) the pitch intervals of the notes must be large, (2) the musical complex must be metrically unaccented, (3) the music must be played at a fast tempo, and (4) all the notes of the same pitch should form a recognizable whole or "gestalt."

11 Collins, J., (2004), p. 55.

12 Bregman, A. S. & Campbell, J., (1971). An in-depth treatment of auditory streaming as well as a variety of auditory illusions and strategies for making rhythmic sense out of sequences of tones may be found in the book by Bregman, A. S., (1990).

13 Kubik, G., (2010b), p. 108, traces the history of *streaming*, its changing terminology, and its application in sub-Saharan African music composition. Wegner, U., (1993), provides a detailed comparison of the psychological aspects of streaming discovered by Bregman and his colleagues with the musicological findings of Kubik. Musical streams are related to Yeston's music-theoretic concept of *strata*; an isochronous stream is a type of stratum. See Parncutt, R., (1994), p. 410 and Yeston, M., (1976) for further discussion.

14 Erickson, R., (1975), p. 116, illustrates the concept of *channeling* with a transcription of a voice and whistle piece performed by the Ba-Benzelé people of Central Africa.

15 Deutsch, D., (1999), p. 313.

16 Although the work of Lerdahl and Jackendoff is primarily theoretical, it has received empirical verification by Deliège, I., (1987).

17 *Ibid.*, p. 115, Figure 30.

18 *Ibid.*, p. 111.

19 McAdams, S. & Bregman, A. S., (1979), pp. 26–43.

20 Zahn, C. T., (1971) proposed using a proximity graph called the *minimum spanning tree* for obtaining gestalt clusters from dot patterns. The minimum spanning tree connects all the points in one tree, and may be constructed by first connecting the closest pair of points, then the second closest pair, and continuing this way until all the points are part of the tree, as long as no cycles are produced, in which case a pair that would create a cycle is skipped. The disadvantage of this method is that, to produce disconnected gestalt components, some parameters must be defined and tuned to identify relatively long edges and delete then from the complete tree.

21 Toussaint, G. T., (1988).

22 Marr, D., (1982). David Marr developed a theory of low-level, bottom-up, vision that he called *early processing*, which involved the construction of a *primal sketch*. Todd, N. P., (1994), proposed an auditory version of a primal sketch in the form of a multiscale model of rhythmic grouping. The SIG is a simple computational-geometric model of a primal sketch that is parameter-free. It is not intended to serve as an alternative theory of rhythmic grouping but merely to illustrate that a simple algorithm is able to extract inherent rhythms.

23 Toussaint, G. T., (2005a), Tzanetakis, G., Kapur, A., Schloss, W. A., & Wright, M., (2007), and Cornelis, O., Lesaffre, M., Moelants, D., & Leman, M., (2010).

24 Roeder, J., (1994), p. 249.

25 Hitt, R. & Zhang, X.-M., (2001), pp. 21–37.

26 Van der Aa, J., Honing, H., & Ten Cate, C., (2015), p. 37.

27 It is believed that the oldest theorem in the world, which marked the birth of deductive reasoning in mathematics, is the theorem of Thales of Miletus (circa 600 BC). See Martin, G. E., (1998), p. 11. See further details in Patsopoulos, D. & Patronis, T., (2006), p. 59, as well as the footnotes in Chapter 8.

28 Kappraff, J., (2002), p. 189. See also Coxeter, H. S. M., (1973).

29 Grahn, J. A. & Brett, M., (2007), p. 893. See the detailed discussion of the ubiquitous use of this pattern in other music as well, including, reggae, South African folk music, Rumanian dances, and protest chants, at the end of Chapter 13.

30 This is the drum ostinato in Chucho Valdez' jazz composition *Invitation*, (EGREM CD0233, Havana, Cuba, 1997).

31 Kenyon, M., (1947), p. 170.

32 If the metric pattern used by Beethoven, [2-2-1-1-2] is rotated by a half cycle, one obtains [1-1-2-2-2], the drumming timeline used in the *bugóbogóbo* music, the symbol of Tanzanian national identity. See Barz, G., (2004), pp. 20–21.

33 Collins, J., (2004), p. 44. The shekere is called *axatse* in Ghana.

CHAPTER 31

Reflection Rhythms, Elastic Rhythms, and Rhythmic Canons

REFLECTION RHYTHMS

WHILE IT IS DIFFICULT to define precisely what is guaranteed to make a "good" rhythm good, it is not hard to list the properties that appear to contribute to a rhythm's goodness in the presence of other appropriate properties. Given that we know a certain rhythm to be good, we may examine its properties to gain insight into what makes it good. However, such properties are, by themselves, neither necessary nor sufficient to make a rhythm good. We have already examined several such properties in previous chapters. Another possible candidate for such a property is that the mirror-symmetric image of a rhythm about some axis of symmetry be equal to its complementary rhythm.[1] I call rhythms that have this geometric property *interlocking reflection* rhythms. These rhythms also exhibit the phenomenological property that if the rhythm and its complement are both played simultaneously, and their acoustic properties such as timbre or intensity differ, the listener has the impression that the rhythm is switching roles, sometimes acting as a figure with its complement as a background, or vice versa. This occurrence is analogous to the unstable perception caused by the cubic wireframe shown in Figure 31.1 (left)

that has no depth cues, and which is sometimes perceived as a cube viewed either from above (center) or below (right). Such figures are known as Necker cubes.[2]

This chapter presents, by means of examples, two simple methods to generate rhythms that exhibit the interlocking reflection property, introduces the notions of elastic rhythms and rhythmic canons, and relates reflection rhythms to rhythmic canons.[3] The first technique uses the well-known snare-drum rudiments called *paradiddles*.[4] These rudiments form the basis of modern drumming technique, and have been shown to increase speed, control, independence, accuracy, creativity, rhythmic innovation, and fluency. They have roots in African *balafon* technique, and have inspired rock drummers as well as modern jazz pioneers.[5] The reader is warned that the mathematics, theory, and algorithms used in paradiddles are not simple. This chapter is more about drumming technique, about left–right independence, and hand–foot independence, than about deep theory. Thus, the reader is advised to use the hands and feet while studying this chapter. Once the simple technique is mastered and has become automatic, a few simple rules thrown in will facilitate the emergence of elegant rhythmic complexity.

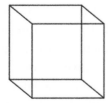 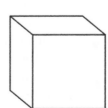 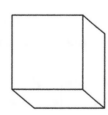

FIGURE 31.1 Unstable perception of the wireframe cube on the left (also Necker cube).

PARADIDDLE METHOD

To illustrate the first method, consider a time span of 16 pulses and refer to Figure 31.2. First, insert an onset at every pulse in the cycle of the rhythm, as in Figure 31.2a, where the label "Right" indicates the rhythm played with the right hand and the label "period" indicates the length of the rhythmic pattern that repeats itself. If we denote by the symbols R and L, the striking of the drum with the right and left hand respectively, then this rhythm repeats the pattern R 16 times. Next, transform this rhythm into a rhythm for two hands by alternating the onsets between the left and right hands, preferably each hand playing on a different drum, as in Figure 31.2b. The idea here is that the right and left hands should produce different sounds either in pitch or timbre so that the listener may perceive both the right-hand and left-hand rhythms simultaneously as two separate streams of pulsations. Note that this operation doubles the length of the period of rhythms played by each hand. The rhythm thus repeats the pattern RL eight times. The next step in the process involves repeating the symbol R after one instance of the RLR pattern to obtain the string RLRR, as in Figure 31.2c. Note that this operation doubles the period again to a length of four. The final step involves alternating between the RLRR pattern and its mirror image LRLL, as in Figure 31.2d, to obtain the pattern RLRRLRLL. This operation again doubles the period of the rhythm, giving it a time span of eight pulses. The rhythm heard on the right-hand drum has the durational pattern [2-1-2-3] shown in polygon notation in Figure 31.3 (right), with its mirror image complementary

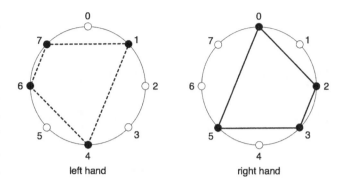

FIGURE 31.3 The left- and right-hand patterns of the *single paradiddle* drum rhythm.

rhythm played on the left-hand drum (left). Note that the rhythm [2-1-2-3] is a deep rhythm, has the rhythmic oddity property, and when played backwards becomes [3-2-1-2], which, as we have seen, is the hand-clapping pattern in Elvis Presley's *Hound Dog*. The complete rhythm played with both hands is one of the three most important rudimentary snare-drum exercises called the *single paradiddle*, and it is not conceived as two separate rhythms, but rather as a single resultant pattern.[6] Here, on the other hand the paradiddle technique is used to generate the individual rhythms played with just the right or left hand. Interestingly, Steve Reich used this rhythm in his composition titled *Phase Patterns*, for four electronic organs, in which the right hand played [2-1-2-3], and the left hand the complement.[7]

For a second example of this method of generating good rhythms, consider the construction of a ternary rhythm with six onsets and 12 pulses. This time we start with a rhythm consisting of 24 onsets and 24 pulses as shown in Figure 31.4a. As in the previous example, this rhythm is first decomposed into its alternating right- and left-hand onsets as in Figure 31.4b. The next step shown in Figure 31.4c is the only step in this process that changes. Instead of repeating the symbol R after the RLR pattern, it is repeated after the RLRLR pattern to create an RLRLRR pattern of period 6. The final step (Figure 31.4d) is the same as before: the RLRLRR pattern is alternated with its mirror image LRLRLL to create a rhythm with period 12. The rhythm played with the right hand, shown in polygon notation in Figure 31.5, has durational pattern [2-2-1-2-2-3], which has the rhythmic oddity property, is a deep rhythm, and is a popular Yoruba bell timeline from West Africa. It is also a Kpelle rhythm of Liberia.[8] The entire rhythm

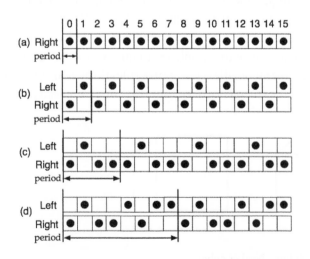

FIGURE 31.2 Constructing a *single paradiddle* reflection rhythm.

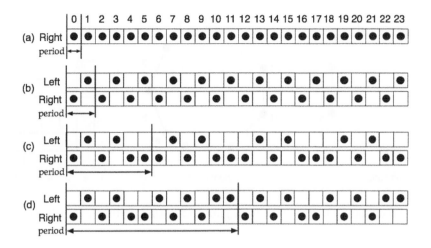

FIGURE 31.4 Constructing a *double paradiddle* reflection rhythm.

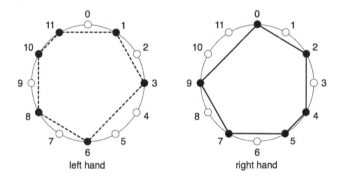

FIGURE 31.5 The left- and right-hand patterns of the *double paradiddle* drum rhythm.

played with both hands is called the *double paradiddle* in rudimentary snare-drum technique.

For the third example that illustrates this method, let us generate a rhythm with eight onsets and 16 pulses.

The process for the first two steps is the same as before, but this time let us start with 32 pulses, as shown in Figure 31.6a and b. The change comes again in the next step (Figure 31.6c), where instead of repeating the symbol R after the RLRLR pattern, it is repeated after the RLRLRLR pattern to create an RLRLRLRR pattern with period 8. The final step (Figure 31.6d) is the same as before: the RLRLRLRR pattern is alternated with its mirror image LRLRLRLL to create a rhythm with period 16. The rhythm played with the right hand, shown in polygon notation in Figure 31.7, has interonset-interval (IOI) vector [2-2-2-1-2-2-2-3], which is a deep rhythm as may be seen from the histogram on the right. This rhythm also has the rhythmic oddity property. The entire rhythm played with both hands is called the *triple paradiddle* in rudimentary snare-drum technique.

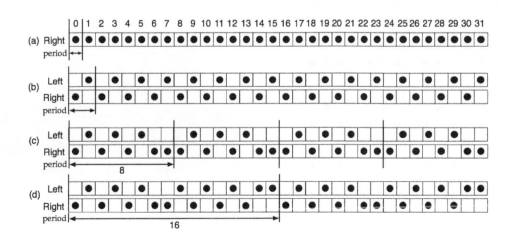

FIGURE 31.6 Constructing a *triple paradiddle* reflection rhythm.

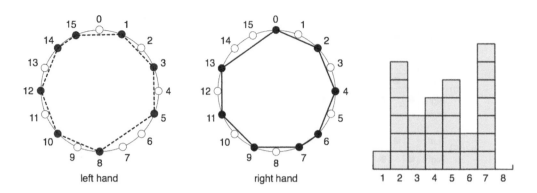

FIGURE 31.7 The left- and right-hand patterns of the *triple paradiddle* drum rhythm and their interval-content histogram.

ALTERNATING-HANDS METHOD

The *paradiddle method* described earlier first puts an onset on every pulse, and then creates a pattern by breaking the alternation between the right and left hands through the repetition of an onset with the same hand, whereas the alternating-hands method maintains the alternation of right and left hands throughout the production of the rhythm, and instead generates a new rhythm by transforming another (usually simpler) seed rhythm.

For the first example that illustrates the alternating-hands method refer to Figure 31.8, and consider the simple four-pulse seed pattern [x x x .] shown in Figure 31.8a with durational pattern [1-1-2]. This rhythm is a universally used simple pattern dating back to at least the *ars antiqua*, associated with prosody, and known as the *short–short–long* pattern. We have seen earlier that it is also a pattern used in the *Baiaó* rhythm of Brazil, a drum rhythm in South Indian classical music, and the *polos* rhythm of Bali. The second step decomposes this cyclic pattern into alternating right- and left-hand

strokes as shown in Figure 31.8b. Finally, this rhythm is alternated with its mirror image creating on the right-hand drum an eight-pulse rhythm with IOI vector [2-3-3], as shown in Figure 31.8c. At the same time, the left hand plays a reflected (mirror image) version of this rhythm. The rhythms in Figure 31.8b and c are typical rhythms played with the *krakebs*, or metal double castanets (illustrated in Figure 31.9) in the *Gnawa* spiritual trance music of the Greater Maghreb region of North Africa, found mainly in Morocco, Algeria, Libya, and Tunisia. One pair is used in each hand to produce a loud and distinctive hypnotic metallic sound that resembles hand clapping.

The reader is invited to apply this alternating-hands method to three more seed patterns given by [1-2-1-2-2] (a version of the cinquillo) in Figure 31.10, the pattern [1-1-1-1-2] in Figure 31.11, and the pattern [2-1-1-2-2] (the popular classical-music rhythmic ostinato) in Figure 31.12. As can be seen by following the procedure

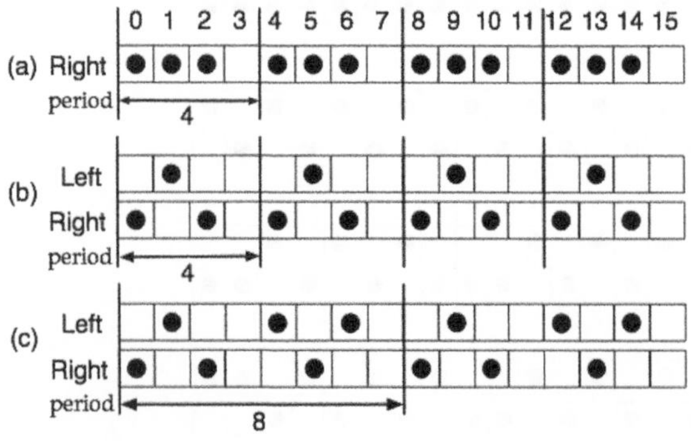

FIGURE 31.8 Constructing the [2-3-3] reflection rhythm using the *alternation* method.

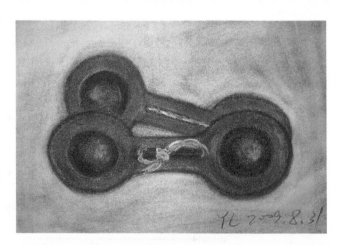

FIGURE 31.9 Greater Maghreb *krakebs* (metal double castanets). (Courtesy of Yang Liu.)

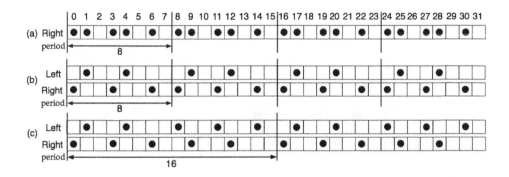

FIGURE 31.10 Constructing the bossa-nova necklace timeline with the alternating hands method.

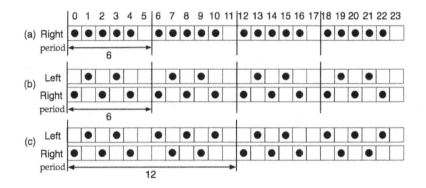

FIGURE 31.11 Constructing the fume–fume timeline with the alternating hands method.

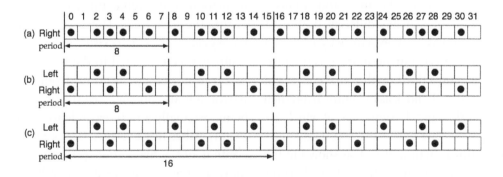

FIGURE 31.12 Generating the *clave son* with the alternating hands method.

illustrated in these figures, the three patterns yield, respectively, the bossa-nova, fume–fume, and *clave son* timelines.

The simple alternating-hands method for rhythm generation that started with the seed pattern [x x x .] produced the pair of right-hand and left-hand rhythms given by the sequences [x.x..x..] and [.x..x.x.], respectively. If we rotate these rhythms in a counterclockwise direction by two pulses, we obtain the rhythms shown in Figure 31.13. The right-hand rhythm is the ubiquitous [3-3-2] *tresillo* pattern, and the left hand produces its reflection about the line through pulses 1 and 5. Here the

right-hand rhythm is shown with solid lines, and the left hand with dashed lines. These two rhythms may be viewed as having other noteworthy geometric relationships to each other. For one, the left-hand rhythm is a rotation of the right-hand rhythm by 180°, that is half a cycle. For another, the left-hand rhythm is the *antipodal* version of the right-hand rhythm. In other words, the left-hand rhythm may be obtained from the right-hand rhythm by replacing every onset by its antipodal onset, that is the onset diametrically opposed to it.

The [3-3-2] tresillo pattern, which constitutes the first half of the *clave son*, by itself, is already imbued with a

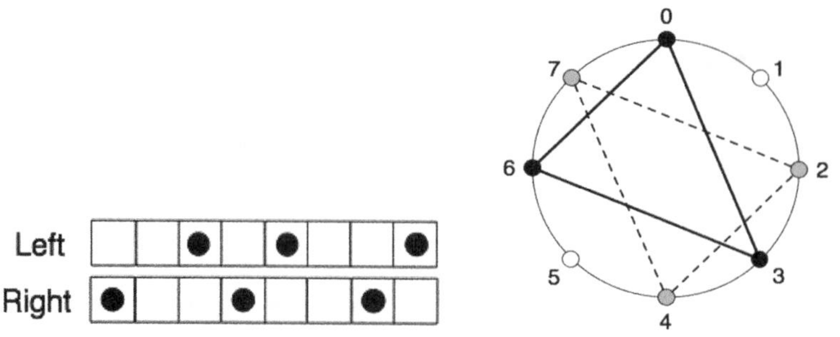

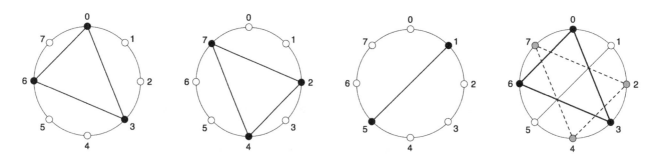

FIGURE 31.13 The tresillo rhythm and its reflection in box and polygon notations.

FIGURE 31.14 Three interlocking rhythms used in Balinese *Kecak* chanting.

great deal of rhythmic power, as evidenced by its widespread popularity all over the world. Playing it alongside its mirror reflection about the line through pulses 1 and 5 that has a distinguishably different timbre, as in the case of the krakebs, may explain why it plays a role in Gnawa trance music. Furthermore, it is interesting to note that if we add a third rhythm that consists of only two onsets at pulses 1 and 5, as in Figure 31.14, then we obtain three interlocking rhythms, two of which are mirror images of each other reflected across a line (mirror) that is itself a third rhythm played, as shown in the rightmost diagram of Figure 31.14. These three rhythms are used in the rhythmic *Kecak* group chanting music of Bali, known for its very powerful trance-inducing effects.[9] Here each of the three rhythms is chanted by a different group of singers. However, it is not clear that rhythms with these geometric properties are in any way necessary to induce trance behavior, as is evidenced from the "self-stabbing" theme of the *Rangda/Barong* ritual. The rhythms used in this piece consist of seven regular perfectly metrically aligned patterns.[10]

ELASTIC RHYTHMS

It has been documented in Chapter 12 that some musicologists believe that the African triple-meter 12-pulse fume–fume rhythm [2-2-3-2-3] was binarized to the

quadruple-meter 16-pulse *clave son* [3-3-4-2-4] by virtue of cultural transformation forces, following the transplantation of the fume–fume from Africa to America during the slave trade. However, it is plausible that the binary *clave son*, which existed in Persia in the thirteenth century, was ternarized to the fume–fume assuming it was transported by the Arabs from Bagdad to Central Africa. It is also possible that the two versions came into being originally and independently among the ancient Pygmy tribes in Central Africa, and that neither binarization nor ternarization processes were responsible for their appearance. The door is still open on this issue. What is certain is that in Cuban music a performer will sometimes switch from playing the binary version to its ternary counterpart within a single piece of music. Furthermore, the two rhythms are sometimes played simultaneously by two musicians, generating a bistable perception that the piece being performed has a binary and ternary feel almost at the same time.[11] More striking perhaps are the examples of music in the world in which a rhythm performed occurs in between a binary and ternary feel. Furthermore, such rhythms need not remain stable in a fixed temporal location between the binary and ternary subdivisions (such as the midpoint), but actually stretch and shrink along a continuum in the space between binary and ternary. I will call such rhythms *elastic* rhythms. Two of the clearest examples of elastic rhythms with IOI patterns that fall in between binary and ternary subdivisions of the time span are manifested in Moroccan *Gnawa* music[12] and Tunisian *Stambēlī* music.[13] In both musical traditions, the elastic rhythms are performed with the iron castanet-like *krakebs* pictured in Figure 31.9.

An idealized circular visualization of the elastic rhythms performed in the Moroccan *Gnawa* and Tunisian *Stambēlī* music, with the *krakeb* castanets, is illustrated in the four diagrams of Figure 31.15. The upper-left diagram shows the binary version using an eight-pulse cycle. One krakeb is held in each hand. The right-hand rhythm [2-3-3] is shown with black-filled circles and solid lines, whereas the left-hand rhythm with onsets at pulses 1, 4, and 6 has gray-filled circles and dashed lines. The letters **R** and **L** are also indicated for right- and left-hand onsets, respectively. The two rhythms are equivalent necklaces and mirror reflections of each other with respect to the axis through pulses 3 and 7. The resultant rhythm has an

IOI pattern [x x x . x x x .]. The two rhythms (left-hand and right-hand) are mapped to a 16-pulse cycle in the upper-right diagram to better illustrate the process. At the other extreme, the ternary version of these rhythms in which both are regular rhythms (regular triangles) is shown in the bottom-right diagram, where the resultant rhythm is the six-onset pulsation [x x x x x x] that lasts the same amount of time as the eight-pulse resultant [x x x . x x x .]. In transforming the binary rhythm to the ternary rhythm, the onset at pulse 10 has been moved to a position close to pulse 11, and the onset at pulse 4 has been moved to a position just after pulse 5. However, the musicians perform the two rhythms by navigating along a continuum in between the perfect binary and ternary versions, yielding rhythms more like the one pictured in the lower-left diagram. Furthermore, although the performers may begin playing the two rhythms close to the binary version, as the tempo of a piece increases, they may transform the rhythms so that they are close to a regular ternary pulsation.

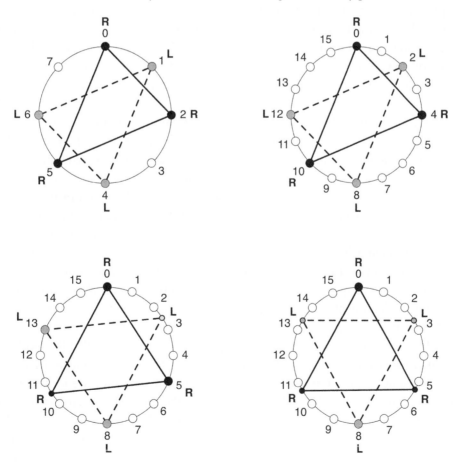

FIGURE 31.15 An idealized circular notation of the elastic process evident in performing Moroccan *Gnawa* and Tunisian *Stambēlī* rhythms with the *krakeb* castanets.

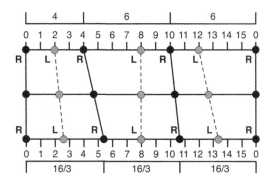

FIGURE 31.16 An idealized linear notation of the elastic process evident in performing Moroccan *Gnawa* and Tunisian *Stambēlī* rhythms with the *krakeb* castanets.

The continuum in which the elastic rhythms are ideally modulated may also be illustrated in the linear fashion shown in Figure 31.16. The binary right-hand rhythm at the top has IOI pattern [4-6-6], whereas the ternary rhythm at the bottom has IOI pattern [16/3-16/3-16/3]. These two versions are connected by solid lines for the right-hand rhythm and by dashed lines for the left-hand rhythm. The time points of the resultant rhythm that lies halfway between the binary and ternary versions are indicated by the horizontal line in the middle. By moving this line upwards or downwards, we may generate other rhythms anywhere in between the perfect binary and perfect ternary versions.

It should be emphasized that the explanations illustrated in Figures 31.15 and 31.16 are simplified idealized versions of the concomitant process in real performances. In practice the perfect mathematical ratios described by this model at any level in between the binary and ternary rhythms are only approximately maintained.

To close this section, it should be noted that elastic rhythms exist between other than binary and ternary rhythms, such as between quintuple and septuple rhythms. Jean During documents examples from the Baluch and Tajik–Uzbek musical traditions, in which rhythms "slide between 3 and 5, and 7 and 8 and 9 "beats." Moreover mathematically, the proportionate difference between 2 + 3 and 3 + 4 is too subtle to be noticed without the necessary skills." He calls such rhythms "ovoid" rhythms.[14]

RHYTHMIC CANONS

A *rhythmic canon* is composed of two or more rotations of a rhythm played at the same time, with the constraint that each rotation (also called a *voice*) uses a tone or timbre that is distinguished from the others, and no two onsets (attacks) of different voices sound in unison. Rhythmic canons provide a popular and effective composition technique that can be traced back to Olivier Messiaen,[15] and has recently received a great deal of attention from the music-theoretical and mathematical points of view.[16] Two of the three interlocking rhythms used in Balinese *Kecak* chanting, shown in the rightmost diagram of Figure 31.14 (the tresillo and its mirror reflection about the axis through pulses 1 and 5, also shown in Figure 31.13), constitute a rhythmic canon in two voices: the first voice (solid lines) starts the rhythm [3-3-2] at pulse 0, and the second voice (dashed lines) starts the same rhythm [3-3-2] half a measure later at pulse 4. In this case the reflection of [3-3-2] about the line through pulses 1 and 5 generates a canon because it is equivalent to a rotation by four pulses (half a measure).

In the earlier example, although no two onsets from different voices sound together, the canon does not use all the onsets available, in particular the onsets at pulses 1 and 5. A rhythmic canon in which every pulse receives exactly one onset is called a *rhythmic tiling canon*. The paradiddle method described at the start of this chapter for producing reflection rhythms yields rhythms performed with the right hand, that are not only reflections but also rotations of the rhythms performed with the left hand. Therefore, if different voices are assigned to the left-hand and right-hand attacks, the paradiddle method may be used to generate rhythmic tiling canons such as [2-1-2-3] and its complement in Figure 31.3, [2-2-1-2-2-3] and its complement in Figure 31.5, and [2-2-2-1-2-2-2-3] and its complement in Figure 31.7.

If you tried to play these rhythms while reading this chapter and find that these are too simple to perform, then you can introduce some additional rules to create more interesting and challenging rhythms. Remember though that the left–right patterns of the paradiddle always remain the same. For example, the single paradiddle is RLRRLRLLRLRRLRLL. Now assume you are playing four congas that have four different timbres, and that they are located in the space around you in the form of a square: two in the front and two in the back. Now you can add the rule that while playing the paradiddle pattern, the attacks should alternate between two consecutive attacks on front congas, and two on back congas, as in the pattern ffbbffbbffbbffbb. The resulting drum attacks you play should then be:

(Rf)(Lf)(Rb)(Rb)(Lf)(Rf)(Lb)(Lb)(Rf)(Lf)(Rb)(Rb)
(Lf(Rf)(Lb)(Lb). If this is still too simple, then apply
an asymmetric front–back rule such as ffbffbffbffbffbf.
With a drum set you can do something similar with the
feet that is different from the hand patterns. As you can
see, the sky is the limit.

NOTES

1 This idea in the pitch domain is called *inversional com-binatoriality*, and was used as a compositional tool by Arnold Schoenberg. See Milstein, S., (1992), p. 6. Thus, it is interesting to investigate its role in the rhythmic domain.

2 Penrose, R., (1992) and Long, G. M. & Olszweski, A. D., (1999).

3 See Toussaint, G. T., (2010), for a more thorough description of the methods described in this chapter.

4 Wanamaker, J. & Carson, R., (1984).

5 Brown, A., (1990).

6 See Koetting, J. & Knight, R., (1986), p. 60, for a discussion on the use of paradiddle structures in traditional African drumming. It is worth pointing out that in creating the paradiddles is it not just a matter of taking any LR rhythm and follow it by its complementary RL rhythm. It has to be only the exact pattern RLRRLRLLRLRRLRLL, etc.

7 Cohn, R., (1992a), Example 1, p. 150.

8 Stone, R. M., (2005), p. 82.

9 McLachlan, N., (2000), p. 63.

10 Becker, J., (1994), p. 48.

11 Vuust, P. & Witek, M. A. G., (2014), p. 8, provide evidence that in bistable perception both rhythms cannot be perceived simultaneously, but rather the percept switches between one rhythm and the other.

12 Sum, M., (2012).

13 Jankowsky, R. C., (2013).

14 During, J., (1997), p. 19. I am grateful to Professor Walter Zev Feldman of New York University for bringing Jean During's reference to my attention.

15 Messiaen, O., (1956).

16 See Agon, C. & Andreatta, M., (2011), Hall, R. W. & Klingsberg, P., (2006), and Tangian, A., (2003), as well as the references therein for some history and a sampling of mathematical properties of rhythmic canons.

CHAPTER 32

Toggle Rhythms

I N THE *ALTERNATING-HANDS* METHOD for generating rhythmic canons composed of "good" rhythms, described in the preceding chapter, the right and left hands continually take turns striking the instrument, analogous to the way our feet do on the ground while we walk, except that, depending on the method employed, the durations between consecutive right- and left-handed strokes may vary. For example, in the rhythm of Figure 31.8, the duration between the first onset (right-hand) and the second onset (left-hand) is one pulse, but the duration between the third onset (right-hand) and the fourth onset (left-hand) is two pulses. In the method described in this chapter, the rhythm emerges from the process of *accenting* the proper onsets with each hand while maintaining all the durations between the left- and right-handed strokes equal to one pulse. In other words, some strokes may be louder than others, or they may differ in timbre, or tonality. Indeed, the soft sounds may even be so muted that they are inaudible or the hand may stop just before coming into contact with the instrument. The important point is that the motion of the hands consists of a continuous pendular alternation of the right and left hands, such that all durations between adjacent pulses are equal. In other words, the downward motions of the hands trace *all* the pulses of the rhythm.[1]

Toggle rhythms are those cyclic rhythms that, when played using the alternating-hands method, have their onsets divided into two consecutive sets, such that the onsets of the first set are played consecutively with one hand, and subsequently the onsets of the second set are played consecutively with the other hand. Thus, playing this way feels as if one hand responds to a question posed by the other hand, analogous to the customary call-and-response method of singing existent in much

of sub-Saharan Africa. The most pleasing and interesting results with this method are obtained when the left and right hands strike drums that are tuned differently, so that they produce sounds of distinct tones or timbres. However, even on a single drum, the left and right hands will almost always produce distinct sounds, since they strike the drum skin at different locations, and thus the effect will still be audible and operative. However, even if all the accented strokes sound the same, the system yields good timelines. Indeed, in some musical practices such as sub-Saharan Africa, timelines by their usual definition have the property that they do not contain accents, that is, all their onsets have equal importance.

The motion of the right and left hands in this manner of playing may be conveniently described with a notation such as RLRLRLRLRLRLRLRL, which indicates all the pulses present in the rhythmic cycle, as well as which pulses are struck with which hand, R standing for the right hand on even-numbered pulses, and L for the left hand on odd-numbered pulses. The rhythm that emerges from this process may be notated using a boldface font for the accented onsets. For example, one possible toggle rhythm with this pattern is **R**L**R**L**R**L**R**LRLRLRL. By accenting the four right-hand and three left-hand strokes, the rhythm that emerges in the form of the accented onsets may be described in box notation as [x . x . x . x . . x . x . x . .]. Note that in this example every pair of consecutive onsets played with the right hand (or left) is separated by one silent pulse. Note also that the transitions between the right-hand and left-hand onsets are separated by an interval of two silent pulses. Toggle rhythms that have this property will be called *smooth* toggle rhythms because this transition is smooth. On the other hand, a 16-pulse pattern such as **RLRLRLRLRLRLRLRL**, which may be expressed in box notation by [x . x . x . x . x x . x . x . x], has no silent pulses between the transition of left-hand and right-hand pulses. This transition is abrupt or sharp, and so toggle rhythms with this property are termed *sharp toggle rhythms*.

The method for generating rhythms in this and the previous chapter are techniques that are useful for drummers and percussion players. By fixing simple repetitive hand-motion patterns that a drummer can learn to play automatically without thinking, and then applying simple repetitive attack patterns on the sequence of drums or other instruments such as bells or cymbals, the drummer can engender the emergence of several complex rhythmic patterns being played simultaneously, without thinking and with little effort other than trying not to listen too carefully to what is being produced, lest one is thrown off guard. These methods may also be used as models for how certain traditional African rhythms on instruments such as the balafon may have come into existence.[2] Figure 32.1 illustrates an algorithm for generating a family of smooth toggle rhythms, by starting with the simplest smooth toggle rhythm that acts as a seed pattern. This seed pattern shown in Figure 32.1a consists of a cycle of eight pulses with two right-hand onsets at pulses 0 and 2, and one left-hand onset at pulse 5. This rhythm, as we have seen earlier, has an interonset-interval structure given by [2-3-3], and is used in the traditional music of several cultures: it is the bell rhythm used in the *nandon*

bawaa music of the Dagarti people of Ghana, as well as a rhythm found in Namibia and Bulgaria. From this seed pattern, we may create new longer rhythms by repeatedly cutting the rhythm in half, and inserting a copy of the *left–right transition segment* in between the resulting two pieces, as illustrated in Figure 32.2. The top of Figure 32.2 shows the eight-pulse seed rhythm with the left–right transition segment shaded. The middle shows the original rhythm cut into two pieces of equal duration. Note that the cut is made at the midpoint that separates the right-hand strokes from the left-hand strokes, which in this case happens between pulses 3 and 4. The bottom shows the final 12-pulse rhythm obtained by splicing the three pieces together, which is the rhythm shown in Figure 32.1b. This process may be iterated by repeatedly inserting the shaded four-pulse transition segment into the preceding rhythm in the same manner. In this way, we may generate the remaining rhythms shown in Figure 32.1, as well as longer ones if desired. However, a cycle of 24 pulses is almost always long enough to serve as a timeline.

The reader will by now be very familiar with the rhythm in Figure 32.1b with interonset-interval structure [2-2-3-2-3], as being the *fume–fume* bell pattern popular in West Africa, and the former Yugoslavia. The rhythm in Figure 32.1c with interonset interval structure [2-2-2-3-2-2-3] is a hand-clapping timeline pattern from Ghana.[3] The rhythm in Figure 32.1e, consisting of 11 onsets among 24 pulses, is perhaps the longest

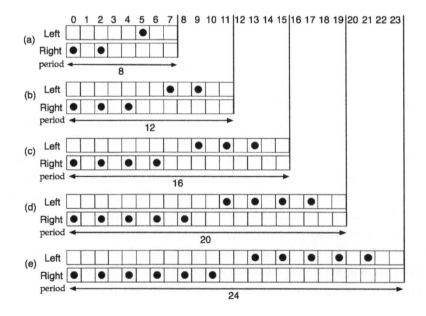

FIGURE 32.1 One method for generating *smooth* toggle rhythms.

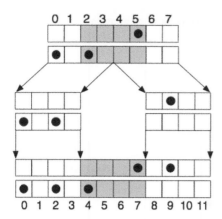

FIGURE 32.2 Splicing a *smooth* toggle rhythm by inserting the right–left transition segment (shaded).

existing smooth toggle timeline, and is played by the Aka pygmies of Central Africa.

Note that the toggle properties are of course rotation invariant, and thus all rotations of toggle rhythms are also toggle rhythms. Consider the seven-onset, 16-pulse rhythm in Figure 32.1c that starts with four downbeats followed by three upbeats. Rotations of this rhythm **RLRLRLRLRLRLRL** that are common in Brazilian samba music include starting with three downbeats instead, as in **RLRLRLRLRLRLRL**, or starting and finishing the sequence with two downbeats each, as in **RLRLRLRLRLRLRL**. The former toggle rhythm is shown in clock notation in Figure 32.3 (left), along with a variant (right) that became well integrated in USA in the 1950s and 1960s. This variant has been used, for example, as a saxophone riff by Bill Haley in *"Rock Around the Clock"* and as a bass riff by the Supremes in *"You Can't Hurry Love"*. On the surface, the change made to obtain this variant appears minimal: only the last onset at pulse 13 has been moved to its adjacent pulse number 12. The consequences of

this small change however are significant. For one, the samba rhythm (left) has the rhythmic oddity property, but the variant violates it with the antipodal pair of onsets at pulses 4 and 12. Furthermore, the samba has four strong double-off-beat onsets at pulses 7, 9, 11, and 13, whereas its variant has only three. These properties, along with the continuous forward locomotive-like motion, endow the samba rhythm with a distinctly African aesthetic. The variant on the other hand swaps the double-off-beat onset at pulse 13 with the last of the four strong downbeats at pulse 12, bringing a strong sense of closure to the pattern. This perhaps explains why the variant rather than the original became so popular in American pop music.

Besides, it is also interesting to compare the interonset-interval histograms of the two rhythms. The samba is missing intervals of duration one and eight. On the other hand, the variant is an all-interval rhythm. Since the number of onsets in a rhythm is a measure of the rhythm's complexity, albeit a weak one, one might be tempted to say that the samba variant is more complex than the original samba. On the other hand, the samba has the rhythmic oddity property and a higher offbeatness value, both of which also measure complexity. Therefore, it seems that the transformation from the original to the variant is exchanging one type of complexity for another.

A similar approach may be used to generate a family of sharp toggle rhythms, as illustrated in Figures 32.4 and 32.5. Figures 32.4 shows a collection of six sharp toggle rhythms ranging in time spans from 4 to 24 pulses, in increments of four pulses. The top of Figure 32.5 shows the shaded four-pulse right–left transition segment that must be inserted into a sharp toggle rhythm to create a new longer sharp toggle rhythm. The remainder of the figure details how the splicing may be done

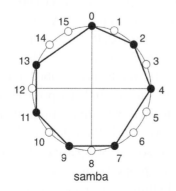
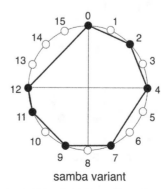

FIGURE 32.3 Brazilian samba timeline (left) and American variant (right).

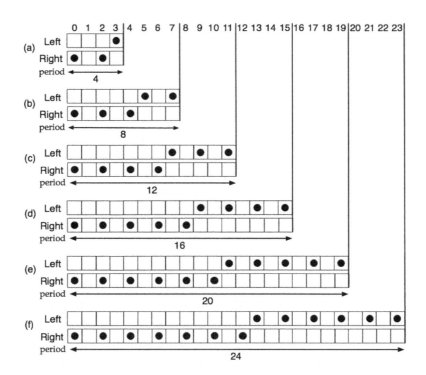

FIGURE 32.4   A method for generating sharp toggle rhythms.

on the sharp toggle rhythm of Figure 32.4b. Note that here the cut made between the right-hand and left-hand onsets occurs between pulses 4 and 5 and partitions the rhythm into two unequal pieces: one with five pulses and three onsets and the other with three pulses and two onsets. The resulting new rhythm timeline at the bottom of Figure 32.5 is the seven-onset 12-pulse timeline of Figure 32.4c. Of course, another simple method

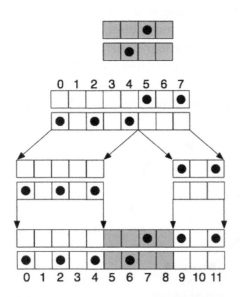

FIGURE 32.5   Splicing a *sharp* toggle rhythm by inserting the right–left transition segment (shaded).

of generating sharp toggle rhythms, if one already has the smooth toggle rhythms to begin with, is to convert a smooth version to a sharp version by adding at appropriate places, one onset to the patterns of each hand. For instance, the smooth toggle rhythm of Figure 32.1b may be converted to the sharp toggle rhythm of Figure 32.4c by adding one right-hand onset at pulse number six and one left-hand onset at pulse number 11.

The 16-pulse toggle rhythm in Figure 32.4d is a timeline played in several parts of Africa.[4] Sometimes this rhythm necklace and its complement are played simultaneously, the rhythm with the right hand, and its complement with the left.[5] One of the longest sharp toggle rhythms is the *bobanji* timeline played on a metal bell by the Aka Pygmies of Central Africa; it has 13 onsets in a cycle of 24 pulses.[6] A rotation of this timeline is shown in Figure 32.4f. The bobanji timeline is actually started on the fourth onset at pulse number 6.

The methods described in the preceding for composing smooth and sharp toggle rhythm timelines generate rhythms with the property that the first set of onsets played with the right hand has one more onset than the second set played with the left hand. Obviously, this property may be easily reversed. Another set of rhythms may be determined by first interchanging the onsets played with each hand, and then playing the rhythms thus obtained in reverse order. For example, such a transformation applied to the sharp toggle rhythm of Figure 32.4c yields the rhythm **RLRLRLRLRLRL**.

Since rhythm timelines are repeated throughout a piece and are thus cyclic, it is natural to represent toggle timelines using two concentric circles as pictured in Figure 32.6, where the outer and inner circles mark the right- and left-hand onsets, respectively, of a rotation of the *standard pattern*. Since the rhythm has seven onsets, seven different timelines may be obtained by starting the cycle at any of the seven onsets. Indeed, as we have seen in previous chapters, all such rhythms are used as timelines in different parts of sub-Saharan Africa.

The representation of cyclic rhythms on a circle permits an alternate definition of toggle rhythms based on the

notion of *linear separability* in geometry. A rhythm (set of integer points on the circle) is a toggle rhythm if there exists a straight line that separates the left-hand onsets (on odd-numbered pulses) from the right-hand onsets (on even-numbered pulses). The rhythm in Figure 32.6, for instance, is a toggle rhythm because there exists a line passing through two points: one being the midpoint between pulses 4 and 5, and another the midpoint between pulses 0 and 11, that leaves all the right-hand pulses on one side of this line, and the left-hand pulses on the other side. Note that although the two circles are drawn as having different sizes for the sake of visualization, they should be considered as one and the same circle for this definition to remain valid in all cases.

The toggle rhythms considered heretofore have the property that the number of right-hand onsets differs by one from the number of left-hand onsets. This is not a requirement for a rhythm to belong to the toggle family. The *clave son* shown in circular toggle notation in Figure 32.7 (left) has only one of its five onsets played with the left hand, and yet it is a toggle rhythm since it admits a line that separates this onset from all others, such as, for example, the line through pulses 1 and 5. On the other hand, the bossa-nova rhythm timeline of Figure 32.7 (right) with three right-hand onsets and two left-hand onsets is not a toggle rhythm since it does not admit any line that separates the onsets at pulses 3 and 13 from the other three.

If the right- and left-hand patterns of the sharp toggle rhythms are combined into single sequences, then the resulting timelines bear a close resemblance to the *pyramid* structure proposed by anthropologist Gerhard Kubik to characterize West and Central

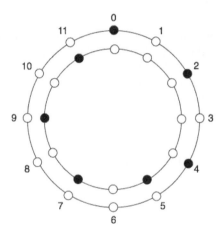

FIGURE 32.6 A double-circle portrayal of a *toggle* rhythm: the right- and left-hand onsets are contained on the outer and inner circles, respectively, fusing together to yield the *standard pattern*.

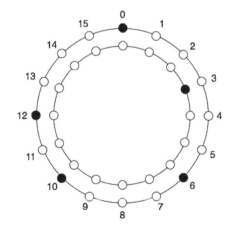 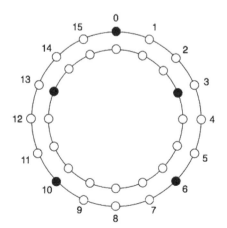

FIGURE 32.7 The *clave son* (left) and the bossa-nova clave (right) in circular toggle notation.

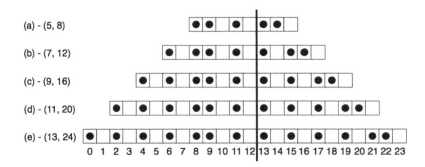

FIGURE 32.8 Kubik's *pyramid* of West and Central African asymmetric timeline patterns.

African asymmetric timeline patterns.[7] Kubik describes the structural relationships between some common African timeline patterns using the trapezoidal shape pictured in Figure 32.8. The pyramid is built from the top down starting with the five-onset, eight-pulse rhythm in Figure 32.8a, which is partitioned into two pieces of five and three pulses given by [x x . x .] and [x x .], respectively. The next rhythm in the pyramid is constructed by inserting the pattern [x .] at the leftmost end of both parts to obtain the rhythm in Figure 32.8b. This approach is continued to create the rhythms in Figure 32.8c–e. Note that putting together the left- and right-hand patterns in Figure 32.4, the rhythms in (b–f) are rotations of the five rhythms listed in

Kubik's pyramid structure of Figure 32.8. The 24-pulse pattern (Figure 32.8e) is the longest asymmetric timeline pattern found in sub-Saharan Africa and was discovered by Kubik and Maurice Djenda in 1966.[8]

NOTES

1 See Toussaint, G. T., (2010), for a more thorough description of the methods described in this chapter.
2 Scherzinger, M., (2010).
3 Kauffman, R., (1980), Table 1, p. 409.
4 Kubik, G., (2010a), p. 45.
5 Kubik, G., (2000), p. 288.
6 *Ibid.*
7 Kubik, G., (1999), p. 54.
8 Kubik, G., (1998), p. 218.

Symmetric Rhythms

Folklore has it that when Napoleon Bonaparte, having been exiled to the island of Elba, set foot there, he exclaimed the words "Able was I ere I saw Elba." Able he probably was, since he managed to escape from Elba within a year, and was soon in power again in France. Whether he knew English well enough to utter these words and whether he truly said them are open to question. However, one thing is irrefutable: this sequence of letters (ABLEWASIEREISAWELBA) reads the same forwards and backwards. This sequence of letters exhibits a property called mirror symmetry (also reflection symmetry), in this case reflection about the letter R, the only letter that occurs once in the sentence. A sequence that exhibits this type of mirror symmetry is called a *palindrome*. Palindromes have been the source of both delightful entertainment, and serious exploration in a wide variety of domains, for thousands of years. Mathematicians study palindromic numbers such as 37485658473.[1] Computer scientists analyze and generate palindromic sequences such as 100001, as well as recursively palindromic sequences such as 101101.[2] In a recursively palindromic sequence, the left and right halves of the sequence are also palindromic. Thus, in the latter sequence, the left and right halves given by 101 are

also palindromic sequences, but in the former sequence, the subsequences 100 and 001 are not. Writers and poets write books, stories, and poems that read the same forwards or backwards using either letters, words, or sentences as units.[3] Artists and designers use decorative frieze patterns that are visual shape palindromes.[4] The frieze patterns in Figure 33.1 are examples of geometric patterns that "read" the same from left to right and from right to left. In other words, each has reflection symmetry about a vertical line through the middle. Note that these patterns also enjoy mirror reflection about horizontal lines through their middles, and they are composed of pairs of adjacent units, each of which also has local mirror vertical and horizontal mirror symmetries. The frieze at the bottom is also recursively palindromic at three levels of mirror reflection about vertical lines.

Symmetry is one of the most consequential features of the world we inhabit. Alexander Voloshinov refers to symmetry as "the most important principle of harmony both in the universe and in art."[5] Our frontal appearance exhibits a strong mirror symmetry about a vertical line through the center of our bodies, even if our internal anatomy is less symmetric. So, it is not surprising that symmetry should play a major role in composing

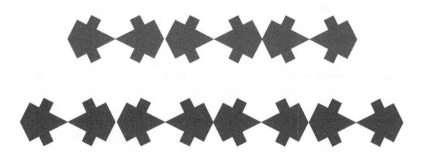

FIGURE 33.1   Palindromic (top) and recursively palindromic (bottom) geometric frieze patterns.

and perceiving music.[6] Simha Arom writes: "One of the features that is most helpful in perceiving metric structure is undoubtedly symmetry."[7] Elsewhere he suggests that symmetry might be a music *universal*.[8]

Many composers have taken advantage of the resources that symmetry has to offer.[9] Johann Sebastian Bach is a prime example. In his *Contrapunctus*, his notes have mirror symmetry about a horizontal line.[10] In *A Musical Offering*, the second half of the piece is the reverse of the first half.[11] His *Goldberg Variations* "is a musical journey through the world of symmetry."[12] Some composers have written pieces of music that sound the same when played forwards or backwards. Such compositions are also called *palindromes* or *crab canons*.[13] A nice example is Joseph Haydn's Symphony number 47 in G major which has a minuet written in 3/4 time (movement number three) that is referred to as *The Palindrome* precisely because it has this property. The melodic contour of the upper notes of this minuet is shown in Figure 33.2, using piano-roll notation. The horizontal axis indicates time in terms of shortest notes. The bold vertical lines indicate the bars of the 3/4 measures. The vertical axis demarcates the pitch of the notes in semitones. Observe that the section in the right diagram is the mirror image reflection of the section on the left. The two sections are shown separated for the

purpose of visualization, but of course in the actual piece, they are contiguous.[14]

A *palindromic* rhythm (also called *nonretrogradable*; because it cannot be used as a different new rhythm by merely reversing it in time) has the same property as Haydn's minuet: it exhibits mirror symmetry about a point in time.[15] Two examples of popular palindromic rhythm timelines that have mirror symmetry about their starting points are given in Figure 33.3. When the rhythms are expressed in polygon notation, this type of mirror symmetry translates to symmetry about a vertical line through the starting onset of the rhythm. The first simple but effective rhythmic ostinato (left) is an electric guitar riff used in the minimalist song titled "*This Must Be the Place*" (subtitled "*Naïve Melody*") from the Talking Heads album, *Speaking in Tongues*, released in 1983. This rhythm is used extensively in sub-Saharan African music. The distinctive lyrics of this song were penned by the band leader David Byrne, and the hypnotic and captivating music was composed in collaboration with his band members Chris Frantz, Jerry Harrison, and Tina Weymouth. The second, more complex palindromic rhythm is the drum "rap" rhythmic ostinato used by Chucho Valdez in his jazz composition "*Invitation*" (EGREM CD0233, Havana, Cuba, 1997). Many other palindromic rhythms are found throughout this book.

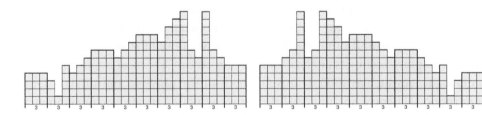

FIGURE 33.2 The palindromic minuet in Haydn's Symphony No. 47.

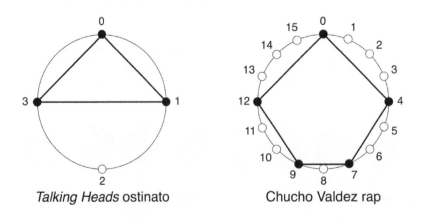

Talking Heads ostinato   Chucho Valdez rap

FIGURE 33.3 A pair of popular *palindromic* rhythms with vertical mirror symmetry.

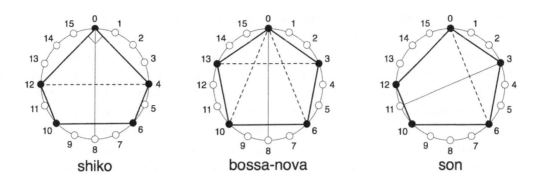

FIGURE 33.4　Three of the six distinguished timelines have reflection symmetry.

Three of the six distinguished five-onset, 16-pulse timelines, revisited throughout this book, and repeated in Figure 33.4 also have mirror symmetry. The shiko and bossa-nova have vertical mirror symmetry, whereas the *clave son* exhibits a more unusual mirror symmetry with respect to the *diagonal* line through pulses 3 and 11. It should be noted in passing that the term symmetric rhythm is sometimes reserved for rhythms that exhibit only vertical symmetry, and the *clave son* would be described as *asymmetric*. According to the terminology used here, however, the *clave son* is symmetrical, whereas the clave rumba with interval vector [3-4-3-2-4] is asymmetric. The distinction reflects whether the cyclic pattern is viewed as a rhythm with a fixed starting point or as a necklace independent of starting point.

If the Chucho Valdez rap rhythm of Figure 33.3 (right) is rotated clockwise by 90°, one obtains the rhythm in Figure 33.5 (left), which is employed in Beijing Opera music, and has horizontal mirror symmetry. Furthermore, if onsets at pulses 2 and 6 are inserted in this rhythm, a sub-Saharan African timeline called *kpatsa* results, as shown in Figure 33.5 (right).[16]

All these mirror-symmetric rhythms contain only one line of reflection. However, rhythms may possess many more such symmetries. Rhythm polygons that possess many axes of reflection will tend to resemble more like regular polygons. Indeed, regular rhythms with *k* onsets have *k* distinct lines of reflection symmetry, as illustrated in Figure 33.6. If a rhythm has an odd number of onsets such as the pentagon on the left, then there is an axis of reflection through every one of its onsets. On the other hand, if the number of onsets is even, as in the hexagon on the right, then there is an axis of reflection through every pair of antipodal onsets as well as through the midpoints of every pair of antipodal edges. In either case, there is an axis of mirror symmetry through every pulse.

It is useful for the purpose of rhythm classification, as well as for measuring the similarity between rhythms, to distinguish between several coarse types of reflection symmetries that may serve as features for characterizing families of rhythms. In addition to the vertical and horizontal reflection lines as well as the lines with positive slope (as in the *clave son*), we may include lines with negative slope, as well as a variety of combinations of these four categories.[17] These combinations permit the distinction between the nine classes of rhythms pictured in Figure 33.7. Traversing from left to right and top

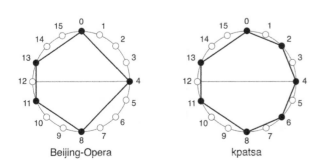

FIGURE 33.5　Two palindromic rhythms with horizontal mirror symmetry.

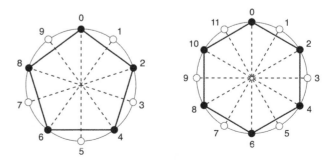

FIGURE 33.6　Regular rhythms possess an axis of mirror symmetry through every pulse.

to bottom, the first rhythm has all four types of mirror symmetry. The second has vertical and horizontal lines of symmetry. The third has diagonal reflection lines of positive and negative slope. Diagrams four to seven have a single diagonal line of symmetry in all four orientations. The eighth rhythm has one vertical and two diagonal lines of symmetry, and the ninth has one horizontal and two diagonal lines of symmetry.

The reader may have observed that Figure 33.7 does not contain the class of rhythms that have exactly two lines of symmetry such that one be either vertical or horizontal, and the other has a negative or positive slope. In short, no rhythms are shown that have exactly two reflection lines that are not orthogonal to each other. There is a good reason for this absence: it turns out that such rhythms do not exist. To see this let us first follow one onset of a rhythm as it is reflected across its lines of

symmetry.[18] As a warm-up exercise, consider the case of two reflection lines, one vertical and the other horizontal, for a cycle of 12 pulses as illustrated in Figure 33.8. Assume that the rhythm we desire has an onset at pulse one. Since the rhythm has a vertical reflection line (left) it must also contain an onset at pulse 11. Furthermore, since the rhythm has a horizontal line of reflection (right) it must also have onsets at pulses 5 and 7. If we reflect these two new onsets about the vertical or horizontal lines of reflection, no new onsets are created, and thus the final rhythm has onsets at pulses 1, 5, 7, and 11, and has only these two lines of reflection.

Now consider applying the same method to create a rhythm that has exactly two lines of reflection such that one is vertical and the other is diagonal (refer to Figure 33.9). Assume that the rhythm contains an onset at pulse 1 and a diagonal reflection line through

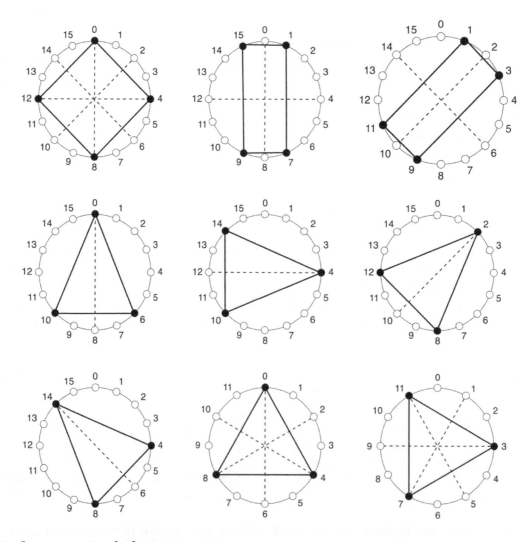

FIGURE 33.7 Some categories of reflection symmetries.

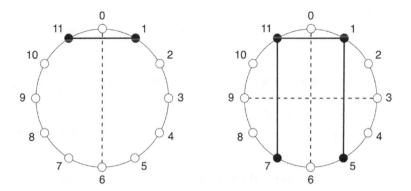

FIGURE 33.8   Generating a symmetric rhythm with two *orthogonal* reflection lines.

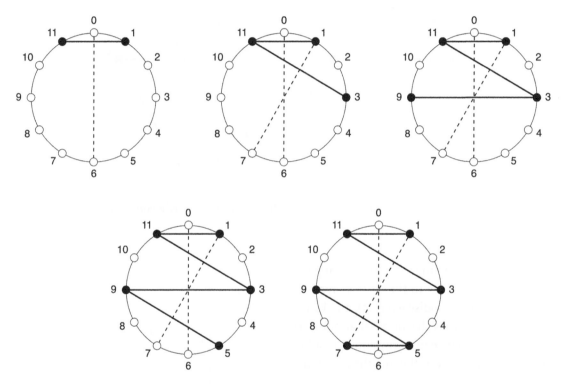

FIGURE 33.9   Attempting to generate a symmetric rhythm with *exactly two nonorthogonal* reflection lines by starting with an onset at pulse one.

pulses 1 and 7. Due to the vertical reflection line, there must be an onset at pulse 11, and because of the diagonal reflection line, this onset is reflected to pulse 3. The onset at pulse 3 is then reflected by the vertical line to pulse nine, which by the diagonal reflection line, goes to pulse 5, finally ending up at pulse 7, by virtue of the vertical reflection line. The rhythm thus created is a regular hexagon with onsets at pulses 1, 3, 5, 7, 9, and 11. However, this rhythm (being regular) has many other reflection lines: one through every pair of antipodal pulses.

As a second example (Figure 33.10), consider starting with an onset at pulse 3 and assume that in addition to

vertical symmetry, it is desired that the rhythm should contain mirror symmetry about the line through pulses 2 and 8. By vertical symmetry, there must be an onset at pulse 9 (left). Because of the diagonal symmetry, there must be onsets at pulses 1 and 7 (middle), and again by vertical symmetry there must also be onsets at pulses 11 and 5 (right). At this stage, no more onsets are induced by these six onsets and two symmetries. Again, the resulting rhythm is a regular hexagon, which has additional mirror symmetries through every pair of antipodal pulses.

As a final example, consider generating a rhythm with symmetry about the vertical line and the line

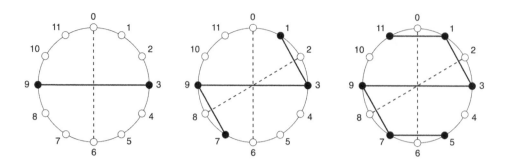

FIGURE 33.10 Attempting to generate a symmetric rhythm with *exactly two nonorthogonal* reflection lines by starting with an onset at pulse three.

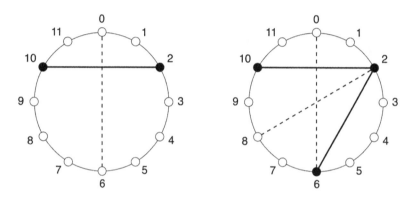

FIGURE 33.11 Attempting to generate a symmetric rhythm with exactly two nonorthogonal reflection lines by starting with an onset at pulse two.

through pulses 2 and 8 by starting with an onset at pulse 2, as in Figure 33.11 (left). Due to the vertical symmetry, there must be an onset at pulse ten (left), and by the diagonal symmetry there must be an onset at pulse six (right), yielding a triangular regular rhythm with additional diagonal symmetry about the line through pulse pair (4,10).

As these concrete examples suggest, no matter how one chooses the nonorthogonal pair of mirror symmetry axes, and no matter which onset one starts with in the iterative process outlined earlier, one will always generate a rhythm that contains a third axis of mirror symmetry. To see this in full generality, let the pair of points (*a,b*) and (*c,d*) in Figure 33.12 (left) denote two arbitrary nonorthogonal axes of symmetry. The symmetry about (*a,b*) implies that we may think of rotating the diagram out of the page about the axis (*a,b*) by 180° to obtain the same image as before rotation. But this rotation brings the axis of symmetry (*c,d*) into the position (*e,f*) on the right. Therefore, the rhythms must also have mirror symmetry about this axis. Furthermore, axis (*c,d*) will coincide with axis (*e,f*) only if the first two

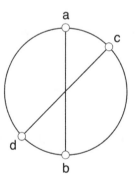 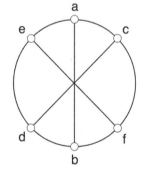

Two Symmetries   Resulting Symmetries

FIGURE 33.12 Two nonorthogonal axes of mirror symmetry (left) always induce a third axis of mirror symmetry (right).

symmetry axes are orthogonal to each other. Therefore, there must in fact be three distinct axes of mirror symmetry in such a situation.

HOURGLASS RHYTHMS

One special case of vertically symmetric rhythms deserves to be highlighted. These are called here symmetric *hourglass* rhythms. To illustrate the motivation

FIGURE 33.13   A Chinese drum in the shape of an *hourglass*. (Courtesy of Yang Liu.)

for adopting this terminology, consider the Chinese drum illustrated in Figure 33.13. It consists of a hollow cylindrical body with skins mounted on both sides, tied together with string. The main feature of this drum is that the shape of the body resembles an hourglass: the diameter of the drum first becomes smaller until the middle of the drum is reached, after which it becomes larger again.[19] An hourglass rhythm has the similar property that its consecutive durations first become smaller and then larger. This terminology has also been used to describe rhythmic phrasings that go from long to short, and then long again, such as the *Clapping Music* pattern of Steve Reich that has groupings of claps that vary as 3-2-1-2. In the Carnatic music of South India such rhythms

are called *damaru yati*, meaning "hourglass shape."[20] Like the symmetric hourglass drum, symmetric hourglass rhythms exhibit vertical mirror symmetry when viewed as rhythms, and other mirror symmetries when considered as necklaces. Apart from the tongue-in-cheek geometrical analogy of shape, no other connection is suggested between hourglass rhythms and hourglass-shaped drums. The examples below are a further special class of hourglass polygons: they are all isosceles triangles. Olivier Messiaen begins his discussion of nonretrogradable rhythms by focusing on such triangles thus: "Outer values identical, middle value free. All rhythms of three values thus disposed are nonretrogradable."[21]

Hourglass rhythms have appeared in many places in this book. For example, the rap rhythm used by Chucho Valdez with durational pattern [4-3-2-3-4] is an hourglass rhythm.[22] Here we give some examples of notable three-attack hourglass rhythms. Such hourglass rhythms are quite common in Chinese,[23] Korean,[24] and Indian[25] music. Two hourglass rhythms with duration interval vectors [2-1-2] and [3-2-3], used in the Beijing Opera, are shown in Figure 33.14. These rhythms are also found in India: [2-1-2] on the left is the *denkhî* and [3-2-3] is the *mathya-tisra* decitala. Three additional Indian hourglass rhythms with $n = 10$, 16, and 20 are shown in Figure 33.15.

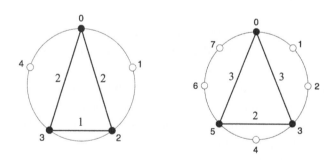

FIGURE 33.14   Two *hourglass* rhythms used in Beijing opera.

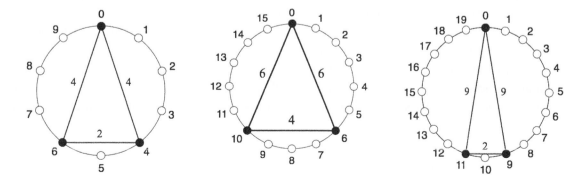

FIGURE 33.15   Three *hourglass* rhythms used in India: *dhenki* (left), *vijaya* (center), and *mathya-samkirna* (right).

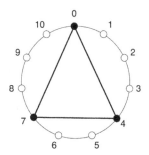

FIGURE 33.16 An *hourglass* Balkan folk rhythm with $n = 11$.

An hourglass rhythm with 11 pulses used in Balkan folk music is pictured in Figure 33.16.

SUBSYMMETRIES

The hourglass rhythms of Figures 33.14–33.16 consist of only three highly structured interonset intervals (IOIs) forming isosceles triangles, and so the reflection symmetry, which is a *global* all-or-none (Yes–No, binary) property, is easily manifested. However, nonregular rhythms with a relatively large number of onsets are unlikely to possess global reflection symmetry. Of the six distinguished timelines, the shiko, bossa-nova, and son (Figure 33.4) enjoy global reflection symmetry, whereas the rumba, soukous, and gahu do not. Therefore, to quantify the amount of symmetry that rhythms possess and obtain a finer graded discrimination between the rhythms, we should measure the number of *local* reflection symmetries they contain. Such symmetries are called *subsymmetries*.[26] In general, even patterns that do not exhibit *global* reflection symmetry may contain an abundance of *local* subsymmetries, which have been found to be more powerful for pattern classification.[27]

The total number of subsymmetries of all lengths of a pattern serves as a measure of the simplicity of the pattern. A pattern containing relatively many subsymmetries will tend to be relatively simple, and a pattern with relatively few subsymmetries will tend to be relatively complex.[28] Figure 33.17 illustrates the calculation of all subsymmetries contained in the *clave son* when the fundamental symbols used in the calculation are pulses. Hence, these subsymmetries are called *pulse*

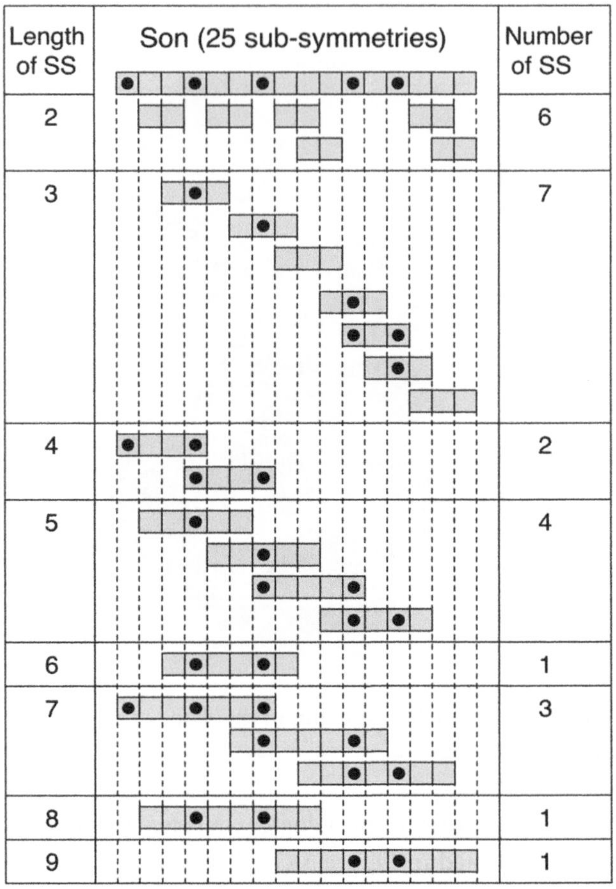

Length of SS	Son (25 sub-symmetries)	Number of SS
2		6
3		7
4		2
5		4
6		1
7		3
8		1
9		1

FIGURE 33.17 Calculation of the number of *pulse subsymmetries* of the *clave son*.

subsymmetries. Large subsymmetries are scarce, as evidenced by the fact that lengths 8 and 9 yield only one subsymmetry. In contrast, short subsymmetries are abundant, and lengths 2 and 3 yield 6 and 7 subsymmetries, respectively. The number of pulse subsymmetries for each of the six distinguished timelines is shown in Figure 33.18 (top row). As one would expect, the son (with 25 pulse subsymmetries) is deemed simpler than the rumba (with 24) but more complex than the shiko (with 29).

For relatively longer rhythms that contain many silent pulses relative to the number of onsets, it is advantageous to compute subsymmetries using symbols that are longer than a pulse. A natural candidate that fits the bill in this context is the IOI, which yields

Property	Rumba	Son	Soukous	Gahu	Bossa	Shiko
Pulse Sub-symmetries	24	25	25	28	28	29
IOI Sub-symmetries	1	2	1	2	4	4

FIGURE 33.18 The number of pulse IOI subsymmetries of the six distinguished timelines.

IOI-subsymmetries. The number of IOI subsymmetries for each of the six distinguished timelines is shown in Figure 33.18 (bottom row). Again, in agreement with expectation, the son (with two IOI subsymmetries) is deemed simpler than the rumba (with one) but more complex than the shiko (with four). Listening experiments with human subjects have provided evidence that the IOI subsymmetries are better predictors of human judgments of rhythm perception and production (performance) complexities than the pulse subsymmetries for several datasets tested.[29]

The number and structure of IOI subsymmetries that the six distinguished timelines possess (pictured in Figure 33.19) provide features for yet another method for rhythm and nonisochronous meter classification. The soukous has only one IOI subsymmetry, and it covers only a small fraction of the rhythm (six pulses). The rumba also contains only one IOI subsymmetry and it covers ten pulses. The gahu has two contiguous nonoverlapping IOI subsymmetries that cover 14 pulses. The remaining three rhythms are completely covered by their IOI subsymmetries. However, the shiko and bossa-nova are each covered by four overlapping IOI subsymmetries. The son is unique among the six-distinguished timelines, in that it is the only rhythm completely covered by precisely two nonoverlapping IOI subsymmetries, namely [3-3] and [4-2-4].

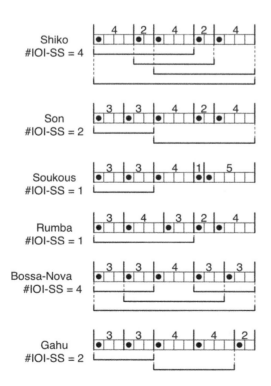

FIGURE 33.19 Calculation of the number of IOI-duration subsymmetries of the six distinguished timelines.

NOTES

1 Guy, R. K., (1989).
2 Ji, K. Q. & Wilf, H. S., (2008).
3 Montfort, N. & Gillespie, W., (2002).
4 Liu, Y. & Toussaint, G. T., (2011, 2010b, 2010c). Shape palindromes have also been called inversions by Kim, S., (1996).
5 Voloshinov, A. V., (1996), p. 109. Just as symmetry is much sought in composing music, so is the breaking of symmetry, leading some to characterize the relation in terms of "love-hate." See Wilson, D., (1986).
6 Donnini, R., (1986) and Liebermann, P. & Liebermann, R., (1990).
7 Arom, S., (1991), p. 188. Christensen, T., (2002), p. 682.
8 Arom, S., (2001), p. 28.
9 Feldman, M., (1981) and Hunter, D. J. & von Hippel, P. T., (2003).
10 Hargittai, I. & Hargittai, M., (1994), p. 14.
11 Harkleroad, L., (2006), p. 37.
12 Du Sautoy, M., (2008), p. 251.
13 Hodges, W. & Wilson, R. J., (2002), p. 83. Hodges, W., (2006), discusses other types of symmetries used by many composers.
14 Handel, S., (1988), p. 315, emphasizes that "the auditory and visual worlds are inherently both temporal and spatial" and "it is a mistake to partial out space and time to different senses." In addition, Schutz, M., (2008), p. 102, outlines considerable evidence that "visual information is an important component of the musical experience," and that "multimodal representations are more accurate." However, Handel, S., (2006), p. 188, provides evidence that it is more difficult to perceive (temporal) symmetry than visual symmetry.
15 Palindromic or nonretrogradable rhythms are strongly associated with the French composer Olivier Messiaen. See Johnson, R. S., (1975) and Messiaen, O., (1956), Chapter 5.
16 Eckardt, A., (2008).
17 Since we are dealing here with rhythms, and not necklaces, the "horizontal," "vertical," and "skew" directions make musical sense because they determine distinct and separate temporal symmetries. Of course, the simplicity of recognizing the visual symmetries present in these polygonal representations of the rhythms does not imply that their corresponding temporal symmetries are easy to perceive. It would be interesting to determine the relative perceptual importance of these symmetries in the temporal domain, as well as the correlation with their visual counterparts.
18 This is a gentle introduction to group theory. For a more advanced treatment on the application of group theory to music, the reader is referred to Noll, T., (2007).

19 A similar hourglass drum called *dhāk* is used in southern India, where the name also refers to a spirit-possession ritual that incorporates the drum (see Roche, D., (2000), p. 63.).

20 Hartenberger, R., (2016), p. 159.

21 Messiaen, O., (1956), p. 20.

22 Chucho Valdez used this rhythm on a drum in his jazz composition *Invitation*, (EGREM CD0233, Havana, Cuba, 1997).

23 Wells, M. St. J., (1991), p. 149.

24 Lee, H.-K., (1981), p. 123.

25 Roche, D., (2000).

26 Alexander, C. & Carey, S., (1968).

27 Toussaint, G. T., (2015a).

28 Toussaint, G. T., & Beltran, J. F., (2013).

29 Toussaint, G. T., (2015b).

CHAPTER 34

Rhythms with an Odd Number of Pulses

IN THE LATTER HALF OF THE NINETEENTH CENTURY, French archaeologists were busy searching for ancient lost cities in North Africa and Greece that had been buried over time by the geological forces of nature. Since 1861, they had been keenly interested in excavating the ancient site of Delphi in Greece, on the southwestern promontory of Mount Parnassus on the mainland. However, they were prevented from moving forward with this project due to bureaucratic stumbling blocks that lasted until 1893. The archaeologists concentrated on a temple dedicated to the god Apollo, best known for being the site of an ancient oracle. Apollo was the god of music, poetry, truth, prophecy, healing, and sunlight, just to name a few of his laurels, but music appears to be his specialization, for statues depicting Apollo frequently show him holding or playing a lyre. The French archaeologists made history when they discovered two hymns inscribed in stone fragments, in decipherable notation unambiguous in its rhythms.[1] Delphi also contained a stadium where every 4 years the Pythian Games were held in honor of Apollo. The discovery of these ancient hymns at the site of the games

inspired the French educator and historian Baron Pierre de Coubertin to attempt to revive the ancient Olympic games.[2] To this end, in 1984, he organized a conference at the Sorbonne University in Paris, with the goal of creating an International Olympics Committee, at which the ancient hymns were played to great effect. The event was a resounding success, and 2 years later, on April 6, 1896, the first modern Olympic games was celebrated before 80,000 spectators in the *Panathenaic* Stadium in Athens.[3]

The performance of the Delphic hymns at the conference in the Sorbonne must have sounded strange to many ears in the audience, since these hymns were composed in the quintuple meter, which was rather unusual in the music that most listeners were accustomed to hearing at that time in Western Europe. Quintuple meter uses measures (or cycles) composed of five pulses (an unusual odd number in that place and time).[4] The first ten measures of the first Delphic hymn are shown in box notation in Figure 34.1.[5] The rhythms of all the measures used in the Delphic hymns come in only four distinct varieties: [2-1-2], [2-1-1-1], [1-1-1-2], and [1-1-1-1-1].

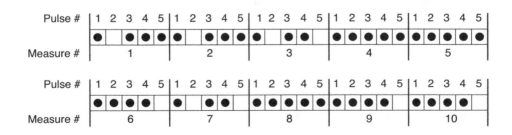

FIGURE 34.1 The rhythms in the first ten measures of the first *Delphic Hymn* to Apollo.

In ancient Greece, quintuple rhythms were common but utilized primarily for the more serious music practiced in religious ceremonies, such as sacred dances and hymns, where the expression of intensity of emotion was a desired outcome. Rhythm was extremely important to the ancient Greeks, as it was deemed to have the power to produce certain specific emotional effects upon the listener. For instance, the rhythm [x . x .] was regarded to be appropriate in religious contexts because it inspired awe, whereas the rhythm [x x x .] was suitable for military marches because it produced energy. C. F. Abdy Williams wrote in 1911, "To the Greek musician the laws of rhythm were as important as those of harmony and counterpoint are to the modern student."[6] Some additional ancient Greek quintuple rhythms not used in the Delphic hymns are shown in Figure 34.2 in polygon notation.

In a tradition that is believed to be hundreds of years old, the Tuareg people, who now live mainly in West Africa, play two rhythms in the performance of their heroic ballads, called *n-geru* and *yalli,* that use the quintuple durational pattern [x . x . .].[7] This rhythm is the same as the second ancient Greek pæonic rhythm shown in Figure 34.2. It is also the metric pattern of the Bulgarian *paidushko horo*[8] and the Macedonian *paiduska* dance.[9] Since the Tuareg people migrated during the past 2,000 years to West Africa from a region closer to Greece in North Africa, it is possible that this rhythm was passed on from Greece to Africa. Perhaps this rhythm forms part of a tradition of the Tuareg people that is older than previously believed. However, this is mere speculation. Indeed, this rhythm is also an ancient Arab-Persian rhythm dating back to at least thirteenth-century Bagdad called *khafif al-ramal*.[10] The quintuple meter [2-1-2] (the Cretic variant in Figure 34.2) is also used in Korean instrumental music.[11]

Leaping forward in time to the twentieth century, the jazz pianist Dave Brubeck was fond of using meters with an odd number of pulses (other than the number three common in the waltz) in some of his best-selling compositions.[12] One of Brubeck's most popular works is the piece "*Take Five,*" written in quintuple time by Brubeck's saxophonist Paul Desmond.[13] The piece was written in a 5/4 time signature with meter [3-2] as shown in Figure 34.3 (left) akin to the ancient Greek two-limbed pæon in Figure 34.2. However, the piano rhythm (and metric feel of the piece) is the six-onset pattern [**X** x . **X** x . **X** . **X** .] with beats at pulses 0, 3, 6, and 8 forming the metric beat pattern [3-3-2-2] shown in Figure 34.3 (right).[14] Scott Murphy provides a list of 38 popular works in quintuple meter, for which the metric patterns [3-2] and [2-3] occurred 31 and 7 times, respectively.[15]

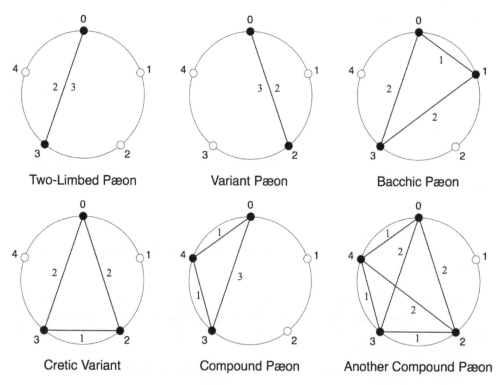

FIGURE 34.2 Six *pæonic* or *quintuple* rhythms of ancient Greece.

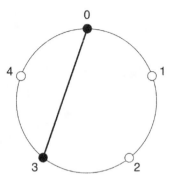
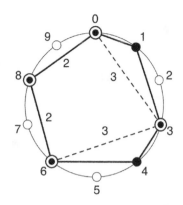

FIGURE 34.3 The quintuple rhythms in Dave Brubeck's "*Take Five*." The 5/4 time signature with beat interonset intervals [3-2] (left), and embedded in a ten-pulse cycle, the piano rhythm [1-2-1-2-2-2], and meter with beats at pulses 0, 3, 6, and 8 forming beat pattern [3-3-2-2] (right).

A second nice example of meters with an odd number of pulses in Brubeck's music is the composition "*Unsquare Dance*." This piece uses a cycle of seven pulses (septuple meter) and is characterized by a three-onset bass line and a four-onset complementary hand-clapping pattern shown in Figure 34.4. The bass line with durational pattern [2-2-3] is the metric pattern of a Bulgarian Easter dance.[16] When started on the third onset, the meter [3-2-2] is obtained, which is the most common rhythm in Macedonia[17] and used in the *makedonsko horo*.[18] Note that both rhythms are *deep*. In the bass rhythm, the interval of duration three occurs once and the interval of duration two twice. In the hand-clapping pattern, the intervals of durations 1, 3, and 2 occur once, twice, and thrice, respectively. If one rotates the bass rhythm by two pulses in a counterclockwise direction, the pattern [2-3-2] results, which is also the meter of the third movement (*Precipitato*) of Prokofieff's "*Piano Sonata Number 7*."[19] Andrew J. Gustar provides a comprehensive and detailed meme-based history of the evolution of septuple meter, across the world starting from ancient Greece, through Western art, ending

with popular music.[20] In addition to the list of quintuple meters, Scott Murphy also provides a list of 70 popular works in septuple meter, for which the metric patterns [2-2-3], [3-2-2], and [2-3-2] occurred 51, 16, and 3 times, respectively.

The next odd number higher than seven is nine. Rhythm timelines or meters that employ nine pulses make their appearance in scattered places that span the planet. Among the San Bushmen of Southern Africa, we find the timelines with duration intervals [2-3-4] and [2-4-3] in Namibia, and [2-2-2-3] in the North–West Kalahari and Botswana.[21] These rhythms are shown in polygon notation in the top row of Figure 34.5.

In Dave Brubeck's "*Blue Rondo à la Turk*," another of his biggest hits, inspired by listening to street musicians while traveling in Turkey, the rhythm alternates between the two nine-pulse rhythmic ostinatos shown in Figure 34.5 (top right and bottom left). The first rhythm with durational pattern [2-2-2-3] is the metric pattern of the Thracian wedding dance *daichovo horo*,[22] and the Turkish *dum-tek* rhythm *tchifty sofian*, in which case the *dum* is sounded on the first and third onsets.[23]

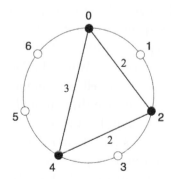
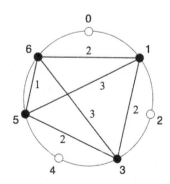

FIGURE 34.4 Seven-pulse rhythms: the bass pattern (left) and the clap pattern (right) in Dave Brubeck's "*Unsquare Dance*."

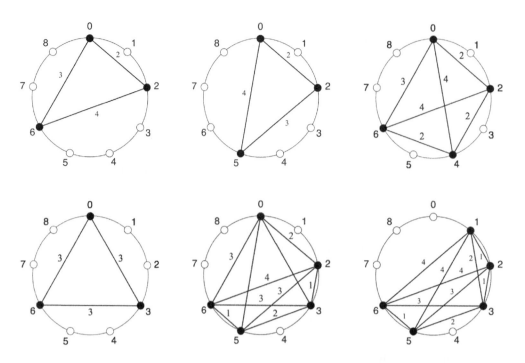

FIGURE 34.5 Nine-pulse rhythms: Top left and top center are the San Bushmen rhythms of Namibia. Top right is the San Bushmen rhythm from the North–West Kalahari and Botswana, as well as one of the two alternating variations in the rhythm timeline of Dave Brubeck's "*Blue Rondo à la Turk*" (the other being the rhythm at the bottom left). At bottom center is the Irish slip jig, and at bottom right is the Transylvanian *invîrtita* dance rhythm.

The other onset rotations of this rhythm, [2-2-3-2], [2-3-2-2], and [3-2-2-2], are also used in Turkey,[24] and the latter meter is used by Don Ellis in "*Strawberry Soup.*"[25] The meter [2-3-2-2] is also the Bulgarian rhythm *chelebishko horo.*[26] These duration patterns are also *deep* rhythms: the intervals of durations 3, 4, and 2 occur once, twice, and thrice, respectively. The second rhythm [3-3-3] is not deep, but it is maximally even. In spite of the similarities in titles, Brubeck's piece should not be confused with Mozart's "*Rondo Alla Turca—Turkish March,*" the Allegretto in his Piano Sonata No. 11, which has a simple 2/4-time signature. Although timelines with nine pulses are rare in sub-Saharan Africa, Robert Kauffman lists this as a common rhythmic pattern.[27] Another perhaps unlikely geographical location in which rhythms with nine pulses are found is Ireland, famous for its *jigs* since at least the seventeenth century.[28] The slip jig (also called hop jig) with metric pattern [2-1-2-1-3] is shown in polygon notation in Figure 34.5 (bottom center). It may be viewed as the eight-pulse cinquillo [2-1-2-1-2], with a silent pulse added at the end. It is not a deep rhythm, since the intervals of duration 1 and 2 both have multiplicity equal to 2. Neither is it maximally even, since five onsets distributed among nine pulses yields a

maximally even rhythm [2-2-2-2-1]. This rhythm is a rotation of [2-2-2-1-2], a Rumanian folk dance rhythm.[29] There are also nine-pulse rhythms that start on a silent pulse such as the Transylvanian *invîrtita* dance rhythm shown on the bottom-right diagram.[30] This rhythm also is not deep, since it contains three intervals of duration 4 and 1.

Figure 34.6 depicts three 11-pulse metric patterns. The first (left) with durational pattern [3-2-3-3] is the meter in Dave Brubeck's "*Countdown.*" If it is started on pulse five, it is the meter of Frank Zappa's "*Outside Now.*"[31] The intervals of durations 2, 5, and 3, occur once, twice, and thrice, respectively, and hence this rhythm is also deep. The second meter (center) with intervals [2-2-2-3-2] comes from the chorus of "*I Say a Little Prayer,*" composed in 1967 by Burt Bacharach and Hal David for singer Dionne Warwick. Bacharach is a prolific music producer as well as a composer and pianist known for his syncopated rhythms and odd-pulse time signatures. The rhythm he used here is also deep: the intervals of durations 3, 5, 4, and 2 occur once, twice, thrice, and four times, respectively. The third meter (right) with interval structure [2-2-3-2-2] is from the Bulgarian folk dance *krivo horo.*[32] Since it is a rotation of the second

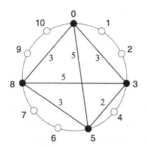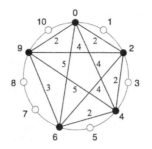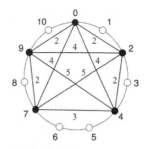

FIGURE 34.6 Eleven-pulse rhythms: the meters of Dave Brubeck's *Countdown* (left), Burt Bacharach and Hal David's chorus of "*I Say a Little Prayer*" (center), and the meter of the Bulgarian folk dance *krivo horo* (right).

meter, it is of course deep. This rhythm as well as [2-2-2-2-3], [2-3-3-3], and [4-3-4] are all meters used in Balkan folk songs.[33]

Three 13-pulse rhythms are pictured in Figure 34.7. The first rhythm (left) with durational pattern [2-3-3-2-3] is a Balkan folk rhythm that was used in Olivier Messiaen's movement No. 20 of his composition "*Vingt Regards sur l'Enfant-Jesus* titled *Regard de l'Église d'Amour*" (or "*Gaze of the Church of Love*").[34] It is another example of a deep rhythm: the interval durations of 6, 2, 3, and 5 pulses occur once, twice, thrice, and four times, respectively. Messiaen was well known for his use of complex rhythms, strongly influenced by the music of India and ancient Greece. He was fond of rhythms with five-pulse cycles, especially the palindromic (nonretrogradable) rhythms such as the Greek *amphimacer* [4-2-4] and the Indian *denkî* [2-1-2].[35]

The second rhythm in Figure 34.7 (center) with durational pattern [3-3-2-2-3] is the meter in the composition "*Connection*" by Don Ellis, a drummer and trumpet player, a composer, and a leader of a big band during the 1960s and 1970s. However, like Dave Brubeck, Ellis is most famous for his compositions with odd meters such as 5, 7, 9, 11, 19, 25, and 33 or more, as well as his polyrhythmic compositions within more conventional time signatures. Some musicologists have used the term "exotic" to describe his rhythms. Don Ellis called them simply "new rhythms" and divided them into two categories for which he is best remembered: (1) complex rhythms within regular meters and (2) "odd meters". The latter category, he eagerly points out, does not refer to "strange" or "weird" rhythms, but rather to those with an odd (as opposed to even) number of pulses in their rhythmic cycle.

In his PhD thesis, Sean Fenlon points out that Ellis divided the exotic meters into two other categories that he called "straight ahead" meters and the more complex "additive" meters. He reserved the term "straight ahead" for those meters that had no accent on any pulse or had an accent only on the first pulse of the cycle. Thus, [x . . . .] is an example of a "straight ahead" five-pulse meter, whereas [x . x . .] is not. By "additive" meters, Ellis meant the construction of meters by the concatenation of groups with at least one group of two pulses and one of three. Among the many noteworthy examples of such complex additive meters used by Ellis are the odd 19-pulse [3-3-2-2-2-1-2-2-2], the odd 25-pulse [2-2-3-2-3-2-2-3-2-2-2] in *How's This for Openers?* as well as the even 18-pulse [3-2-2-2-3-2-2-2] employed in *Strawberry Soup*, one of his most well-liked and sophisticated composition.[36]

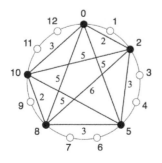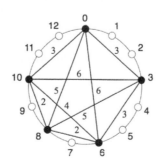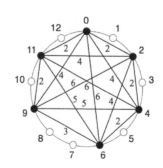

FIGURE 34.7 Thirteen-pulse rhythms: (left) "*Vingt Regards sur l'Enfant-Jesus: movement 20*," Messiaen, (center) "*Connection*" by Don Ellis, and (right) Bulgarian meter *krivo plovdivsko horo*.

The third rhythm in Figure 34.7 (right) with interval sequence [2-2-2-3-2-2] is a Bulgarian dance rhythm called *krivo plovdivsko horo*, meaning *"Crooked dance from the Plovdiv region."*[37] A rotation of this rhythm with durational pattern [2-2-2-2-2-3] is the meter of a Balkan song.[38] This particular 13-pulse rhythm is also deep: the intervals of durations 3, 5, 6, 4, and 2, occur once, twice, thrice, four times, and five times, respectively.

In Bulgarian music, there also exists a common 15-pulse meter named *buchimish*, with interval structure [2-2-2-2-3-2-2].[39] As the examples illustrated suggest, the characteristic feature of Bulgarian dance music, and Balkan music in general, is the use of asymmetric meters composed of groups of two-pulse and three-pulse beats. These rhythms are often referred to as aksak rhythms, already introduced in Chapter 8. Simha Arom provides a detailed classification of aksak rhythms and a long list of authenticated aksak rhythms that contain additional numbers of pulses including 17, 19, 21, 25, 27, 33, and 35.[40]

NOTES

1 Reinach, T., (1893). See also Weil, H., (1893).
2 Macaloon, J. J., (2007). See also Guttmann, A., (1992).
3 Smith, M. L., (2004).
4 Odd here refers to the mathematical number contrasted with *even* numbers. One of the most fundamental mathematical two-class disjoint categorizations of rhythms is into those with an even number of pulses, and those with an odd number of pulses. Within these two categories, further bifurcations may arise as a result of applying musical or additional mathematical features. The word "odd" is also used to suggest that such rhythms were considered strange (or artificial) among the population at large (see Frampton, J. R., (1926), p. 400). This is not to imply that measures of five pulses were absent from classical music of the nineteenth century. A noteworthy example is the second movement of Tchaikovski's Symphony No. 6 in B minor, Op. 74, titled *Pathétique*, composed in 1893. I am indebted to the Harvard University mathematician Noam Elkies for this example.
5 Abdy Williams, C. F., (2009), p. 39.
6 Abdy Williams, C. F., (2009), p. 1.
7 Wendt, C. C., (2000), p. 221.
8 Petrov, B., (2012), p. 160.
9 Singer, S., (1974), p. 386.
10 Wright, O., (1978), p. 216. The cycle of the *khafif al-ramal* actually has ten pulses, and so the complete rhythm is [2-3-2-3].
11 Ku, L. H., (1981), p. 123.
12 Tracey, A., (1961), p. 113. In spite of Dave Brubek's legendary successful career and his many fans, some critics were not too sympathetic with his rhythmic explorations. Andrew Tracey criticizes Brubek for not being African enough in the choice of his rhythms, while admitting that time signatures such as [2-2-2-3] look promising. However, this time signature is more Turkish than African.
13 Lamb, E., (2012), p. 3.
14 Keith, M., (1991), p. 132.
15 Murphy, S., (2016). Appendix A.
16 Rice, T., (2004), p. 51. Fracile, N., (2003), p. 199, lists both [2-2-3] and [3-2-2] in their compilation of Balkan rhythms. The meter [2-2-3] is also the Bulgarian *ruchenitza horo* (see Petrov, B., (2012), p. 160.).
17 Singer, S., (1974), p. 386.
18 *Ibid.*, p. 43.
19 Sicsic, H.-P., (1993).
20 Gustar, A. J., (2012).
21 Poole, A., (2018).
22 *Ibid.*, p. 33 and Petrov, B., (2012), p. 160.
23 Hagoel, K., (2003), p. 115.
24 Cler, J., (1994), p. 194.
25 Keith, M., (1991), p. 126.
26 Petrov, B., (2012), p. 160.
27 Kauffman, R., (1980), Table 1, p. 409.
28 Hast, D. E. & Scott, S., (2004), p. 66.
29 Proca-Ciortea, V., (1969), rhythm No. 51 in group X, p. 188.
30 *Ibid.*, p. 183.
31 Keith, M., (1991), p. 126.
32 London, J., (1995), p. 68. The generic name for Bulgarian dances that use 11-beat meters is *kopanitsa*. See Rice, T., (2004), p. 82 and Petrov, B., (2012), p. 160.
33 Fracile, N., (2003), p. 199.
34 Walt, van der S., (2007).
35 Šimundža, M., (1987, 1988).
36 Fenlon, S. P., (2002).
37 Rice, T., (2004), p. 67.
38 Fracile, N., (2003), p. 199.
39 Petrov, B., (2012), p. 160.
40 Arom, S., (2004), p. 45.

CHAPTER 35

Visualization and Representation of Rhythms

In this book, I have up to this point used predominantly three notation systems that are most convenient for the types of analyses undertaken: the box notation (in several variations), the convex polygon notation, and the interonset interval (IOI) vector notation (numerical durational patterns). The first two approaches are geometric and emphasize visualization. The third method indicates duration with numbers. Another noteworthy numerical system is *gongche* notation, the traditional Chinese system that uses numbers to indicate pitches, and dots and lines for rhythm.[1] The musical information and the notation systems that encode this information are interdependent. They have their unique advantages and drawbacks, and different applications may benefit more from notation systems that are tailored to them.[2] Music performers, ethnomusicologists, and music psychologists may also benefit more from notation systems that are tailored to their particular needs.[3] In this chapter several additional notation systems are reviewed, and the contexts in which they are used are indicated. Standard Western music notation is notably left out of this discussion because its descriptions are ubiquitous.

ALTERNATING-HANDS BOX NOTATION

In Chapter 35, a technique using alternating hands was described for generating toggle rhythms. The technique is also ideal for teaching and learning hand drumming. Recall that it consists of continuously, and gently, striking the drum by alternating with the right and left hand, much like walking, making sure that all the durations between two consecutive strikes are equal, and that all the sounds made are identical to each other, much like a metronome.[4] The box notation system can be modified so that it provides a clearer visualization of the alternating-hands process and that it is easy to read while practicing the technique, as illustrated in Figure 35.1. This notation also helps the student understand and remember the structure of the rhythms more easily. In addition, it helps the student acquire coordination, left–right independence, and timing precision, by embodying a metronomic sense of pulse that is essential for mastering rhythmic performance.

The idea is simple: use two side-by-side sequences of boxes, one for each hand, and rotate them to a vertical position so that the left and right hands play the left and right columns, respectively. In Figure 35.1 time flows from the bottom to the top. On the left is the *clave son*, and on the right the clave rumba. The differences between the two rhythms are highlighted by the onsets played by the left hand. The *clave son* contains only one left-hand onset at pulse three, whereas the clave rumba contains an additional left-hand onset at pulse seven. In order to develop proper left–right independence, playing with the reversal of the roles of the left and right hands should also be mastered.

SPECTRAL NOTATION

Consider the IOI durations of the *clave son* [3-3-4-2-4] and plot these numbers, in the order in which they occur, as heights of bars in a bar graph, such as that illustrated in Figure 35.2 (left). This type of notation has been used by researchers in the linguistics field interested in the

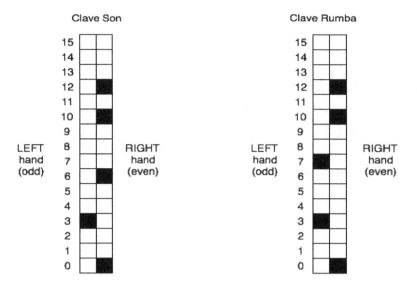

FIGURE 35.1   The *alternating-hands* box notation.

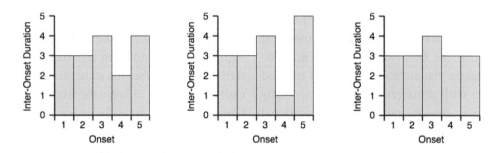

FIGURE 35.2   From left to right, respectively, the *son*, *soukous*, and *bossa-nova* timelines in spectral notation.

analysis of speech rhythms. This notation provides a compact view of the *spectrum* of the IOIs in the order in which they appear in time, and it accentuates the irregularity (or regularity) of the rhythm. Hence, we call it spectral notation. In Figure 35.2 it may be perceived quite compellingly that the bossa-nova (right) is more regular (uniform) than the soukous (middle).

Any rhythm notation system suggests a variety of new methods for measuring the distance (dissimilarity) between rhythms, and spectral notation is no exception. A natural way to measure distance in this case is by overlaying one spectrum over the other and calculating the area of the region in between the two curves, as shown in Figure 35.3.

On the left, the spectra of the *clave son* and bossa-nova are superimposed over each other, and the area difference between the two curves, highlighted in dark gray, is 2. On the right, the bossa-nova spectrum is superimposed over the soukous spectrum to reveal an area between the curves equal to 4. Thus, we conclude

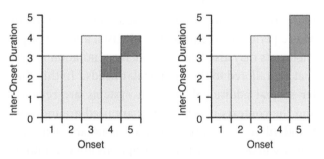

FIGURE 35.3   The area difference between the spectra of the *son* and the *bossa-nova* (left), and between the spectra of the *bossa-nova* and the *soukous* (right).

that the bossa-nova is more similar to the son than to the soukous.

TEDAS NOTATION

A disadvantage of the spectral notation elucidated in Figure 35.3 is that the veracity of the relative time durations of the IOIs along the time axis is lost because these intervals are displayed in the vertical direction. In an attempt to recover this valuable lost visual information, while at the same time maintaining some of the advantages of the spectral notation, in 1987 Kjell Gustafson, a researcher at the Phonetics Laboratory of the University of Oxford interested in displaying speech rhythm, proposed an original and simple method that places time on both the vertical and horizontal axes.[5] Since an element that displays the durations along two orthogonal directions determines a square, Gustafson called his notation TEDAS notation, an acronym for *Time Elements Displayed as Squares.* Figure 35.4 shows (top to bottom) the son, soukous, and bossa-nova timelines in TEDAS notation.

The TEDAS notation also suggests measuring the distance between rhythms by the area difference of their graphs.[6] In Figure 35.5 (left), the TEDAS graphs of the son and bossa-nova superimposed over each other yield an area difference between the two curves, highlighted in dark gray, equal to 6. On the right, the TEDAS graph of the son superimposed over the TEDAS graph of the soukous yields an area between the curves equal to 8. Thus, by this measure, we may conclude that the son is more similar to the bossa-nova than to the soukous. By representing the IOI durations as squares of the durations, the TEDAS notation places greater emphasis on longer IOIs, since two-dimensional area grows as the square of the one-dimensional duration. How the application of such a weighting scheme of the IOIs to the design of similarity measures affects the predictability of human judgments is still an open problem.

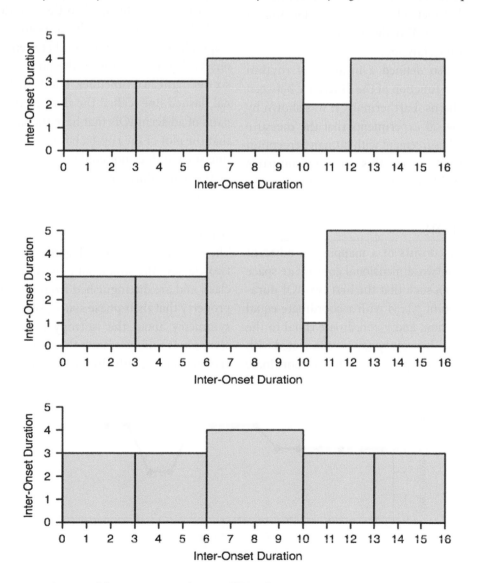

FIGURE 35.4 The *son, soukous,* and *bossa-nova* timelines in TEDAS notation.

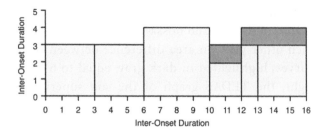 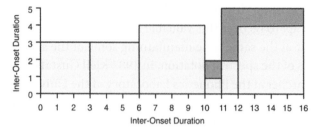

FIGURE 35.5   The area difference between the *son* and the *bossa-nova* (left), and between the *son* and the *soukous* (right).

CHRONOTONIC NOTATION

A variant of the TEDAS notation was later discovered by the music psychologist Ludger Hofmann-Engl who christened it *chronotonic* notation. In this notation, every onset of duration *k* pulses receives *k atoms* (pulses) of height *k*. The sequence of atoms of a rhythm connected by line segments between adjacent atoms determines a *chonotonic chain*. Figure 35.6 illustrates the *chronotonic chain* obtained for the *clave son*.

Hofmann-Engl also defined a measure of rhythm similarity based on a function of the difference between two chronotonic chains. Furthermore, it was shown by means of psychological experiments that the measure exhibits considerable agreement with human perception of rhythmic similarity.[7] How this measure compares with other more well-known measures, such as the edit distance, is yet to be determined.

PHASE SPACE PLOTS

A phase space plot consists of a mapping of adjacent IOI durations onto a two-dimensional coordinate space (the Euclidean plane), such that the first two IOI durations determine a point $p_1(x,y)$ with x-coordinate equal to the first IOI duration, and y-coordinate equal to the second IOI duration. This point $p_1(x,y)$ is connected with a straight edge to a point $p_2(x,y)$ that has x-coordinate

equal to the second IOI duration and y-coordinate equal to the third IOI duration, and so forth.[8] In general, if the rhythm has *n* onsets (attacks), the phase space plot has *n* points $p_1(x,y)$, $p_2(x,y)$, ..., $p_n(x,y)$ corresponding to *n* pairs of adjacent IOI durations, such that $p_i(x,y)$ has an x-coordinate equal to the duration of IOI_i and a y-coordinate equal to the duration of IOI_{i+1}, for *i* = 1, 2, ..., *n*. Since the rhythms are cyclic, IOI_{n+1} corresponds to the IOI duration between the last and first onsets. Figure 35.7 shows the phase space plots of the six distinguished timelines. Points that lie on the diagonal dashed line (called the *isotropy line*) correspond to pairs of adjacent IOIs that have the same duration. Note that the plot of the rumba has no vertex on the isotropy line since it does not contain any pair of adjacent IOIs of the same duration. The plots are general polygons (phase space polygons also called embedding graphs[9]) whose vertices and edges may fall on top of each other (self-intersect), as is the case with five of these rhythms. These six phase space polygons may be classified into two classes. The shiko, son, and bossa-nova fall in one class, and are distinguished from the other three by the property that their phase space polygons have reflection symmetry about the isotropy line. Furthermore, the *clave son* is unique among the six by virtue of the property that it is the only rhythm whose phase space polygon

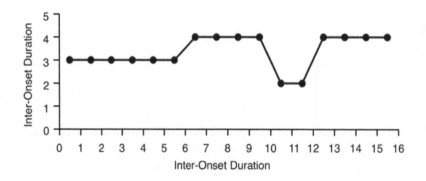

FIGURE 35.6   The chronotonic chain of the *clave son*.

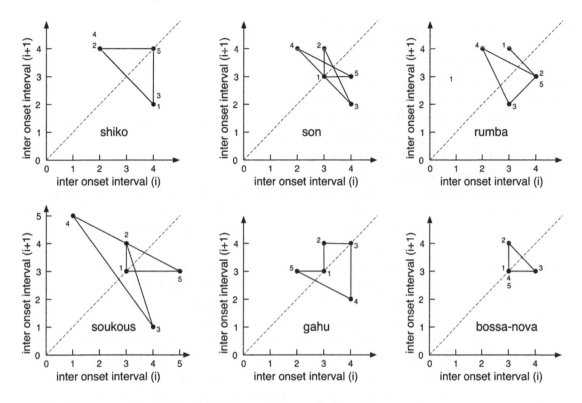

FIGURE 35.7 The phase space plots (polygons) of the six distinguished timelines.

has reflection symmetry and no two vertices lie on top of each other. The gahu is also uniquely distinguishable from the other five, since it is the only rhythm that has a phase space plot that is a *simple* polygon. A polygon is simple provided that all its vertices are unique, and it has no self-intersections.[10] Only the gahu has a phase space plot composed of a simple polygon.

Phase space plots provide a tool for artistic and analytical exploration of rhythmic structures. In the acoustic domain, they have been applied to audio-based music visualization for music structure analysis, with the goal of facilitating users to browse through music collections, in the context of music information retrieval.[11] The application of phase space plots to the analysis of symbolically represented rhythms is still an area in its infancy waiting to be mined.

Any representation of rhythms suggests methods to measure rhythm similarity, and phase space plots are no exception. Two phase space plots may be compared quantitatively by measuring their musically relevant geometric features and then calculating the distance between the resulting feature vectors. To illustrate the usefulness of phase space plots and the measurement of their similarity for the case of rhythm timelines, let us return to the problem of binarization of the ternary fume–fume timeline [2-2-3-2-3] by means of the natural (proximity Gestalt inspired) algorithm that snaps the onsets of the fume–fume in a 12-pulse span to their nearest pulses in an overlaid 16-pulse span. This problem, left unsettled in Chapter 12, will be resolved here geometrically. The nearest pulse snapping algorithm generated three different possible binarizations of [2-2-3-2-3], namely [3-2-4-3-4], [3-3-4-2-4], and [2-3-4-3-4], raising the question of explaining the claims made by ethnomusicologist Rolando Pérez-Fernández that the *clave son* [3-3-4-2-4] is the "correct" one.[12] In Chapter 12 it was attempted to resolve this problem by calculating the total quantization errors obtained with the global Minkowski metrics, for its three binarizations, for the cases of the Manhattan, Euclidean, and Maximum metrics. The best result (obtained with the Maximum metric) resulted in a tie that was broken by invoking the psychological Gestalt spatial theory of rhythmic resolution.[13] In more concrete terms, for our special case, the fume–fume begins with two IOIs of the same duration [2-2], and the only binarization that also starts with two IOIs of equal duration, the *clave son* with [3-3]. However, it is possible to resolve the problem with purely geometrical global similarity measures using the phase space plots, as shown later using two different approaches.

Two musically relevant geometric features of the plots are (1) the number of polygon vertices that lie on the isochrony line and (2) the presence or absence of reflection symmetry of the polygon about the isochrony line. The fume–fume polygon has one vertex on the isochrony line as well as reflection symmetry with respect to this line. By contrast, the phase space polygons of the binarizations [3-2-4-3-4] and [2-3-4-3-4] do not share either of these properties, whereas the polygon corresponding to the *clave son* enjoys both properties. By this comparison metric, it follows that the *clave son* is the binarization most similar to the fume–fume (Figure 35.8).

For the second geometric resolution of the binarization problem, we need some additional terminology and definitions. The similarity between two phase space polygons may also be measured using the areas of their *convex hulls*. A polygon is *convex* if the line segment connecting every pair of points that lie in the polygon also lies completely in the polygon. Figure 35.9 (left) illustrates a convex polygon, where *x* and *y* represent any two points in the polygon. The polygon on the right in this figure is not convex because there exists a pair of points *x* and *y* such that the line segment connecting them

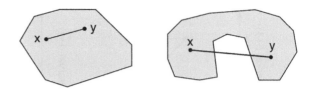

FIGURE 35.9 A convex polygon (left) and a nonconvex polygon (right).

passes through the exterior of the polygon. The area of the convex hull of a set in the Euler lattice has been used previously as a feature for characterizing chords and scales in terms of their preference. It turns out that this feature is also helpful in the rhythm domain.

A polygon is said to be *star-shaped* if there exists a point *x* in the polygon such that for every point *y* in the polygon the line segment connecting *x* and *y* also lies completely in the polygon. Intuitively, if the star-shaped polygon is considered to be the floor plan of an art gallery, then one guard suitably located can see the entire gallery from that spot.[14] The polygon in Figure 35.10 (left) shows one such location *x* from which the entire polygon is visible. Recall that a polygon is called *simple*

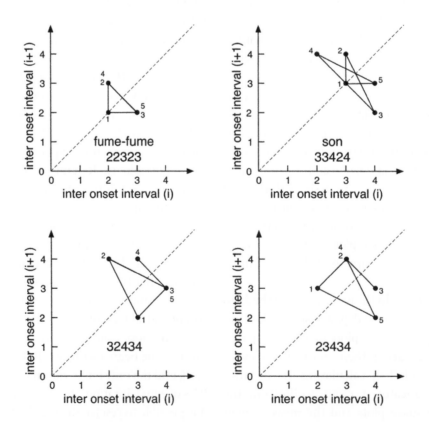

FIGURE 35.8 The phase space plots of the ternary fume–fume and its three binarizations resulting from the nearest onset snapping rule.

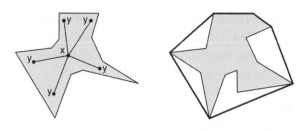

FIGURE 35.10 A star-shaped polygon (left) and the convex hull of a simple polygon (right).

provided that its boundary does not intersect itself. In Figure 35.7 the only phase space polygon that is simple is the one that represents the gahu timeline. All the other five polygons are self-intersecting. Furthermore, the gahu polygon is star-shaped, since it is entirely visible from its vertex labeled number 1. The *convex hull* of a polygon is the convex set of minimum area that encloses it. Figure 35.10 (right) shows a polygon (shaded) and its convex hull (bold lines). Intuitively, if an elastic band is wrapped around the polygon, it will take on the shape of a convex polygon that is precisely the convex hull of the polygon.[15] These notions of convexity and star-shapedness (also called star-convex) have been successfully applied to the analysis of spaces of chords, scales,

and harmony in the pitch domain.[16] Indeed, when the intervals of traditional musical scales are represented on an Euler lattice, all of them form star-shaped structures.[17] This property may thus qualify as a universal in music and provides inspiration to search for similar rhythmic universals by means of geometric methods.

With this additional terminology, let us return to the binarization problem of the fume–fume [2-2-3-2-3]. Figure 35.11 shows the convex hulls (shaded) of the phase space polygons for the fume–fume as well as its three binarization candidates. The areas of their convex hulls are: 0.5 for the fume–fume, 1.5 for the *clave son*, and 2.0 for the two additional binarizations [3-2-4-3-4] and [2-3-4-3-4]. Therefore, by comparing the magnitude of the areas, it follows that the *clave son* is again the binarization most similar to the fume–fume. Whether phase space plots of rhythms will be useful for measuring rhythm similarity in more general settings is a research problem yet to be explored.

TANGLE DIAGRAMS

The visualization techniques described in the preceding are useful for all types of rhythms. In addition, there are some specific visualization techniques

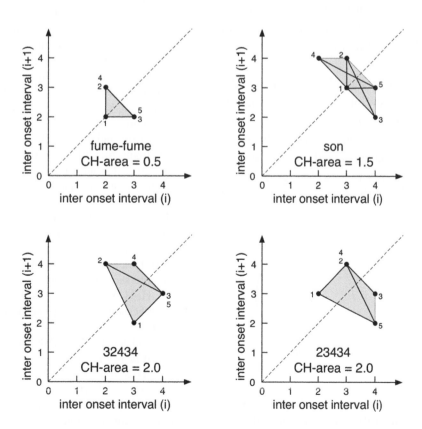

FIGURE 35.11 The convex hulls (shaded) of the phase space plots of the ternary fume–fume and its three binarizations.

that emerge from the structure of restricted classes of rhythms. The application of *tangle theory*, a branch of topology closely related to knot theory, to the visualization of Euclidean rhythms is a case in point.[18] We have seen in Chapter 21 titled Euclidean rhythms, that given two numbers *k* and *n*, where *n* is the number of pulses and *k* is the number of onsets, the Euclidean algorithm generates the Euclidean rhythm *E*(*k,n*). The Euclidean algorithm may also be used to generate the continued fraction of the rational number *k/n*. Furthermore, there is a one-to-one correspondence between a continued fraction of a rational number and a tangle.[19] A *tangle diagram* consists of a circle with two vertices near the top and two near the bottom. The top vertices are connected to the bottom vertices with arcs that may intertwine with each other. Figure 35.12 shows the tangle diagrams for the tresillo Euclidean rhythm [2-1-2-1-2] (left), and the 13-pulse rhythm [3-2-3-2-3] (right). The tangle diagram on the right is clearly visually more complex than the one on the left, suggesting that the 13-pulse Euclidean rhythm is more complex than the cinquillo. There is thus the possibility that there exists a correlation between the visual complexity of the tangle diagram of a rhythm, and the aural perceptual complexity of the rhythm. However, this question has not been investigated. How useful tangle diagrams will be for the analysis of rhythm in the future only time will tell.

In closing this chapter, it should be noted that the new technologies being developed around the application of computer software and hardware, for a variety of purposes such as notation, analysis, education, games, and composition, have spurred an explosion of new visualization techniques that explore graph techniques in two dimensions (such as tree languages[20]), and

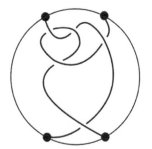

FIGURE 35.12 The *tangle* diagrams corresponding to the Euclidean cinquillo rhythm (left) and the 13-pulse rhythm [3-2-3-2-3] (right).

three-dimensional spaces.[21] There is now an annual conference called TENOR[22] (International Conference on Technologies for Music Notation and Representation) that explores this exploding field of research.

NOTES

1 Lau, F., (2008). p. 52.
2 Cohen, D. & Katz, R., (1979). Ellingson, T., (1992), focuses on notation systems developed in several cultures. See also Ellingson, T., (1992) and Kaufman Shelemay, K., (2000). Koetting, J. & Knight, R., (1986), p. 60, find that box notation is convenient for their "fastest pulse analysis" of rhythm. Brinkman, A. R., (1986), describes a binomial notation system for pitch that codes additional information. Benadon, F., (2009a), provides tools for notating microtiming in jazz while minimizing the complexity of notation. For a comparison of several mathematical notation systems, see Liu, Y. & Toussaint, G. T., (2010a).
3 For instance, Perkins, D. N. & Howard, V. A., (1976), p. 76, propose a "psychological" notation system that visually reflects the "clustering" or figural perceptual grouping of the attack points of a rhythm.
4 Konaté, F. & Ott, T., (2000), p. 20. With this technique, also called "hand-by-hand" drumming, the performer playing a beat pattern also feels all the pulses in between the beats, which aids the acquisition of a metronomic precision.
5 Gustafson, K., (1987, 1988).
6 The application of the difference in area between two curves has also been applied to geometric measures of melodic similarity, where the melody curves represent pitch as a function of time. See ÓMaidín, D., (1998), Aloupis, G., Fevens, T., Langerman, S., Matsui, T., Mesa, A., Nuñez, Y., Rappaport, D., & Toussaint, G., (2006), and Lubiw, A. & Tanur, L., (2004).
7 Hofmann-Engl, L., (2002).
8 Ravignani, A., (2017). See also Ravignani, A. & Norton, P., (2017), for additional references and applications of the phase space plots to the analytic exploration of music.
9 Monro, G. & Pressing, J., (1998).
10 Toussaint, G. T., (1991). See also Meisters, G. H., (1975), for additional interesting properties of simple polygons, and Grünbaum, B., (1998), for enumeration of a class of self-intersecting polygons.
11 Wu, H.-H. & Bello, J. P., (2010).
12 Pérez-Fernández, R. A., (1986) and Pérez Fernández, R. A., (2007).
13 See McLachlan, N., (2000), for his spatial theory of rhythmic resolution.
14 Avis, D. & Toussaint, G. T., (1981).
15 Toussaint, G. T., (1985).
16 See the papers by Honingh, A.K. & Bod, R., (2005), for a treatment of convexity and the well-formedness of musical objects, Honingh, A. K., (2006), for convexity

and compactness as models for the preferred intonation of chords, and Honingh, A. K., (2009), for finding automatic modulation using convex sets of notes.

17 Honingh, A. K., & Bod, R., (2011).

18 Kirk, J. & Nicholson, N., (2016).

19 Conway, J., (1970).

20 Jacquemard, F., Ycart, A., & Sakai, M., (2017).

21 See the online papers by Ham, J. J., (2017), for 3-D spatial drum notation, and by Kim-Boyle, D., (2017), for 3-D score notation systems.

22 The conference proceedings of TENOR may be found here: www.tenor-conference.org/proceedings.html.

CHAPTER 36

Rhythmic Similarity and Dissimilarity

I N SEVERAL CHAPTERS IN THIS BOOK, the topic of rhythm similarity has been mentioned briefly in different contexts. In this chapter, this topic is considered in greater detail. Before examining the complex problem of measuring sequence similarity, let us look at the simpler problem of comparing only two shapes. Consider the three pairs of shapes ordered left to right in Figure 36.1. Everyone would without doubt agree that the two shapes on the left are very hard to distinguish from each other and that the two in the center are more similar to each other than the pair on the right. We are all experts at judging shape similarity. Even newly born infants can distinguish between squares and circles, and they can tell the difference between the faces of their mother and father. The similarity between two objects is one of the most important features for distinguishing between them and for pattern recognition in general. Indeed, the survival of the human species depends crucially on this skill, and therefore, it is not surprising that humans are masters at telling the difference between the myriad of shapes they encounter in the world, even when their differences are

quite subtle. Can we construct a mathematical formula for measuring the similarity between two shapes, and more importantly, for the topic of this book, between two musical rhythms? The artificial intelligence pioneer Minsky (1981) frames the question this way: "What are the rules of musical resemblance? I am sure that this depends a lot on how melodies are 'represented' in each individual mind."[1] To give a more complete account of this field of research, and because no single method works well in all applications, in this chapter, several popular methods used to measure rhythm similarity are compared and illustrated with examples. For pedagogical reasons, we begin with the simplest possible measure (the Hamming distance) and then move on to more complex measures, highlighting their strengths and weaknesses along the way.

HAMMING DISTANCE

The Hamming distance is perhaps the most natural way to measure the dissimilarity between two rhythms represented as sequences of symbols. Consider the

FIGURE 36.1   Pairs of (from left to right) identical, similar, and dissimilar shapes.

FIGURE 36.2 The Hamming distance between these two symbol sequences is 4.

two horizontal sequences of eight symbols each in Figure 36.2. They differ in their third, fourth, sixth, and eighth positions. The Hamming distance between two sequences is defined as the number of corresponding locations where the two sequences differ.[2] In this example the Hamming distance is equal to four.

This distance measure begets its name from the mathematician and computer pioneer Richard Hamming. During World War II, Hamming was programming early versions of computers to calculate equations that physicists were using to determine whether or not exploding an atomic bomb would ignite the whole planet's atmosphere and thus possibly destroy the entire human race. After the war, he worked at Bell Laboratories using punched card readers. One of the problems with these cumbersome machines was that they introduced many errors when reading the sequences of input bits (ones and zeros). As a result of this frustration, he invented several coding methods to detect and correct errors, and published them in a seminal paper in which he introduced the Hamming distance. The Hamming distance may be computed very efficiently,[3] and works well in computer applications where one is interested in detecting whether two sequences are identical or not and in correcting errors when they differ.

Measuring the similarity between two symphonies, songs, melodies, or rhythms is a challenging problem in music analysis and technology, and has many applications ranging from generating playlists to copyright infringement resolution, music information retrieval,[4] phylogenetic analysis, and discovering the evolution of rhythmic patterns and motifs in a style of music.[5] There exist two broad approaches to measuring the similarity between objects: *feature*-based methods and *transformation*-based methods. In feature-based methods, objects are compared in terms of the number of traits they have in common.[6] In transformation-based approaches, similarity is measured by how little effort is required to transform one object to another.[7]

The Hamming distance described earlier has been applied to the study of cultural evolution through biological phylogenetic analyses of Turkmen textiles[8] as well as musicology.[9] Although the Hamming distance is appropriate for the study of carpet similarity based on the presence or absence of visual motifs, for applications to temporal sequences it has a serious drawback. Consider the two pairs of sequences in Figure 36.3. In Figure 36.3a the two sequences differ in their fourth and fifth positions. In Figure36.3b the sequences differ in their fourth and eighth positions. Thus, in both Figure 36.3a and b, the Hamming distance is equal to 2. If the sequences are viewed as strictly visual patterns arranged on a carpet, it is possible that some viewers might consider that the pair of sequences in Figure 36.3a are more similar to each other than those in (b), because the symbols that do not match are closer together in (a) than in (b). If the sequences are considered as mere collections of objects such that the order of the symbols is unimportant, then this variation in separation may not matter much. However, if the symbols make up temporal sequences (as in speech or music), then their relative positions in the sequence may be crucial. Consider for instance the two pairs of sequences in Figure 36.3 as representing onsets of musical notes in a cycle of eight pulses. In the pair (Figure 36.3b), the onsets that differ occur at the fourth and eighth pulse positions, whereas

FIGURE 36.3 A weakness of the Hamming distance for musical rhythm analysis.

in (a) they differ at the fourth and fifth pulse positions. In the former case, the two onsets divide the cycle into two equal half cycles, and in both half cycles, the onset occurs at the end of the half cycle. On the other hand, in the latter case, the two onsets occur in the last and first positions of the half cycles. Thus, by this reckoning, the sequences in Figure 36.3b may be considered to be more similar to each other than the sequences in (a). This ambivalence in possible judgments highlights the importance of context for measuring sequence similarity.

SWAP DISTANCE

The *swap distance* does not suffer from the problem outlined in the preceding, inherent in the Hamming distance. To define the swap distance, we first need a definition of the *swap operation*. One of the oldest and most fundamental operations in computer science is the interchange of elements in strings of symbols made up of numbers and letters, in the context of the design of algorithms for sorting, dating back to 1974.[10] When the elements in a sequence that are interchanged are required to be adjacent to each other, the swap operation is called a *miniswap* or *primitive swap*,[11] or *transposition*.[12] Here, on the other hand, the shorter term, *swap* is used to mean an interchange of two *adjacent* elements. For rhythms represented as binary sequences in box notation, this means that the swap operation interchanges an empty box with a filled box when the two boxes lie next to each other. Thus, in Figure 36.4 the sequence in (a) may be converted to that in (b) with one swap. On the other hand, to convert the sequence in (c) to that in (d) requires four swaps. Note that this number of swaps is equivalent to the distance that the source attack in (c) must travel to reach the target position in (d). The *swap distance* between two rhythms is defined as the *minimum* total number of swaps necessary to transform one rhythm into the other, and is equivalent to the minimum value of the sum of distances (measured in number of pulses) traveled by all the attacks during this transformation.

Figure 36.5 illustrates the calculation of swap distance between the shiko and gahu rhythm timelines. Since each rhythm has five attacks, it follows that the first attack of shiko must move to the first attack of gahu, the second to the second, and so on. The arrows indicate the positions to which each attack of the shiko must move in order for it to match its corresponding position of the gahu. The bottom row indicates the cost in terms of the number of swaps that each attack incurs in reaching its goal. The swap distance between the two sequences is simply the sum of all the individual costs. Thus, the swap distance between the shiko and gahu timelines is equal to 3.

The swap distance is a special case of the Minkowski metric described earlier in Chapter 12 on the topic of binarization and quantization of rhythms. To see this, represent the rhythms as d-dimensional vectors, where d is the number of attacks, and each element in the vector is the x-coordinate of the attack. Consider two rhythms X and Y consisting of d attacks each. Let the coordinates of the attacks of rhythm X be $x_1, x_2, ..., x_d$, and of rhythm Y be $y_1, y_2, ..., y_d$. Recall that the Minkowski metric between rhythms X and Y is given by

$$d_p(X,Y) = \left(|x_1 - y_1|^p + |x_2 - y_2|^p + \cdots + |x_d - y_d|^p \right)^{1/p}$$

where $1 \leq p \leq \infty$. In our example $d = 5$, Shiko = X has the vector (0, 4, 6, 10, 12) and Gahu = Y has the vector (0, 3, 6, 10, 14). Now letting $p = 1$, the swap distance is equal to

$$d_1(X,Y) = |x_1 - y_1| + |x_2 - y_2| + |x_3 - y_3| + |x_4 - y_4| + |x_5 - y_5|$$

$$= |0 - 0| + |4 - 3| + |6 - 6| + |10 - 10| + |12 - 14|$$

$$= 0 + 1 + 0 + 0 + 2$$

$$= 3$$

This first-order Minkowski metric, also called the L_1 norm, because $p = 1$, has been used frequently in pattern

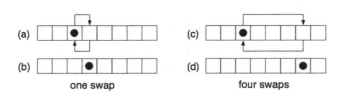

FIGURE 36.4 Illustrating the definition of the *swap* operation.

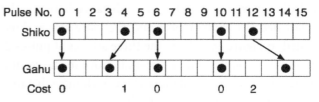

FIGURE 36.5 The swap distance between the *shiko* and *gahu* timelines.

recognition applications since the 1950s, and in 1970, it was shown to be optimal for some applications with suitable probability density functions governing the data.[13] In the context of music, it has recently been applied to voice leading.[14] One issue that comes up when using p-Minkowski metrics for either voice leading or measuring similarity, in general, is the selection of the value of p. Hall and Tymoczko (2012) have developed a novel and powerful approach to deal with this issue by means of the concept of submajorization.[15] Submajorization is a tool that permits the avoidance of total commitment to a particular metric (even beyond Minkowski metrics), while still being able to make similarity judgments that remain true regardless of the metric used.

DIRECTED SWAP DISTANCE

The swap distance described in the preceding works well when both rhythms have the same number of attacks. However, if the two rhythms have a different number of attacks, then it is impossible to convert the sequence with fewer attacks to the sequence with a higher number of attacks, since there are not enough of them. To solve this more general problem the swap distance may be modified in the following manner. Let us call the rhythm that has a higher number of attacks the *denser* rhythm, and the other the *sparser* rhythm. The *directed swap distance* is then defined as the *minimum* total number of swaps needed by all the attacks of the denser rhythm to convert it to the sparser rhythm, with the constraints that every attack of the denser rhythm must move to an attack position of the sparser rhythm, and every attack of the sparser rhythm must receive at least one attack of the denser rhythm. The term "directed" comes from the fact that the mapping used here is directional, from the denser rhythm to the sparser rhythm. Note that now the basic operation is not a swap between an occupied box and an empty box, but simply a *shift* of an attack by a distance of one pulse. In other words, an occupied box may be moved on top of another occupied box, leaving the previous box empty. An example calculation of the directed swap distance between the renowned Bo Diddley beat and the *clave son* is illustrated in Figure 36.6. Here the three attacks at pulses 2, 3, and 4 of the Bo Diddley beat all move to the second attack of the *clave son* at pulse three, yielding a distance of two, between this pair of rhythms.

This distance measure between sequences is also used in computational biology to measure the similarity

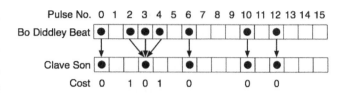

FIGURE 36.6 The directed swap distance between the *Bo Diddley beat* and the *clave son*.

between long molecules such as DNA, which are modeled as symbol sequences. In this context, the calculation of this distance measure is called the *restriction scaffold assignment* problem.[16] Several variants and generalizations of the directed swap distance have been proposed in the literature on information retrieval and combinatorial pattern matching, including the *fuzzy* Hamming distance[17] and the *generalized* Hamming distance.[18] More recently this distance measure has also been applied to voice leading.[19]

MANY-TO-MANY ASSIGNMENT DISTANCE

Although the swap and directed swap distances may work well in some situations where the rhythms are not too different from each other, in some cases, they yield unsatisfactory results in the sense that they may not agree with the judgments of musicologists. It is reasonable to expect that a more realistic distance measure should allow fission as well as fusion of attacks in the transformation process of one rhythm to another. The *many-to-many assignment* distance serves this function well. Consider the comparison of two popular Baroque Siciliani rhythms in Figure 36.7 given by duration interval vectors [3-1-2-3-1-2] and [3-1-1-1-4-2]. Both rhythms have six attacks, and so the swap distance assigns the attack at pulse nine of the first rhythm to the distant attack at pulse six of the second rhythm, yielding a rather large distance value equal to 4. However, the attack at pulse nine is a preparatory (anticipation syncopation) attack for the attack at pulse ten. Therefore,

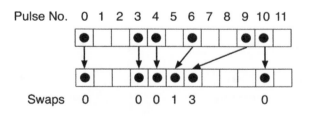

FIGURE 36.7 The swap distance between two Baroque Siciliani rhythms is equal to 4.

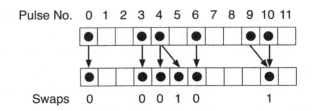

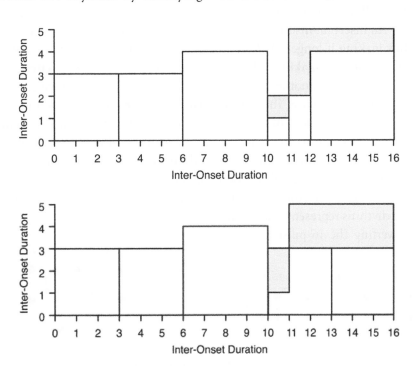

FIGURE 36.8 The many-to-many assignment distance between the rhythms of Figure 36.7 is equal to 2.

a better assignment would match the attack at pulse nine of the first rhythm to the attack at pulse ten of the second rhythm.

A many-to-many assignment between these two rhythms is shown in Figure 36.8. Note that in this assignment there is a *fission* operation in which the attack at pulse 4 in the first rhythm is assigned to two attacks at pulses 4 and 5, and there is a *fusion* operation in which the two attacks at pulses 9 and 10 of the first rhythm are mapped to the attack at pulse 10 of the second rhythm. In this assignment, the resulting distance is two. An efficient algorithm for computing this distance measure also exists.[20] This distance measure has also been recently applied to voice leading.[21]

GEOMETRIC DISTANCE

Just as the spectral notation may be employed to define an area distance between two rhythms by overlaying

their spectral representations, the Time Elements Displayed As Squares (TEDAS) and chronotonic notations may be used in the same way to obtain a geometric area distance between two rhythms that is called *chronotonic distance*.[22] Figure 36.9 illustrates the TEDAS distances (shaded areas) between the son and soukous (distance = 8 at the top), and between the bossa-nova and soukous (distance = 12 at the bottom).

There exist many additional mathematical methods for measuring rhythm similarity, including measures designed for comparing time series,[23] histograms,[24] chords, and scales,[25] which may be directly applied to rhythm,[26] raising the question of which measure one should use in practice. The answer is of course that it depends on the application at hand. There are numerous considerations to be taken into account before choosing a measure, such as the computational complexity of the algorithm available for computing it, how well it correlates with human judgments, and how well it performs in phylogenetic analyses of rhythms.[27] The Hamming distance may be computed very efficiently in time linearly proportional to the length of the rhythms, but as we have seen, it has musicological limitations. A generalization of the Hamming distance that allows shifting operations circumvents some of its drawbacks and may also be computed efficiently.[28] On the other hand, the *edit* distance may be calculated

FIGURE 36.9 The area distance (shaded) in TEDAS notation.

using dynamic programming,[29] which takes time proportional to the square of the lengths of the rhythms.[30] There is experimental evidence that the edit distance correlates well with human judgments of rhythm similarity.[31] For the specific problem of content-based polyphonic music retrieval, there exist generalizations of the edit distance, such as the earthmover's distance and the transportation distance,[32] with some evidence that for such applications the geometric methods are superior to the edit-distance approach.[33] And of course there is Larry Polansky's survey, already mentioned in a previous chapter, for an in-depth comparative analysis of a plethora of metric equations for measuring distance and similarity between a variety of musical objects such as rhythms, scales, contours, tuning systems, and so on.[34]

EDIT DISTANCE

The swap distance described in the preceding works well when the two rhythms being compared have the same number of pulses. However, in general, it is required to compare two rhythms that have different numbers of pulses. The *edit* distance between two sequences of symbols, also called Levenshtein distance, after its inventor Vladimir Levenshtein, the father of Russian information theory, is defined as the minimum number of *edits* (or *mutations*) necessary to convert one sequence to the other.[35] The edit operations allowed are *insertions*, *deletions*, and *substitutions*.[36] An insertion inserts one symbol into a sequence, thus making it longer. A deletion removes one symbol from a sequence, making it shorter. A substitution replaces one symbol by another and thus does not change the length of the sequence. These operations allow the comparison of sequences of different lengths. For example, one way to convert the sequence **WAITER** to **WINE** is by means of two deletions (**A** and **R**) to obtain **WITE**, followed by one substitution of **T** by **N**, resulting in a total of three edit operations. In the context of musical rhythms represented as binary sequences, consider converting the 16-pulse *clave son*

to the 12-pulse fume–fume, illustrated in Figure 36.10. One rather inefficient set of edit operations is obtained by aligning one sequence over the other and performing the required edits in left to right order. This computation yields four substitutions of silent pulses by attacks at pulses 2, 4, 7, and 9, three substitutions of attacks by silent pulses at locations 3, 6, and 10, one deletion of an attack at pulse 12, and three deletions of silent pulses at pulses 13, 14, and 15, for a grand total of 11 edits. However, the edit distance between these two rhythms is equal to 4. Since the fume–fume is four pulses shorter than the *clave son*, the minimum possible number of edits is 4. Furthermore, this minimum is realizable if four silent pulses are deleted judiciously. In particular, as shown in Figure 36.11, deletion of silent pulses 2, 5, 9, and 15 will accomplish the task.

It may be observed that if one rhythm has many more pulses than another, then the edit distance grows as a function of the *difference* in lengths between the two rhythms being compared, since this many pulses must be either removed from the longer rhythm or added to the shorter rhythm to make them the same length. It may be argued that, in general, this difference should contribute to the rhythms' dissimilarity rating. However, if this is deemed undesirable for some applications, then the rhythms may be mapped to a common number of pulses before applying the edit distance. For example, in the case of the 16-pulse *clave son* and the 12-pulse fume–fume, they may both be mapped to a 48-pulse cycle. This is akin to a discrete simulation of placing both rhythms on a single continuous circle. Since the maximum value that the edit distance can take when comparing two sequences of length n and m is equal to the sum of the lengths of the two sequences, another method used for normalizing the edit distance is to divide the edit distance by the sum of their lengths. Experimental comparisons of the edit distance with other measures of distance and dissimilarity have demonstrated its

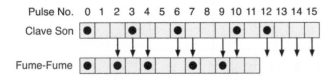

FIGURE 36.10 Eleven edit operations for converting the *clave son* to the *fume–fume* rhythm.

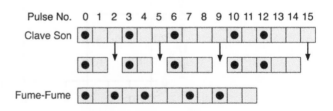

FIGURE 36.11 The edit distance between the *clave son* and the *fume–fume* rhythms is 4.

superiority in terms of predicting human judgments, for the case of melodies[37] as well as rhythms.[38]

NOTES

1 Minsky, M., (1981), p. 35.

2 Hamming, R. W., (1986).

3 Toussaint, G. T., (2004b). Given two binary sequences with n symbols in each, the Hamming distance may be computed using a number of simple operations that is proportional to n.

4 Bello, J. P., (2011) and Antonopoulos, I., Pikrakis, A., Theodoridis, S., Cornelis, O., Moelants, D., & Leman, M., (2007).

5 Crawford, T., Iliopoulos, C. S., Winder, R., & Yu, H., (2001), p. 56 and Mont-Reynaud, B., & Goldstein, M., (1985).

6 Tversky, A., (1977).

7 Hahn, U., Chater, N., & Richardson, L. B., (2003).

8 Tehrani, J. & Collard, M., (2002), p. 443. See also Collard, M. & Tehrani, J., (2005), p. 108.

9 Paiement, J.-F., Bengio, S., & Eck, D., (2008).

10 de Bruijn, N. G., (1974).

11 Biedl, T., Chan, T., Demaine, E. D., Fleischer, R., Golin, M., & Munso, J. I., (2001).

12 Navarro, G., (2004), p. 32.

13 Toussaint, G. T., (1970).

14 Callender, C., Quinn, I., & Tymoczko, D., (2008).

15 Hall, R. W. & Tymoczko, D., (2012).

16 Ben-Dor, A., Karp, R. M., Schwikowski, B., & Shamir, R., (2003) and Colannino, J., & Toussaint, G., (2005a, 2005b). For an efficient algorithm to compute the directed swap distance, see Colannino, J., Damian, M., Hurtado, F., Iacono, J., Meijer, H., Ramaswami, S., & Toussaint, G. T., (2006). See also Gusfield, D., (1997).

17 Bookstein, A., Klein, S. T., & Raita, T., (2001).

18 Bookstein, A., Kulyukin, V. A., & Raita, T., (2002).

19 Callender, C., Quinn, I., & Tymoczko, D., (2008).

20 Colannino, J., Damian, M., Hurtado, F., Langerman, S., Meijer, H., Ramaswami, S., Souvaine, D., & Toussaint, G. T., (2007). Efficient many-to-many point matching in one dimension. *Graphs and Combinatorics.* Vol. 23, *Computational Geometry and Graph Theory,* The Akiyama-Chvatal Festschrift, pp. 169–178. See also Mohamad, M., Rappaport, D., & Toussaint, G. T., (2011).

21 Callender, C., Quinn, I., & Tymoczko, D., (2008).

22 ÓMaidín, D., (1998), proposed a similar measure, and Aloupis, G., Fevens, T., Langerman, S., Matsui, T., Mesa, A., Nuñez, Y., Rappaport, D., & Toussaint, G., (2006), showed that it could be computed efficiently.

23 Kulp, C. W. & Schlingmann, D., (2009). Berenzweig, A., Logan, B., Ellis, D. P. W., & Whitman, B., (2004), provide a survey of acoustic music-similarity measures.

24 Cha, S.-H. & Srihari, S. N., (2002).

25 See Morris, R., (1979–1980), Chrisman, R., (1971), and the references therein.

26 See for example, Scott, D. & Isaacson, E. J., (1998).

27 Toussaint, G. T., (2006b).

28 Jiang, M., (2009).

29 Smith, L. A., McNab, R. J., & Witten, I. H., (1998). See also Bradley, D. W. & Bradley, R. A., (1999), p. 196, for the application of dynamic programming to the comparison of bird songs.

30 Wagner, R. A. & Fischer, M. J., (1974) and Wagner, R. A., (1999).

31 Post, O. & Toussaint, G. T., (2011) and Toussaint, G. T., Campbell, M., & Brown, N., (2011).

32 Typke, R., Giannopoulos, P., Veltkamp, R. C., Wiering, F., & van Oostrum, R., (2003).

33 Lemström, K. & Pienimäki, A., (2007), p. 148.

34 Polansky, L., (1996).

35 Levenshtein, V. I., (1966). Mutation operations are commonly employed to compare molecular sequences in computational biology. In the music domain, similar operations are called rhythmic or *metric modulation.* Arlin, M. I., (2000), provides an excellent discussion of a variety of metric modulation operations. See also Mongeau, M. & Sankoff, D., (1990), p. 165.

36 Substitutions are sometimes also called *replacements.* Other operations may also be added to generalized versions of the edit distance. For example, Lowrance, R. & Wagner, R. A., (1975), added the *swap* operation to the three standard operations. See also Takeda, M., (2001), Orpen, K. S. & Huron, D., (1992), and Oomen, B. J., (1995).

37 Müllensiefen, D. & Frieler, C., (2004).

38 Post, O. & Toussaint, G. T., (2011). See Toiviainen, P. & Eerola, T., (2003) for an empirical comparison of Finnish and South-African listeners' judgments regarding rhythm similarity.

CHAPTER 37

Grouping and Meter as Features of Rhythm Similarity

In Chapter 36 it was pointed out that the edit distance between two sequences is a method of choice for measuring the similarity between two rhythms represented as sequences of symbols. Not only does the edit distance apply to rhythms with different numbers of pulses and onsets but also for rhythms composed of symbols that take multiple values. In addition, experiments with human listeners have shown that the edit distance is among the best measures for predicting human judgments of rhythm similarity.[1] However, while admirably exploiting the purely durational information contained in the interonset intervals (IOIs) of a rhythm, the edit distance ignores two important psychological gestalts that influence rhythm perception: grouping and meter. For example, the edit distance, as used here, assigns equal weights to all edits, and thus does not distinguish between onsets that are syncopated and those that are strongly metrical. Therefore it is reasonable to expect that the performance of the edit distance might be improved by modifying it so as to incorporate grouping and/or meter information about the rhythms being compared. Indeed, it has been shown in the *acoustic* domain that incorporating meter information in rhythm similarity measures improves their ability to predict human judgments of rhythm similarity.[2] In this chapter, modifications of the edit distance are explored that incorporate grouping and/or meter information in the *symbolic* domain. In addition, experiments with human listeners and a dataset consisting of nine Middle-Eastern and Mediterranean rhythms are reported, which demonstrate that combining appropriate grouping and/or

meter information with the edit distance may significantly improve the power of the edit distance to predict human judgments of rhythm similarity.

The listening tests were carried out in two locations with subjects that belonged to distinct cultural groups, and differed in terms of musical experience, to test whether the edit distance, grouping, and meter are culture-dependent parameters. One test was carried out at Harvard University with undergraduate students majoring in music that had experience in classical music and were unfamiliar with traditional Middle Eastern music.[3] A similar listening test was subsequently replicated with predominantly Emirati and other Middle-Eastern undergraduate students and administrative staff at New York University Abu Dhabi (NYUAD) in Abu Dhabi (United Arab Emirates), who were not musically trained, but were familiar with traditional Middle-Eastern music as listeners. The experiments used nine *dum-tak* rhythms that consisted of two sounds, a low-pitched (*dum*) and a high-pitched (*tak*).[4] These rhythms are shown in box notation in Figure 37.1, where a *dum* sound is denoted by a square containing a black-filled circle, a *tak* sound by a white-filled circle, and a silent pulse by an empty square. Note that there is considerable variability among the nine rhythms with respect to the numbers of pulses, *dums*, and *taks* that they possess. The number of pulses varies between six and nine, and the numbers of *dums* and *taks* both vary between one and three. These rhythms also differ in their complexity: for instance, rhythms with eight pulses are considered to be simpler than those with seven or nine pulses.[5]

Rhythms		#Pulses	#Dums	#Taks
Baladii	●●● ○●● ○	8	3	2
Darb as-sarl	● ○●● ○○	8	2	3
Dawr-hindii	●○○● ○	7	2	3
Grantchasko	● ● ● ○	9	3	1
Laz	● ○ ○	7	1	2
Maqsuum	●○ ○● ○	8	2	3
Samaaii	●○○○●○	6	2	3
Sayyidii	●○ ●● ○	8	3	2
Sombati	● ○○● ○	8	2	3

FIGURE 37.1 The nine Middle-Eastern and Mediterranean *dum-tak* rhythms used in the listening tests and symbolic calculations.

The acoustic sound samples of the nine rhythms were created using Apple *Garageband* available in the MacBook Pro laptop computer. The rhythms, entered in Musical Instrument Digital Interface format, were exact. Each onset triggered one of two sounds: either a low-pitched *dum* sound (similar to a hand striking the center of a drum) or a high-pitched *tak* sound (akin to a hand striking the edge of a drum). Each rhythm was played four times in succession at a constant tempo of 200 pulses per minute, resulting in sound samples that lasted between 8 and 11 s depending on the number of pulses in the rhythm. Thus all the rhythms contained interpulse intervals of the same duration of clock time.

The participants in the tests listened to the rhythms while sitting on a chair using noise-canceling Bose headphones (model *QuietComfort* 15). The headphones were connected to a MacBook Pro laptop computer (at Harvard) and iMac desktop computer at NYUAD, on which was displayed the graphical user interface of the *Sonic Mapper* sound comparison software developed by Gary P. Scavone, using *Qt* for the user interface and *RtAudio* for audio output. The *Sonic Mapper* software offers three alternatives for comparing sounds: two-dimensional similarity mapping, sorting, and the more traditional pairwise comparison tests.[6] The method of randomized pairwise comparisons was used in all the listening tests.[7] The resulting ratings yield a dissimilarity matrix that can be compared to dissimilarity matrices obtained from the mathematical models, via Mantel tests of correlation.[8] The participants did not listen to the rhythms before the experiment began, they received no training on how to judge the range of similarity and dissimilarity judgments, and they were not provided with a definition of similarity (which they sometimes requested). Furthermore, the participants did not have a practice trial before the test. By sliding a cursor on a bar, listeners were permitted to judge two rhythms as equal in terms of similarity or dissimilarity, thereby reducing the possibility of circular errors in judgments.[9]

THE EDIT DISTANCE WITH GROUPING INFORMATION

In Chapter 19 the mutual nearest-neighbor (MNN) graph was shown to be a simple yet powerful grouping algorithm of the onsets of a rhythm, possessing several advantages over other formal mathematical models of grouping that use only durational information of a rhythm's IOIs. Recall that the MNN graph is obtained by joining a pair of onsets O_i and O_j with an edge if and only if both onsets are nearest neighbors of each other.[10] An onset O may have two nearest neighbors (one succeeding and the other preceding O) if the two distances between O and its nearest neighbors are equal. Each connected component of the MNN graph corresponds to a separate group in a rhythm's grouping structure. Two versions of the MNN graph are obtained depending on whether the algorithm is applied to the linear box notation or the circular cyclic representations of a rhythm. The groups obtained in the former case are referred to as *linear* groups, whereas in the latter case, they are called *cyclic* groups.

Linear and cyclic groups may differ from each other according to their cardinality and structure, as illustrated in Figure 37.2 with two *dum-tak* rhythms expressed in both box and circle notation: the nine-pulse grantchasko (left) and the eight-pulse sombati (right). In the box notation, the distances measured between pairs of onsets are Euclidean (straight lines), and in the circle notation, the distances are geodesic (shortest arcs along the circle representing durations). In the circle notation, the white-filled circles denote silent pulses, the single black-filled circles denote *tak* sounds, and the annuli with black interiors denote *dum* sounds. The MNNs are indicated with solid bidirectional lines, such as the onset pairs (0, 2), (5, 7), and (7, 0) in the grantchasko. The nearest neighbors (not mutual) of an onset are indicated with unidirectional dashed lines. For example in the sombati, the onset at pulse 2 is the nearest neighbor of the onset at pulse 0. The MNN graph yields different groupings depending on whether it is calculated linearly

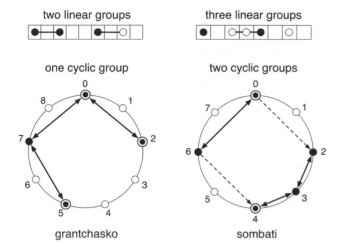

two linear groups   three linear groups

one cyclic group   two cyclic groups

grantchasko   sombati

FIGURE 37.2 Contrasting the *linear* and *cyclic* groups obtained with the *MNN* graph of two *dum-tak* rhythms: the grantchasko (left) and sombati (right).

or circularly. From Figure 37.2 it is observed that for the grantchasko the MNN graph yields two linear groups but only one cyclic group, and for the sombati, the MNN graph specifies three linear groups but only two cyclic groups. Thus the MNN graph reflects the assertion of Cooper and Meyer (1960) regarding grouping when they write that "Rhythmic groups are not respecters of bar lines. They cross them more often than not; and one of the first things that the reader must learn is that the bar line will tell him little about rhythmic grouping."[11] The MNN graph of the nine rhythms in the dataset (in box notation) is shown in Figure 37.3, along with the numbers of linear and cyclic groups generated by the algorithms.

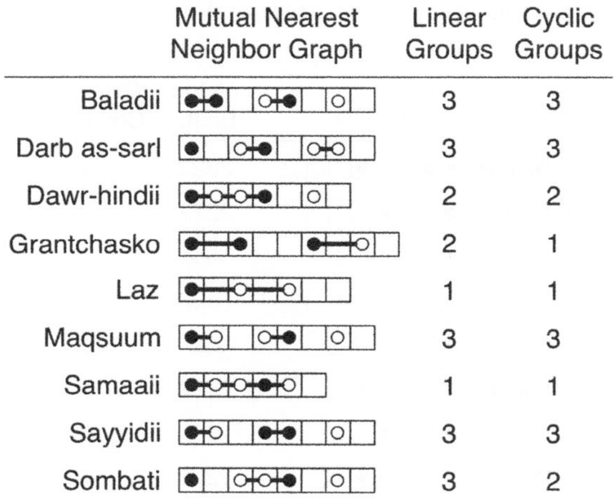

Mutual Nearest Neighbor Graph		Linear Groups	Cyclic Groups
Baladii		3	3
Darb as-sarl		3	3
Dawr-hindii		2	2
Grantchasko		2	1
Laz		1	1
Maqsuum		3	3
Samaaii		1	1
Sayyidii		3	3
Sombati		3	2

FIGURE 37.3 The *MNN* graph of the nine *dum-tak* rhythms in box notation, and the number of *linear* and *cyclic* groups contained in each rhythm.

One question that arises before deciding how to modify the edit distance to improve its ability to predict human judgments of rhythm similarity, by incorporating grouping information, concerns which features of grouping should be employed. There are a variety of possible features of the structure of groupings that could be brought to bear on the problem. The simplest feature that might be a good candidate for characterizing a rhythm is the group cardinality, that is, the number of groups that the onsets of a rhythm contains. This property suggests that the dissimilarity between two rhythms may be measured by the absolute value of the difference in the number of groups that the two rhythms contain. Note that this measure is not guaranteed to work in both directions (similar and dissimilar). If the difference between the group cardinalities of each of the two rhythms is nonzero, then the rhythms will differ in their grouping structure, and thus tend to sound different, but if they have the same group cardinalities, it does not imply that the rhythms will be perceived to be similar, since rhythms with equal group cardinalities need not possess similar grouping structures.

Table 37.1 lists the correlations between the human judgments of rhythm similarity and the linear and cyclic group cardinality differences, for the NYUAD and Harvard listeners, obtained with the Mantel test. All the correlations are highly statistically significant, as indicated by the low p values listed under each correlation coefficient. Note that the grouping features measure dissimilarity (or distance) and are thus correlated negatively with similarity ratings. The correlations for the cyclic grouping feature are considerably higher than for the linear feature, for both NYUAD and Harvard participants, suggesting that both groups of listeners perceive groups across bar lines, in support of the aforementioned assertion of Cooper and Meyer that "Rhythmic groups are not respecters of bar lines." In addition, the correlations are substantially higher for the Harvard listeners than for the NYUAD listeners, suggesting that listeners unfamiliar with these rhythms are more likely to use the grouping cardinality difference as a perceptual feature in their ratings of rhythm similarity than those subjects familiar with the rhythms.

A second question that arises once the grouping feature has been selected is how to combine it with the edit distance. In the interest of simplicity, the edit distance modified to incorporate grouping information is defined as the sum of the edit distance plus the group cardinality

TABLE 37.1 Comparisons of *Linear* and *Cyclic* Group Cardinality Differences, as Predictors of Human Judgments of Rhythm Similarity (Via Mantel Test Correlations)

Participants	Linear Group Cardinality Difference	Cyclic Group Cardinality Difference
NYUAD	$r = -0.369$	$r = -0.581$
	$p = 0.038$	$p = 0.0012$
Harvard	$r = -0.511$	$r = -0.639$
	$p = 0.019$	$p = 0.0006$

TABLE 37.2 Comparisons of the Edit Distance Alone, and the Edit Distance Plus Linear and Cyclic Group Cardinality Differences, as Predictors of Human Judgments of Rhythm Similarity (Via Mantel Test Correlations)

Participants	Edit Distance Alone	Edit Distance + Linear Group Cardinality Difference	Edit Distance + Cyclic Group Cardinality Difference
NYUAD	$r = -0.689$	$r = -0.716$	$r = -0.792$
	$p = 0.0004$	$p = 0.002$	$p = 0.0002$
Harvard	$r = -0.686$	$r = -0.804$	$r = -0.824$
	$p = 0.0011$	$p = 0.0002$	$p = 0.0003$

difference. Table 37.2 lists the correlations obtained with the Mantel test, for the NYUAD and Harvard listeners, for the edit distance alone, for the edit distance plus the linear group cardinality difference, and the edit distance plus the cyclic group cardinality difference, as predictors of human judgments of rhythm similarity. For both sets of listeners, the correlations for the edit distance + group cardinality difference are considerably higher than for the edit distance alone. Although the edit distance by itself is already a measure of the dissimilarity of the grouping structure of two rhythms being compared, the results in Table 37.2 suggest that there is additional discrimination information contained in the group cardinality difference that the edit distance does not capture.[12]

THE EDIT DISTANCE WITH METER INFORMATION

Chapter 18 introduced Michael Keith's mathematical measure of meter complexity, defined for the family of asymmetrical meters that consist of underlying metrical beats that have durational patterns consisting of strings of 2's and 3's.[13] Since complex rhythms and meters sound different from simple ones, and the perceived meter induced by a rhythm influences the rhythm's perceived complexity, the difference in the meter complexity of two rhythms is a reasonable candidate for a feature of rhythm dissimilarity. As with the group cardinality difference, this measure is also not guaranteed to work in both directions. If the meter complexities of the two rhythms differ, then the rhythms will tend to sound different, but identical complexities do not imply

that the rhythms will be perceived to be similar. In this section the edit distance is modified by incorporating metric information, and results are described with the same dataset of *dum-tak* rhythms used in the preceding section. These results then permit the comparison of grouping information with meter information, in terms of their relative ability to predict human judgments of rhythm similarity for this dataset. The metric beats and resulting Keith meter complexities of the nine *dum-tak* rhythms are listed in Figure 37.4, where the complexity values vary between 2 and 7.

Table 37.3 lists the correlations obtained with the Mantel test for the group-cardinality difference and Keith meter-complexity difference, as predictors of human judgments of rhythm similarity. The correlations for the meter complexity feature are notably higher than for

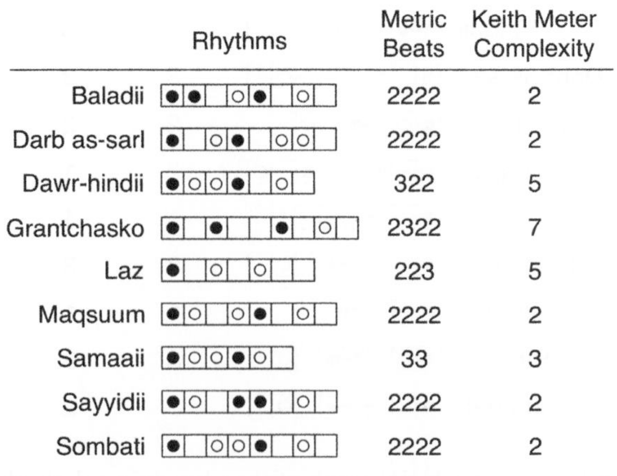

Rhythms	Metric Beats	Keith Meter Complexity
Baladii	2222	2
Darb as-sarl	2222	2
Dawr-hindii	322	5
Grantchasko	2322	7
Laz	223	5
Maqsuum	2222	2
Samaaii	33	3
Sayyidii	2222	2
Sombati	2222	2

FIGURE 37.4 The metric beats and Keith meter complexity of the nine *dum-tak* rhythms.

TABLE 37.3   Comparisons of the Group Cardinality and Keith Meter Complexity Differences as Predictors of Human Judgments of Rhythm Similarity (Via Mantel Test Correlations)

Participants	Group Cardinality Difference	Keith Meter Complexity Difference	Group Card Diff + Keith Meter Complexity Difference
NYUAD	$r = -0.369$	$r = -0.693$	$r = -0.702$
	$p = 0.038$	$p = 0.004$	$p = 0.001$
Harvard	$r = -0.511$	$r = -0.693$	$r = -0.754$
	$p = 0.019$	$p = 0.007$	$p = 0.002$

TABLE 37.4   Comparisons of the Keith Meter Complexity Difference, the Edit Distance Alone, and the Edit Distance Plus 0.5 (Keith Meter Complexity Difference) as Predictors of Human Judgments of Rhythm Similarity (Via Mantel Test Correlations)

Participants	Keith Meter Complexity Difference	Edit Distance Alone	Edit Distance + 0.5 (Keith Meter Complexity Difference)
NYUAD	$r = -0.693$	$r = -0.689$	$r = -0.877$
	$p = 0.004$	$p = 0.0004$	$p = 0.0001$
Harvard	$r = -0.693$	$r = -0.686$	$r = -0.875$
	$p = 0.007$	$p = 0.0011$	$p = 0.0001$

the grouping feature, suggesting that both NYUAD and Harvard listeners used meter information to a greater degree than grouping information in their rhythm similarity ratings. In addition, the correlations of the similarity ratings with meter complexity difference for the NYUAD and Harvard subjects are identical ($r = -0.693$), suggesting that both groups of listeners gave equal weight to this perceptual feature, in their similarity ratings.

Table 37.4 shows the correlations of the Keith meter complexity difference, the edit distance alone, and the edit distance plus 0.5 (Keith meter complexity difference) as predictors of human judgments of rhythm similarity (using Mantel tests). For all three measures, the correlations are almost identical for both groups of listeners.

To conclude, these experiments indicate that, for this complex rhythm dataset, both grouping information and meter information improve the power of the edit distance to predict human judgments of rhythm similarity. Furthermore, the meter feature, as used here, provides greater improvement to the edit distance than the grouping feature. These results are encouraging because the group cardinality and Keith meter complexity differences between two rhythms are rather simple and crude features of grouping and metric structure. Moreover, the manner in which the values of these features were combined with the edit distance (by simple addition) is

also rudimentary. It is possible that even better results may be obtained by using more sophisticated features of grouping and metrical structure and by combining them with the edit distance in a more discerning way. Such questions are left for future investigation.

NOTES

1 Post, O. & Toussaint, G. T., (2011), Toussaint, G. T., Matthews, L., Campbell, M., & Brown, N., (2012), and Toussaint, G. T. & Oh, S. M., (2016).
2 Gómez-Marín, D., Jordá, S., & Herrera, P., (2015). See also Smith, L. M., (2010).
3 Toussaint, G. T., Campbell, M., & Brown, N., (2011).
4 Shiloah, A., (1995). See also Touma, H. H., (1996).
5 Einarson, K. M. & Trainor, L. J., (2015), p. 93.
6 Scavone, G. P., Lakatos, S., & Harbke, C. R., (2002).
7 Kendall, M.G. & Smith, B. B., (1940).
8 Refer to Mantel, N., (1967) for a definition and justification of the Mantel test with dissimilarity matrices that contain dependencies, and to Bonnet, E. & Van de Peer, Y., (2001) for software freely available for computing the Mantel test.
9 Parizet, E., (2002).
10 Ozaki, K., Shimbo, M., Komachi, M., & Matsumoto, Y., (2011).
11 Cooper, G. & Meyer, L. B., (1960), p. 6.
12 Cao, E., Lotstein, M., & Johnson-Laird, P. N., (2014), p. 446, consider that the "Edit distance is a natural metric for comparing rhythms with different perceptual groups."
13 Keith, M., (1991).

CHAPTER 38

Regular and Irregular Rhythms

ONCE UPON A TIME, a Chinese Buddhist monk traveled to India in search of religious scriptures to take back to China. On his way, he came upon a flooded river where a large fish offered to help him get across. In return for the favor, the fish requested that while in India the monk should find out how to achieve salvation. On his return trip to China, having spent many years in India collecting the scriptures, the monk encountered the same flooded river and the same fish to help him across. In the middle of the river the fish asked the monk about his request, but the monk had forgotten about it. In anger, the fish dumped the monk into the river. Luckily a fisherman passing by saved the monk, but the scriptures were completely destroyed. Back in China, the monk was extremely angry at the fish. Therefore, he carved a wooden sculpture in the form of a fish and beat its head using a wooden stick. To the monk's surprise, the fish disgorged one character from the lost scriptures he had collected during his travels in India. The monk continued hammering the fish head every day until several years later he had recovered the entire scripture.

This story describes how the hollow wooden sculpture of the fish head with a slit representing the mouth became a percussion instrument called *muyü*, which is used in traditional Chinese music. The muyü comes in many different sizes and shapes. The one illustrated in Figure 38.1 looks more like a true fish than the more common designs.

Traditionally, the muyü blocks were used in Buddhist temples to accompany the chanting of the monks. However, today they are employed widely in Beijing Opera. The rhythms played on this instrument are mostly *regular* rhythms.[1] Regular rhythms have been imbued with emotional power. K. M. Wilson writes:

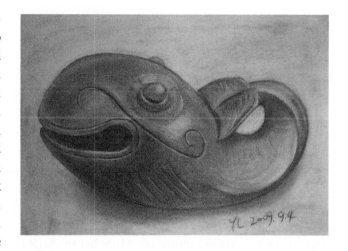

FIGURE 38.1   Chinese *muyü* temple block. (Courtesy of Yang Liu.)

"If we estimate emotion by its dynamic quality, the regular rhythms are the stronger."[2] Regular rhythms may be described on the rhythm circle as regular polygons, that is polygons with all their sides equal and all their angles equal. The rhythm most frequently played on the muyü has durational pattern [4-4-4-4] shown as a polygon in Figure 38.2 (left). However, the pieces usually end with a more syncopated rhythm such as that shown in the middle, which is quite irregular, has durational pattern [4-5-2-1-4], and uses seven of the eight possible interonset intervals (IOIs), as can be seen from its interval histogram on the right.

A. M. Jones writes: "The simplest rhythmic background to a song is a steady succession of regular claps."[3] This is probably why regular rhythms are present in almost all the music of the world, especially if they are used for dancing or marching. In modern electronic dance music, the mixing of regular and irregular

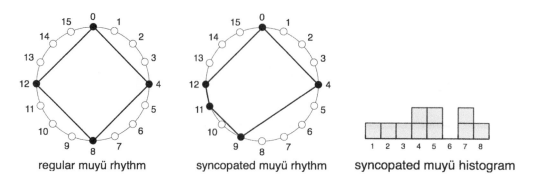

FIGURE 38.2 Typical *muyü* rhythms.

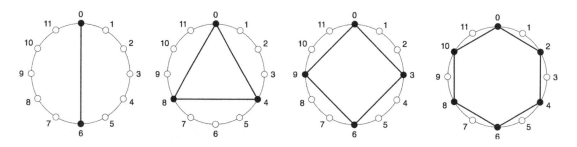

FIGURE 38.3 Four common regular rhythms played sometimes, simultaneously, on different instruments, in sub-Saharan African music.

(asymmetric) rhythms is one of its key features.[4] In the 12-pulse polyrhythmic music of sub-Saharan Africa, it is common to play simultaneously the four regular rhythms shown in Figure 38.3, each on a different drum, a percussion instrument, or with hand clapping. These four rhythms represent all the possible divisions of the 12-pulse cycle into regular polygons, save the 12-sided polygon obtained by playing every pulse.

All the rhythms of the world may thus be classified into two categories: *regular* and *irregular*.[5] Since the latter category is much broader than the former, it is useful to refine it by measuring the degree of irregularity possessed by a rhythm. The *Oxford Dictionary* definition of the term *irregular* is clear enough: "occurring at uneven or varying rates or intervals." Irregular is the opposite of *regular*, which in the context of time in musical rhythm means that onsets occur with the same duration between individual adjacent instances. What we need then to quantify rhythm irregularity is a measure of the amount of unevenness present in the adjacent IOI durations. It is easy enough to distinguish visually at a glance between a regular and an irregular polygon, at least when the polygon has a relatively small number of sides. Similarly, for a sounded rhythm it is not difficult to tell whether or not it consists of regularly spaced

beats. Thus, there is little doubt in any listener's mind that the seven rhythms in Figure 38.4 are all irregular.

What is more difficult to gauge is which of two given rhythms is more irregular than the other. Consider for example the rhythms labeled one and seven in Figure 38.4, with IOIs [1-1-1-1-1-1-2-2-2] and [1-1-2-1-1-2-1-1-2], respectively. At first glance, rhythm seven appears more regular because the long durations are evenly spaced out amongst the short ones.[6] Furthermore the overall shape of the rhythm appears to be a regular triangle with clipped-off corners. It may also be viewed as the IOI pattern [1-2-1] repeated three times. However, rhythm one is composed of two perfectly regular equal-duration chains: the left half composed of three long edges, and the right half made up of six short edges. It is as if a regular six-sided polygon and a regular 12-sided polygon were each cut vertically in half, and the left side of the hexagon was glued to the right side of the dodecagon. How does one weigh one of these properties against the other? A psychological solution to the problem would be to ask human subjects which of the two rhythms appears to be more regular. On the other hand, a mathematical approach might either measure the irregularity with a local contrast measure, such as the normalized pairwise variability index (nPVI) used to measure language rhythm irregularity,[7] or produce a list

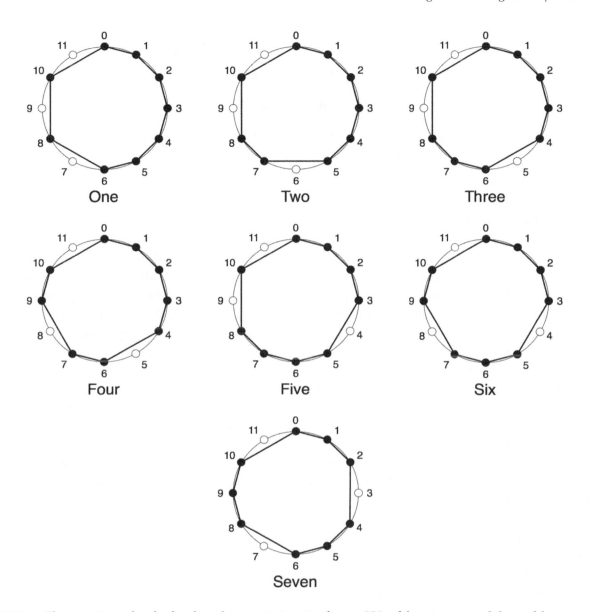

FIGURE 38.4 The seven irregular rhythm *bracelets* containing six adjacent IOIs of duration one and three of duration two.

of geometric properties possessed by the family of regular polygons, and then count how many of these properties are present in each of the two polygons being compared. However, finding such properties is not a trivial task. In addition to being able to distinguish among the different polygons, such properties should correlate well with human perception of the irregularity of the corresponding acoustic rhythms. Interestingly, the nPVI values for rhythms one and seven are 8.33 and 57.8, respectively, and thus, by this measure, rhythm one is considered to be more regular than rhythm seven.

As for a list of possible geometric properties that one might hope to use to measure regularity, consider a fundamental geometric property of a polygon: its area. One of the earliest theorems proved in geometry by the ancient Greeks states that of all polygons with a given number of sides and total length of its edges, the regular polygons maximize the area. Unfortunately, this theorem does not help to produce a ranking of the seven polygons because they all have the same area. To see this, consider the pair of polygons labeled one and seven in Figure 38.5. Connect each vertex of the polygon to the center of the circle creating nine triangles: three shaded equilateral triangles and six white smaller isosceles triangles. The three shaded triangles are equal, and all six small triangles are equal. Since all the polygons may be so partitioned into three of the shaded triangles and six of the smaller triangles, it follows that the two polygons have the same area. Indeed, all seven polygons consist of cyclic permutations of these same nine triangles.

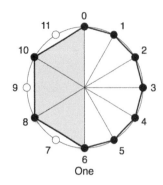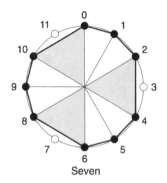

FIGURE 38.5 The areas of all seven polygons are the same.

Another geometric property of regular polygons is that they exhibit a large number of mirror symmetries. This suggests that perhaps the more regular a polygon is, the greater the number of symmetries it will possess. Polygon one has one axis of symmetry about the line through pulses 3 and 9. However, polygon seven has three axes of symmetry about the lines through pulse pairs (1,7), (5,11), and (9,3). This suggests that what seems quite plausible perceptually, that is, that polygon seven is more regular than polygon one, appears to be also true from the point of view of geometric symmetry. Furthermore, rhythm seven is a Euclidean necklace, and thus distributes the nine onsets among the 12 pulses in a maximally even way.

The reader may have observed upon viewing the seven polygons in Figure 38.4 that some of them, such as polygons 1, 4, 5, and 7, contain axes of mirror symmetry at various inclinations, whereas the others do not. Now it is known that, in the domain of visual perception, vertical and horizontal symmetries are detected more easily than their counterparts by the human perceptual system. Therefore, one may wonder if the problem of ranking these polygons by the amount of regularity they possess is made easier when the polygons are rotated so as to make their symmetries more evident. Of course, the number of these symmetries possessed by the polygons is not affected by rotating them.

Figure 38.6 shows the seven bracelets of Figure 38.4, where each is rotated so as to better reveal its regularity. This has been accomplished by rotating the polygons that contain an axis of symmetry, so that the axis is vertical, and rotating the remaining polygons so as to maximize the number of their edges that are vertical and horizontal. The perceptual difference between the two sets of figures is striking, and it appears easier to say from Figure 38.6, which polygons are more regular

than the others. It is fair to say that polygon seven is the most regular of them all. Furthermore, polygons 2, 3, and 6 are most irregular since they possess no axes of symmetry. To rank these three amongst themselves, we may resort to the arrangement of their three long edges. Polygon six may be considered more regular than the other two since all three long edges are separated by short edges, i.e., the long edges are more regularly spaced. Finally, we may classify polygon three as being more regular than polygon two, because in the former, the two adjacent long edges are more evenly separated from the third long edge (4 and 2) than in the latter (5 and 1). In this way, we may conclude that polygon two is the most irregular among this collection.

The preceding discussion focused on measuring the irregularity of polygons. However, as already emphasized, one must be careful in carrying over this visual analysis from the domain of polygons to the aural perception of musical rhythms. For example, in the case of polygons, a geometric feature described in the preceding, such as the presence of axes of mirror symmetry, is rotation invariant. On the other hand, the visual perception of these polygons may change markedly when the polygons are rotated. The same is true for the aural perception of rhythms when they are rotated. Therefore, when measuring rhythm irregularity, it is important to stick to the rhythms themselves rather than their more general necklace or bracelet representations, which ignore the starting onset of the rhythmic cycles.

Several chapters in this book have already dealt with concepts that may be interpreted as implicit or indirect measures of irregularity. They include the measures of syncopation, metrical complexity, offbeatness, and rhythmic oddity. Here we turn our attention to measuring irregularity directly. One rather obvious way to quantify the irregularity of a rhythm is simply to

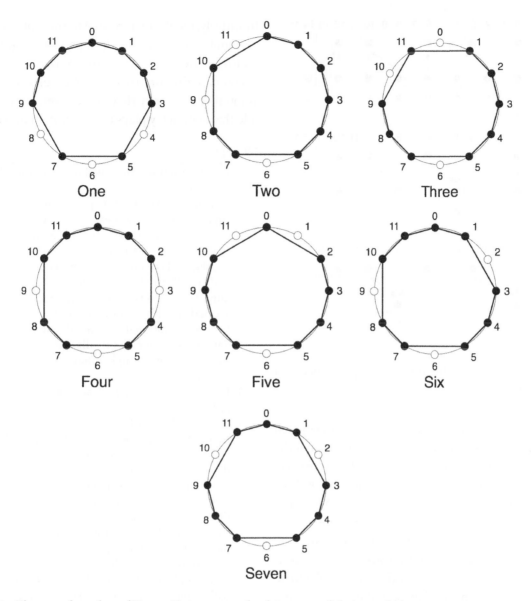

FIGURE 38.6 The seven bracelets of Figure 38.4 are rotated to better reveal their regularity.

calculate some suitable distance measure between the given rhythm and a suitable perfectly regular rhythm. This sounds simple enough, but how should the regular rhythms be placed in the cycle? One natural solution is to place the regular rhythms to align with the meter. For example, for some applications, a square in a 16-pulse cycle could be placed so as to create the rhythm [x . . . x . . . x . . . x . . .], rather than say [. x . . . x . . . x . . . x . .], or [. . x . . . x . . . x . . . x .]. For other applications, one may want to calculate the minimum distance over all rotations of this pattern. One may even consider rotations of the rhythm in a continuous circle.

Another question is what should be done if a rhythmic cycle admits more than one perfectly regular rhythm. Consider for example the 16-pulse *clave son*

timeline. It admits three regular rhythms with 2, 4, and 8 beats, determined by the durational patterns [8-8], [4-4-4-4], and [2-2-2-2-2-2-2-2], respectively. This scenario suggests several possibilities. If it is desired to measure the irregularity relative to one of these specific underlying meters, then that is the regular rhythm that should be used in the calculation. Figure 38.7 illustrates the calculations for the *clave son* timeline with 8, 4, and 2 beats per cycle using the *directed swap* distance measure,[8] giving costs equal to 6, 5, and 11 (top, middle, and bottom, respectively). If a measure of irregularity is desired that is independent of these three meters, or in the context of polyrhythmic music in which all three meters are present simultaneously, then we may take the average of these three values to obtain (6 + 5 + 11)/3 =

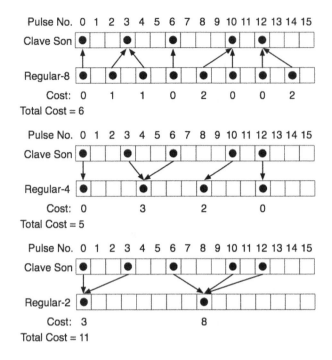

FIGURE 38.7 Converting the directed swap distance to a measure of irregularity.

22/3. For a comparison consider the clave rumba given by [x . . x . . . x . . x . x . . .], in which the third onset of the *clave son* is moved forward by one pulse. The reader may verify that the distances of the clave rumba to the three regular meters are 6, 4, and 10, respectively. Averaging over the three meters yields a value of $(6 + 4 + 10)/3 = 20/3$. Thus, according to this mathematical measure, the *clave son* is more irregular than the clave rumba. By contrast, most musicologists consider the clave rumba to be more syncopated than the *clave son*, thus highlighting the difference between the two related notions of syncopation and irregularity. The *clave son* more clearly separates the "call" or first half of the measure [x . . x . . x .] from the "response" or second half of the measure [. . x . x . . .], than the rumba with counterparts [x . . x . . . x] and [. . x . x . . .]. Thus, by this reckoning the *clave son* is indeed more irregular than the clave rumba.

This chapter highlights the variety of methods for measuring the irregularity of a musical rhythm.

Intuitively, regularity is related to concepts such as evenness and symmetry. Clearly, a perfectly regular rhythm is also a perfectly even rhythm. What is more difficult is to compare rhythms that are not regular in terms of the amount of regularity they possess and to determine the role that symmetry plays in the creation of regularity.

NOTES

1 Regular rhythms are called *plain* rhythms by Rahn, J., (1983), p. 245.

2 Wilson, K. M., (1927), p. 4. Concerning irregular rhythms, Wilson writes: "We may say that the more irregular is the rhythm of music, the more intellectual it will be." I think, however, that to characterize regular and irregular rhythms as "merely emotional" and "intellectual" appears to be a rather facile dichotomy. The world is replete with irregular rhythms that are laden with emotional content. The distinction between emotional and intellectual rhythms is perhaps a useful one, but its discriminating feature must surely be more complex than irregularity considered in isolation.

3 Jones, A. M., (1954a), p. 27 and Jones, A. M., (1973/1974), p. 45.

4 Butler, M. J., (2001).

5 In the real world of acoustic rhythms played by performers, there are of course no perfectly regular rhythms, if regularity is defined in precise mathematical terms, as is done here. In real performances, rhythms live in a continuous space, no two IOIs have exactly the same duration, and hence *all* rhythms are irregular in this strict sense. This is further exacerbated in real performances that exhibit expressive timing. Regularity here is analyzed in an abstract music-theoretical setting much as pitches are studied with set theory. In practice, with acoustic input, the concept of regularity should be replaced with *approximate* regularity. In any case, rhythms, that are almost regular, are often perceived by a listener to be perfectly regular, and that is the important point for this discussion.

6 This rhythm is the bell pattern of the Afro-Peruvian *Samba Malató*. See Feldman, H. C., (2006), p. 112.

7 See Chapter 17 on rhythm complexity for a discussion of the nPVI.

8 Another distance measure that has been explored recently in this context is the edit distance (Toussaint, G. T., 2012b).

CHAPTER 39

Evolution and Phylogenetic Analysis of Musical Rhythms

PHYLOGENETIC ANALYSIS OF CULTURAL OBJECTS

GENEALOGICAL ANALYSIS VIA PHYLOGENETIC trees is a mathematical and computational tool originally developed for analyzing the evolutionary relatedness among a group of biological organisms.[1] More recently, it has been applied to the study of the evolution of a growing variety of cultural objects including, but not limited to, stone projectile points, helmets, swords, pottery compositions, pottery designs, baskets, languages, puberty rituals, marriage patterns, textile designs, written texts, and musical instruments.[2] There are differences and similarities between the methods used and inferences that can be made from phylogenetic analysis, when applied to either biology or culture. This technically difficult and vast topic will not be addressed here. It suffices to note that one of the main controversial issues in the evolution of culture is whether it manifests itself vertically (phylogenesis: parental inheritance) or horizontally (ethnogenesis: borrowing). It may be tempting to hypothesize that, when it comes to culture, in contrast to biology, ethnogenesis is the primary factor in determining the evolution of cultural objects. However, there are counterexamples to this hypothesis. Phylogenetic analysis of Turkmen textiles based on decorative motifs has provided no evidence that supports the hypothesis "that ethnogenesis has always been a more significant process in cultural evolution than phylogenesis."[3] The same issue has been explored in the domain of religious studies, where phylogenetic analysis has been used to investigate whether religious violence is inherited from parent congregations (phylogenesis) or whether it diffuses horizontally between groups (ethnogenesis).[4]

In the domain of music, phylogenetic analysis has been applied in several ways to different aspects of the music under investigation. It can be applied to a single piece of music to gain insight into its compositional structure[5] or to the analysis of the transmission history of copied manuscripts (or variants) of individual pieces of music, such as Orlando Gibbons' Prelude in G.[6] Performance analysis has also been explored with phylogenetic analysis. Whereas earlier algorithmic analyses of performance restricted their attention to tempo and dynamics, phylogenetic analysis has more recently focused on including durations of notes or rhythmic interonset intervals (IOIs), such as in the comparative study of 29 performances of Bach sonatas for violin solo,[7] as well as seven recorded performances of Steve Reich's *Clapping Music*.[8]

When applied to the historical evolution of musical rhythm, phylogenetic analysis raises several questions at different stages of the process.[9] Some of the fundamental issues at stake are (1) How should musical rhythms be represented mathematically in the first place? (2) What is a good model (in evolutionary terms) of a rhythm mutation operation? (3) How should the distance (or dissimilarity) between two rhythms be measured? (4) How should the geographical (or cultural) distance between two different rhythms be measured, specifically when one or both of these rhythms have multiple appearances in different locations in the world? These questions beckon a multidisciplinary approach to search for answers. In this chapter we illustrate the application of phylogenetic analysis to the historical evolution and prototypicality of rhythm timelines in two widely different genres of music: the historical evolution of 12-pulse meters used in the flamenco music of Southern Spain, and the historical evolution of quintuple rhythms in the music of ancient Greece.

PHYLOGENETIC ANALYSIS OF FLAMENCO COMPÁS

Anyone who has seen flamenco dancing, or heard classical music favorites such as Georges Bizet's *Carmen*, Claude Debussy's *Iberia Suite*, or Carl Orff's *Carmina Burana*, has surely been enchanted by the marked, fresh, high-pitched sound of castanets, such as those pictured in Figure 39.1. Castanets are clam-shaped clappers traditionally made of hardwoods, such as olive trees, that are hung on the thumbs of each hand to allow the fingers to create rhythms by causing the two halves to strike against each other. Although the castanets are visually and aurally compelling, and add spice to the music and dance, the rhythms produced on the castanets play a decorative rather than a central structuring role. The most important rhythm in nearly all flamenco music is the underlying metric pattern used, or *compás* as it is called in Spain, which is often played by hand clapping, and is always felt in an embodied manner by the musician or dancer, even when it is not sounded. These metric patterns play a similar role in flamenco music as timelines do in sub-Saharan African music, *wazn* in Arabic music, and *talas* in Indian music.

FIGURE 39.1 Spanish *castanets*. (Courtesy of Yang Liu.)

By some accounts there are more than 70 different styles of flamenco music in Spain, determined by factors such as the location where the music is popular, whether it is instrumental or voice only, what kinds of instruments are used, in what festivals it is employed, what scales are used, what kinds of stories they tell (love, war, poverty, persecution, etc.), and what types of rhythms or compás accompany the song. All these different styles use predominantly a ternary cycle of 12 pulses that is typically executed by clapping all the pulses, and marking the compás by accenting certain claps by means of loudness, timbre, or a combination of the two. In the case of timbre, the clap is usually performed either high pitched and crisp or low pitched and muffled. Other flamenco styles exist that use exclusively binary cycles of four or eight pulses. These include such notables as the *tango* and its variants that include the *tanguillo, rumba, farruca, garrotín, zambra,* and *mariana.* However, all these binary styles use one and the same metric pattern given by [. x x x], where "." denotes in this context not a silent pulse (as in the rest of the book), but rather a soft clap, and "x" denotes a loud clap. Here we are concerned with the phylogenetic analysis of the more characteristic ternary 12-pulse metric patterns employed in flamenco music.

At this point the reader may be wondering if there is any connection between the 12-pulse rhythms used in flamenco music, and the 16-pulse *clave son* timeline frequently examined in this book. A slight digression here is in order. First, recall that there is a great deal of mathematical, musicological, and perceptual similarity between the binary *clave son* given by [3-3-4-2-4] and its ternary counterpart with interval structure [2-2-3-2-3]. Furthermore, as described in Chapter 10, there exists evidence that the ternary version may have mutated to the binary version by a process that ethomusicologist Rolando Pérez Fernández calls *binarization*, in the context of intercultural transplantation (ethnogenesis). As for the relation between flamenco music and Afro-Cuban music, historically, there has been a great deal of cultural transplantation between the geographical triangle consisting of Cuba, Southern Spain, and sub-Saharan Africa. More than 1,000 years ago the Arabs who lived in Southern Spain and North Africa traveled south, exploring much of the African continent, as far south as Mali, Ghana, and beyond. They have also written accounts describing the music they encountered in their travels. A common metric pattern (compás) used in Flamenco music, such as the guajira given by [x . . x . . x . x . x .], is a rhythm played on the musical bow by the

San people of Botswana, Namibia, and Angola.[10] After the Spanish conquest of Cuba, the slave trade created cultural exchanges between all three regions. The popular *punto cubano* timeline in Cuba uses the same metric pattern as the guajira, and is often accompanied by the wooden claves that play the derived version [x . x x . x x . x . x .]. In addition, musicologists have certain beliefs based on both historical knowledge and oral tradition, about the evolution of flamenco music. Therefore, flamenco music provides a convenient dataset on which to test phylogenetic analysis tools to determine what can be garnered from analyzing the metric patterns. At the same time, such an analysis may shed light on the evolution of the *clave son* and its migrations between Ghana and Bagdad, keeping in mind that the simpler a rhythmic timeline is, the higher is the probability that it was born independently in different places, without necessarily migrating from one place to another.

In the present analysis, all dimensions of flamenco music other than the durational patterns of the meters are ignored. These radical assumptions are made to simplify the analysis and to determine how much information can be obtained from such a minimalist approach. The method illustrated is akin to the methods used in bioinformatics that study human evolution by focusing on the DNA molecules, which are sequences of symbols in an alphabet of four letters, while ignoring all the other rich biological, psychological, neurological, and cultural dimensions that characterize human beings.

There is consensus among flamenco musicologists that the fountain of flamenco music is the fandango, which uses the meter [x . . x . . x . . x . .].[11] This rhythm is symmetric and periodic, repeating the waltz-like metric pattern [x . .] four times. There are in flamenco music four additional (asymmetric) 12-pulse meters shown later.[12]

[. . x . . x . x . x . x]—soleá
[. . x . . . x x . x . x]—bulería
[x . x . x . . x . . x .]—seguiriya
[x . . x . . x . x . x .]—guajira

These four patterns along with the fandango are depicted in convex polygon notation in Figure 39.2, where the "0" marks the time at which the meter starts, but which may be different from the position at which the rhythm is sometimes "launched" for the convenience of the dancers. Before proceeding further, a caveat should be noted concerning the labels attached to these rhythms. The soleá and bulería patterns are used in a large variety of flamenco styles. Furthermore, the pattern here referred to as soleá is sometimes used in a style named *bulería tradicional*, and the pattern called bulería is sometimes employed in the *bulería moderna*. The guajira rhythm is used in fewer styles but is also most notably used in the petenera. The names adopted here for these rhythmic timelines are used mainly for convenience, but they also reflect the most notable styles that use the meters.

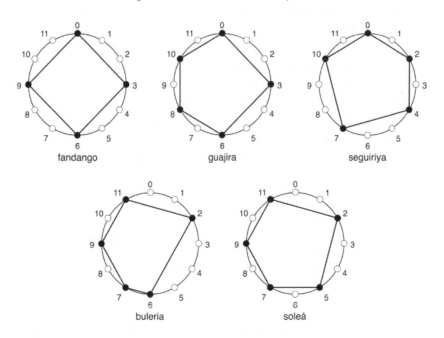

FIGURE 39.2 The flamenco ternary (12-pulse) meters in polygon notation.

It is known that the bulería pattern is a relatively recent mutation that evolved from the soleá pattern in the twentieth century. It is also interesting to note that, of all the five rhythms, only the bulería has the rhythmic-oddity property, a geometric measure of the rhythm's complexity. This accords with the theory that, as patterns evolve, their complexity tends to increase,[13] which highlights the importance of the rhythmic oddity property in isolating bulería from the older patterns. The more obvious difference between the bulería and all the other rhythms is that it is the only rhythm that contains intervals of lengths 1, 2, 3, and 4. The other rhythms have intervals of lengths 2 and 3 only. There are several measures of rhythm complexity that depend on the variability of these interonset durations, the most noteworthy being the standard deviation (SD) and the normalized pairwise variability index (nPVI). The soleá has SD = 0.55 and nPVI = 10, whereas the bulería has SD = 1.14 and nPVI = 53.8. According to both measures, the bulería is more complex than the soleá, again supporting the theory that evolution favors an increase in complexity. The guajira, often described as having a significant Cuban influence, is also distinguished from the other four rhythms in that it is the only five-onset rhythm with an offbeatness value of zero.

Since the fandango has four onsets and the other rhythms have five, the directed swap distance discussed earlier was used to compute the distance matrix shown in Figure 39.3.[14] The bottom row in this figure, labeled with the word "Total" indicates the sum of the distances of the rhythm listed at the top of the column, to all the other rhythms in the set. This figure in itself already reveals some interesting information. It shows for example that the seguiriya meter with (Total = 31) is the most different from the collection, and the fandango and guajira meters with (Total = 21) are the most similar to the rest of the group.

Ethnomusicologists are beginning to embrace mathematical and computational methods, such as phylogenetic analysis, in their exploration of music.[15] Phylogenetic trees belong to the family of *proximity graphs*[16] that provide not only an effective method for visualizing the distance matrix obtained from rhythm datasets but also yield additional structures such as clustering relationships and information for the reconstruction of possible ancestral rhythms. They may also be used to discover "evolutionary chains" of minimal mutations in motifs frequently used in a particular style or period of music.[17] A variety of techniques have been developed for generating phylogenetic trees from distance matrices.[18] Some of these methods yield *graphs* containing cycles rather than trees, when the underlying proximity structure between the rhythms is not tree-like. One notable example in this class is a program called *SplitsTree*.[19] Like the more traditional phylogenetic tree programs, *SplitsTree* computes a drawing in the plane with the property that the distance between any two nodes reflects as closely as possible the true distance, between the corresponding two rhythms, that is contained in the distance matrix. If the tree structure does not match the data perfectly, then new nodes in the graph may be introduced. Such is the case, for example, in the *SplitsTree* of Figure 39.4, which contains two new nodes indicated with white circles. Such nodes may suggest implied "ancestral" rhythms from which their "offspring" may be derived. In addition, edges may split to form parallelograms, as is also evident in Figure 39.4. The relative sizes of these parallelograms are proportional to an *isolation* index that indicates the significance of the clustering relationships inherent in the distance matrix. *SplitsTree* also computes the *splitability* index, a measure of the goodness-of-fit of the entire splits graph. This fitness value is obtained by dividing the sum of all the approximate distances in the splits graph by the sum of all original distances in the distance matrix.

The SplitsTree of Figure 39.4 suggests a clear clustering structure in which the bulería and soleá form one tight cluster, the fandango and guajira form a

	Soleá	Bulería	Seguiriya	Guajira	Fandango
Soleá	0	1	11	7	7
Bulería	1	0	12	8	8
Seguiriya	11	12	0	4	4
Guajira	7	8	4	0	2
Fandango	7	8	4	2	0
Total	**26**	**29**	**31**	**21**	**21**

FIGURE 39.3 The directed swap distance matrix for the five flamenco ternary meters.

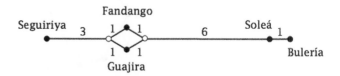

FIGURE 39.4 The *SplitsTree* obtained with the directed-swap distance matrix.

second tight cluster, and the seguiriya is off by itself. The phylogenetic tree supports the historical knowledge that the bulería pattern evolved from the soleá pattern. Furthermore, if we believe that the rhythms situated near the center of the tree correspond to earlier ancestral rhythms, this result lends support to the musicological tenet that the fandango is the fountain from which spring all flamenco rhythms. Indeed, in the genealogical trees that musicologists constructed based on historical evidence, the fandango is located low and at the center of the main trunk of such trees. That these trees do not place the fandango at the bottom of the trunk of the tree is not surprising, given that these genealogical trees were constructed with the entire music in mind, including voice and instrumentation, and not the metric rhythm in isolation. Musicologists also believe that the bulería pattern is a recent mutation of the soleá. The phylogenetic tree also suggests that the bulería, being a leaf of the tree, is an offshoot of the soleá, thus lending support to this theory.

One of the two white nodes on either side of the fandango and guajira nodes is closer to the *center* of the tree than are the fandango and guajira. The center of a tree may be considered to be the point that minimizes its maximum distance to any node. This suggests that the rhythm corresponding to this white node is an ancestor of all these rhythms. However, we do not know what this rhythm is, and therefore it must be reconstructed. Since the output of the *SplitsTree* program provides the lengths (distances) of all the edges in the graph, from these lengths, we may infer the following graph distances, illustrated in Figure 39.5, between the unknown ancestral rhythm and all other rhythms:

d(ancestor, guajira) = 1
d(ancestor, fandango) = 1

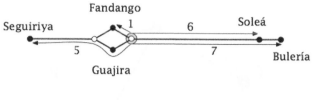

FIGURE 39.5 Reconstructing the ancestral rhythm using distance information.

d(ancestor, seguiriya) = 5
d(ancestor, soleá) = 6
d(ancestor, bulería) = 7

Examining these distances in the graph leads one to suspect that the ancestral rhythm should be quite similar to both the guajira and the fandango, and would therefore consist of either four or five onsets. An exhaustive computer search revealed the five-onset rhythm given by [x . . x . . x . . x x .] illustrated in polygon notation in Figure 39.6 (left). The reader may verify that this rhythm satisfies the five required distance constraints using the directed swap distance. Furthermore, this rhythm, which may be viewed as the composition of the fandango with one additional onset at pulse number ten, is the only five-onset 12-pulse rhythm that satisfies these distance constraints.

A search of the literature to determine if this "ancestral" rhythm makes an appearance anywhere yielded the following three observations worth noting. First, it has been established that this accented metric pattern belongs to the genre of flamenco music known as Fandango de Huelva.[20] Chuck Keyser mentions in his book that this pattern is used often in flamenco music as a closure or resolution phrase.[21] Furthermore, it is almost identical, and has the same feel, as a rhythm Nan Mercader calls the *Fandango de Huelva*. This rhythm is the composition of the "ancestral" rhythm of Figure 39.6 (left) with an additional onset halfway

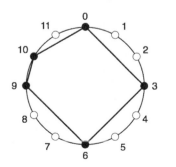
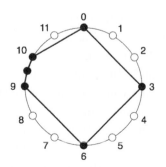

FIGURE 39.6 The "ancestral" rhythm of flamenco (left) and the *fandango de Huelva* (right).

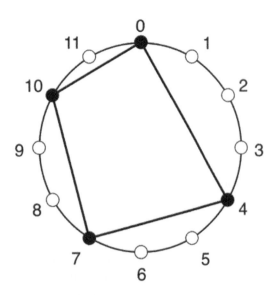

FIGURE 39.7 The Arab-Persian rhythm *al-hazaj*.

between onset number nine and onset number ten. In other words, the fandango de Huelva may be described in a time span of 24 units as [x . . . . . x . . . . . x . . . . . x x x . . .]. Removing the middle decorative onset from the triplet in the fandango de Huelva yields the "ancestral" rhythm obtained from the phylogenetic tree. This phylogenetic analysis contradicts the theory put forth by Molina and Mairena that the seguiriya rhythm is the ancestral compás of flamenco music.[22] For an alternative departure point in the phylogenetic studies of flamenco music, the reader is referred to the detailed study by Bernat J. de Cisneros Puig.[23]

It is also worth pointing out the close relationship that exists between the seguiriya and an ancient thirteenth century Arab-Persian rhythm called *al-hazaj*. The latter, shown in polygon notation in Figure 39.7, has a durational pattern [4-3-3-2] and results when one silences the second onset of the seguiriya.[24]

THE GUAJIRA

According to Nan Mercader, among others, the guajira rhythm was strongly influenced by Cuban music (the punto Cubano in particular) and has the flavor of nineteenth century colonial Spain.[25] The pattern is also the underlying rhythm of the Mexican *son*.[26] It was notably used by Leonard Bernstein as the metric pattern for his piece "America" in the hit movie *West Side Story*.[27] Supposedly the rhythm traveled from Cuba to Spain via the Canary Islands, where it mixed with some indigenous elements.[28] The name comes from the word *guajiro*, for a

creole campesino. The guajira pattern, like the fandango, is located near the center of the graph, and the two have a twin relationship with each other. They both have the same set of directed-swap distances to the other three rhythms. Does this suggest that the guajira, like the fandango, plays a significant ancestral role in the evolution of flamenco? There is evidence to suggest that it does. For almost 800 years, from 711 to 1492, Andalusia, the motherland of flamenco music, was ruled by the Arabs, and North African Moors, who travelled extensively in the African continent, going as far as South Africa. In the Tuareg region of Africa, there exist several 12-pulse patterns that are used for dancing and informal entertainment. In particular, an old rhythm given by [x . . x . . x . x . x .] with the name of *abakkabuk* with IOIs [3-3-2-2-2] is the same as the guajira rhythm.[29] Furthermore, the *San* people of Southern Africa play the same rhythm on a musical bow. Thus, it is possible that the guajira rhythm, employed in flamenco music, traveled from sub-Saharan Africa to Andalusia by way of the Moors. Conversely, there is no doubt that sub-Saharan African music has been influenced by Arabic music.[30] There is also evidence that in the fourth century BC, before the Arabs explored West Africa, the Carthaginians set sail across the Straigts of Gibraltar and traveled down the African coast.[31] So perhaps this rhythm traveled to Europe a 1,000 years before the Arabs conquered North Africa.

This is all well for the ternary rhythms, but is there any connection between the ternary 12-pulse guajira rhythm and the binary 16-pulse *clave son*? If the guarira pattern on a 16-h clock is binarized by snapping its onsets to the nearest onsets of a superimposed 16-h clock, as shown in Figure 39.8, then it becomes the

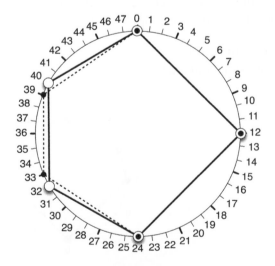

FIGURE 39.8 The binarization of the guajira metric pattern.

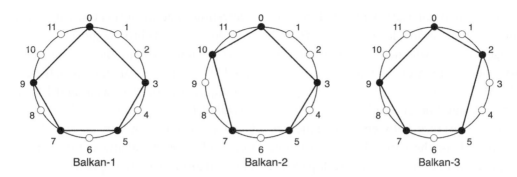

FIGURE 39.9   Three Balkan folk rhythms with five onsets among 12 pulses.

kpatsa rhythm of Ghana with durational pattern [4-4-3-2-3].[32] Figure 39.8 employs a 48-h clock to accommodate both 16-pulse and 12-pulse cycles. The *clave son* is a rotation and *permutation* of these intervals, in which the interval of duration two flanked by intervals of duration three is moved to the location in between the intervals of duration four. This type of operation is the topic of Chapter 40.

As a final note, one cannot ignore the close relationship that exists between the Balkan folk rhythms shown in Figure 39.9 and the flamenco and dub-Saharan African rhythms. The first Balkan rhythm (left) is a rotation of the guajira. The other two are rotations of the fume–fume or ternary version of the *clave son*, and have the same rhythmic contour as the *clave son*. Since the Balkan region lies geographically in between Bagdad and Spain, it is conceivable that these rhythms migrated either intact or in mutated forms from one place to the other.

PHYLOGENETIC ANALYSIS OF ANCIENT GREEK PAEONIC RHYTHMS

In the history of the evolution of musical rhythms, there is a paucity of documented examples that yield evidence of the gradual development across centuries, and of a prototype rhythm evolving into a set of its variants. The ancient Greek rhythmic paeonic (quintuple time) genus discussed by Aristotle and Aristoxenus[33] is one such notable example, thus providing a unique and much needed opportunity to test the evolutionary efficacy of the mathematical phylogenetic tools presently available. A paeonic rhythm has durational ratio 2:3, and can be notated succinctly using binary sequences. The rhythm [x . x x .] is the root of the paeonic genus, and is called the *cretic*.[34] Martin L. West documents seven variants of this prototype rhythm that appeared gradually over a period ranging from the seventh century BC to the second century AD.[35] The eight rhythms listed in their chronological order of appearance in Table 5.1 of West's book are reformatted in box notation in Figure 39.10.

	1 2 3 4 5	7th c. BC	5th c. BC	3rd-2nd c. BC	2nd c. AD
1	● ●●	✔	✔	✔	✔
2	● ●	✔	?		
3	● ●●●		✔	✔	✔
4	●●●●		✔	✔	✔
5	●●●●●			✔	✔
6	● ● ●				✔
7	●●● ●				✔
8	●● ●●				✔

FIGURE 39.10   Historical record of the increasing variety of quintuple rhythms in use from the seventh century BC to the second century AD.

In the following, we apply correlation and phylogenetic analyses using the tools available in the *SplitsTree-4* software package[36] to test whether mathematical evidence supports the evolution of the paeonic genus, as documented by M. L. West.

Consider first the question of whether a mathematical measure of the distance between two sequences can serve as a model for the chronological evolution (or historical appearance) of the ancient Greek paeonic rhythms. To answer this question, the order of appearance of the eight paeonic rhythms in Figure 39.10 is tested for association with the *edit* distance calculated from the root of the paeonic genus (rhythm No. 1) to all the other seven rhythms. The use of the edit distance is motivated in part by the fact that it correlates well with human judgments of perception.[37] However, there is no a priori reason why such is a necessity, since evolutionary mutations may be directed by properties other than perceptual similarity, such as changes in increased rhythm complexity. Recall from Chapter 36 that the edit distance between two sequences (rhythms) is defined as the minimum number of mutations required to transform one sequence to the other. There are three types of mutations: (1) a symbol may be deleted, shortening the rhythm, (2) a symbol may be inserted, lengthening the rhythm, and (3) one symbol may be substituted by another. A natural test to measure the association between two orderings is the Spearman rank correlation coefficient.[38] The edit distances from rhythm No. 1 to the eight rhythms are 0, 1, 1, 1, 2, 2, 2, and 2. The Spearman rank correlation between this sequence of distances and the chronological order is $r = 0.91$, with a two-tailed value of $p = 0.00155$.[39] This suggests the hypothesis that the edit distance between the ancestral rhythm and its progenitors predicts the chronological evolutionary trajectory of the rhythms. It would thus be interesting to perform this test with other datasets that have documented chronological histories.

Let us now turn to the second question that asks whether a mathematical measure of the distance between two sequences can serve as a model for identifying the *central prototype* rhythm from which all the other ancient Greek paeonic rhythms are derived. To this end we perform a phylogenetic analysis of these rhythms using the *BioNJ* algorithm in the *SplitsTree-4* software package. The input to the *SplitsTree* program is in general any distance matrix that contains the distances (edit distance in the present case) between every pair of rhythms in the family. The *BioNJ* algorithm is a neighbor-joining algorithm that constructs a tree from the distance matrix using an agglomerative process consisting of iteratively picking a pair of rhythms, creating a new node that represents these rhythms, and updating the distance matrix by replacing both rhythms with the new node, while minimizing the difference between the sum of distances in the matrix and the sum of geodesic distances in the tree.

The edit distance matrix of the eight rhythms documented by M. L. West yields the sum of the distances from each rhythm (a measure of centrality) to all the other seven rhythms, shown in Figure 39.11 (bottom row labeled **Total**). Thus, the rhythm with the minimum total sum is considered to be the ancestral root or central prototype, from which all the other rhythms may be generated with the minimum number of edit distance mutations. From Figure 39.11 it is observed that the oldest ancestral rhythm does have the minimum total edit distance of 11, and the most recent offspring has the largest distance of 14. However, the third and fifth rhythms are tied with the first, and the second rhythm is tied with the last, yielding a less than definitive outcome.

However, the book by M. L. West does not contain all the ancient Greek quintuple rhythms. In the fourth

No.	Rhythm	x.xx.	x.x..	x.xxx	xxxx.	xxxxx	x.x.x	xxx.x	xx.xx
1	x.xx.	0	1	1	1	2	2	2	2
2	x.x..	1	0	2	2	3	1	2	3
3	x.xxx	1	2	0	2	1	1	2	2
4	xxxx-	1	2	2	0	1	3	2	2
5	xxxxx	2	3	1	1	0	2	1	1
6	x.x.x	2	1	1	3	2	0	1	2
7	xxx.x	2	2	2	2	1	1	0	2
8	xx.xx	2	3	2	2	1	2	2	0
	Total	11	14	11	13	11	12	12	14

FIGURE 39.11 The edit distance matrix for the eight rhythms in West's table in Figure 39.10.

No.	Rhythm	x.xx.	x.x..	x..x.	x.xxx	xxxx.	x.x.x	xx.x.	xxx.x	xx.xx
1	x.xx.	0	1	1	1	1	2	2	2	2
2	x.x..	1	0	2	2	2	1	2	2	3
3	x..x.	1	2	0	2	2	2	1	3	2
4	x.xxx	1	2	2	0	2	1	3	2	2
5	xxxx.	1	2	2	2	0	3	1	2	2
6	x.x.x	2	1	2	1	3	0	2	1	2
7	xx.x.	2	2	1	3	1	2	0	2	1
8	xxx.x	2	2	3	2	2	1	2	0	2
9	xx.xx	2	3	2	2	2	2	1	2	0
	Total	12	15	15	15	15	14	14	16	16

FIGURE 39.12 The edit distance matrix for the nine rhythms obtained by deleting [x x x x x] from West's list, and inserting the two rhythms numbered 3 [x . . x .] and 7 [x x . x .] from the book by Abdy Williams.

century BC, Aristoxenus, the Greek music theorist, student of Aristotle, and writer of the *Elements of Rhythm*, contains, according to C. F. Abdy Williams,[40] two additional quintuple rhythms, namely [x . . x .] and [x x . x .]. Furthermore, the rhythm [x x x x x] listed in West's table was used infrequently in ancient Greece. To quote Abdy Williams, "a continuous use would not be made of such a rhythm; for its character is quite alien to the paeon. The reason Aristoxenus objects to giving five primary times (onsets) to the paeon is that the ancients considered a succession of short notes mean and vulgar." It is therefore of interest to perform a phylogenetic analysis with nine rhythms, including the two rhythms documented in the work of Aristoxenus, and excluding the rhythm with five primary times (successive onsets). With this new list on nine rhythms, one obtains the distance matrix in Figure 39.12, where the two additional rhythms are numbered 3 and 7 in the list. With these nine rhythms, the results provide good agreement with the historical evolutionary data. The most ancient rhythm now is the unique rhythm with the smallest total sum of distances equal to 12.

Figure 39.13 shows the *BioNJ* tree calculated from the matrix of Figure 39.12 The order and numbering in which the rhythms appear historically in West's table in Figure 39.10 is indicated with bold numbers, and the two additional rhythms listed by Abdy Williams are circled. The most ancient (ancestral prototype) rhythm is clearly situated in the center of the tree. The correlation and phylogenetic analyses using the rhythms documented by M. L. West thus provide support of the chronological evolution of those rhythms. Furthermore, if the two rhythms listed by Abdy Williams are included, and the five-onset rhythm number 5 in West's table [x x x x x]

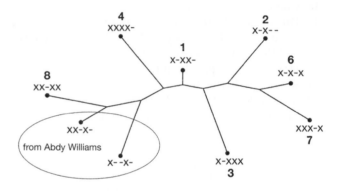

FIGURE 39.13 The *BioNJ* tree calculated from the edit distance matrix of Figure 39.12.

is deleted from the group, then the phylogenetic analysis of the resulting nine rhythms provides support for the historical account of the evolution of the ancient Greek quintuple rhythms.

NOTES

1 Baum, D. A. & Smith, S. D., (2012).
2 The book by Lipo, C. P., O'Brien, M. J., Collard, M., & Shennan, S. J., Eds., (2006), provides a fascinating and illuminating collection of chapters dedicated to the application of phylogenetic analysis to a wide variety of cultural objects (other than music), that serves as a useful introduction to the subject. See also Mace, R., Holden, C. J., & Shennan, S., Eds., (2005).
3 Collard, M. & Tehrani, J., (2005).
4 Matthews, L. J., Edmonds, J., Wildman, W. J., & Nunn, C. L., (2012).
5 Colannino, J., Gómez, F., & Toussaint, G. T., (2009).
6 Windram, H., Charlston, T., & Howe, C. J., (2014).
7 Liebman, E., Ornoy, E., & Chor, B., (2012).
8 Sethares, W. A. & Toussaint, G. T., (2014).
9 Toussaint, G. T. (2003b) and Toussaint, G. T., (2012a).
10 Kaemmer, J. E., (2000), p. 314.

11 De Cisneros Puig, B. J., (2017).

12 See Cano, D. M., (1983), Keyser, C. H., (1993), Fernández-Marin, L., (2001), Gamboa, J. M., (2002), Rossy, H., (1966), Manuel, P., (2004), p. 102, and Parra, J. M., (1999).

13 Crofts, A. R., (2007), p. 25.

14 See Guastavino, C., Gómez, F., Toussaint, G., Marandola, F., & Gómez, E., (2009), for a comparison of the directed swap distance with other distance measures.

15 See Toussaint, G. T., (2012) for a survey of computational tools for phylogenetic analysis of musical rhythms.

16 There are many other types of proximity graphs that may be used to visualize data. See Toussaint, G. T., (1980, 1988), as well as Jaromczyk, J. W. & Toussaint, G. T., (1992).

17 Crawford, T., Iliopoulos, C. S., Winder, R., & Yu, H., (2001) and Miranda, E. R., (2004).

18 Gonnet, G. H., (1994).

19 Dress, A., Huson, D., & Moulton, V., (1996), Huson, D. H., (1998), Bryant, D. & Moulton, V., (2004), and Huson, D. H. & Bryant, D., (2006).

20 De Cisneros Puig, B. J., (2017).

21 Keyser, C. H., (1993).

22 Molina, R. & Mairena, A., (1963). For further discussion, see also Díaz-Báñez, J. M., (2017).

23 De Cisneros Puig, B. J., (2017), p. 2.

24 Wright, O., (1978). p. 216.

25 Mercader, N., (2001), p. 26.

26 Stanford, E. T., (1972), p. 77.

27 See London, J., (2004), p. 129 and London, J., (1995), p. 65, for an analysis of Bernstein's "America," and a discussion of other complex meters.

28 See Manuel, P., (2004), for an account of the complex musical interactions between Cuba and Spain, relevant to the *guajira*, over the past few centuries.

29 Wendt, C. C., (2000), p. 222. Rotating the guajira by a half cycle yields the durational pattern [2-2-2-3-3], which is a common African rhythmic pattern. See Kauffman, R., (1980), Table 1, p. 409, Stone, R. M., (2005), p. 82. Kwabena Nketia, J. H., (1962), p. 132, identifies this rhythm as a hand-clapping pattern used in *Kwasi Dente*, a recreational maiden song of the Akan people of Ghana. It is also the rhythmic structure of the drumming and hand-clapping music that accompanies the *sōt silām* dance from *Mirbāt* in the south of Oman. See El-Mallah, I. & Fikentscher, K., (1990).

30 Jones, A. M., (1954b).

31 Fryer, P., (2003), p. 106 and Migeod, F. W. H. & Johnston, H., (1915), p. 414.

32 Eckardt, A., (2008), p. 64.

33 West, M. L., (1992), p. 140.

34 Abdy Williams, C. F., (1911), p. 43.

35 West, M. L., (1992), p. 143.

36 Huson, D. H. & Bryant, D., (2006).

37 Post, O. & Toussaint, G. T., (2011). See also Toussaint, G. T., (2013b).

38 Spearman, C., (1904).

39 This correlation value was obtained with Spearman's Rho Calculator on the Social Science Statistics web page: www.socscistatistics.com/tests/spearman/Default2.aspx.

40 Abdy Williams, C. F., (1911).

CHAPTER 40

Rhythm Combinatorics

A S WE HAVE SEEN SEVERAL TIMES ALREADY, the *clave son* pattern in binary notation consists of five onsets among 16 pulses. It has also been pointed out early on in this book that the number of ways one may select five items from among a collection of 16 items is given by the formula (16!)/(5!)(11!) that yields the number 4,368. This is a large number of patterns. However, as pointed out in earlier chapters, most of these are unsuitable for use as timeline rhythms. So far in this book, a variety of systematic methods have been described for reducing the size of this large set by discarding rhythms that do not possess certain properties deemed desirable. One disadvantage of proceeding in this way is that the list of possible properties to examine has to be created in the first place, and may be unbounded. A different approach, and one that proceeds in the opposite direction, starts out with a well-known rhythm as a *seed* from which to generate a family of close relatives that hopefully inherit "goodness" by virtue of their proximity to the seed rhythm. In fact, an example of this approach has already been discussed in the case in which a maximally even rhythm was used to generate a family of almost maximally even rhythms. In that approach, the rhythm was viewed as a binary sequence, and each onset was swapped with its neighboring silent pulse on either side. In this chapter, the rhythm is viewed instead as a durational pattern, or a sequence of interonset intervals (IOIs), and a family of rhythms is obtained from one "good" rhythm by swapping the positions of the IOIs of the "good" rhythm, that is, by generating all the permutations of these intervals.[1] Let us illustrate this approach with several seed patterns, and consider first the *clave son*.

The *clave son* pattern in binary sequence representation is [x..x..x...x.x...], which yields the durational pattern [3-3-4-2-4]. The first question is how many permutations of this pattern of five numbers exist? Note that these numbers form *multisets*, since repetitions of elements are permitted. We have a total of five IOIs that belong to three different classes: one IOI of duration 1, two IOIs of duration 2, and two IOIs of duration 3. Therefore the total number of permutations of the interval set [3-3-2-4-2] is (5!)/(1!)(2!)(2!) = 30. All 30 patterns are shown in box notation in Figures 40.1 and 40.2 with the *clave son* placed first in the list.

The reader is encouraged to play all 30 rhythms for comparison. Some sound better than others, and some are more difficult to play than others, but none are bad,

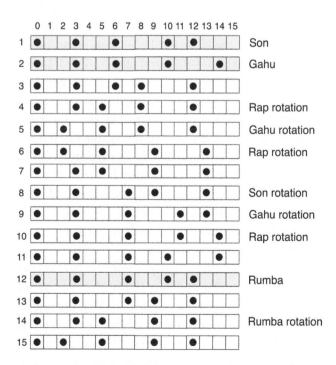

FIGURE 40.1   The first 15 interval permutations of the *clave son*.

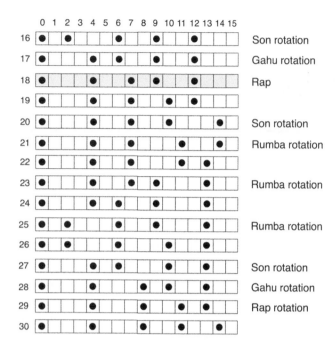

FIGURE 40.2 The second 15 interval permutations of the *clave son*.

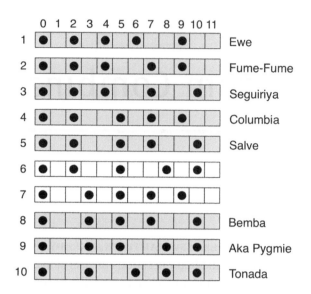

FIGURE 40.3 The ten IOI permutations of the fume–fume timeline [2-2-3-2-3].

and all of them sound interesting. Among these 30 permutations of the *clave son*, we find the rumba, the gahu, and the rap timelines, as well as, of course, their onset rotations. If one rhythm may be obtained from another by a permutation of its interval vector, the two rhythms will be said to belong to the same *interval combinatorial class*. Thus, the *clave son* [3-3-4-2-4], clave rumba [3-4-3-2-4], gahu [3-3-4-4-2], and rap [4-3-2-3-4] belong to the same interval combinatorial class, whereas shiko [4-2-4-2-4], soukous [3-3-4-1-5], and bossa-nova [3-3-4-3-3] each belong to their own distinct classes. Returning to the 30 rhythms of the son–rumba–gahu–rap interval combinatorial class, the remaining 26 rhythms resemble each other to one degree or another but tend to sound more modern, more "jazzy." Indeed, any one of them could be successfully incorporated in new music.

For the second example, consider the fume–fume timeline with binary sequence representation [x . x . x . . x . x . .] and durational pattern [2-2-3-2-3]. This pattern has five IOIs and two IOI multisets: the IOI of duration 3 occurs twice, and the IOI of duration 2 occurs thrice. Therefore the number of permutations is (5!)/(2!)(3!) = 10. The ten permutations are shown in Figure 40.3. At least eight of these (shown shaded) are used in music from Ghana, Spain, Cuba, South Africa, and Central Africa, and are identified by some of their names.

For a third example, consider the cinquillo timeline with binary sequence representation [x . x x . x x .] and durational pattern [2-1-2-1-2]. This pattern also has five IOIs and two multisets: the IOI of duration 1 occurs twice and the IOI of duration 2 occurs thrice. Therefore the number of permutations is again (5!)/(2!)(3!) = 10. The ten permutations are shown in Figure 40.4. At least seven of these (shown shaded) are used in music from Senegal, Surinam, Rumania, Cuba, Turkey, Guinea, and Classical music, and are identified by their names.

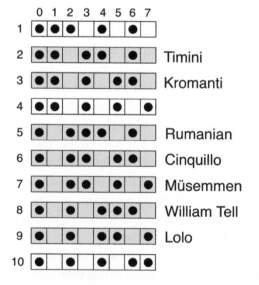

FIGURE 40.4 The ten IOI permutations of the cinquillo timeline [2-1-2-1-2].

Combinatorial methods such as those described earlier, for generating musical rhythms, have been used by composers,[2] as well as by church-bell ringers to create interesting nonrepetitive patterns.[3] It is a tool that facilitates composition. Furthermore, the various permutations of a rhythm can be selected either at random or according to deterministic musicological rules. A recent example of this approach is the recombination of rhythm midi loops by Sioros and Guedes (2011), who first order the rhythms by decreasing complexity.[4] Indeed, one might say that the application of combinatorics to music shifts the creative process from the ineffable domain of creative invention to the more mechanical realm of choice.

NOTES

1 In describing his principal method for creating variation in compositions Schillinger, J., (2004), p. xi, called these types of operations *general permutations* to distinguish them from *circular* permutations. The combinatorial method for generating and analyzing permutations and combinations of elements in sets is a technique that has been used frequently in music theory. Although Schillinger had many useful ideas, some of his terminology and notation were not well received at the time of his writing. See Backus, J., (1960), for a critique.

2 Babbitt, M., (1962), used combinatorial methods as a means to obtain maximal diversity in his compositions. See also Coxeter, H. S. M., (1968), Forte, A., (1973), Reiner, D. L., (1985), Duncan, A., (1991), Read, R. C., (1997), Dabby, D. S., (2008), and Benson, D. J., (2007), Chapter 9. Knobloch, E., (2002), is a nice survey of combinatorial methods popular in the seventeenth century. Toussaint, G. T., (2002, 2003a) has applied combinatorial methods to generate "good" rhythm timelines. See also De Souza, R. C., (2016).

3 Roaf, D. & White, A., (2006), p. 123.

4 Sioros, G. & Guedes, C., (2011), p. 381.

CHAPTER 41

What Makes the *Clave Son* Such a Good Rhythm?

THE PREFERRED PROPERTIES OF MUSIC, in general, and rhythm, in particular, may be explored at many levels of generality, and at any point in the objective–subjective continuum. In Plato's *Republic*, Socrates tells Glaucon that "Good rhythm depends on simplicity."[1] For Jeff Pressing: "The effectiveness of a groove from the standpoint of reception is assessed by its ability to engage human movement and attention."[2] Bruce F. Katz proposes measuring musical preference at the neurological level: "A musical passage will be preferred to the extent that it creates synchrony in the neurons that are responsible for processing the passage."[3] The approach taken in this book to measure rhythmic preference follows in the footsteps of Socrates and Pressing. In this chapter, stock is taken of the variety of properties that are possessed by the *clave son* rhythm, in particular, and "good" rhythms, in general. The role that these properties play is examined, individually and as a whole, in contributing to the salience of this particular rhythm. Any one of these properties, by itself, is perhaps not sufficient to establish a rhythm as a candidate for a good timeline, let alone characterize the *clave son* uniquely among all rhythms. However, a group of them in combination can be quite compelling. In the final analysis, it is a question of striking the right equilibrium within and between these properties.

EVENNESS AND BALANCE

Perfect evenness, discussed in detail in Chapter 20, is obtained only when all the adjacent interonset duration intervals of a rhythm are equal. This happens only when

a rhythm is perfectly regular, such as, for instance, the rhythm [x . x . x . x . x . x . x . x .]. But as has been pointed out more than once, "much of our enjoyment of music comes from its balance of predictability and surprise."[4] Rhythms with identical interonset intervals (IOIs) contain too much regularity and symmetry to be musically interesting. This is not to say they have no psychological power or social usefulness. It is well known that perfectly regular rhythms have been used for thousands of years to embolden soldiers into battle. Indeed, according to the musicologist Curt Sachs, humanity's sensation of regular rhythms is acquired from walking and marching.[5] No doubt such rhythms used in warfare also provide soldiers with the requisite energy for long marches. Furthermore, Sachs points out that the rhythmic tempo of much music matches the rhythm of a person's stride. To succeed as a musical timeline, however, a rhythm must be somewhat asymmetrical, it must possess some irregularity, and it must contain an element of surprise. In the words of Alexander Voloshinov: "Exact symmetry in art is perceived as static and frigid, and only the introduction of asymmetrical elements into symmetry brings about a kind of enlivening."[6] Recent scientific discoveries suggest that such "broken" symmetries are relevant to more than art, and have "deep cosmological implications."[7] Marvin Minsky, an artificial intelligence pioneer at The Massachusetts Institute of Technology, expressed the need for deviation from perfect repetition as follows: "Good music germinates from tiny seeds. How cautiously we handle novelty, sandwiching the new between repeated sections of familiar stuff! The clearest kind of change is near-identity, in thought just as in vision. Slight shifts in view may best reveal an object's form or even show us whether it is there at all."[8]

As we have already seen in Chapter 6, there are 4,358 different ways of arranging five onsets among 16 pulses, thus yielding as many rhythms. However, only six note-worthy members of this class have been used more extensively than others as timelines in traditional music in different cultures around the world, and it is fair to say that only the *clave son* has overshadowed the others to take center stage by capturing the human imagination. Even though 4,358 is a large number, the fraction of uninteresting patterns in this family of rhythms is also large. For instance, the timelines that have all their five onsets occupying adjacent pulses such as [x x x x x . . . .], or [. . . x x x x x . . . .], or [x x x x x . . . .], etc., are not nearly as interesting. For a rhythm to be a good timeline, the five onsets should be distributed *almost* as evenly as possible within the 16-pulse time span. A perfectly even rhythm with five onsets among 16 pulses is not possible since five does not divide evenly into 16 on a digital clock. Of course, it does so on a continuous clock, but then the rhythm becomes regular and loses all its uniqueness. The only numbers on either side of five to divide evenly into 16 are four and eight. Four is the closest to five, and yields the rhythm [x . . . x . . . x . . . x . . .]. Superimposing this rhythm on the *clave son* yields [x̲ . . x . x̲ . . x . x̲ . . x . x̲ . . .], where the underscore sign indicates the positions of four perfectly evenly distributed beats. A distance measure, such as the swap distance, between a given rhythm and the perfectly even rhythm may serve as a measure of evenness of the given rhythm. The reader may verify that the swap distance between the *clave son* and the four-beat regular rhythm is 5. Similarly, the swap distances between the four-beat rhythm and the five other distinguished timelines are given as follows: shiko = 4, rumba = 4, soukous = 6, bossa-nova = 6, and gahu = 7. Thus, according to this measure of maximal evenness, the son has a relatively low score of 5, and is thus a fairly regular rhythm. Furthermore, the score is not an extreme value in either direction, indicating that the *clave son* strikes a balance between too much and too little regularity, but leaning closer to regularity than irregularity.

Chapter 20 explored the concept of centroid-balanced rhythms. The amount of centroid balance possessed by a rhythm is measured by the distance between the center of gravity of the onsets of a rhythm (in its circular representation) and the center of the circle. If these two centers coincide for a rhythm, then the rhythm is perfectly centroid balanced. A relatively smaller distance implies a relatively greater balance. The distances from the centers of gravity of the six distinguished timelines and the center of the circle are given as follows: son = 1.6, shiko = 8.0, bossa-nova = 8.0, soukous = 8.0, rumba = 9.0, and gahu = 14.7 (where the radius of the circle is 100). The *clave son* has the distinction of being much more balanced than the other five timelines, and is almost perfectly balanced.

RHYTHMIC ODDITY

As pointed out in Chapter 15, a rhythm has the rhythmic oddity property if no two of its onsets are located diametrically opposite to each other on the circle. As already pointed out, this property by itself is not sufficient to establish the saliency of a timeline, as is clearly evident from a rhythm such as [x x x x x . . . .], which is not a good timeline but nevertheless does exhibit the property. Figure 41.1 illustrates the presence or absence of the rhythmic oddity property in the six distinguished timelines. It can be seen that gahu, soukous, and shiko do not have the property. For rhythms with attacks that are fairly evenly spaced, such as these six timelines, adding this property appears to contribute to their saliency. However, by itself the property is not sufficient to highlight the *clave son* uniquely from among the group of six time-lines, since the rumba and the bossa-nova also possess this property.

As we have seen earlier, one way to generalize the binary-valued rhythmic oddity property to a multi-valued one, so as to discriminate between rhythms that have the property, is to measure how far a rhythm is from losing the property. Recall that the set of *antipodal* pulses of a rhythm are the pulses diametrically opposite to the rhythm's onset. Figure 41.1 shows the six distinguished timelines with lines connecting their onsets to their antipodal pulses. One way to measure the amount of rhythmic oddity in a rhythm is with a version of the swap distance in which, for each onset of the rhythm, the minimum distance to its nearest antipodal pulse is calculated. The sum of these distances over all the onsets of a rhythm is a measure of how little oddity the rhythm possesses; a higher value indicates greater rhythmic oddity because the rhythm is then further away from losing the property.

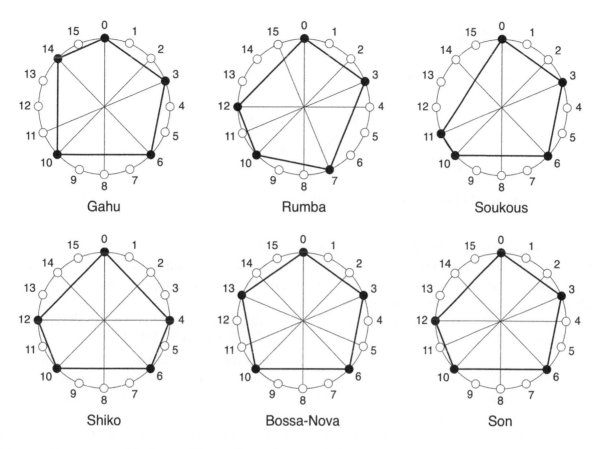

FIGURE 41.1   The amount of rhythmic oddity in the six distinguished timelines.

As an illustrative example, consider the calculation of the amount of rhythmic oddity contained in the gahu rhythm, and refer to Figure 41.1. The distance from the first onset to its nearest antipodal pulse is two, realized by either pulse 2 or 14. The distance of the second onset to its nearest antipodal pulse (pulse two) is one. The distance of the third onset to its nearest antipodal pulse is zero, since it is itself an antipodal pulse of the fifth onset. The fourth onset is at distance one from pulse 11. Finally, the fifth onset has distance zero since it lies at the antipodal pulse of the third onset. Therefore, the overall oddity score of gahu is four. The oddity values of the remaining five timelines increase from left to right with the following values: five for rumba, five for soukous, six for shiko, six for bossa-nova, and seven for son. Thus, according to this measure of rhythmic oddity, the son may be discriminated from the rumba and bossa-nova. Indeed, the *clave son* has the highest value of rhythmic oddity among this family of rhythms. Surprisingly, this is one of the few properties for which the *clave son* takes on a complete extreme value when compared with the other five distinguished timelines.

OFFBEATNESS

Offbeatness was the subject matter of Chapter 16. In the 16-pulse cycle in which the six distinguished timelines live there are four main beats occurring at pulses 0, 4, 8, and 12 that divide the cycle into four equal parts. Relative to these main beats, the remaining onsets may be considered to be off the beat. There is a difference, however, between onsets that lie at pulse two, for example, rather than pulses one and three, since the former lies exactly at the midpoint between two adjacent beats, whereas the latter lies halfway between a beat and the midpoint between two beats. Indeed, Reinhard Flatischler calls the former kind of onsets *offbeats*, and the latter kind *double-time offbeats*.[9] Thus, being off the beat has something to do with the mathematical concept of divisibility of durations into the cycle time span. It was pointed out in Chapter 16 that in a cycle of 16 pulses there is a group of special positive integers that have the property that they are not divisible by 8, 4, or 2. These numbers are 1, 3, 5, 7, 9, 11, 13, and 13. Onsets occurring at these locations may be considered to be strongly off the beat. The *offbeatness* property measures the number of such

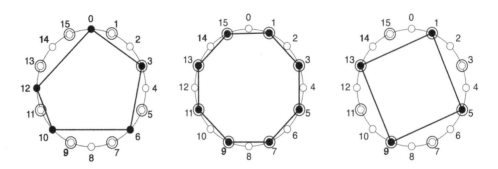

FIGURE 41.2 The *clave son* (left) has one *strongly offbeat* onset. Regular rhythms (center and right) may be composed entirely of offbeats onsets.

pulse positions occupied by onsets. Thus, offbeatness is a mathematical definition of a property related to the concept of syncopation, and a rhythm that has this property is usually considered to be more interesting or lively. The *clave son* in Figure 41.2 (left) has an offbeatness value of one since it has only one onset at one of these positions (pulse 3).

It should be pointed out, however, that the offbeatness measure only makes sense when the rhythm is viewed in the context of an underlying regulative beat structure. The rhythms in Figure 41.2 (center and right) are composed entirely of offbeat onsets, and would thus receive a high value of offbeatness according to this measure. Yet when viewed in isolation (without a regulative structure), they are perfectly regular rhythms that should be characterized as having no offbeatness. The six distinguished timelines have the following offbeatness values that vary in the range between 0 and 2: shiko = 0, gahu = 1, son = 1, soukous = 2, bossa-nova = 2, and rumba = 2. According to this measure, the *clave son* falls squarely in the middle of this range of values.

WEIGHTED OFFBEATNESS

The offbeatness measure counts the number of onsets that lie at the odd-numbered pulses, which in this context are the onsets *strongly* off the beat, or in the terminology of Flatischler, *double-time* offbeats. The *weighted offbeatness* measure counts the total number of onsets that are both offbeats and double-time offbeats, but places different weights on each type of offbeat. In the absence of any *extra*mathematical knowledge about the relative importance of these two kinds of offbeats, either psychological, musicological, or music-theoretic, the double-time offbeat onsets may be counted as two offbeat onsets. Then the resulting values fall in the range between 2 and 6: shiko = 2, son = 4, rumba = 5, gahu = 5,

soukous = 6, and bossa-nova = 6. Again, by this more general measure of offbeatness, the *clave son* value falls in the middle of the range of values.

METRICAL COMPLEXITY AND KEITH'S MEASURE OF SYNCOPATION

It was pointed out in Chapter 13 on mathematical measures of syncopation, that the Lerdahl-Jackendoff metrical hierarchy assigns to the 16 pulses of a 16-pulse rhythm, the weights (5, 1, 2, 1, 3, 1, 2, 1, 4, 1, 2, 1, 3, 1, 2, 1), respectively. If for a given rhythm the weights corresponding to the onset locations are summed, we obtain a measure of the metrical expectedness (or simplicity) of the rhythm. For the *clave son*, the metrical simplicity is 13. The maximum value that five onsets in a cycle of 16 pulses may take is 17. Subtracting 13 from 17 yields the value 4 as a measure of the metrical complexity of the *clave son*. The six distinguished timelines in increasing value of metrical complexity are as follows: shiko = 2, son = 4, rumba = 5, gahu = 5, soukous = 6, and bossa-nova = 6. Once more, the range of values is between 2 and 6, and son has a value that falls in the middle of this range.

Chapter 13 also introduced Keith's mathematical measure of syncopation, which is, however, restricted to rhythms that have pulse numbers equal to a power of two, such as $2^4 = 16$. The six 16-pulse distinguished timelines in increasing value of Keith syncopation are as follows: shiko = 1, son = 2, rumba = 2, gahu = 3, bossa-nova = 3, and soukous = 4. For this measure, the values range between 1 and 4, and son has a value that avoids these two extremes.

MAIN-BEAT ONSETS AND CLOSURE

The six distinguished timelines have four main beats at pulses 0, 4, 8, and 12, as indicated in Figure 41.3 by means of double circles. The number of onsets of a 16-pulse

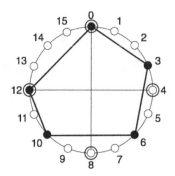

FIGURE 41.3 The *clave son* has two onsets at main beats on pulses 0 and 12.

rhythm that coincide with these four beats is a measure of the rhythm's synchronicity with the underlying beat. One might call it a *beatness* measure. Furthermore, a rhythm that has an onset at the last of these main four beats (pulse 12) contains *closure*, a property that appears to be attractive to many listeners. Of the six distinguished timelines, the shiko, son, and rumba have closure, whereas the bossa-nova, soukous, and gahu do not. The number of main beats contained in each of the six distinguished timelines is as follows: shiko = 3, son = 2, rumba = 2, soukous = 1, bossa-nova = 1, and gahu = 1. The range of values is from 1 to 3, and the son moreover falls squarely in the middle.

CARDINALITY OF DISTINCT DURATIONS

Several measures of rhythm complexity have been explored in earlier chapters. One such measure was the entropy of the *full* IOI histograms, where the latter are viewed as discrete probability distributions. This entropy is an *implicit* measure of the number of distinct durations present in the rhythm. A higher value of entropy results from a flatter histogram, which in turn implies the presence of a wider range of distinct durations. The number of distinct durations may also be measured *explicitly* by just counting them. From the histograms in Figure 8.3, the following data may be readily observed: shiko = 4, bossa-nova = 4, son = 5, rumba = 6, soukous = 7, and gahu = 7. The values range from 4 to 7, and the *clave son* falls in the lower-middle range. Furthermore, the son and rumba values constitute the medians of the six rhythms.

CARDINALITY OF DISTINCT ADJACENT DURATIONS

It has already been pointed out Chapter 8 that the *full* interonset histograms sometimes fail to distinguish

between different rhythms because the latter may have identical histograms. In those cases, it may help to also examine the *adjacent* interonset histograms of Figure 8.4. Yet another measure of rhythm complexity counts the number of distinct *adjacent* IOIs present in the rhythm. The histograms in Figure 8.4 yield the following values: shiko = 2, bossa-nova = 2, son = 3, rumba = 3, gahu = 3, and soukous = 4. The values range from 2 to 4, with son falling midway between these two extremes.

ONSET COMPLEXITY AND SUM OF DISTINCT DISTANCES

In addition to counting the number of distinct distances between all its pairs of onsets that a rhythm may possess, one may focus on a *single* onset and measure its contribution to the total number of distinct distances. One might say that the number of distinct distances that are incident on a given onset is a measure of that *onset's complexity* relative to the other onsets. Then the sum of the onset distances summed over all the onsets in a rhythm provides an onset-based measure of the rhythm's complexity. Figure 41.4 shows next to each onset, the number of distinct distances to the other onsets that it determines. The number in the shaded circle at the center of each polygon indicates the sum of all these onset values. According to this measure of complexity, the shiko, son, and bossa-nova are all considered equal, and the rumba has the highest complexity. Thus, this measure does not discriminate between the son and the shiko or bossa-nova.

DEEP RHYTHMS, DEEPNESS, AND SHALLOWNESS

Recall from Chapter 27 that a rhythm is deep if its full IOI histogram has the property that no two columns have the same height (not counting the columns of height zero). According to this binary-valued measure, among the six distinguished timelines, only the shiko and bossa-nova are deep. However, this binary measure may be converted into a *multivalued* measure of deepness by calculating the distance between the histogram of a given rhythm and that of a deep rhythm. But then, since we are trying to measure the deepness property itself, without regard to which distances actually realize the necessary unique heights of the histogram, the histogram columns should first be sorted by increasing height, before calculating the distances between them.

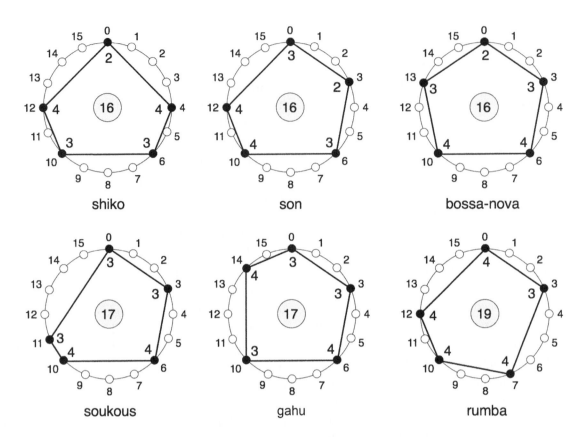

FIGURE 41.4   Onset complexity and distinct distances.

Sorting the histograms in Figure 8.3 yields the histograms in Figure 41.5.

Note that the shiko and bossa-nova have perfectly deep histograms. Now the distance between two histograms may be measured by the sum of the absolute values of the differences in heights of the corresponding columns.[10] The reader may readily verify that the deepness distances of the six rhythms are as follows: shiko = 0, bossa-nova = 0, son = 4, rumba = 4, soukous = 6, and gahu = 6. A larger value of distance implies a smaller value of deepness. Therefore, this distance may also be considered as a measure of *shallowness*. Again, in terms of deepness, two rhythms are deeper than the son (bossa-nova and shiko), and two are shallower (soukous and gahu). Thus, in terms of deepness (or shallowness), the son follows the middle path.

Shiko	0	0	0	0	1	2	3	4
Son	0	0	0	1	2	2	2	3
Rumba	0	0	1	1	1	2	2	3
Soukous	0	1	1	1	1	2	2	2
Gahu	0	1	1	1	1	2	2	2
Bossa	0	0	0	0	1	2	3	4

FIGURE 41.5   The histogram values sorted in increasing order.

TALLNESS

The *tallness* property measures the maximum height of the columns in the IOI histogram. A larger tallness value suggests the tendency that there is a larger concentration of IOIs, which is somewhat related to the deepness property. Also, a tall histogram has less chance of realizing many distinct distances, and so it is also related inversely to rhythm complexity. From Figure 8.3 the following values of tallness are evident: soukous = 2, gahu = 2, son = 3, rumba = 3, shiko = 4, and bossanova = 4. Once more, the *clave son* avoids the extreme values that this property takes.

PROTOTYPICALITY AND PHYLOGENETIC TREE CENTRALITY

The chapter on flamenco meters introduced the application of *phylogenetics trees* to the analysis of families of rhythms, to obtain a clustering of the rhythms as well as to infer a possible evolutionary phylogeny. Simultaneously, such an analysis sometimes constructs nodes in the tree that may determine ancestral rhythms. Applying similar techniques to the six distinguished timelines yields additional insight into the special status

Rhythm	Shiko	Son	Soukous	Rumba	Bossa	Gahu
Shiko	0	1	2	2	2	3
Son	1	0	1	1	1	2
Soukous	2	1	0	2	2	3
Rumba	2	1	2	0	2	3
Bossa	2	1	2	2	0	1
Gahu	3	2	3	3	1	0
Σ	10	**6**	10	11	8	12

FIGURE 41.6 The *swap-distance matrix* for the six distinguished timelines.

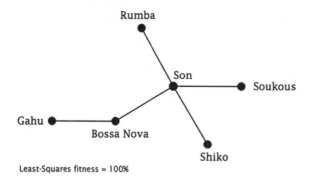

FIGURE 41.7 The phylogenetic tree *SplitsTree* computed using the swap-distance matrix.

enjoyed by the *clave son*. The first step in such an analysis is the computation of a distance matrix. Computing the swap distance between all pairs of the six timelines yields the distance matrix shown in Figure 41.6.

In addition to these distances, the bottom row of Figure 41.6 shows, for each rhythm named at the top, the *sum* of all its distances to the other rhythms. This number is a measure of how dissimilar a rhythm is to the others. Note that the *clave son*, with the lowest value of six, is the unique rhythm most similar to all others. This suggests that it is the central rhythm of this family of rhythms. In other words, the *clave son* is the rhythm that minimizes the total number of swap mutations needed from which to generate all the other rhythms. Therefore, we may consider this number as a measure of the *centrality* of a rhythm with respect to both a group of rhythms and a specific distance measure. A smaller number reflects a greater degree of centrality. Figure 41.7 shows the *SplitsTree* calculated from the distance matrix of Figure 41.6. This tree visually encapsulates at a glance all the information contained in the distance matrix.

MIRROR SYMMETRY

In Chapter 33 on symmetric rhythms, it was shown that rhythms notated on a circle may be classified according to a variety of mirror (reflection) symmetries that depend on the orientation of the line of symmetry: vertical, horizontal, and diagonal being the three main categories. Furthermore, symmetry is often cited as a contributing factor to making music sound good. Indeed, as we have seen earlier, Simha Arom has suggested that symmetry might be a music universal. Only shiko, son, and bossa-nova possess mirror symmetry, and only *clave son* possesses diagonal mirror symmetry. As with many other geometric properties of rhythm, symmetry by itself is no guarantee that a rhythm that possesses it will be a successful timeline. After all, regular polygons are replete with mirror symmetries, but fail to make interesting timelines. However, a rhythm that has only one line of mirror symmetry may stand a better chance. Furthermore, a line of symmetry that has a diagonal orientation perhaps yields rhythms that are more interesting or surprising than those possessing a vertical line of mirror symmetry. Note that the very popular 8-pulse tresillo (3-3-2), 12-pulse fume–fume (2-2-3-2-3), and 12-pulse *bembé* (2-2-1-2-2-2-1) timelines also have a single axis of reflection symmetry, and it is oriented in a diagonal direction. At present this is mere speculation, and psychological experiments would have to be carried out to determine whether such symmetry preferences exist.

PULSE SUBSYMMETRIES

As pointed out in Chapter 33, few rhythms possess the all-or-none feature of global mirror (reflection) symmetry. One weakness of such a coarse measure of symmetry is that it does not provide a finer than binary discrimination between rhythms. For example, both the son and shiko contain one axis of reflection symmetry, and thus reflection symmetry considers both rhythms equally symmetric, although we can see and hear that the shiko is more symmetric than the son. One way to relax global symmetry, by quantifying it, is to count the number of local symmetries (subsymmetries) instead. Recall that a subsymmetry consists a contiguous subsequence of pulses (sounded or silent) that are palindromic. The numbers of such pulse subsymmetries for the six distinguished timelines are as follows: shiko = 29, son = 25, soukous = 25, rumba = 24, bossa-nova = 28, and gahu = 28. According to this measure, the shiko is more symmetric than the son (29 versus 25, respectively), in agreement with human perception. Also, the son has a value that avoids the two extremes.

IOI DURATION SUBSYMMETRIES

When rhythms have large IOI durations, strings of contiguous silent pulses tend to inflate the count of subsymmetries present. This has the effect that the rumba, which is much less symmetric than the son, has almost as many pulse subsymmetries (24 versus 25, respectively). Since there are in general fewer contiguous IOIs in a rhythm, counting the IOI subsymmetries can compensate for the inflation of pulse subsymmetries. Recall from Chapter 33 that an IOI subsymmetry consists of a sequence of contiguous IOIs that are palindromic. The numbers of such IOI subsymmetries for the six distinguished timelines are as follows: shiko = 4, son = 2, soukous = 1, rumba = 1, bossa-nova = 4, and gahu = 2. According to this measure, the shiko is more symmetric than the son (4 versus 2, respectively), also in agreement

with human perception. Furthermore, the son has two IOI subsymmetries compared with one for the rumba, which is also more appropriate. Finally, the son not only has a value that avoids the two extremes, but it is the only rhythm among the distinguished six that is composed of the union of two disjoint IOI subsymmetries. In contrast, the gahu rhythm also contains two disjoint IOI subsymmetries, but they do not span the entire rhythm. This unique feature of the binary *clave son* is shared with the ternary fume–fume that has an IOI structure [2-2-3-2-3] composed of the disjoint union of [2-2] and [3-2-3].

AREA OF PHASE SPACE PLOTS

Figure 41.8 shows the convex hulls (with their areas shaded) of the phase space polygons of the six distinguished timelines. Note that the areas range from the small 0.5 for the bossa-nova to the large 5.0 for the soukous. On the other hand, the area of the son polygon is 1.5 and thus falls in between the extremes.

FRACTAL METRIC HIERARCHY

In Chapter 18 the workings of the hierarchical meter-based model of rhythm perception proposed by Longuet-Higgins and Lee (1982) were illustrated using the six distinguished timelines as examples. The analysis not only

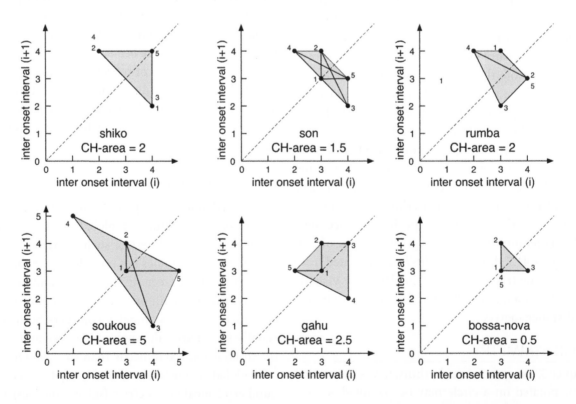

FIGURE 41.8 The areas (shaded) of the convex hulls of the phase space polygons of the six distinguished timelines.

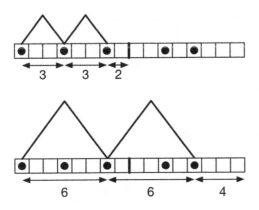

FIGURE 41.9 Illustration of the fractal nature of the metrical hierarchy of the *clave son*.

revealed the fractal nature of the two-level metrical hierarchy, but also showed that, among the six rhythms, only the *clave son* possesses this property. The fractal property of the *clave son* is clarified visually in Figure 41.9. The 16-pulse *clave son* rhythm consists of the concatenation of two equal eight-pulse sections each. The first section of eight pulses has three onsets at pulses 0, 3, and 6 that form the hypothesized meter and the first level of the metric hierarchy (top diagram). It is in fact the tresillo rhythm with duration pattern [3-3-2]. The second level of the

metric hierarchy (bottom diagram in Figure 41.9) updates the beat length to have a duration of six pulses with onsets at pulses 0 and 6, and predicts an onset at pulse 12. The *clave son* rhythm has onsets precisely at these pulse locations, and thus the meter at this level has duration pattern [6-6-4], which is precisely a magnification (the double) of [3-3-2], establishing the fractal property.

SHADOW CONTOUR ISOMORPHISM

Of the six distinguished timelines, the *clave son* is the only one that has a cyclic rhythmic contour that is the same as the contour of its shadow rhythm. Figure 41.10 shows the six timelines along with their shadow rhythms and the interonset durations of the shadow rhythms. The rhythmic contours of both the rhythms and their shadows are shown in Figure 41.11 for comparison. The rhythmic contour of the *clave son* is [0 + − + −] and that of its shadow is [+ −0 + −] which is a rotation of the former. Interestingly, the shadow of the shiko rhythm is a rotation of the bossa-nova rhythm.

Although this property succeeds in uniquely selecting the *clave son* from among the six distinguished timelines, like the other properties, it falls short of characterizing good timelines in a more general setting,

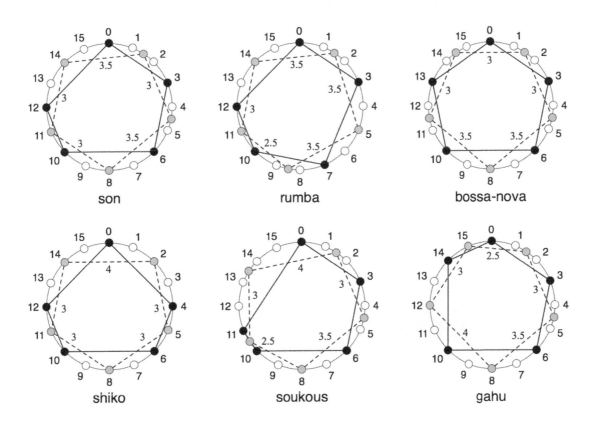

FIGURE 41.10 The six distinguished timelines and their *shadow* rhythms.

Timeline	Rhythm IOIs	Rhythm Contour	Shadow IOIs	Shadow Contour
Son	3, 3, 4, 2, 4	0 + - + -	3, 3.5, 3, 3, 3.5	+ - 0 + -
Rumba	3, 4, 3, 2, 4	+ - - + -	3.5, 3.5, 2.5, 3, 3.5	0 - + + 0
Bossa -Nova	3, 3, 4, 3, 3	0 + - 0 0	3, 3.5, 3.5, 3, 3	+ 0 - 0 0
Shiko	4, 2, 4, 2, 4	- + - + 0	3, 3, 3, 3, 4	0 0 0 + -
Soukous	3, 3, 4, 1, 5	0 + - + -	3, 3.5, 2.5, 3, 4	+ - + + -
Gahu	3, 3, 4, 4, 2	0 + 0 - +	3, 3.5, 4, 3, 2.5	+ + - - +

FIGURE 41.11   The rhythmic contours, their shadows, and the contours of their shadows.

as the example in Figure 41.12 illustrates. Here the time-line with interval vector [1-2-2-1-10] is too skewed to be a successful timeline, and yet it possesses a rhythmic contour that is a rotation of the contour of its shadow (dashed line). Whether this property has any psychological or neurological weight in restricted contexts is yet to be determined experimentally. However, as a mathematical property, it is clearly useful for characterizing rhythms, in general, and the *clave son*, in particular.

For convenience, all the properties discussed in this chapter are listed together for the six distinguished timelines, in the table of Figure 41.13. A property that a rhythm

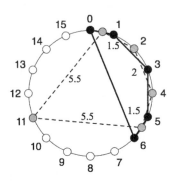

FIGURE 41.12   A timeline (solid lines) with a rotated contour of its shadow (dashed lines).

Property	Shiko	Son	Soukous	Rumba	Bossa	Gahu
Maximal Evenness	-	-	-	-	Yes	-
Centroid Balance	8.0	1.6	8.0	9.0	8.0	14.7
Swap-distance Evenness	4	5	6	4	6	7
Rhythmic Oddity	-	Yes	-	Yes	Yes	-
Rhythmic Oddity Amount	6	7	5	5	6	4
Off-Beatness	0	1	2	2	2	1
Weighted Off-Beatness	2	4	6	5	6	5
Metrical Complexity	2	4	6	5	6	5
Keith Syncopation	1	2	4	2	3	3
Pressing Complexity	6	14.5	15	17	19.5	22
Entropy-Adjacent	0.97	1.52	1.92	1.52	1.52	0.72
Entropy-Full	1.84	2.24	2.72	2.44	2.72	1.84
nPVI	66.7	40.5	70.5	41.0	23.8	14.3
Closure	Yes	Yes	-	Yes	-	-
Main-Beat Onsets	3	2	1	2	1	1
Distinct Durations	4	5	7	6	4	7
Distinct Adjacent Durations	2	3	4	3	2	3
Onset Distinct Distances	16	16	17	19	16	17
Deep	Yes	-	-	-	Yes	-
Shallowness	0	4	6	4	0	6
Tallness	4	3	2	3	4	2
Swap-Distance Centrality	10	6	10	11	8	12
Mirror Symmetry	Yes	Yes	-	-	Yes	-
Diagonal Mirror Symmetry	-	Yes	-	-	-	-
Pulse Sub Symmetries	29	25	25	24	28	28
IOI Sub Symmetries	4	2	1	1	4	2
Area of Phase Space Polygons	2.0	1.5	5.0	2.0	0.5	2.5
Fractal Metric Hierarchy	-	Yes	-	-	-	-
Shadow Contour Isomorphism	-	Yes	-	-	-	-

FIGURE 41.13   List of 29 properties possessed by the six distinguished timelines.

does not possess is indicated by the dash (—) for easy visualization. Of the 14 properties that have numerical values, 11 of them yield values that for the *clave son* fall, if not near the middle of the range spanned by the six rhythms, avoid extreme values. This list provides strong evidence that at least for the *clave son*, and perhaps for other rhythms as well, in order for a rhythm to be "good," these 11 properties should in general not take on extreme values, but rather should follow the doctrine of the "golden mean." On the other hand, there are three properties that take on extreme values for the *clave son*: the centroid balance, the amount of rhythmic oddity, and the swap distance centrality. In addition, there are three properties unique to the *clave son* among the six distinguished timelines: diagonal mirror symmetry, fractal metric hierarchy, and shadow contour isomorphism. It has been established that the *clave son* is the most centroid-balanced rhythm not just among these six distinguished timelines, but among all rhythms with five onsets in a 16-pulse cycle.[13]

Although the shadow contour isomorphism property and the extreme centroid balance property are obtained by only the *clave son*, among the six distinguished timelines, like the other properties for which the *clave son* is preferred over the other five timelines on the basis of *rotation-invariant* properties, the question arises as to why the particular rotation of the *clave son* necklace is preferred over any of its other four rotations that start on an onset. To answer this question, refer to Figure 41.14 that shows the *clave son* (labeled as son-1) and its four rotations that start on the other four onsets.

Two desirable musicological properties of rhythm timelines that have not yet been discussed are the *call-response* property and *metric ambiguity* (see Figure 41.15). Call-response refers to the property that the rhythm evokes the feeling that it consists of two discernible parts (groups of onsets), the first of which plays the role of "asking a question," followed by the second, which "provides an answer." Asking a question opens a perceptual space and answering it closes that space. This call-response property may also be expressed in terms of the neo-Platonic model of cosmology, which ascribes a "specific affiliation of the numbers 2 and 3 with centripetal and centrifugal forces, respectively."[11] The *clave son* starts with two triple units and ends with a duple unit. The triple units thus create a centrifugal (outward or opening) force, whereas the duple unit provides a centripetal (inward closing) force. This property of musical rhythm is inherent in much of sub-Saharan African diasporic music and beyond. It is related

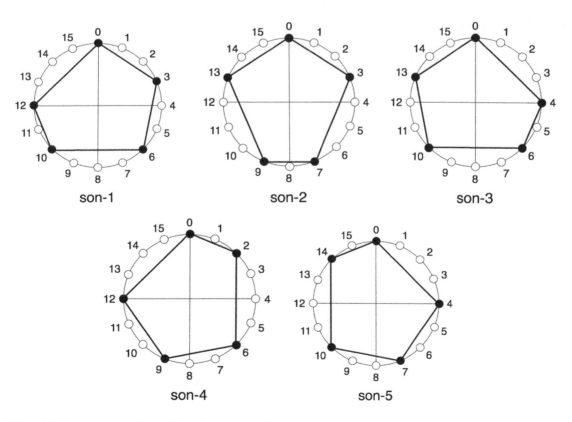

FIGURE 41.14   All five onset rotations of the *clave son* timeline.

Property	Son-1	Son-2	Son-3	Son-4	Son-5
Closure	Strong	-	-	Weak	-
Platonic Centrifugal Centripetal Forces	Yes	-	-	-	-
Metric Ambiguity	Yes	-	-	-	-

FIGURE 41.15 List of metric properties of the five rotations of the *clave son* timeline.

to the property of *closure*. Call-response provides closure, but closure does not necessarily convey call response. Of the five rotations of the *clave son* shown in Figure 41.14, only son-1 (the true son) and son-4 have closure. Of these, son-1 has *strong* closure, provided by the fourth onset, which lies at pulse ten, at the midpoint of the third quarter cycle. By contrast, the closure property of son-4 may be called *weak* because the fourth onset is situated at pulse nine, which is close to the midpoint of the cycle and far from the last onset.

As we have seen in Chapter 13, an effective method by which to add spice and surprise to a rhythm is through syncopation. Syncopation introduces a touch of cognitive insecurity, or metrical ambiguity, or what Neil McLachlan calls *gestalt despatialization*.[12] In the case of the *clave son* timeline, which has four strong beats at pulses 0, 4, 8, and 12, metrical ambiguity can be introduced by first misguiding the listener's brain into predicting a sequence of equal-duration intervals of three pulses each. For this to happen, there must be

at least two such intervals at the start of the sequence. This means that the first three onsets must occur at pulses 0, 3, and 6. However, this is not sufficient to cause the listener to experience metrical ambiguity. To do so, the last interval must have duration equal to four units, and hence be determined by onsets at pulses 12 and 0. In this way, the rhythm ends with a clear and contradictory four-pulse interval. Only son-1 has this property.

NOTES

1 Impagliazzo, J., (1989), p. 16.
2 Pressing, J., (2002), p. 290.
3 Katz, B. F., (2004), p. 30.
4 Levitin, D. J., Chordia, P., & Menon, V., (2012), p. 3716.
5 Sachs, C., (1953). See also De Leew, T., (2005), p. 41 and Bonus, A. E., (2010), p. 371.
6 Voloshinov, A. V., (1996), p. 111.
7 Wade, D., (2006), p. 38 and Anderson, P. W., (1972).
8 Minsky, M., (1981), p. 39.
9 Flatischler, R., (1992).
10 Distance measures between histograms have a variety of applications in music information retrieval, such as feature selection, indexing, pattern classification, and cluster analysis. Beltran, J. F., Liu, X., Mohanchandra, N., & Toussaint, G. T., (2015) and Cha, S.-H. & Srihari, S. N., (2002), propose a variety of alternate techniques to measure the distance between histograms.
11 Cohn, R., (2016a), paragraph [8.7].
12 McLachlan, N., (2000).
13 Refer to the discussion of balanced rhythms in Chapter 20.

CHAPTER 42

On the Origin, Evolution, and Migration of the *Clave Son*

L ET US RETURN TO the Spanish sailor that we encountered in Chapter 12. Recall that in the sixteenth century he travels from Sevilla to Havana on a Spanish fleet of galleons on its way to pick up gold from Mexico and silver from Bolivia. During one of his wanderings in Havana, he comes upon a group of black former slaves drumming and dancing passionately in the street, and is captivated by one of the crisp invigorating rhythms that one of the musicians is playing with a pair of wooden sticks. The sailor's brain interprets a signal relayed by his ear, which is stimulated by a sound wave traveling from the sticks to the ear, causing his eardrum to vibrate. This sound wave is the *acoustic signal* that embodies the rhythm that the musician is playing. The acoustic signal is a complex waveform that contains much information about the sound as well as the environment in which the sound is produced, in addition to the points in time at which the sticks are struck together. However, a greatly simplified and idealized representation of this acoustic signal, after being recorded by electronic equipment, and graphically displayed on a computer screen, might look something like the waveform in Figure 42.1 (left). Of course, the sailor is not privy to this graphical information.

It is well known that human perception does not result from a mere *bottom–up* processing of the objective physical stimulus presented to the perceiving mechanism. It is rather a partly subjective and constructive interactive process that also involves *top–down* processing, in which the perceiver projects a medley of competing hypotheses about what is perceived. These hypotheses emerge from a variety of biological as well as cultural expectations possessed by the listener.[1] In the words of David Huron, "Rhythms are perceived categorically."[2] The brain has, as it were, a collection of rhythmic templates, and when it encounters an unknown rhythm it perceives it as one of its close rhythmic templates.[3] Therefore the rhythm that our sailor perceives in the situation described may correspond to an idealized waveform different from that shown in Figure 42.1 (left), and closer to a rhythm that is much more familiar to the sailor.

Back onboard ship on his return to Sevilla, the sailor, while trying to entertain his shipmates one evening, tries to remember and reconstruct the rhythm with a pair of wooden spoons. However, had there been a recording of it, the rhythm performed that night might have looked more like the rhythm on the right in Figure 42.1. Note that in this rhythm the second onset is closer to the

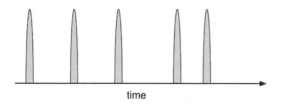
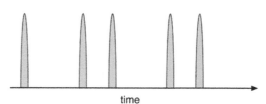

FIGURE 42.1   Two idealized acoustic signals or waveforms in real time.

third onset than the first. In the rhythm on the left, the corresponding three onsets are equally spaced in time.

The scenario just described illustrates a *mutation* step in the type of evolution known as *Lamarckian* evolution.[4] The original rhythm heard by the sailor has been changed slightly for whatever reason (maybe misperception in the first place, due to aural illusions or to cultural expectations, perhaps bad memory during reconstruction, possibly poor performance capability, or perchance purposeful creativity). Another sailor on board may later try to play this mutated rhythm for his friends back home upon arrival in Huelva, only to make a further slight change to it. Thus, in Lamarckian evolution the progressive changes or errors introduced into the original rhythm may accumulate rapidly, resulting in the speedy extinction of the original. In the process of transmitting the rhythm in this way from one person to another a dozen times, the final rhythm obtained may bear little resemblance to its progenitor. In the context of biology, Lamarckian evolution, which is named after the French biologist Jean-Baptiste Lamarck, espouses the idea that an organism may transfer to its offspring characteristics that are acquired by the organism during its lifetime.

Consider on the other hand a culture in which the village master drummer teaches a young pupil to play a rhythm on a drum using the following instructions. First strike the drum with both hands 16 times starting with the right hand, to make a steady pulse alternating between right and left hands, much like walking. When this steady pulse is well established, play 5 of these pulses loudly and the remaining 11 softly. In particular, play the first stroke loud, the next two strokes soft, another loud, two more soft, another loud followed by three soft, a loud, a soft, a loud, and finally three soft. Finally, when the pattern of loud beats is clear, skip all the soft beats altogether by stopping the hands that play soft beats just before they touch the drum. The resulting rhythm will

have the second onset played with the left hand, and the other four with the right hand. Were we to represent this scenario graphically, the acoustic signal would look like the waveform in Figure 42.2 (left), with the added information of pulse numbers implicitly encoded in either the oral instructions set forth by the master teacher, or in the alternating right–left hand motions that mark out the pulses. The onsets would be played on pulses 1, 4, 7, 11, and 13. These instructions would make it virtually impossible for a student, once having attained a steady motion, to produce the rhythm on the right in which all the onsets are played with the right hand on pulses 1, 5, 7, 11, and 13.

This second scenario described earlier differs considerably from the first. In the first, a new instance of a rhythm is brought about by means of an attempt to *copy the rhythm*. The information transmitted from one generation to another consists of a *mutated* facsimile of the rhythm itself. This replica of the rhythm is a complicated continuous acoustic signal susceptible to the smallest of perturbations, and thus easily morphed into other rhythms. By contrast, in the second scenario, a new instance of a rhythm is brought about by means of *executing a set of instructions* for producing it. Here the information transmitted across generations is not the rhythm itself, but rather a list of *instructions* for *creating* the rhythm. As long as the instructions are passed along intact, the rhythm produced with them will likely remain the same, barring other drastic outcomes such as a lack of performance skill. In contrast to the *concrete continuous* acoustic signal, these instructions are *abstract discrete* logical entities. As such they are more robust and stable than the malleable continuous objects that make up acoustic signals. Hence, rather than a small quantitative error, a large qualitative error would have to be made in order to mutate the rhythm. Alternately, the rules themselves would have to be changed in order to mutate the rhythm. This type of evolution is called

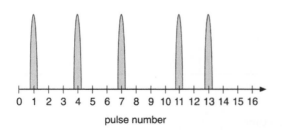

FIGURE 42.2 Two rhythms notated in pulse-measured time.

Weismannian evolution[5] after the German evolutionary biologist Friedrich Leopold August Weismann, who was strongly opposed to Lamarckism. It is analogous to the modern view in biology that holds that an organism is produced by executing the discrete "instructions" encoded in DNA molecules.

Both types of evolution described earlier may be at work in a culture based on oral tradition, such as sub-Saharan Africa.[6] In the case of Lamarckian evolution, the acoustic signal itself is passed on aurally. In Weismannian evolution, the instructions may be passed on orally and memorized for subsequent oral transmission. A third oral approach that a master may use to teach rhythms to a pupil is by means of language mnemonics. As poets and linguists know very well, words, phrases, and sentences possess rhythm. In recent years much research has been done to uncover the fascinating relationships between music and language.[7] It is therefore not surprising that many cultures have passed on rhythms from one generation to another by means of mnemonics. In the majority of languages, the vowels sounding closest to [i] and [u] (spoken with a relatively closed mouth) take less time to articulate than the other vowels.[8] Furthermore, certain syllables in a language sound stronger than others, thus automatically introducing accents and strong beats. Such is the case for example with *ta* and *na*, *ta* being stronger than *na*. Also, some syllables are naturally longer than others when sounded, such as *tan* and *ta*, the former being longer than the latter. By stringing a group of such syllables together, rhythmic patterns are automatically created.[9] Thus, mnemonic systems, sometimes referred to as "nonsense syllables," are in fact fairly universally accessible, and make a great deal of sense.

One of the earliest mnemonic syllable systems for teaching and transmitting musical rhythms was developed in the eighth century by the well-known prosodist al-Khalīl, who wrote one of the first books on the theory of rhythm, *Kitāb al-iqā*. According to this system the shortest beat, corresponding to one pulse, was *ta*, represented here in box notation in Figure 42.3 (row 1). The syllable *tan* was used for a sounded pulse that is followed by an unsounded pulse, as in row two. The combinations *tana*, *tanan*, and *tananan* correspond to the rhythms in rows three, four, and five, respectively, where a pulse containing a black circle represents the strong beat, the pulse with a gray circle denotes a secondary (optional, or ornamental) beat, and the empty box is a silent pulse.

In the early thirteenth century, in the year 1258, at a time when Bagdad was one of the most brilliant intellectual centers of the world, a leading musician and music theorist (also a calligrapher and physicist) living there, by the name of Safi al-Din al-Urmawi (AD 1216–AD 1294), managed to survive the destruction of Bagdad at the hands of the Mongol invaders. Safi al-Din produced a seminal book, by some accounts estimated to be written in the year 1252, which included a theory of rhythm, titled *Kitāb al-adwār* (The Book of Musical Modes).[10] One of the most important rhythms in this book identified as *al-thaquīl al-awwal* is described using the mnemonic syllable system as *tanan tanan tananan tan tananan*. Figure 42.3 shows this rhythm in box notation. If we disregard the secondary onsets (labeled as optional) and retain only the fundamental *ta* beats, we obtain the *clave son* shown directly underneath the *al-thaquīl al-awwal* rhythm. A mnemonic system such as this provides better copying fidelity than merely attempting to replicate an acoustic signal. However, in

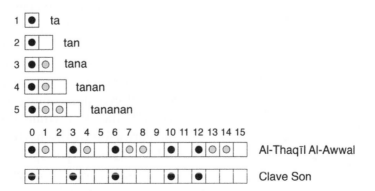

FIGURE 42.3   A mnemonic system for the rhythm *al-thaqīl al-awwal*.

practice there may still be some inexactness present in the pronunciation of the syllables. Hence this system of cultural inheritance could be said to lie somewhere in between Lamarckism and strict Weismannian evolution.

In addition to the mnemonic syllable system, Safi al-Din's book includes an exact written notation system for describing the rhythms within lines of text, as well as a geometric version that uses circles to represent the rhythmic cycles. In both notations, the pulses are denoted by dots, and circles denote the partition of the pulses into groups that start with a main beat. Thus the *al-thaqīl al-awwal* rhythm is written by Safi al-Din in running text as O . . . O . . . O . . . . O . . O . . . . using this notation. He also includes a circular representation illustrated in Figure 42.4. The circle is divided into 16 pulses indicated by black dots. A smaller circle inside the large circle contains the name of the rhythm, and the lines connecting the inner circle to the small white-filled circles on the large outer circle indicate the partition of the cycle into groups. An arrow indicates where the rhythmic cycle starts and in which direction time flows (counterclockwise). Hence this diagram denotes the rhythm [3-3-4-2-4].

Safi al-Din's book appears to contain the earliest historical records of the *clave son* rhythm. Written notations such as these are examples of sets of instructions that provide an even higher copying fidelity than mnemonic syllable systems, and therefore they greatly facilitate Weismannian evolution. Therefore, we can be pretty certain that the *clave son* in its binary form [3-3-4-2-4] existed in Baghdad as early as the thirteenth century. However, it was not until the twentieth century that this rhythm took center stage on the world scene. During the 1960s in the West African country

of Ghana, there appeared a dance and music called *kpanlogo* that uses this rhythmic pattern played on an iron bell. Furthermore, in Central Africa there is a music named *gome*, popular during the 1920s, that also uses this bell pattern as a timeline. Some musicologists believe that the migration of gome from Central to West Africa is responsible for the adoption of this bell pattern in the kpanlogo dance. Others believe that this rhythm made its way more than once back and forth between South America and West Africa via the slaves, and that it returned to Africa in the 1880s in a mutated form with West Indian regimental bands stationed at the Portuguese slave trading post of Elmina on the coast of Ghana. It is known that 1,000 years earlier, while the Arabs lived in North Africa, they traveled extensively southward to Ghana and beyond.[11] Whether this rhythmic pattern originated in Bagdad and made its way to Ghana, or vice versa, or whether it emerged independently in both regions, or whether it originated among the Pygmies in central Africa is an open question.

There is debate among ethnomusicologists concerning the types of mutations that may transform one rhythm to another as it migrates from one culture to another. On the one hand, there are those such as Rolando Pérez Fernández who believe that African ternary rhythms were transformed into binary rhythms in America when they were exposed to Spanish music.[12] In an oral tradition that is governed by Lamarckian transmission of acoustic signals, it is plausible that a ternary rhythm such as the fume–fume of Figure 42.5 could be transformed to its binary version of Figure 42.2 (left) by virtue of cultural expectation. After all, from the purely acoustic point of view, the binary and ternary versions are almost the same, especially when played at fast tempos. Comparing their ternary and binary polygon representations of Figure 10.2, it is clear that both rhythms have the same perceptual grouping: one group of three onsets followed by a space, followed by one group of two onsets followed by another space. Experiments by Stephen Handel demonstrated that "Two rhythms that had the same perceptual grouping were judged as being identical, even if the timing between the groups was different."[13] The two rhythms also have identical *rhythmic contours*, a feature that has been shown by psychologists to be more easily perceptible than precise quantitative relations. It is easier for subjects to judge whether an interonset-interval duration is equal to, greater than, or less than the preceding or the following duration, than to

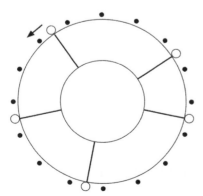

FIGURE 42.4 Safi al-Din's circular representation of the rhythm *al-thaqīl al-awwal*.

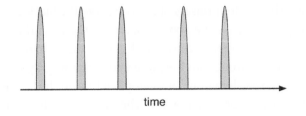 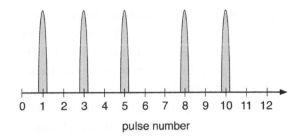

FIGURE 42.5   The fume–fume rhythm notated in real time and pulse-measured time.

judge quantitative relations such as twice as long or half as long.[14] Furthermore, there is evidence that rhythms are perceived *categorically*, meaning that rhythms that are slightly different from each other may be perceived as being identical.[15]

A contrary view is held by the anthropologist Gerhard Kubik who holds that social factors are not sufficient to explain the presence or absence of asymmetric timeline patterns in the music of the New World. Kubik believes that rhythmic timeline patterns possess discrete mathematical structures that are *cultural invariables*, and as a consequence, they either do not migrate at all or they migrate intact. In Kubik's words: "Accentuation, mnemonic syllables, and instrumentation of the timeline patterns are all highly variable, but no one can change their intrinsic mathematical structure. Any such attempt simply destroys the pattern."[16] If the instructions for teaching rhythmic timelines in a culture make use of mnemonics, then a rhythm such as the fume–fume, implicitly encoded as in Figure 42.5 (right), would enjoy stronger copying fidelity, and be governed by Weismannian evolution. It would then likely migrate intact to another culture if the people carried the code with them and continued to use it in their teaching. However, the fact that the *clave son* existed in Bagdad hundreds of years before the slave migrations from West Africa to the New World, and that in the West African oral musical tradition, rhythms were coded mnemonically, suggests that both the binary and ternary versions of the *clave son* probably existed simultaneously in West Africa before the advent of the slave trade and that they migrated independently to the New World. One must be careful of ascribing evolutionary influences on the basis if similarity alone, since such similarity may arise independently as a result of functional constraints, especially where universals are involved.[17] Of course, the Europeans living in the New World would probably have been unaware of the mnemonics in a foreign

language used by the African slaves, and thus the process of binarization in a Lamarckian manner could have easily occurred among the Europeans. Mathematical structures per se are not immune to change; it depends on the type of mathematics employed. Durational patterns expressed in terms of continuous mathematics are quite malleable (Lamarckian) and can easily change, as has been repeatedly demonstrated with listening experiments in the laboratory. In general, discrete mathematics is more resistant to continuous change, it is more Weismannian, but asymmetric timeline patterns need not be coded mathematically in terms of the number of pulses, as suggested by Kubik. They could be coded in the discrete mathematical relational terms such as equal, greater than, or less than, as reflected in rhythmic contours. Mathematics alone does not necessarily impose cultural invariability.

The *clave son* rhythmic pattern acquired its name from the type of Cuban music knows as the *son*. The cradle of the son music is the eastern Cuban province of Oriente, where it may have taken preliminary forms in the seventeenth century. However, in several accounts, its verified existence comes later in the nineteenth century in the cities of Guantánamo, Baracoa, Manzanillo, and Santiago de Cuba, where it took on its more present-day form in the early twentieth century. On the other hand, Peter Manuel dates the practice of the clave concept earlier to the 1850s.[18] From Oriente it was taken to Havana in 1909 by the soldiers of the Permanent Army, where in the early part of the twentieth century it slowly acquired a faster tempo, and musicians incorporated other instruments such as the trumpet, congas, and piano.[19] In the 1940s and 1950s, Havana was the playground of the United States, and Cuban music such as the *son*, *mambo*, and *cha-cha-cha* made its way to New Orleans and New York. In New York, with the influence of the Puerto Rican community it morphed into *salsa*, which in the last half of the twentieth century conquered the world. The question of when the *clave son* rhythmic pattern was incorporated into son music is more difficult to ascertain. Some musicologists believe that the son borrowed the pattern from the rumba music when it migrated to Havana.

In the rock-and-roll period of the 1950s and 1960s in the United States, there was one rhythm that was used by so many musicians in so many songs that it stuck out from among all others. It was called the *Bo-Diddley Beat*, named after the singer and songwriter Bo Diddley who made it popular in several of his recordings. In 1955 on the nationally televised *Ed Sullivan Show*, Bo Diddley performed one of his own songs called *Bo Diddley*, violating the contract he had signed to perform another song made popular by Tennessee Ernie Ford, titled *Sixteen Tons*. Bo Diddley's song, sometimes referred to as a "freight-train" stomp, had a hard driving guitar and drum rhythm, also known by the mnemonic "*shave and a hair-cut, two bits*," often used as a door-knock pattern, and referred to by H. C. Longuet-Higgins as the "cliché rhythm."[20] It is shown in box notation in Figure 42.5 (line two). Much music inspired by the Bo Diddley Beat has a similar feel but uses variations of this basic rhythmic pattern. A small sample of these is listed in Figure 42.6 starting with the *clave son* itself (on line 1), which does not have the Bo Diddley Beat onsets at pulses 2 and 4.

The third example in Figure 42.6 shows the floor-tom rhythm (actually played on a cardboard box) used in Buddy Holly's "*Not Fade Away*" recorded in 1957. It has dropped the second onset of the Bo Diddley Beat and inserted an onset at pulse nine. This version has the feel of the *shiko* timeline [4-2-4-2-4] with anticipatory onsets at pulses 3 and 9. In addition to this floor-tom rhythm, another variant is heard on backup vocals sung by *The Crickets*, shown on line four. Here the gray circles indicate the hummed consonant sound "*mmm*" and the black circles the plosive syllable sound "*pa.*"

Also in 1957 Buddy Holly released the song "*Bo Diddley*" that makes use of the strong drum timeline shown on line five. Rather surprisingly, in this version of the *clave son* Buddy Holly moved the second onset from pulse 3 to pulse 2, converting it in effect to a rotation of the *shiko* timeline. In another Buddy Holly song, the chorus sings another variant with the mnemonic: "DUM DE DUM DUM," "OH BOY."[21]

In the 1958 release of Duane Eddy's "*Cannon Ball*" the song starts with a strong high-pitched sound marking

out the last four onsets of the *clave son* (the black circles in line six). As the song progresses, these synthesized sounds are replaced by hand claps. The first onset is sounded throughout the piece with a drum (gray circle). Line seven shows one of the top ten hits on the U.S. Pop Chart in 1958: "Willie and The Hand Jive" by Johnny Otis. It may be obtained by removing the third onset at pulse four in Buddy Holly's drum pattern in "*Not Fade Away*." The eighth line shows the guitar rhythm in Elvis Presley's 1961 rendition of "*Marie's the Name*" (*of his latest flame*). The guitar riff has five strong downward strums that coincide with the *clave son* onsets and two (anticipatory) upward strums on pulses 2 and 5, in effect converting the first three onsets of the *clave son* to the *cinquillo* rhythm.

In 1964 Bo Diddley and Chuck Berry released a song called "*Bo's Rhythm*" on the album *Two Great Guitars*. The guitar rhythm at times plays the pattern on line nine, which consists of the *clave son* with two additional filling onsets at pulses 4 and 5. When the rhythm is strummed on a guitar at a fast pace, it is convenient for a performer to add these two onsets so that the hand strums the pattern *up-down-up-down*.

In 1982 the group *Bow Wow Wow* released the song "*I want Candy,*" in which the rhythm shown on line ten is played on the *timbales*. This rhythm consists of the *clave son* with two fill onsets added at pulses 5 and 11. George Michael's "Faith" released in 1987 has a bass line that follows the first six onsets of the pattern in line 11. It embellishes the *clave son* with two onsets, one anticipatory bass note at pulse 9, and a high-pitched snapping sound at pulse 14 that helps to propel the cycle forwards by breaking the two-onset closing response section of the *clave son*. Much more recently in 2009, the Los Angeles group *The Black Eyed Peas* released the album *The E.N.D.* (short for *The Energy Never Dies*). The song *Electric City* in this album uses a strong bass sound in the form of the *clave son* with a snapping high-pitched onset at pulse 14, as in George Michael's "*Faith*" shown on line 12.

It is worth noting that almost all the "variants" inspired by the *clave son*, pictured in Figure 42.6, contain the *clave son* as a subset of their onsets, with some additional onsets inserted in various pulse locations. Only two of the rhythms (lines 4 and 5) move the second onset of the *clave son* from pulse position 3 to location 2. Thus all these rhythms are examples of shellings of the *clave son* by addition (the topic of Chapter 28), but none

FIGURE 42.6   A small sample of the variants of the *clave son* rhythmic pattern.

of them are the result of subtracting onsets from the *clave son*. In contrast, the Mbuti Pygmies of the eastern part of the Democratic Republic of Congo incorporate a timeline in their music that is obtained by a shelling of the *clave son* by subtraction. The *clave son* has mirror symmetry about the axis through pulses 2 and 11, as shown in Figure 42.7 (left). The Mbuti Pygmies play the shelling obtained by deleting the second onset on the third pulse.[22] By deleting an onset on the axis of symmetry, the resulting rhythm also has a mirror symmetry along the same axis of the *clave son*. This is a nice example of symmetry-preserving shelling, a technique used by Steve Reich in his composition of *Drumming*

(see Chapter 28). The *clave son* is part of the repertoire of the BaYaka Pygmies of the Central African Republic. It is estimated that these two cultural groups diverged at least 70,000 years ago.[23] It is possible that we have here a couple of extremely ancient rhythm timelines and one of the earliest examples of shelling. It makes one wonder whether one of these two rhythms generated the other or whether the two appeared independently of each other.

Figure 42.8 shows the histogram of the frequencies with which each onset is used in the variants of the Bo Diddley rhythm listed in Figure 42.6. It is clear from this histogram that the indispensable beats are the five onsets that make up the *clave son*, whereas the

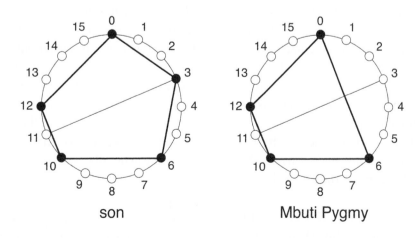

FIGURE 42.7   The *clave son* timeline (left) and a symmetry-preserving shelling of the *clave son* played by the Mbuti Pygmies of the Democratic Republic of Congo (right).

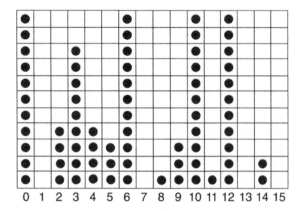

FIGURE 42.8 The histogram of the onsets in the Bo Diddley variants of Figure 42.6.

FIGURE 42.11 The *guiro*. (Courtesy of Yang Liu.)

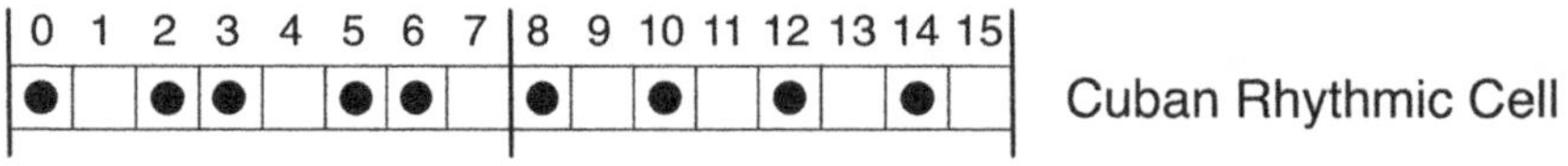

| 0 | 1 | 2 | 3 | 4 | 5 | 6 | 7 | 8 | 9 | 10 | 11 | 12 | 13 | 14 | 15 |

Cuban Rhythmic Cell

FIGURE 42.9 The 16-pulse rhythmic cell that characterizes Cuban music.

| 0 | 1 | 2 | 3 | 4 | 5 | 6 | 7 | 8 | 9 | 10 | 11 | 12 | 13 | 14 | 15 |

Kassa timeline

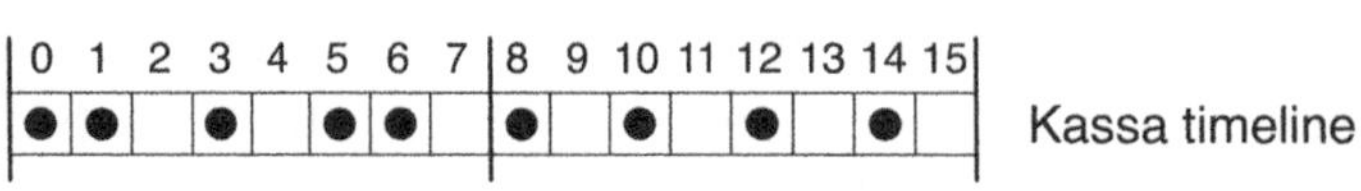

FIGURE 42.10 The *kassa* timeline from Guinea.

embellishment onsets occur mainly between the second and third onsets of the *clave son*, and just before the fourth onset.

In the *Dictionary of Cuban Music* Helio Orovio writes that the 16-pulse pattern shown in Figure 42.9 forms the rhythmic cell that characterizes Cuban music. It consists of the syncopated cinquillo pattern [2-1-2-1-2] followed by the regular pattern [2-2-2-2]. The former he calls *compás fuerte* (strong timeline), and the latter *compás débil* (weak timeline), and he emphasizes that the strong and weak parts always alternate. This rhythmic cell contains the *clave son* as a subset. Comparing this rhythmic cell to the *clave son* variants listed in Figure 42.6 shows that it contains several others as well. This pattern is also used as a bell timeline rhythm in West African music. The further slight variant shown in Figure 42.10 is the timeline of the *kassa* rhythm of Guinea.

Both rhythms shown in Figures 42.9 and 42.10 are rhythms played on the *guiro* in the Danzón style of Cuban music. The guiro, shown in Figure 42.11, is made from a hollowed-out gourd, on which parallel ridges are carved out at small intervals. It is played by means of scraping a thin stick made of wood or metal over the ridges.

Rhythm timelines are a type of *cultural object*. A rhythm that is considered successful from an evolutionary perspective will survive and multiply. The power of a rhythm timeline to disseminate itself among different cultural communities is the property of *fecundity*. Abundant evidence has been put forward that the *clave son* is such a rhythm, and thus it is only fitting that it should be investigated to determine the musicological and mathematical reasons for its fertility.[24] By analyzing the reasons for the success of the *clave son*, we should be able to gain insight into what makes a "good" rhythm good more generally.

NOTES

1 Morrison, S. J. & Demorest, S. M., (2009).
2 Huron, D., (2007), p. 191. See also Schulze, H.-H., (1989), Honing, H., (2002), and Patel, A. D., (2008), p. 112.

3 The mental templates may be partly culturally acquired in the form of the "perception norms" discussed by Anku, W., (1997), p. 212. They may also be determined by the perceptual figural grouping mechanisms in the brain, as suggested by experiments performed by Handel, S., (1992, 1998b).

4 Blackmore, S., (2000), p. 59. The term "Lamarckian" comes from the evolutionary theories of Jean-Baptiste de Lamarck and today refers exclusively to the principle of the inheritance of acquired features. See also Diettrich, O., (1992).

5 *Ibid.*, See also Weismann, A., (1889).

6 Kaufman Shelemay, K., (2000), p. 24.

7 See Longuet-Higgins, H. C. & Lee, C. S., (1982) and Longuet-Higgins, H. C., Webber, B., Cameron, W., Bundy, A., Hudson, R., Hudson, L., Ziman, J., Sloman, A., Sharples, M., & Dennett, D., (1994), for early analogies between music and natural languages, and Patel, A. D., (2006, 2008) and Cohen, D. & Katz, R., (2008), p. 18, for more recent work. There is ongoing debate about which came first, music or language. Brown, S., (2001), offers a possible resolution to this issue with his "musilanguage" model of music evolution which hypothesizes that music and language both evolved simultaneously from a communication method he calls "musilanguage." For an introduction to evolutionary musicology, see Brown, S., Merker, B., & Wallin, N. L., (2001). See also Fitch, W. T., (2005) and McGowan, R. W. & Levitt, A. G., (2011).

8 Hughes, D. W., (2000), p. 105.

9 Ester, D. P., Scheib, J. W., Kimberly, & Inks, J., (2006), describe a syllabic system called *Takadimi* that has found successful application to the teaching of rhythm, and Colley, B., (1987), compares several syllabic methods for improving rhythm literacy.

10 Wright, O., (1995).

11 Stone, R. M., (2000), p. 14.

12 Pérez-Fernández, R. A., (1986, 2007). See Chapter 12.

13 Handel, S., (1992), p. 497.

14 Roederer, J. G., (2008), p. 10. Cohen, J. E., (2007), p. 142, writes: "In recognizing the pattern represented by the score, the brain compares relations between successive events rather than absolute values." Rahn, J., (1995), develops music-theoretical analytical tools based on such nonnumerical predicates. A related approach to soften the hardness of absolute values employs *fuzzy* set theory, and Quinn, I., (1997), provides an example of their usefulness when applied to the theory and practice of musical contours.

15 Desain, P. & Honing, H., (2003).

16 Kubik, G., (1998), p. 218.

17 Blench, R., (1982) argues on the basis of the similarity of musical instruments that Indonesia had a cultural influence on Central Africa through colonization. Jeffreys, M. D. W., (1966/1967), on the other hand, suggests that the influence traveled in the converse direction.

18 Manuel, P., (2009), p. 189.

19 Orovio, H., (1992), p. 456.

20 Longuet-Higgins, H. C., Webber, B., Cameron, W., Bundy, A., Hudson, R., Hudson, L., Ziman, J., Sloman, A., Sharples, M., & Dennett, D., (1994), p. 105.

21 Brady, B., (2002), p. 82.

22 Poole, A., (2018).

23 Grauer, V., (2006) and Grauer, V., (2009).

24 Toussaint, G. T., (2011).

CHAPTER **43**

Epilogue

IN THIS BOOK, I have presented some of the computational and mathematical foundations for the characterization of "good" musical rhythms, as well as for their phylogenetic analysis, by focusing on the structure and the roots of one rhythm in particular, that has captured the imagination of the world, and withstood the test of time, the *clave son.* The attempt to discover the origin of this rhythm raises a more general question. How old is musical rhythm? This question is difficult to answer in absolute terms, but I can venture a guess that musical rhythm is probably as old as the human mind. Even before humans were creating rhythm with musical instruments, the sound of rhythm was abundantly inherent in nature as the effects of wind, rain, rivers, waterfalls, thunderstorms, and galloping horses, not to mention the "music" produced by animals.[1] Although many of these sounds may lack a discernible cyclic pattern with precise small interval ratios determined by integers, once in a while perhaps by chance nature will provide one. Furthermore, the perception of rhythm, as we have seen, is very much a constructive process in which the mind projects a categorical pattern on an input stimulus. Vijay Iyer considers music perception and cognition to be *embodied activities.*[2] Therefore even if nature produces an arrhythmic (continuous) sound pattern, the human mind will mold it into a categorical (discrete) rhythm. Thus, in the human mind perceiving rhythm is likely to be as old as the mind's ability to recognize perceptual categories.

The perception of rhythm in nature is one thing, but what about the creation and production of rhythm with either the human body (hands and voice) or external musical instruments? Some archeologists have claimed that the first deliberately built musical instrument

is a flute made out of bone during the late Stone Age, found in Germany.[3] There is still debate about whether the holes found on this bone were made judiciously by the human hand or serendipitously by some animal's teeth. In any case, the question of designing and building sophisticated musical instruments by hand, such as the flute, violin, or harpsichord may be dispensed with outright, since human hands themselves become musical instruments when clapped, as exemplified so plainly in the *hambone* rhythms used by the slaves in African–American plantations, the *flamenco* music of Southern Spain, or the twentieth-century minimalist musical composition *Clapping Music* by Steve Reich. Elevating the flute to the status of the earliest musical instrument reflects a longstanding bias against rhythm as the fundamental property of music, which is echoed in the writings of some scholars. For example, in discussing the relationship between the characteristics of tone (pitch, volume, and timbre) and rhythm, George C. Gow writes "one must remember that rhythm is the *means of handling the material,* not the material itself."[4] In contrast, Jeremy Montagu has suggested that during the early Stone Age when people were making stone tools, stones may have become the first nonbody musical instruments used for making concussive sounds.[5] The stones may have given way to bone or hard wooden sticks such as the *clappers* used in ancient China, or the *claves* so popular today in many parts of the world. Stone chimes have been used in ancient China for thousands of years, and have made their way to most parts of the world. In many places stones are still used in music today. In the Basque country of the northeastern part of Spain, for instance, *Xalaparta* music incorporates wooden sticks pounded on either wooden planks or stone slabs (lithophones) of different thickness and composition that produce sounds with a wide variety of different pitches and timbres.

In this book, I have focused on more specific questions. How old is the rhythmic pattern with interonset interval structure [3-3-4-2-4]; what is so special about this particular permutation of the numbers 2, 3, and 4; and how did it manage to conquer the world, to become a household rhythm, a popular door knock pattern? There are surely cultural, sociological, technological, economic, military, and political factors that may be partially responsible for the migrations of this rhythm from one culture to another, and from one continent to another. In addition to such diffusion or *ethnogenesis*, this rhythm may have been discovered in different places, and at different times, completely independently, a process referred to as *polygenesis*.[6] This is almost certainly true for the ubiquitous [3-3-2] rhythm timeline, and it is not unreasonable to suppose that the concatenation of two instances of this rhythm, given by [3-3-2–3-3-2] may have been independently transformed to [3-3-4-2-4] in geographically isolated cultures. However, the emphasis in this book has been the exploration of which mathematical, musicological, and psychological properties have made this pattern the salient universal cultural meme that it is today? To analyze this rhythm in a meaningful context, the related more general question that has been considered is: what makes a "good" rhythm good? This question led to the consideration of a wide variety of mathematical properties that good rhythms possess, thus narrowing down the infinitude of possible candidate rhythms to smaller groups of rhythms that could then be narrowed down further so as to characterize the pattern [3-3-4-2-4] as distinct from all others, in terms of these properties. In the end, the combination of the mathematical property of *shadow contour isomorphism* with the psychological property of *gestalt despatialization* accomplished the task of reducing the smaller groups to the unique rhythm: the *clave son*. Furthermore, it has been established that the *centroid-balance* property uniquely characterizes the *clave son* cyclic interonset-interval pattern among all other rhythms with five onsets and 16 pulses, by rendering it the most centroid-balanced pattern of all. Nevertheless, since this property is invariant under rotations, it does not isolate the rotation [3-3-4-2-4] from among all other rotations of this pattern.

An implicit question repeatedly considered in this book has been: What can *mathematics* tell us about musical rhythm, and is what it tells us useful? From the examination of all the mathematical, musicological, and psychological properties discussed and applied to rhythm timelines, the general pattern that emerges in attempting to characterize the *clave son* rhythm may be described by the philosophy of the *golden mean*, taught to children as the *Goldilocks* principle.[7] This philosophy has surfaced in several different parts of the world at different times in history. The Chinese philosopher Confucius espoused it, the Greek philosopher Aristotle promoted it, and Buddhist philosophy embraced it, calling it the *middle path*. This philosophy proposes that for achieving success in life no single property or attribute of humanity should be present in excess; the properties should exist in moderation between that of excess and deficiency. From the 31 rhythmic properties listed in Figures 41.13 and 41.15, it may be observed that for almost all these properties, the *clave son* takes on a value between the possible extremes, and indeed, close to the middle of the range of values afforded by each of these properties.

One of the main goals of this book has been to promote the application of phylogenetic analysis tools, first successfully applied to DNA analysis by biologists, and later to the evolution of cultural objects such as language, to the analysis of musical rhythms. Although in the past few decades phylogenetic techniques have been used by archaeologists and anthropologists to study the evolution of a wide variety of cultural objects besides language, such as stone projectile points, helmets, swords, pottery compositions, pottery designs, baskets, puberty rituals, marriage patterns, written texts, textile designs, and even musical instruments, music itself has been conspicuously left out of these studies. In the first edition of this book, I wrote that "the phylogenetic analysis of musical rhythm is just beginning." Since its publication in 2013 the application of phylogenetic tools to the analysis of specific families of rhythms from several musical traditions has been fruitful. A challenging problem remaining for future research is whether we can obtain a phylogeny of *all* the world's rhythms? The phylogenetic analysis of musical rhythm, especially in the form of timelines and ostinatos, has much to contribute not only to musicology but also to the study of human migrations, as has already been done successfully with language and DNA molecules.

There is an ongoing debate about which came first, music or language.[8] A new model of the evolution of music put forward by Steven Brown suggests that *both* music and language evolved from a common ancestral cultural object that he calls *musilanguage*. Musilanguage has the properties of lexical tone, combinatorial phrase formation, and expressive phrasing mechanisms. Just as rhythm is the most important and fundamental aspect of music, it is probably also the case that rhythm is the fountain of musilanguage. Indeed, Bruce Richman has suggested that groups of sounds evolve into definite patterns with meaning by means of the creation of expectancies based on the repetition and regularity of rhythms.[9]

The nature of the studies on the relationship between music and culture has been mostly qualitative, descriptive, and ethnographic rather than quantitative. The folklorist and ethnomusicologist Alan Lomax was one of the pioneers in steering ethnomusicological research away from such an approach and into a more quantitative direction. Over many years, he carried out an extensive cross-cultural project to determine to what extent a song style reflects other cultural patterns. To carry out this project, Lomax developed a system he called *Cantometrics* for measuring song styles and correlating the resulting features with cultural data. To label his data consisting of thousands of songs from all over the world, he trained human subjects to detect a variety of features of song style. Among these many features, there were five rather coarse measures of rhythm type: (1) perfectly regular rhythms he called *one-beat rhythms*, (2) *simple* measures consisting of 2, 4, 6, 8, 9, or 12 pulses per cycle, in which the accent falls on the first pulse, (3) *complex* meters in which the cycle time span is not divisible by two or three, (4) *irregular* rhythms in which the accents occur at irregular intervals, and (5) rhythms in which no accent pattern may be distinguished, which he called *free rhythms* or *parlando rubato*. With the resulting data, the song styles were submitted to classical cluster analyses to obtain a categorization of songs that could then be correlated with both culture and geographical location. The approach proposed to study the evolution of music in this book differs from *Cantometrics* on several fronts. For one it concerns rhythm only, on the grounds that rhythm is the most fundamental aspect of music, and should be explored and understood in detail first, before analyzing the other parameters of music. The features of rhythm proposed here are more numerous and more detailed than the coarse rhythmic features employed in *Cantometrics*. The rhythmic features used here are calculated objectively by quantitative mathematical formulas rather than subjectively by trained human subjects. The measures of similarity between pairs of rhythms employed here are more sophisticated than those used in the classical cluster analysis programs. Finally, and perhaps most importantly, the new phylogenetic analysis tools developed in the field of computational evolutionary biology are being applied in the new approach proposed here. These phylogenetic techniques have been illustrated with two examples of their application to musical rhythm: flamenco meters and ancient Greek pæonic (quintuple) rhythms. In both cases, interesting and promising results are evident.

The systematic computational methods presented here for generating and classifying good rhythm timelines within the broad context of designing tools for composition, and fitness functions for genetic algorithms, fall under the general umbrella that includes structural and generative methods for the analysis of the rhythm timelines of West and Central Africa in a cultural context. It is hoped that the structural properties of the rhythm timelines explored here, their mathematical formulations, and the algorithms used to generate these rhythms will help not only in the quest to determine a characterization of what makes a "good" rhythm good but also provide useful ideas for the ethnomusicologist and the composer.

It is both surprising and encouraging that much useful information may be garnered from a purely mathematical analysis of the durational patterns of musical rhythm, an analysis that ignores all the other less mathematical parameters associated with musical rhythm. Nevertheless, to characterize the uniqueness of the *clave son* rhythm, the psychological property of *gestalt despatialization* was enlisted. Perhaps the moral of this story, at least at this point in time, is that since the perception of rhythm is in part intrinsically subjective and constructed in our minds, a purely mathematical solution to this characterization problem is not attainable. On the other hand, it may be that the right mathematical model has not yet been found, and that to be successful such a model must take into account the neurobiology of rhythm. An ideal candidate for such a mathematical model is the notion of centroid-balance explored in Chapter 20, since among all possible rhythms with 5 onsets and 16 pulses, the *clave son* is, although not perfectly balanced, the most balanced of them all. A tantalizing open problem is the determination of whether the centroid-balance property has any neurological correlates. Thus, the geometrical foundations laid out in this book beckon the infusion of higher-level quantitative musicological knowledge into the mathematical structures considered thus far, in order to determine the "correlation between external stimuli and internal structures," and thus to more fully and accurately understand musical rhythm.[10] Such a venture continues to offer an ever-expanding minefield for systematic, comparative, and computational musicology.[11]

NOTES

1 Gray, P. M., Krause, B., Atema, J., Payne, R., Krumhansl, C., & Baptista, L., (2001), p. 52, describe the structural similarities in the rhythms and scales used by animals and humans. See Ravignani, A., (2014) for the evolutionary origins of rhythm.
2 Iyer, V., (2002).
3 Kunej, D. & Turk, I., (2000), p. 235.
4 Gow, C. G., (1915), p. 638.
5 Montagu, J., (2004), p. 171.
6 List, G., (1978), p. 46.
7 Murphy, S., (2016), [4.3].
8 Gray, P. M., Krause, B., Atema, J., Payne, R., Krumhansl, C., & Baptista, L., (2001), p. 54, describe evidence that "seems to signal … that music has a more ancient origin even than human language."
9 Richman, B., (2001), p. 304.
10 Goldenberg, J., Mazursky, D., & Solomon, S., (2001), p. 2433. See also Thaut, M. H., (2008) regarding rhythm and the brain.
11 Parncutt, R., (2007).

References

BOOKS, BOOK CHAPTERS AND THESES

Abdy Williams, C. F. (2009). *The Aristoxenian Theory of Musical Rhythm. Cambridge Library Collection - Music.* Cambridge University Press, Cambridge, UK.

Acquista, A. (2009). *Tresillo: A Rhythmic Framework Connecting Differing Rhythmic Styles, Master of Music Project Report.* California State University, Long Beach, CA.

Agawu, K. (1995a). *African Rhythm: A Northern Ewe Perspective.* Cambridge University Press, Cambridge, MA.

Agawu, K. (2003). *Representing African Music: Postcolonial Notes, Queries, Positions.* Routledge, London.

Agawu, K. (2016). *The African Imagination in Music.* Oxford University Press, Oxford, UK.

Allouche, J.-P. & Shallit, O. (2002). *Automatic Sequences.* Cambridge University Press, Cambridge, UK.

Amiot, E. (2009). Eine kleine Fourier musik. In: *Mathematics and Computation in Music,* T. Klouche & T. Noll (eds), CCIS 37. Springer-Verlag, Berlin, Heidelberg. pp. 469–476.

Anku, W. (1995). Towards a cross-cultural theory of rhythm in African drumming. In: *Intercultural Music,* C. T. Kimberlin & A. Euba (eds), vol. 1. E. Breitinger, Bayreuth. pp. 167–202.

Anku, W. (2002a). *Structural Set Analysis of African Music 1: Adowa.* Soundstage Production, Legon, Ghana.

Anku, W. (2002b). *Structural Set Analysis of African Music 2: Bawa.* Soundstage Production, Legon, Ghana.

Arom, S. (1991). *African Polyphony and Polyrhythm.* Cambridge University Press, Cambridge, UK.

Arom, S. (2001). Prolegomena to a biomusicology. In: *The Origins of Music,* N. L. Wallin, B. Merker, & S. Brown (eds). MIT Press, Cambridge, MA. pp. 27–29.

Arom, S. (2009). *La Fanfare de Bangui.* Éditions La Découverte, Paris.

Ascher, M. (2002). *Mathematics Elsewhere: An Exploration of Ideas Across Cultures.* Princeton University Press, Princeton, NJ and Oxford.

Ashton, A. (2007). *Harmonograph: A Visual Guide to the Mathematics of Music.* Walker & Company, Ontario.

Bach, E. & Shallit, J. (1996). *Algorithmic Number Theory - Volume I: Efficient Algorithms.* MIT Press, Cambridge, MA.

Bamberger, J. (2000). *Developing Musical Intuitions.* Oxford University Press. New York.

Barz, G. (2004). *Music in East Africa.* Oxford University Press, Oxford, UK.

Baum, D. A. & Smith, S. D. (2012). *Tree Thinking: An Introduction to Phylogenetic Biology.* W. H. Freeman, New York.

Beament, J. (2005). *How We Hear Music: The Relationship between Music and the Hearing Mechanism.* The Boydell Press, Woodbridge, UK.

Beckenbach, E. & Bellman, R. (1961). *An Introduction to Inequalities, New Mathematical Library.* Random House, New York.

Beckman, P. (1971). *A History of Pi.* The Golem Press, New York.

Belinga, M. S. E. (1965). *Litterature et Musique Populaire en Afrique Noire.* Ed. Cujas, Paris.

Benson, D. J. (2007). *Music: A Mathematical Offering.* Cambridge University Press, Cambridge, UK.

Berendt, J.-E. (1987). *Nada Brahma: The World is Sound,* Translated by H. Bredigkeit. Inner Traditions, Rochester, VT.

Berry, W. (1987). *Structural Functions in Music.* General Publishing Company, Ltd., Don Mills, Canada.

Bhattacharya, C. & Hall, R. W. (2010). *Geometrical representations of North Indian thaats and raags.* In: *Bridges: Mathematical Connections in Art, Music, and Science,* Sarhangi, R. (ed.), Pecs, Hungary. pp. 341–346.

Blacking, J. (1973). *How Musical Is Man?* Washington Press, Seattle, Washington, DC.

Blackmore, S. (2000). *The Meme Machine.* Oxford University Press, Oxford, UK.

Blades, J. (1992). *Percussion Instruments and Their History,* 4th edn. Bold Strummer Ltd, Westport, CT.

Blum, S. (1992). Analysis of musical style. In: *Ethnomusicology - An Introduction.* H. Myers (ed.). W. W. Norton & Company, New York. pp. 165–218.

Bonus, A. E. (2010). The metronomic performance practice: A history of rhythm, metronomes, and the mechanization of musicality. PhD Thesis, Department of Music, Case Western Reserve University.

Bouchet, A. (2010). *Imparité rythmique.* ENS, Culturemath, Paris.

Bradley, D. W. & Bradley, R. A. (1999). Application of sequence comparison to the study of bird songs. In: *Time Warps, String Edits, and Macromolecules: The Theory and Practice of Sequence Comparison,* D. Sankoff & J. Kruskal (eds). CSLI Publications, Stanford University. pp. 189–207.

Bregman, A. S. (1990). *Auditory Scene Analysis: The Perceptual Organization of Sound*. MIT Press, Cambridge, MA.

Breiman, L., Friedman, J. H., Olshen, R. A., & Stone, C. J. (1984). *Classification and Regression Trees*. Wadsworth & Brooks/Cole Advanced Books & Software, Monterey, CA.

Brown, S. (2001). The *"musilanguage"* model of music evolution. In: *The Origins of Music*, N. L. Wallin, B. Merker, & S. Brown (eds). MIT Press, Cambridge, MA. pp. 271–300.

Brown, S., Merker, B., & Wallin, N. L. (2001). An introduction to evolutionary musicology. In: *The Origins of Music*, N. L. Wallin, B. Merker, & S. Brown (eds). MIT Press, Cambridge, MA. pp. 3–24.

Butler, M. J. (2006). *Unlocking the Groove - Rhythm, Meter, and Musical Design in Electronic Dance Music*. Indiana University Press, Bloomington, IN.

Cambouropoulos, E. (2005). Musical rhythm: A formal model for determining local boundaries, accents and metre in a melodic surface. In: *Music, Gestalt, and Computing*, LNCS Vol. 1317. Springer, Berline. pp. 277–293.

Cano, D. M. (1983). *Cante y Baile Flamencos*. Everest, León.

Cargal, J. M. (1988a). *Discrete Mathematics for Neophytes: Number Theory, Probability, Algorithms, and Other Stuff*. BibSonomy, Montogomery, AL.

Cargal, J. M. (1988b). *Discrete Mathematics for Neophytes: Number Theory, Probability, Algorithms, and Other Stuff*, Chapter 3. BibSonomy, Montogomery, AL. p. 2.

Chashchina, S. (2016). Before and after metre: To the issue of "out-of-metrical" rhythm research. In: *Music: Transitions/Continuities*, M. Veselinović, et al., (eds). Belgrade. pp. 145–159.

Chazelle, B. (2006). *The Algorithm: Idiom of Modern Science*. Princeton University. www.cs.princeton.edu/~chazelle/pubs/algorithm-print.html

Chatman, S. (1965). *A Theory of Meter*. Mouton & Co., The Hague.

Chemillier, M. (2002). Ethnomusicology, ethnomathematics. The logic underlying orally transmitted artistic practices. In: *Mathematics and Music*, G. Assayag, H. G. Feichtinger, & J. F. Rodrigues (eds). Springer, Berlin. pp. 161–183.

Chernoff, J. M. (1979). *African Rhythm and African Sensibility*. The University of Chicago Press, Chicago, IL.

Christensen, T. (2002). *The Cambridge History of Western Music Theory*. Cambridge University Press, Cambridge, UK.

Clayton, M. (2000). *Time in Indian Music*. Oxford University Press, Oxford, UK.

Clough, J., Conley, J. & Boge, C. (1999). *Scales, Intervals, Keys, Triads, Rhythm, and Meter*. W. W. Norton & Company, New York.

Collard, M. & Tehrani, J. (2005). Phylogenesis versus ethnogenesis in Turkmen cultural evolution. In *The Evolution of Cultural Diversity: A Phylogenetic Approach*, R. Mace, C. J. Holden, & S. Shennan (eds). UCL Press, London, UK. pp. 108–130.

Collins, J. (2004). *African Musical Symbolism in Contemporary Perspective*. Pro Business, Berlin, (ISBN 3-938262-15-X).

Conway, J. (1970). An Enumeration of knots and links, and some of their algebraic properties. In: *Problems in Computational Algebra*. J. Leech (ed.). Pergamon Press, Oxford, UK. pp. 329–358.

Cook, N. (1990). *Music, Imagination and Culture*. Oxford University Press, Oxford, UK.

Cooper, G. W. & Meyer, L. B. (1960). *The Rhythmic Structure of Music*. University of Chicago Press, Chicago, IL.

Cover, T. M. (1987). Figure-ground problem for sound. In: *Open Problems in Communication and Computation*, T. Cover & B. Gopinath (eds). Springer-Verlag, New York. p. 171.

Coxeter, H. S. M., (1973). Regular Polytopes. (Third ed.). Dover Publications.

Crawford, R. (2004). George Gershwin's 'I Got Rhythm' (1930). In: *The George Gershwin Reader*, R. Wyatt & J. A. Johnson (eds). Oxford University Press, Oxford - New York. pp. 156–172.

Cross, I. (1999). Is music the most important thing we ever did? Music, development and evolution. In: *Music, Mind and Science*, S. W. Yi (ed.). Seoul National University Press, Seoul. pp. 10–39.

Dannenberg, R. B. & Hu, N. (2002). Discovering musical structure in audio recordings. In: *Music and Artificial Intelligence*, C. Anagnostopoulou, M. Ferrand, & A. Smaill (eds). Springer, Berlin. pp. 43–57.

De Leew, T. (2005). *Music of the Twentieth Century: A Study of its Elements and Structure*. Amsterdam University Press, Amsterdam.

Dembski, S. & Straus, J. N. (eds) (1986). *Milton, Babbitt, in Words about Music*. University of Wisconsin Press, Madison, WI, p. 105.

Deutsch, D. (1999). Grouping mechanisms in music. In: *The Psychology of Music*, D. Deutsch (ed.), 2nd edn. Academic Press, San Diego, CA.

D'Indy, V. (1902). *Cours de Composition Musicale*. Durand, Paris.

Dominguez, M., Clampitt, D., & Noll, T. (2009). WF scales, ME sets, and Christoffel words. In: *Mathematics and Computation in Music*, T. Klouche & T. Noll (eds), CCIS 37. Springer-Verlag, Berlin, Heidelberg. pp. 477–488.

Du Sautoy, M. (2008). *Symmetry: A Journey into the Patterns of Nature*. Harper Collins Publishers, New York.

Duda, R. O. & Hart, P. E. (1973). *Pattern Classification and Scene Analysis*. John Wiley, New York.

Dworsky, A. & Sansby, B. (1999). *A Rhythmic Vocabulary*. Dancing Hands Music, Minnetonka.

Eades, P. (1988). Symmetry finding algorithms. In: *Computational Morphology*, G. T. Toussaint (ed.). Elsevier, Amsterdam. pp. 41–51.

Eckardt, A. (2008). Kpatsa: An examination of a Ghanaian dance in the United States. Masters Thesis, Bowling Green State University.

Eglash, R. (2005). *African Fractals - Modern Computing and Indigenous Design*. Rutgers University Press, New Brunswick, NJ.

Ellingson, T. (1992). Notation. In: Ethnomusicology – *An Introduction*, H. Myers (ed.). W. W. Norton & Company, New York. pp. 165–218.

Erickson, R. (1975). *Sound Structure in Music*. University of California Press, Berkeley, CA.

Feldman, H. C. (2006). *Black Rhythms of Peru*. Wesleyan University Press, Middletown, CT.

Fenlon, S. P. (2002). The exotic rhythms of Don Ellis. Doctor of Musical Arts Dissertation, Peabody Institute, Johns Hopkins University.

Fernandez, L. (2004). *Flamenco Music Theory: Rhythm, Harmony, Melody, Form*. Acordes Concert, Madrid, Spain.

Fernandez, R. A. (2006). *From Afro-Cuban Rhythms to Latin Jazz*. University of California Press, Berkeley, CA.

Fitch, W. T. (2012). The biology and evolution of rhythm: Unravelling a paradox. In: *Language and Music as Cognitive Systems*, P. Rebuschat, M. Rohmeier, J. A. Hawkins, & I. Cross (eds). Oxford University Press, Oxford, UK. pp. 73–95.

Flatischler, R. (1992). *The Forgotten Power of Rhythm*. LifeRhythm, Mendocino, CA.

Gamboa, J. M. (2002). *Cante por Cante: Discolibro Didactico de Flamenco*. New Atlantis Music and Alia Discos, Madrid.

Garland, T. H. & Kahn, C. V. (1995). *Math and Music: Harmonious Connections*. Dale Seymour Publications, Palo Alto, CA.

Gerard, G. (1998). *Salsa! The Rhyhtm of Latin Music*. White Cliffs Media, Tempe, AZ.

Glass, L. & Mackey, M. C. (1988). *The Rhythms of Life*. Princeton University Press, Princeton, NJ.

Goines, L. & Ameen, R. (1990). *Funkifying the Clave - Afro-Cuban Grooves for Bass and Drums*. Manhattan Music Publications, New York.

Gregory, B. (2005). Entropy and complexity in music: Some examples. Master of Science Thesis, University of North Carolina at Chapel Hill.

Gusfield, D. (1997). *Algorithms on Strings, Trees, and Sequences: Computer Science and Computational Biology*. Cambridge University Press, Cambridge.

Gustafson, K. (1987). A new method for displaying speech rhythm, with illustrations from some Nordic languages. In: *Nordic Prosody* IV, K. Gregersen & H. Basboll (eds). Odense University Press, Denmark. pp. 105–114.

Gustafson, K. (1988). The graphical representation of rhythm. In: *Progress Reports from Oxford Phonetics*, vol. 3. University of Oxford, Oxford, UK. pp. 6–26.

Guttmann, A. (1992). *The Olympics: A History of the Modern Games*. University of Illinois Press, Urbana-Chicago.

Hagoel, K. (2003). *The Art of Middle Eastern Rhythm*. OR-TAV Music Publications, Kfar Sava, Israel.

Hall, A. C. (1998). *Studying Rhythm*. Prectice Hall, Upper Saddle River, NJ.

Hamilton, A. (2007). *Aesthetics and Music*. Continuum International Publishing Group, New York.

Hamming, R. W. (1986). *Coding and Information Theory*. Prentice-Hall, Englewood Cliffs, NJ.

Handel, S. (2006). *Perceptual Coherence: Hearing and Seeing*. Oxford University Press, New York.

Hargittai, I. & Hargittai, M. (1994). *Symmetry: A Unifying Concept*. Shelter Publications, Inc., Bolinas, CA.

Harkleroad, L. (2006). *The Math behind the Music*. Cambridge University Press, Cambridge, UK.

Hartenberger, R. (2016). Performance Practice in the Music of Steve Reich, Chapter 11. In: *Clapping Music*. Cambridge University Press, Cambridge, UK, pp. 153–167.

Hartigan, R., Adzenyah, A., & Donkor, F. (1995). West African Rhythms for Drum Set. Manhattan Music, Miami, FL.

Hast, D. E. & Scott, S. (2004). *Music in Ireland*. Oxford University Press, Oxford, UK.

Hasty, C. F. (1997). *Meter as Rhythm*. Oxford University Press. Oxford, UK.

Herrera, P., Yeterian, A., & Gouyon, F. (2002). Automatic classification of drum sounds: A comparison of feature selection methods and classification techniques. In *Music and Artificial Intelligence*, C. Anagnostopoulou, M. Ferrand, & A. Smaill (eds). Springer-Verlag, Berlin. pp. 69–80.

Hijleh, M. (2008). Toward a practical theory of world rhythm. M.A. Thesis, Department of Music, University of Sheffield.

Hodges, W. (2006). The geometry of music. In: *Music and Mathematics: From Pythagoras to Fractals*, J. Fauvel, R. Flood, & R. Wilson (eds). Oxford University Press, Oxford, UK. pp. 91–112.

Hodges, W. & Wilson, R. J. (2002). Musical patterns. In: *Mathematics and Music*, G. Assayag, H. G. Feichtinger, & J. F. Rodrigues (eds). Springer-Verlag, Berlin. pp. 79–87.

Hoffman, P. (1999). *The Man who Loved only Numbers: The Story of Paul Erdős and the Search for Mathematical Truth*. Hyperion, New York.

Hollos, S. & Hollos, J. R. (2014). *Creating Rhythms*. Abrazol Publishing, Longmont, CO.

Honingh, A. K. (2009). Automatic modulation finding using convex sets of notes. In: *Mathematics and Computation in Music*, T. Klouche & T. Noll (eds), MCM 2007, CCIS 37. Printer-Verlag, Berlin Heidelberg. pp. 88–96.

von Hornbostel, E. M. & Sachs, C. (1992). Classification of musical instruments. In: *Ethnomusicology - An Introduction*, H. Myers (ed.). W. W. Norton & Company, New York. pp. 165–218.

Huron, D. (2007). *Sweet Anticipation: Music and the Psychology of Expectation*. MIT Press, Cambridge, MA.

Huron, D. (2009). Is music an evolutionary adaptation? In: *The Cognitive Neuroscience of Music*, I. Peretz & R. Zatorre (eds). Oxford University Press, Oxford, UK. pp. 56–75.

Jan, S. (2007). *The Memetics of Music: A Ne-Darwinian View of Musical Structure and Culture*. Ashgate Publishing Company, Burlington, VT.

Jeans, J. (1968). *Science & Music*. Dover Publications, Inc., New York.

Jedrzejewski F. (2017). New investigations on rhythmic oddity. In: *Mathematics and Computation in Music. MCM 2017. Lecture Notes in Computer Science*, O. Agustin-Aquino, E. Lluis-Puebla, & M. Montiel (eds), vol. 10527. Springer-Verlag. pp. 227–237.

Jensen, K. (2010). On the inherent segment length in music. In: *Machine Audition: Principles, Algorithms and Systems: Principles, Algorithms and Systems*, W. Wang (ed.). Publisher Idea Group Inc. (IGI), Hershey, Pennsylvania.

Johnson, I. (2009). *The Birth of Tragedy by Friedrich Nietzsche*. Richer Resources Publications, Arlington, VA.

Johnson, R. S. (1975). *Messiaen*. University of California Press, Berkeley and Los Angeles, CA.

Johnson, T. A. (2003). *Foundations of Diatonic Theory - A Mathematically Based Approach to Music Fundamentals*. Key College Publishing, Emeryville, CA.

Jones, A. M. (1959). *Studies in African Music*. Oxford University Press, Amen House, London.

Jones, I. M. (1937). Gershwin analyzes science of rhythm. *Los Angeles Times*, February 7, p. 5. In: *The George Gershwin Reader*, R. Wyatt & J. A. Johnson (eds). Oxford University Press, Oxford, New York. pp. 244–246.

Kaemmer, J. E. (2000). Southern Africa: An introduction. In: *The Garland Handbook of African Music*, R. M. Stone (ed.). Garland Publishing, Inc. and Taylor & Francis Group, New York. pp. 310–331.

Kappraff, J. (2002). *Beyond Measure: Essays in Nature, Myth, and Number*. World Scientific, Singapore.

Kartomi, M. J. (1990). *On Concepts and Classifications of Musical Instruments*. The University of Chicago Press, Chicago, IL.

Kaufman Shelemay, K. (2000). Notation and oral tradition. In: *The Garland Handbook of African Music*, R. M. Stone (ed.). Garland Publishing, Inc. and Taylor & Francis Group, New York. pp. 24–42.

Keith, M. (1991). *From Polychords to Pólya: Adventures in Musical Combinatorics*. Vinculum Press, Princeton, NJ.

Kernfeld, B. (1995). *What to Listen For in Jazz*. Yale University Press, New Haven, CT.

Keyser, C. H. (1993). *Introduction to Flamenco: Rhythmic Foundation and Accompaniment*. Santa Barbara, CA.

Kim, S. (1996). *Inversions: A Catalog of Calligraphic Cartwheels*. Byte Books, Peterborough, NH.

Klőwer, T. (1997). *The Joy of Drumming: Drums and Percussion Instruments from Around the World*. Binkey Kok Publications, Diever.

Knobloch, E. (2002). The sounding algebra: Relations between combinatorics and music from Mersenne to Euler. In: *Mathematics and Music*, G. Assayag, H. G. Feichtinger, & J. F. Rodrigues (eds). Springer-Verlag, Berlin. pp. 27–48.

Konaté, F. & Ott, T. (2000). *Rhythms and Songs from Guinea*. Lugert Verlag, Oldershausen.

Krause, E. F. (1975). *Taxicab Geometry*. Dover Publications, New York.

Kubik, G. (1999). *Africa and the Blues*. University Press of Mississippi, Jackson, MS.

Kubik, G. (2000). Central Africa: An introduction. In: *The Garland Handbook of African Music*, R. M. Stone (ed.). Garland Publishing, Inc. and Taylor & Francis Group, New York. pp. 260–290.

Kubik, G. (2010a). *Theory of African Music - Vol. I*. University of Chicago Press, Chicago, IL and London.

Kubik, G. (2010b). *Theory of African Music - Vol. II*. University of Chicago Press, Chicago and London.

Kulp, C. W. & Schlingmann, D. (2009). Using Mathematica to compose music and analyze music with information theory. In: *Mathematics and Computation in Music*, T. Klouche & T. Noll (eds), CCIS 37. Springer-Verlag, Berlin, Heidelber. pp. 441–448.

Kunej, D. & Turk, I. (2000). New perspectives on the beginnings of music: Archeological and musicological analysis of a middle Paleolithic bone flute. In: N. L. Wallin, B. Merker, & S. Brown, *The Origins of Music*. MIT Press, Cambridge, MA. pp. 235–268.

Kwabena Nketia, J. H. (1962). *African Music in Ghana: A Survey of Traditional Forms*. Longmans, Green and Co. Ltd., Accra.

Kwabena Nketia, J. H. (1963). *Folk Songs of Ghana*. University of Ghana, Legon.

Kwabena Nketia, J. H. (1974). *The Music of Africa*. Norton, New York.

Langer, S. K. (1957). *Problems of Art: Ten Philosophical Lectures*. Charles Scribner's Sons, New York.

Langevin, A. & Riopel, D. (eds). (2005). *Logistics Systems Design and Optimization*. Springer-Verlag, New York.

Lau, F. (2008). *Music in China*. Oxford University Press, Oxford, UK.

Lehmann, B. (2002). The syntax of 'Clave' - perception and analysis of meter in Cuban and African Music. M.A. Thesis, Tufts University, Medford, MA.

Lerdahl, F. & Jackendoff, R. (1983). *A Generative Theory of Tonal Music*. MIT Press, Cambridge, MA.

Levine, M. (1995). *The Jazz Theory Book*. Sher Music Co., Petaluma, CA.

Levitin, D. J. (2006). *This is Your Brain on Music: The Science of a Human Obsession*. Dutton, New York.

Levitin, D. J. (2008). *The World in Six Songs*, Viking Canada, Toronto.

Lewin, D. (2007). *Generalized Musical Intervals and Transformations*, Oxford University Press, New York.

Lewis, A. C. (2005). *Rhythm: What it Is and How to Improve Your Sense of it*. Eightseigh Press, San Francisco, CA.

Lipo, C. P., O'Brien, M. J., Collard, M., & Shennan, S. J. (eds) (2006). *Mapping Our Ancestors: Phylogenetic Approaches in Anthropology and Prehistory*. Transaction Publishers, New Brunswick, NJ.

Locke, D. (1998). *Drum Gahu: An Introduction to African Rhythm*. White Cliffs Media, Gilsum, NH.

Lomax, A. (1968). *Folk Song Style and Culture*. American Association for the Advancement of Science, Washington, DC, Publication no 88.

London, J. (2003). Rhythm. In: *The New Grove Dictionary of Music and Musicians*, J. Tyrrell (ed.), 2nd edn, vol. 21, Oxford University Press. p. 277.

London, J. (2004). *Hearing in Time: Psychological Aspects of Musical Meter*, Oxford University Press, New York.

Lothaire, M. (2002). *Algebraic Combinatorics on Words*. Cambridge University Press, Cambridge, UK.

Loy, G. (2006). *Musimathics, Vol. 1: The Mathematical Foundations of Music*. MIT Press, Cambridge, MA.

Loy, G. (2007). *Musimathics, Vol. 2: The Mathematical Foundations of Music*. MIT Press, Cambridge, MA.

Macaloon, J. J. (2007). *This Great Symbol. Pierre de Coubertin and the Origins of the Modern Olympic Games*. Routledge, London.

Mace, R., Holden, C. J., & Shennan, S., (eds) (2005). *The Evolution of Cultural Diversity: A Phylogenetic Approach*. UCL Press, London.

Mahner, M. (ed.) (2001). *Scientific Realism: Selected Essays of Mario Bunge*. Prometheus Books, Amherst, New York.

Malabe, F. & Weiner, B. (1990). *Afro-Cuban Rhythms for Drumset*. Manhattan Music, Inc., Miami, FL.

Mandelbrot, B. B. (1982). *The Fractal Geometry of Nature*. W. H. Freeman & Co., San Francisco, CA

Manuel, P. (2006). Flamenco in focus - An analysis of a performance of soleares. In: *Analytical Studies in World Music*, M. Tenzer (ed.). Oxford University Press, New York. pp. 92–119.

Manuel, P., Bilby, K. & Largey, M. (2006). *Caribbean Currents - Caribbean Music from Rumba to Reggae*. Temple University Press, Philadelphia, PA.

Marr, D. (1982). *Vision: A Computational Investigation into the Human Representation and Processing of Visual Information*. Freeman, New York.

Martin, G. E. (1998). *Geometric Constructions*. Springer-Verlag, New York.

Martineau, J. (2008). *The Elements of Music: Melody, Rhythm, and Harmony*. Walker & Company, New York.

Mauleón, R. (1997). The Cuban clave: Its origins and development in world musics. Masters Thesis, Mills College, Oakland, CA.

Mazzola, G. (2002). The topos geometry of musical logic. In: *Mathematics and Music*, G. Assayag, H. G. Feichtinger, & J. F. Rodrigues (eds). Springer-Verlag, Berlin, pp. 199–213.

Mazzola, G. (2003). *The Topos of Music: Geometric Logic of Concepts, Theory, and Performance*. Birkhäuser, Basel.

McKay, C. & Fujinaga, I. (2006). Musical genre classification: Is it worth pursuing and how can it be improved? *Proceedings of the International Conference on Music Information Retrieval*, Victoria, Canada, pp. 101–106.

Mercader, N. (2001). *La Percusión en el Flamenco*. Nueva Carisch España, S.L., Madrid.

Messiaen, O. (1956). *The Technique of my Musical Language*, Translated by J. Satterfield. Alphonse Leduc, Paris.

Milne, A. J., Bulger, D., Herff, S., & Sethares, W. A. (2015). Perfect balance: A novel principle for the construction of musical scales and meters. In: *Mathematics and Computation in Music—MCM*, T. Collins, D. Meredith, & A. Volk (eds), vol. 9110. LNAI and Springer-Verlag, Heidelberg. pp. 97–108.

Milstein, S. (1992). *Arnold Schoenberg: Notes, Sets, Forms*. Cambridge University Press, Cambridge, UK.

Mocquereau, A. (1932). *"Le Nombre Musical Grégorien"* - A Study of Gregorian Musical Rhythm, vol. I, part I, English Translation by A. Tone. Desclée & Co., Paris.

Moles, A. (1966). *Information Theory and Esthetic Perception*. The University of Illinois Press, Urbana, IL and London.

Molina, R. & Mairena, A. (1963). *Mundo y Formas del Cante Flamenco*. Ed. Revista de Occidente, Madrid.

Montfort, N. & Gillespie, W. (2002). *2002: A Palindrome Story*. Spineless Books, Urbana, IL.

Moore, R. & Sayre, E. (2006). An Afro-Cuban Bata piece for Obatala, king of the white cloth. In: *Analytical Studies in World Music*, M. Tenzer (ed.). Oxford University Press, New York. pp. 120–160.

Morales, E. (2003). *The Latin Beat - The Rhythms and Roots of Latin Music from Bossa Nova to Salsa and Beyond*. Da Capo Press, Cambridge, MA.

Morrison, S. J. & Demorest, S. M. (2009). Cultural constraints on music perception and cognition. In: *Progress in Brain Research*, J. Y. Chiao (ed.), vol. 178. Elsevier, The Netherlands. pp. 67–77.

Muchimba, F. (2008). *Liberating the African Soul: Comparing African and Western Christian Music*. Authentic Publishing, Colorado Springs, CO.

Murphy, J. P. (2006). *Music in Brazil*. Oxford University Press, New York.

Nettl, B. (2005). *The Study of Ethnomusicology - Thirty-One Issues and Concepts*. University of Illinois Press, Urbana, IL.

Nietzsche, F. (1889). *Die Götzen-Dämmerung (Twilight of the Gods)*. Leipzig: Verlag von C. G. Naumann.

Niven, I. (1981). *Maxima and Minima without Calculus*, vol. 7. The Mathematical Association of America, Washington, DC.

Novotney, E. D. (1998). The 3:2 relationship as the foundation of timelines in West African Musics. Doctor of Musical Arts Thesis, University of Illinois at Urbana-Champaign.

Orovio, H. (1992). *Diccionario de la Música Cubana: Biográfico y Técnico*. Editorial Letras Cubanas, La Habana, Cuba.

Ortiz, F. (1995). *La Clave*. Editorial Letras Cubanas, La Habana, Cuba.

Palmer, P. (2009). *Blues & Chaos: The Music Writing of Robert Palmer*. Scribner, New York.

Parra, J. M. (1999). *El Compás Flamenco de Todos los Estilos*. Ediciones Apóstrofe, Barcelona.

Patel, A. D. (2008). *Music, Language, and the Brain*. Oxford University Press, Oxford, UK.

Peñalosa, D. (2009). *The Clave Matrix – Afro-Cuban Rhythm: Its Principles and African Origins*. Bembe Books, Redway, CA.

Peck, R. W. (2015). All-interval structures. In: *Mathematics and Computation in Music*, MCM 2015, T. Collins, D. Meredith, A. Volk (eds), LNAI 9110. Springer, Switzerland. pp. 279–290.

Percival, H. W. (1946). *Thinking and Destiny*. The Word Foundation, Inc., Rochester, NY.

Pérez-Fernández, R. A. (1986). *La binarización de los ritmos ternarios africanos en América Latina*. Casa de las Américas, Havana.

Piccard, S. (1939). *Sur les ensembles de distances des ensembles de points dun espace euclidien*, vol. 13. Mem. Univ. Neuchatel, Neuchatel, Switzerland.

Polster, B. (2004). *Q.E.D. – Beauty in Mathematical Proof*. Wooden Books and Walker & Company, New York.

Potter, P. (2000). *Four Musical Minimalists*. Cambridge University Press, Cambridge, UK.

Rahn, J. (1980). *Basic Atonal Theory*. Schirmer Books, New York.

Rahn, J. (1983). *A Theory for All Music: Problems and Solutions in the Analysis of Non-Western Forms*. University of Toronto Press, Toronto, Canada.

Randel, D. M. (ed.) (2003). *The Harvard Dictionary of Music*, 4th edn. The Belknap Press of Harvard University Press, Cambridge, MA.

Redmond, L. (1997). *When the Drummers Were Women - A Spiritual History of Rhythm*. Three Rivers Press, New York.

Reich, S. (1980). *Clapping Music for Two Performers*. Universal Edition Ltd., London.

Reich, S. (2002). *Writings on Music 1965–2000*. Oxford University Press, New York.

Reingold, E. M. & Dershowitz, N. (2001). *Calendrical Calculations: The Millenium Edition*. Cambridge University Press, Cambridge, UK.

Rentink, S. (2003). Kpanlogo: Conflict, identity crisis and enjoyment in a Ga drum dance. M. A. Thesis, Department of Musicology, University of Amsterdam.

Rice, T. (2004). *Music in Bulgaria*. Oxford University Press, New York and Oxford, UK.

Richman, B. (2001). How music fixed "nonsense" into significant formulas: On rhythm, repetition, and meaning. In: *The Origins of Music*, N. L. Wallin, B. Merker, & S. Brown. MIT Press, Cambridge, MA. pp. 301–314.

Richards, E. G. (1998). *Mapping Time: The Calendar and its History*. Oxford University Press, Oxford, UK.

Ritchie, S. (2012). Baroque Clichés. In: *Before the Chinrest*. Indiana Press, Bloomington. pp. 76–82.

Roaf, D. & White, A. (2006). Ringing the changes: Bells and mathematics. In: *Music and Mathematics: From Pythagoras to Fractals*, J. Fauvel, R. Flood, & R. Wilson (eds). Oxford University Press, New York. pp. 113–130.

Roederer, J. G. (2008). *The Physics and Psychophysics of Music: An Introduction*, 4th edn. Springer-Verlag, New York.

Rosalia, R. V. (2002). Migrated Rhythm: The Tambú of Curaçao, CaribSeek.

Ross, C. (2007). *Architecture and Mathematics in Ancient Egypt*. Cambridge University Press, Cambridge, UK.

Rossy, H. (1966). *Teoria del Cante Jondo*. Punt Groc and Associats, Breda, Barcelona.

Rowe, R. (2001). *Machine Musicianship*. MIT Press, Cambridge, MA.

Sachs, C. (1943). *The Rise of Music in the Ancient World - East and West*. W. W. Norton & Company, New York.

Sachs, C. (1953). *Rhythm and Tempo: A Study in Music History*. W. W. Norton & Company, New York.

Sacks, O. (1998). *The Man Who Mistook His Wife for a Hat: And Other Clinical Tales*. Touchstone, New York.

Safi al-Din al-Urmawî, Kitâb al-Adwâr 1252 (1938). *La Musique Arabe*, Translated by R. Erlanger. Paul Geuthner, Paris.

Scheirer, E. D. (2000). Music-Listening Systems. PhD Thesis, Massachusetts Institute of Technlogy, Cambridge, MA.

Schillinger, J. (2004). Theory of rhythm. In: *The Schillinger System of Musical Composition*, vol. I. Clock and Rose Press, Harwitch Port, MA. pp. 1–95.

Sethares, W. A. (2007). *Rhythm and Transforms*. Springer-Velag, London.

Shiloah, A. (1995). *Music in the World of Islam: A Socio-Cultural Study*. Scolar Press, Aldershot, England.

Shmulevich, I. & Povel, D.-J. (2000b). Complexity measures of musical rhythms. In: *Rhythm Perception and Production*, P. Desain & L. Windsor (eds). Swets & Zeitlinger, Lisse, The Netherlands. pp. 239–244.

Sicsic, H.-P. (1993). Structural, dramatic and stylistic relationships in Prokofiev's Sonatas no. 7 and no. 8. D.M.A. Thesis, Rice University.

Smith, L. M. (2010). Rhythmic similarity using metrical profile matching. *Proceedings of the International Computer Music Conference*, New York, pp. 177–182.

Smith, M. L. (2004). *Olympics in Athens 1896: The Invention of the Modern Olympic Games*. Profile Books Ltd, London.

Song, C. (2014). Syncopation: Unifying music theory and perception. PhD Thesis, Queen Mary, University of London.

Steinhardt, P. J. & Ostlund, S., (eds) (1987). *The Physics of Quasicrystals*. World Scientific Publishing Co. Pte. Ltd., Singapore.

Stone, R. M. (2000). Exploring African music. In: *The Garland Handbook of African Music*, R. M. Stone (ed.). Garland Publishing, Inc. and Taylor & Francis Group, New York. pp. 13–21.

Stone, R. M. (2005). *Music in West Africa*. Oxford University Press, Oxford, UK.

Streich, S. (2006). Music complexity: A multi-faceted description of audio content. PhD Thesis, Universitsat Pompeu Fabra, Barcelona.

Sum, M. (2012). Music of the Gnawa of Morocco: Evolving spaces and times. PhD Thesis, University of British Columbia, Vancouver, Canada.

Takeda, M. (2001). String resemblance systems: A unifying framework for string similarity with applications to literature and music. In: *Combinatorial Pattern Matching*, A. Amir & G. M. Landau (eds), vol. 2089. LNCS, Jerusalem, Israel. pp. 147–151.

Tan, S.-L., Pfordresher, P., & Harré, R. (2010). *Psychology of Music: From Sound to Significance*. Psychology Press, Hove, UK.

Tang, P. (2007). *Masters of the Sabar - Wolof Griot Percussionists of Senegal*. Temple University Press, Philadelphia, PA.

Tanguiane, A. S. (1993). *Artificial Perception and Music Recognition*. Springer-Verlag, Berlin, Heidelberg.

Temperley, D. (2001). *The Cognition of Basic Musical Structures*. MIT Press, Cambridge, MA.

Temperley, D. (2002). A Bayesian approach to key-finding. In: *Music and Artificial Intelligence*, C. Anagnostopoulou, M. Ferrand, & A. Smaill (eds). Springer-Verlag, Berlin. pp. 195–206.

Temperley, D. (2007). *Music and Probability*. MIT Press, Cambridge, MA.

Tenzer, M. (2006). Analysis, categorization, and theory of musics of the world. In: *Analytical Studies in World Music*, M. Tenzer (ed.). Oxford University Press, Oxford, UK. pp. 3–38.

Thaut, M. H. (2008). *Rhythm, Music, and the Brain*. Taylor & Francis Group, New York and Abingdon, UK.

Touma, H. H. (1996). *The Music of the Arabs*. Amadeus Press, Portland, OR.

Toussaint, G. T. (1988). A graph-theoretical primal sketch. In: *Computational Morphology*, G. T. Toussaint (ed.). North-Holland, Amsterdam. pp. 229–260.

Toussaint, G. T. (2012a). Phylogenetic tools for evolutionary musicology. In: *Mathematical and Computational Musicology*, T. Klouche (ed.), vol. I. Berlin National Institute for Music Research and Springer-Verlag, Berlin.

Toussaint, G. T. (2013a). *The Geometry of Musical Rhythm*. Chapman & Hall and CRC Press, Boca Raton, FL.

Trehub, S. (2001). Human processing predispositions and musical universals. In: *The Origins of Music*, N. L. Wallin, B. Merker, & S. Brown (eds). MIT Press, Cambridge, MA. pp. 428–448.

Tymoczko, D. (2011). *A Geometry of Music: Harmony and Counterpoint in the Extended Common Practice*. Oxford University Press, New York.

Unruh, A. J. (2000). Kpanlogo: A detailed description of one arrangement of a West-African Music and Dance Genre. Masters Thesis, Bowling Green State University, Bowling Green, OH.

Uribe, E. (1993). *The Essence of Brazilian Persussion and Drum Set*. CCP/Belwin Inc., Miami, FL.

Uribe, E. (1996). *The Essence of Afro-Cuban Persussion and Drum Set*. Warner Brothers Publications, Miami, FL.

van der Walt, S. (2007). Rhythmic techniques in a selection of Olivier Messiaen's Piano works. Master of Music Thesis, University of Pretoria.

Veltman, J. (2006). Syllable placement and metrical hierarchy in sixteenth-century motets. In *Music Analysis East and West, Computing in Musicology*, W. B. Hewlett & E. Selfridge-Field (eds), vol. 14. MIT Press, Cambridge, MA. pp. 73–89.

Varney, J. (1999). Colombian Bambuco: The evolution of a national music style, PhD Thesis, Grifith University, South Brisbane, Australia.

Vinke, L. N. (2010). Factors affecting the perceived rhythmic complexity of auditory rhythms. M.A. Thesis, Bowling Green State University.

Wade, B. C. (2004). *Thinking Musically*. Oxford University Press, Oxford, UK.

Wade, D. (2006). *Symmetry: The Ordering Principle*. Wooden Books and Walker & Company, New York.

Wagner, R. A. (1999). On the complexity of the extended string-to-string correction problem. In: *Time Warps, String Edits, and Macromolecules: The Theory and Practice of Sequence Comparison*, D. Sankoff & J. Kruskal (eds). CSLI Publications, Stanford University, Stanford, CA. pp. 215–235.

Wanamaker, J. & Carson, R. (1984). *International Drum Rudiments*. Alfred Publishing Co., Inc., Sherman Oaks, CA.

Washburne, C. (1995). Clave: The African roots of salsa. In: *Kalinda! Newsletter for the Center for Black Music Research, Fall Issue*. Columbia University, New York, pp. 7–11.

Watkins, A., Scheaffer, R., & Cobb, G. (2004). *Statistics in Action: Understanding a World of Data*. Key College Publishing, Emeryville, CA.

Weismann, A. (1889). *Essays Upon Heredity*, vols. 1 and 2. The Clarendon Press, Oxford, UK.

Wendt, C. C. (2000). Tuareg music. In: *The Garland Handbook of African Music*, R. M. Stone, (ed.). Garland Publishing, Inc. and Taylor & Francis Group, New York. pp. 206–227.

West, M. L. (1992). *Ancient Greek Music*. Oxford University Press, New York.

Wright, O. (1978). *The Modal System of Arab and Persian Music AD 1250–1300*. Oxford University Press, Oxford, UK.

Wright, M. J. (2008). The shape of an instant: Measuring and modeling perceptual attack time with probability density functions. PhD Thesis, Department of Music, Stanford University.

Wright, D. (2009). *Mathematics and Music*. American Mathematical Society, Providence, RI.

Yeston, M. (1976). *The Stratification of Musical Rhythm*. Yale University Press, New Haven, CT.

JOURNAL ARTICLES AND CONFERENCE PROCEEDINGS

Agawu, K. (1987). The rhythmic structure of West African music. *The Journal of Musicology*, 5/3:400–418.

Agawu, K. (1995b). The invention of "African Rhythm." *Journal of the American Musicological Society*, 48/3:380–395.

Agawu, K. (2006). Structural analysis or cultural analysis? Competing perspectives on the "standard pattern" of West African rhythm. *Journal of the American Musicological Society*, 59/1:1–46.

Agmon, E. (1997). Musical durations as mathematical intervals: Some implications for the theory and analysis of rhythm. *Music Analysis*, 16/1:45–75.

Agon, C. & Andreatta, M. (2011). Modelling and implementing tiling rhythmic canons in OpenMusic visual programming language. *Perspectives of New Music*, 1–2/49:66–92.

Ahmad, W. & Kondoz, A. M. (2011). Analysis and synthesis of hand clapping sounds based on adaptive dictionary. *Proceedings of the International Computer Music Conference*, University of Huddersfield, Huddersfield, UK, pp. 257–263.

Aichholzer, O., Caraballo, L. E., Díaz-Báñez, J. M., Fabila-Monroy, R., Ochoa, C., & Nigsch, P. (2015). Characterization of extremal antipodal polygons. *Graphs and Combinatorics*, 31/2:321–333.

Akhtaruzzaman, Md. (2008). Representation of musical rhythm and its classification system based on mathematical and geometrical analysis. *Proceedings of the International Conference on Computer and Communication Engineering*, Kuala Lumpur, Malaysia, pp. 466–471.

Akhtaruzzaman, Md., Rashid, M. M., & Ashrafuzzaman, Md. (2009). Mathematical and geometrical analysis and representation of North Indian musical rhythms based on multi polygonal model. *Proceedings of the International Conference on Signal Processing Systems*, Kuala Lumpur, Malaysia, pp. 493–497.

Akpabot, S. (1972). Theories on African music. *African Arts*, 6/1:59–62.

Akpabot, S. (1975). Random music of the Birom. *African Arts*, 8/2:46–47.

Alén, O. (1995). Rhythm as duration of sounds in Tumba Francesa. *Ethnomusicology*, 39/1:55–71.

Alexander, C. & Carey, S. (1968). Subsymmetries. *Perception and Psychophysics*, 4:73–77.

Aloupis, G., Fevens, T., Langerman, S., Matsui, T., Mesa, A., Nuñez, Y., Rappaport, D., & Toussaint, G. T. (2006). Algorithms for computing geometric measures of melodic similarity. *Computer Music Journal*, 30/3:67–76.

Alperin, R. C. & Drobot, V. (2011). Golomb rulers. *Mathematics Magazine*, 84:48–55.

Amiot, E. (2007). David Lewin and maximally even sets. *Journal of Mathematics and Music*, 1/3:157–172.

Amiot, E. & Sethares, W. A. (2011). An algebra for periodic rhythms and scales. *Journal of Mathematics and Music*, 5/3:149–169.

Ammann, R. (1997). Le rythme kanak. *Cahiers de Musiques Traditionnelles*, 10:237–247.

Anand, A., Wilkinson, L., & Tuan, D. N. (2009). An L-infinity norm visual classifier. *Proceedings of the Ninth IEEE International Conference on Data Mining*, IEEE Computer Society, Washington, DC.

Anderson, P. W. (1972). More is different - broken symmetry and the nature of the hierarchical structure of nature. *Science*, 177/4047:393–396.

Andreatta, M., Noll, T., Agon, C., & Assayag, G. (2001). The geometrical groove: Rhythmic canons between theory, implementation and musical experiment. *Les Actes des 8e Journées d'Informatique Musicale*, Bourges, 93–97.

Anku, W. (1997). Principles of rhythm integration in African drumming. *Black Music Research Journal*, 17/2:211–238.

Anku, W. (2000). Circles and time: A theory of structural organization of rhythm in African music. *Music Theory Online*, 6/1.

Anku, W. (2007). Inside a master drummer's mind: A quantitative theory of structures in African music. *Transcultural Music Review*, 11, (article No. 5) [Accessed October 27, 2009].

Antonopoulos, I., Pikrakis, A., Theodoridis, S., Cornelis, O., Moelants, D., & Leman, M. (2007). Music retrieval by rhythmic similarity applied on Greek and African traditional music. *Proceedings of the 8th International Conference on Music Information Retrieval*, Vienna, Austria, pp. 297–300.

Apel, W. (1960). Vier plus vier=drei plus drei plus zwei. *Acta Musicologica*, 32:29–33.

Apostol, T. M. & Mnatsakanian, M. A. (2000). Finding centroids the easy way. *Math Horizons*, 8/1:7–12.

Arlin, M. I. (2000). Metric mutation and modulation: The nineteenth-century speculations of F.-J. Fétis. *Journal of Music Theory*, 44/2:261–322.

Arom, S. (1984). The constituting features of Central African rhythmic systems: A tentative typology. *The World of Music*, 26/1:51–67.

Arom, S. (1988). Systèmes musicaux en Afrique Subsaharienne. *Intersections*, 9/1:1–19.

Arom, S. (1989). Time structure in the music of Central Africa: Periodicity, meter, rhythm and polyrhythmics. *Leonardo*, 22/1:91–99.

Arom, S. (1994). More on rhythmical marking: A reply to Herve Rivière. *Ethnomusicology*, 38/2:321–322.

Arom, S. (2004). L'aksak: Principes et typologie. *Cahiers de Musiques Traditionnelles*, 17:12–48.

Asada, M. & Ohgushi, K. (1991). Perceptual analyses of Ravel's "Bolero." *Music Perception: An Interdisciplinary Journal*, 8/3:241–249.

Asch, M. I. (1975). Social context and the musical analysis of Slavey drum dance songs. *Ethnomusicology*, 19/2:245–257.

Assar, G. R. F. (2003). Calendars at Babylon and Seleucia on the Tigris. *Iran*, 41:171–191.

Avis, D. & Toussaint, G. T. (1981). An efficient algorithm for decomposing a polygon into star-shaped polygons. *Pattern Recognition*, 13/6:395–398.

Avorgbedor, D. (1987). The construction and manipulation of temporal structures in "Yeve" cult music: A multidimensional approach. *African Music*, 6/4:4–18.

Ayres, B. (1972). Effects of infant carrying practices on rhythm in music. *Yearbook of the International Folk Music Council*, 4:142–145.

Bååth, R. (2012). A new look at subjective rhythmisation. *Abstracts of the Conference on Perspectives on Rhythm and Timing*, Glasgow, Scotland, July 19–21, 2012.

Babbitt, M. (1962). Twelve-tone rhythmic structure and the electronic medium. *Perspectives of New Music*, 1/1:49–79.

Backer, E. (2005). On musical stylometry – a pattern recognition approach. *Pattern Recognition Letters*, 26:299–309.

Backus, J. (1960). Re: Pseudo-science in music. *Journal of Music Theory*, 4/2:221–232.

Bainbridge, D. (2001). The challenge of optical music recognition. *Computers and the Humanities*, 35/2:95–121.

Barber, J. J. & Ogle, K. (2014). To *P* or Not to *P*? *Ecology*, 95/3:621–626.

Barton, S., Getz, L., & Kubovy, M. (2017). Systematic variation in rhythm production as tempo changes. *Music Perception*, 34/3:303–312.

Becker, J. (1968). Percussive patterns in the music of mainland Southeast Asia. *Ethnomusicology*, 12/2:173–191.

Becker, J. (1994). Music and trance. *Leonardo Music Journal*, 4:41–51.

Béhague, G. (1973). Bossa & bossas: Recent changes in Brazilian urban popular music. *Ethnomusicology*, 17/2:209–233.

Bello, J. P. (2011). Measuring structural similarity in music. *IEEE Transactions on Speech, Audio and Language Processing*, 19/7:2013–2025.

Bello, J. P., Daudet, L., Abdallah, S., Duxbury, C., Davies, M., & Sandler, M. B. (2005). A tutorial on onset detection in music signals. *IEEE Transactions on Speech and Audio Processing*, 13/5:1035–1047.

Beltran, J. F., Liu, X., Mohanchandra, N., & Toussaint, G. T. (2015). Measuring musical rhythm similarity: Statistical features versus transformation methods. *International Journal of Pattern Recognition and Artificial Intelligence*, 29/2:1–23.

Benadon, F. (2007). A circular plot for rhythm visualization and analysis. *Music Theory Online*, 13/3.

Benadon, F. (2009a). Gridless beats. *Perspectives of New Music*, 47/1:135–164.

Benadon, F. (2009b). Time warps in early jazz. *Music Theory Spectrum*, 31/1:1–25.

Benadon, F. (2010). Expressive timing via metric hybrids. *Proceedings of the 11th International Conference on Music Perception and Cognition*, Seattle, WA, August 23–27.

Ben-Dor, A., Karp, R. M., Schwikowski, B., & Shamir, R. (2003). The restriction scaffold problem. *Journal of Computational Biology*, 10/2:385–398.

Benzon, W. L. (1993). Stages in the evolution of music. *Journal of Social and Evolutionary Systems*, 16/3:273–296.

Berendt, J.-E. (1987). *Nada Brahma: The World is Sound*, Translated by H. Bredigkeit. Rochester, VT.

Berenzweig, A., Logan, B., Ellis, D. P. W., & Whitman, B. (2004). A large-scale evaluation of acoustic and subjective music-similarity measures. *Computer Music Journal*, 28/2:63–76.

Berry, W. (1985). Metric and rhythmic articulation in music. *Music Theory Spectrum*, 7:7–33.

Bettermann, H., Amponsah, D., Cysarz, D., & Van Leeuwen, P. (1999). Musical rhythms in heart period dynamics: A cross-cultural and interdisciplinary approach to cardiac rhythms. *The American Journal of Physiology*, 277/5:H1762–H1770.

Biamonte, N. (2014). Formal functions of metric dissonance in rock music. *Music Theory Online*, 20/2.

Bianchi, I., Burro, R., Pezzola, R., & Savardi, U. (2017). Matching visual and acoustic mirror forms. *Symmetry*, 9/39:1–21.

Biedl, T., Chan, T., Demaine, E. D., Fleischer, R., Golin, M., & Munso, J. I. (2001). Fun-Sort—Or the chaos of unordered binary search. *Discrete Applied Mathematics*, 144:231–236.

Bishop, A. J. (1990). Western mathematics: The secret weapon of cultural imperialism. *Race and Class*, 32/2:51–65.

Bispham, J. (2006). Rhythm in music: What is it? Who has it? And why? *Music Perception*, 24/2:125–134.

Blacking, J. (1955). Some notes on a theory of African rhythm advanced by Erich von Hornbostel. *African Music*, 1/2:12–20.

Blau, S. K. (1999). The hexachordal theorem: A mathematical look at interval relations in twelve-tone composition. *Mathematics Magazine*, 72/4:310–313.

Blench, R. (1982). Evidence for the Indonesian origins of certain elements of African culture: A review, with special reference to the arguments of A.M. Jones. *African Music*, 6/2:81–93.

Block, S. & Douthett, J. (1994). Vector products and intervallic weighting. *Journal of Music Theory*, 38/1:21–41.

Bloom, G. S. (1977). A counterexample to a theorem of S. Piccard. *Journal of Combinatorial Theory, Series A*, 22:378–379.

Bolton, T. L. (1894). Rhythm. *The American Journal of Psychology*, 6/2:145–238.

Boltz, M. G. (1999). The processing of melodic and temporal information: Independent or unified dimensions? *Journal of New Music Research*, 28/1:67–79.

Boltz, M. G. (2011). Illusory tempo changes due to musical characteristics. *Music Perception*, 28/4:367–386.

Bonnet, E. & Van de Peer, Y. (2001). Zt: A software tool for simple and partial Mantel tests. *Journal of Statistical Software*, 7/10:1–12.

Bookstein, A., Klein, S. T., & Raita, T. (2001). Fuzzy Hamming distance: A new dissimilarity measure. *Proceedings 12th Annual Symposium on Combinatorial Pattern Matching, Volume LNCS 2089*, Springer-Verlag, Berlin, pp. 86–97.

Bookstein, A., Kulyukin, V. A., & Raita, T. (2002). Generalized Hamming distance. *Information Retrieval*, 5/4:353–375.

Boveiri, H. R. (2010). On pattern classification using statistical moments. *International Journal of Signal Processing, Image Processing, and Pattern Recognition*, 3/4:15–24.

Bradby, B. (1987). Symmetry around a centre: Music of an Andean community. *Popular Music*, 6:197–218.

Brady, B. (2002). Oh, Boy! (Oh, Boy!): Mutual desirability and musical structure in the Buddy Group. *Popular Music*, 21/1:63–91.

Brăiloiu, C. (1951). Le rythme aksak. *Revue de Musicologie*, 33:71–108.

Bregman, A. S. & Campbell, J. (1971). Primary auditory stream segregation and perception of order in rapid sequences of tones. *Journal of Experimental Psychology*, 89/2:244–249.

Breslauer, P. (1988). Diminutional rhythm and melodic structure. *Journal of Music Theory*, 32/1:1–21.

Brewer, R. (1999). The use of habanera rhythm in rockabilly music. *American Music*, 17/3:300–317.

Brinkman, A. R. (1986). A binomial representation of pitch for computer processing of musical data. *Music Theory Spectrum*, 8:44–57.

Brown, A. (1990). Modern jazz drumset artistry. *The Black Perspective in Music*, 18/1–2:39–58.

Brun, V. (1964). Euclidean algorithms and musical theory. *Enseignement Mathématique*, 10:125–137.

Bryant, D. & Moulton, V. (2004). NeighborNet: An agglomerative algorithm for the construction of phylogenetic networks. *Molecular Biology and Evolution*, 21/2:255–265.

Buerger, M. J. (1976). Proofs and generalizations of Patterson's theorems on homometric complementary sets. *Zeitschrift für Kristallographie*, 143:79–98.

Buerger, M. J. (1978). Interpoint distances in cyclotomic sets. *The Canadian Mineralogist*, 16:301–314.

Burton, A. R. & Vladimirova, T. (1999). Generation of musical sequences with genetic techniques. *Computer Music Journal*, 23/4:59–73.

Butler, M. J. (2001). Turning the beat around: Reinterpretation, metrical dissonance, and asymmetry in electronic dance music. *Music Theory Online*, 7/6.

Butterfield, M. W. (2006). The power of anacrusis: Engendered feeling in groove-based musics. *Music Theory Online*, 12/4.

Callender, C., Quinn, I., & Tymoczko, D. (2008). Generalized voice leading spaces. *Science*, 320:346–348.

Cameron, D., Potter, K., Wiggins, G., & Pearce, M. T. (2012). Perception of rhythmic similarity in Reich's Clapping Music: Factors and models. *Proceedings of the 12th International Conference on Music Perception and Cognition and the 8th Triennial Conference of the European Society for the Cognitive Sciences of Music*, Thessaloniki, Greece, July 23–28, pp. 195–196.

Cao, E., Lotstein, M., & Johnson-Laird, P. N. (2014). Similarity and families of musical rhythm. *Music Perception*, 31/5:444–469.

Carey, N. & Clampitt, D. (1989). Aspects of well-formed scales. *Music Theory Spectrum*, 11/2:187–206.

Carey, N. & Clampitt, D. (1996). Self-similar pitch structures, their duals, and rhythmic analogues. *Perspectives of New Music*, 34/2:62–87.

Casey, M., Veltkamp, R., Goto, M., Leman, M., Rhodes, C., & Slaney, M. (2008). Content-based music information retrieval: Current directions and future challenges. *IEEE Proceedings*, 96/4:668–696.

Cemgil, A. T., Desain, P., & Kappen, B. (2000). Rhythm quantization for transcription. *Computer Music Journal*, 24/2:60–76.

Cha, S.-H. & Srihari, S. N. (2002). On measuring the distance between histograms. *Pattern Recognition*, 35:1355–1370.

Chaitin, G. J. (1974). Information theoretic computational complexity. *IEEE Transactions on Information Theory*, IT-20:10–15.

Chemillier. M. (2004). Periodic musical sequences and Lyndon words. *Soft Computing*, 8:611–616.

Chemillier, M. & Truchet, C. (2003). Computation of words satisfying the "rhythmic oddity property" (after Simha Arom's work). *Information Processing Letters*, 86:255–261.

Chen, H. C. & Wong, A. K. C. (1983). Generalized texture representation and metric, *Computer Vision, Graphics, and Image Processing*, 23:182–206.

Chernoff, J. M. (1991). The rhythmic medium in African music. *New Literary History*, 22/4:1093–1102.

Chew, E., Volk, A., & Lee, C.-Y. (2005). Dance music classification using inner metric analysis. *Proceedings of the 9th INFORMS Computer Society Conference*, Kluwer, pp. 255–370.

Childs, A. P. (2006). Structural and transformational properties of all-interval tetrachords. *Music Theory Online*, 12/4.

Chrisman, R. (1971). Identification and correlation of pitch-sets. *Journal of Music Theory*, 15/1–2:58–83.

Cilibrasi, R., Vitányi, P., & de Wolf, R. (2004). Algorithmic clustering of music based on string compression. *Computer Music Journal*, 28/4:49–67.

Clayton, M. R. L. (1996). Free rhythm: Ethnomusicology and the study of music without metre. *Bulletin of the School of Oriental and African Studies*, 59/2:323–332.

Cler, J. (1994). Pour une theorie de l'aksak. *Revue de Musicologie*, T. 80e/2e:181–210.

Clough, J. & Douthett, J. (1991). Maximally even sets. *Journal of Music Theory*, 35:93–173.

Clough, J., Engebretsen, N., & Kochavi, J. (1999). Scales, sets, and interval cycles: A taxonomy. *Music Theory Spectrum*, 21/1:74–104.

Cohen, J. E. (2007). Information theory and music. *Systems Research and Behavioral Science*, 7/2:137–163.

Cohen, D. & Katz, R. (1979). The interdependence of notation systems and musical information. *Yearbook of the International Folk Music Council*, 11:100–113.

Cohen, D. & Katz, R. (2008). Rhythmic patterns reflecting cognitive constraints and aesthetic ideals. *Journal of New Music Research*, 37/1:15–35.

Cohn, R. (1991). Bartók's octatonic strategies: A motivic approach. *Journal of the American Musicological Society*, 44/2:262–300.

Cohn, R. (1992a). Transpositional combination of beat-class sets in Steve Reich's phase-shifting music. *Perspectives of New Music*, 30/2:146–176.

Cohn, R. (1992b). The dramatization of hypermetric conflicts in the scherzo of Beethoven's Ninth Symphony. *19th Century Music*, 15/3:188–206.

Cohn, R. (2000). Weitzmann's regions, my cycles, and Douthett's dancing cubes. *Music Theory Spectrum*, 22/1:89–103.

Cohn, R. (2001). Complex hemiolas, ski-hill graphs, and metric spaces. *Music Analysis*, 20/3:295–326.

Cohn, R. (2003). A tetrahedral model of tetrachordal voice leading space. *Music Theory Online*, 9/4.

Cohn, R. (2015). Why we don't teach meter and why we should. *Journal of Music Theory Pedagogy*, 29:5–24.

Cohn, R. (2016a). A Platonic model of funky rhythms. *Music Theory Online*, 22/2.

Cohn, R. (2016b). Teaching atonal and beat-class theory, modulo small. *Brazilian Journal of Music and Mathematics*, 1/1:15–24.

Colannino, J. & Toussaint, G. T. (2005a). An algorithm for computing the restriction scaffold assignment problem in computational biology. *Information Processing Letters*, 95:466–471.

Colannino, J. & Toussaint, G. T. (2005b). Faster algorithms for computing distances between one-dimensional point sets. In: *Proceedings of the XI Encuentros de Geometría Computacional*, F. Santos & D. Orden (eds). Servicio de Publicaciones de la Universidad de Cantabria, Santander, Spain. pp. 189–198.

Colannino, J., Damian, M., Hurtado, F., Iacono, J., Meijer, H., Ramaswami, S., & Toussaint, G. T. (2006). An O(n log n)-time algorithm for the restriction scaffold assignment problem. *Journal of Computational Biology*, 13/4:979–989.

Colannino, J., Damian, M., Hurtado, F., Langerman, S., Meijer, H., Ramaswami, S., Souvaine, D., & Toussaint, G. T. (2007). Efficient many-to-many point matching in one dimension. *Graphs and Combinatorics*. Vol. 23, *Computational Geometry and Graph Theory*, The Akiyama-Chvatal Festschrift, pp. 169–178.

Colannino, J., Gómez, F., & Toussaint, G. T. (2009). Analysis of emergent beat-class sets in Steve Reich's Clapping Music and the Yoruba bell timeline. *Perspectives of New Music*, 47/1:111–134.

Colley, B. (1987). A comparison of syllabic methods for improving rhythm literacy. *Journal of Research in Music Education*, 35/4:221–235.

Condit-Schultz, N. (2016). Deconstructing nPVI. *Proceedings of the International Conference on Music Perception and Cognition*, San Francisco, CA, July 5–9, pp. 800–804.

Conley, J. K. (1981). A psychophysical investigation of judgments of complexity in music. *Psychomusicology*, 1/2:59–71.

Coons, E. & Kraehenbuehl, D. (1958). Information as a measure of structure in music. *Journal of Music Theory*, 2/2:127–161.

Cornelis, O., Lesaffre, M., Moelants, D., & Leman, M. (2010). Access to ethnic music: Advances and perspectives in content-based music information retrieval. *Signal Processing*, 90:1008–1031.

Correa, D. C., Saito, J. H., & Costa, L. F. (2010). Musical genres: Beating to the rhythms of different drums. *New Journal of Physics*, 12:053030.

Courlander, H. (1942). Musical instruments of Cuba. *The Musical Quarterly*, 28/2:227–240.

Cox, T. F. (1981). Reflexive nearest neighbors. *Biometrics*, 37/2:367–369.

Coxeter, H. S. M. (1968). Music and mathematics. *Mathematics Teacher*, 61/3:312–320.

Coyle, E. J. & Shmulevich, I. (1998). A system for machine recognition of music patterns. *Proceedings of the IEEE International Conference on Acoustics, Speech, and Signal Processing*, Seattle, WA.

Craizer, M., Teixeira, R. C., & da Silva, M. A. H. B. (2013). Polygons with parallel opposite sides. *Discrete and Computational Geometry*, 50:474–490.

Crawford, T., Iliopoulos, C. S., Winder, R., & Yu, H. (2001). Approximate musical evolution. *Computers and the Humanities*, 35:55–64.

Crofts, A. R. (2007). Life, information, entropy, and time. *Complexity*, 13/1:14–50.

Cronin, C. (1997–1998). Concepts of melodic similarity in music-copyright infringement suits. *Computing in Musicology*, 11:187–209.

Crook, L. (1982). A musical analysis of the Cuban rumba. *Latin American Music Review*, 3/1:92–123.

Cross, I. (1998). Music and science: Three views. *Revue Belge de Musicologie*, LI I:207–214.

Cross, I. (2001). Music, cognition, culture and evolution. *Annals of the New York Academy of Sciences*, 930:28–42.

Cuthbert, M. C. (2006). Generalized set analysis of Sub-Saharan African rhythm? Evaluating and expanding the theories of Willie Anku. *Journal of New Music Research*, 35/3:211–219.

Dabby, D. S. (2008). Creating musical variation. *Science*, 320:62–63.

Dahlig-Turek, E. (2009). Studying rhythm morphology. *Lithuanian Musicology*, 10:127–137.

Dalla Bella, S. & Peretz, I. (2005). Differentiation of classical music requires little learning but rhythm. *Cognition*, 96:B65–B78.

Davies, M., Madison, G., Silva, P., & Gouyon, F. (2012). The effect of microtiming deviations on the perception of groove in short rhythms. *Music Perception*, 30/5:497–510.

De Bruijn, N. G. (1974). Sorting by means of swapping. *Discrete Mathematics*, 9:333–339.

De Cisneros Puig, B. J. (2017). Discovering flamenco metric matrices through a pulse-level analysis. *Analytical Approaches to World Music*, 6/1:2.

De Fleurian, R., Blackwell, T., Ben-Tal, O., & Müllensiefen, D. (2016). Information-theoretic measures predict the human judgment of rhythm complexity. *Cognitive Science*, 41/3:800–813.

De Souza, R. C. (2016). Discrete and combinatorial mathematics, geometry and mathematics of continuous functions used in some of my compositional problems. *Brazilian Journal of Music and Mathematics*, 1/1:102–115.

De Valpine, P. (2014). The common sense of p values. *Ecology*, 95/3:617–621.

Deliège, I. (1987). Grouping conditions in listening to music: An approach to Lerdahl & Jackendoff's grouping preference rules. *Music Perception*, 4:325–360.

Demaine, E. D., Erickson, J., Krizanc, D., Meijer, H., Morin, P., Overmars, M., & Whitesides, S. (2008). Realizing partitions respecting full and partial order information. *Journal of Discrete Algorithms*, 6/1:51–58.

Desain, P. & Honing, H. (1999). Computational models of beat induction: The rule-based approach. *Journal of New Music Research*, 28/1:29–42.

Desain, P. & Honing, H. (2003). The formation of rhythmic categories and metric priming. *Perception*, 32:341–365.

Diaz, J. D. (2017). Experimentations with timelines in Afro-Bahian Jazz: A strategy of rhythm complication. *Analytical Approaches to World Music*, 6/1:1–34.

Díaz-Báñez, J. M. (2017). Mathematics and flamenco: An unexpected partnership. *The Mathematical Intelligencer*, 39/3:27–39.

Dibben, N. (1994). The cognitive reality of hierarchic structure in tonal and atonal music. *Music Perception*, 12/1:1–25.

Diettrich, O. (1992). Darwin, Lamarck and the evolution of science and culture. *Evolution and Cognition*, 1st Series, 2/3.

Dixon, S. (1997). Beat induction and rhythm recognition. *Proceedings of the Australian Joint Conference on Artificial Intelligence*, Perth, Australia, pp. 311–320.

Don, G. W., Muir, K. K., Volk, G. B., & Walker, J. S. (2010). Music: Broken symmetry, geometry, and complexity. *Notices of the American Mathematical Society*, 57/1:30–49.

Donnini, R. (1986). The visualization of music: Symmetry and asymmetry. *Computers and Mathematics with Applications*, 12B/1–2:435–463.

Douthett, J. & Krantz, K. (2008). Dinner tables and concentric circles: A harmony in mathematics, physics, and music. *College Mathematics Journal*, 39:203–211.

Dress, A., Huson, D., & Moulton, V. (1996). Analysing and visualizing sequence and distance data using SPLITSTREE. *Discrete Applied Mathematics*, 71:95–109.

Duffy, S. & Pearce, M. (2018). What makes rhythms hard to perform? An investigation using Steve Reich's Clapping Music. *PLoS ONE*, 13/10:e0205847.

Duke, R. A. (1994). When tempo changes rhythm: The effect of tempo on nonmusicians' perception of rhythm. *Journal of Research in Music Education*, 42/1:27–35.

Duncan, A. (1991). Combinatorial music theory. *Journal of the Audio Engineering Society*, 39:427–448.

During, J. (1997). Rythmes ovoïdes et quadrature du cycle. *Cahiers de Musiques Traditionnelles*, 10:17–36.

Eerola, T., Himberg, T., Toiviainen, P., & Louhivuori, J. (2006). Perceived complexity of Western and African folk melodies by Western and African listeners. *Psychology of Music*, 34/3:337–371.

Einarson, K. M. & Trainor, L. J. (2015). The effect of visual information on young children's perceptual sensitivity to musical beat alignment. *Timing and Time Perception*, 3:88–101.

Ellis, J., Ruskey, F., Sawada, J., & Simpson, J. (2003). Euclidean strings. *Theoretical Computer Science*, 30/1:321–340.

El-Mallah, I. & Fikentscher, K. (1990). Some observations on the naming of musical instruments and on the rhythm in Oman. *Yearbook for Traditional Music*, 22:123–126.

Elschekova, A. (1966). Methods of classification of folktunes, *Journal of the International Folk Music Council*, 18:56–76.

Entsua-Mensah, T. E. (2015). Time in Indian and West African music. *Research on Humanities and Social Sciences*, 5/6:166–171.

Erdős, P. & Turán, P. (1941). On a problem of sidon in additive number theory and some related problems. *Journal of the London Mathematical Society*, 16:212–215.

Essens, P. (1995). Structuring temporal sequences: Comparison of models and factors of complexity. *Perception and Psychophysics*, 57/4:519–532.

Ester, D. P., Scheib, J. W., & Inks, K. J. (2006). Takadimi: A rhythm system for all ages. *Music Educators Journal*, 93/2:60–65.

Fagg, B. (1956). The discovery of multiple rock gongs in Nigeria. *African Music*, 1/3:6–9.

Farbood, M. M. & Schoner, B. (2009). Determining feature relevance in subject responses to musical stimuli. In: *Mathematics and Computation in Music*, E. Chew, A. Childs, & C.-H. Chuan (eds), CCIS 38. Springer-Verlag, Berlin, Heidelberg. pp. 115–129. *Proceeding of Second International Conference, John Clough Memorial Conference*, New Haven, CT.

Feldman, M. (1981). Crippled symmetry. *Anthropology and Aesthetics*, 2:91–103.

Feldman, H. C. (2005). The black pacific: Cuban and Brazilian echoes in the Afro-Peruvian revival. *Ethnomusicology*, 49/2:206–231.

Fernández-Marin, L. (2001). El Flamenco en las aulas de música: De la transmision oral a la sistematización de su estudio. *Musica y Educación*, 45:13–30.

Fisher, G. H. (1967). Measuring ambiguity. *The American Journal of Psychology*, 80/4:541–557.

Fitch, W. T. (2005). The evolution of music in comparative perspective. *Annals of the New York Academy of Sciences*, 1060:1–21.

Fitch, W. T. & Rosenfeld, A. J. (2007). Perception and production of syncopated rhythms. *Music Perception*, 25/1:43–58.

Fitch, W. T. & Rosenfeld, A. J. (2016). Dance, music, meter and groove: A forgotten partnership. *Frontiers in Human Neuroscience*, 10/64. doi: 10.3389/fnhum.2016.00064.

Flanagan, P. (2008). Quantifying metrical ambiguity. *Proceedings of the International Symposium on Music Information Retrieval*, Philadelphia, PA, pp. 635–640.

Floyd, S. A., Jr. (1999). Black music in the circum-Caribbean. *American Music*, 17/1:1–38.

Forte, A. (1964). A theory of set-complexes for music. *Journal of Music Theory*, 8/2:136–183.

Forte, A. (1973). The basic interval patterns. *Journal of Music Theory*, 17/2:234–272.

Fracile, N. (2003). The Aksak rhythm, a distinctive feature of the Balkan folklore. *Studia Musicologica Academiae Scientiarum Hungaricae*, 44/1–2:197–210.

Frampton, J. R. (1926). Evidence of the naturalness of the less usual rhythms. *The Musical Quarterly*, 12/3:400–405.

Franklin, J. N. (1974). Ambiguities in the X-ray analysis of crystal structures. *Acta Crystallographica*, A-30:698–702.

Freedman, E. G. (1999). The role of diatonicism in the abstraction and representation of contour and interval information. *Music Perception: An Interdisciplinary Journal*, 16/3:365–387.

Fryer, P. (1998). The 'discovery' and appropriation of African music and dance. *Race and Class*, 39:1–20.

Fryer, P. (2003). Our earliest glimpse of West African music. *Race and Class*, 45/1:105–110.

Gabrielson, A. (1973a). Similarity ratings and dimension analyses of auditory rhythm patterns. I, *Scandinavian Journal of Psychology*, 14:138–160.

Gabrielson, A. (1973b). Similarity ratings and dimension analyses of auditory rhythm patterns. II, *Scandinavian Journal of Psychology*, 14:161–176.

Gamer, C. (1967). Deep scales and difference sets in equal-tempered systems. *American Society of University Composers: Proceedings of the Second Annual Conference*, St. Louis, MO, pp. 113–122.

Gascuel, O. (1997). BIONJ: An improved version of the NJ algorithm based on a simple model of sequence data. *Molecular Biology and Evolution*, 14:685–695.

Gatty, R. (1912). Syncopation and emphasis. I. *The Musical Times*, 53/832:369–372.

Gerson-Kiwi, E. (1952). Migrations and mutations of oriental folk instruments. *Journal of the International Folk Music Council*, 4:16–19.

Gerstin, J. (2017). Comparison of African and diasporic rhythm: The Ewe, Cuba, and Martinique. *Analytical Approaches to World Music*, 6/1, 90.

Gibson, D. (1993). The effects of pitch and pitch class content on the aural perception of dissimilarity in complementary hexachords. *Psychomusicology*, 12:58–72.

Gibson, M. & Byrne, J. (1991). Neurogen, musical composition using genetic algorithms and co-operating neural networks. *Proceedings of the Second International Conference on Artificial Neural Networks*, Institute of Electrical Engineers, Stevenege, England, pp. 309–313.

Gilman, B. I. (1909). The science of exotic music. *Science*, 30/772:532–535.

Gjerdingen, R. O. (1993). "Smooth" rhythms as probes of entrainment. *Music Perception: An Interdisciplinary Journal*, 10/4:503–508.

Goldberg, D. (2015). Timing variations in two Balkan percussion performances. *Empirical Musicology Review*, 10/4:305–328.

Goldenberg, J., Mazursky, D., & Solomon, S. (2001). Structures of the mind and universal music. *Science*, 292/5526:2433.

Gómez, E. & Herrera, P. (2008). Comparative analysis of music recordings from Western and non-Western traditions by automatic tonal feature extraction. *Empirical Musicology Review*, 3/3:140–156.

Gómez, F., Khoury, I., Kienzle, J., McLeish, E., Melvin, A., Pérez-Fernandez, R., Rappaport, D. & Toussaint, G. T., (2007). Mathematical models for binarization and ternarization of musical rhythms. *Proceedings of BRIDGES: Mathematical Connections in Art, Music, and Science*, San Sebastián, Spain, pp. 99–108.

Gómez-Marín, D., Jordá, S., & Herrera, P. (2015). PAD and SAD: Two awareness-weighted rhythmic similarity distances. *Proceedings of the 16th International Conference for Music Information Retrieval*, Málaga, Spain, pp. 666–672.

Gonnet, G. H. (1994). New algorithms for the computation of evolutionary phylogenetic trees. In: *Computational Methods in Genome Research*, S. Suhai (ed.). Plenum Press, New York. pp. 153–161.

Gotham, M. (2013). Review of Godfried Toussaint, The geometry of musical rhythm: What makes a "good" rhythm good? *Music Theory Online*, 19/2.

Gotham, M. (2015). Meter metrics: Characterizing relationships among (mixed) metrical structures. *Music Theory Online*, 21/2:1–21.

Gow, G. C. (1915). Rhythm: The life of music. *The Musical Quarterly*, 1/4:637–652.

Goyette, J. (2011). Pumping the all-interval tetrachords: Some algorithms for generating the Z-related sets. *Society for Music Theory Annual Meeting*, Minneapolis, MN, October.

Grahn, J. A. & Brett, M. (2007). Rhythm and beat perception in motor areas of the brain. *Journal of Cognitive Neuroscience*, 19/5:893–906.

Grahn, J. A. (2009). Neuroscientific investigations of musical rhythm: Recent advances and future challenges. *Contemporary Music Review*, 28/3:251–277.

Grauer, V. A. (1965). Some song-style clusters: A preliminary study. *Ethnomusicology*, 9:265–71.

Grauer, V. (2006). Echoes of our forgotten ancestors. *The World of Music*, 48/2:5–58.

Grauer, V. (2009). Concept, style, and structure in the music of the African Pygmies and Bushmen: A study in cross-cultural analysis. *Ethnomusicology*, 53/3:396–424.

Grauer, V. (2017). Commentary on comparing timeline rhythms in pygmy and bushmen music by Adrian Poole. *Empirical Musicology Review*, 12/3–4: 199–204.

Gray, P. M., Krause, B., Atema, J., Payne, R., Krumhansl, C., & Baptista, L. (2001). The music of nature and the nature of music. *Science*, 5:52–54.

Grünbaum, B. (1998). Selfintersections of polygons. *Geombinatorics*, 8:37–45.

Guastavino, C., Gómez, F., Toussaint, G. T., Marandola, F., & Gómez, E. (2009). Measuring similarity between flamenco rhythmic patterns, *Journal of New Music Research*, 38/2:129–138.

Gustar, A. J. (2012). The closest thing to crazy: The shocking scarcity of septuple time in Western music. *Journal of the Royal Musical Association*, 137/2:351–400.

Guy, R. K. (1989). Conway's RATS and other reversals. *The American Mathematical Monthly*, 96/5:425–428.

Haack, J. K. (1998). The mathematics of Steve Reich's Clapping Music. *Proceedings of BRIDGES: Mathematical Connections in Art, Music and Science*, Winfield, Kansas, pp. 87–92.

Haack, J. K. (1991). *Clapping Music - a combinatorial problem*. *The College Mathematics Journal*, 22:224–227.

Hahn, U., Chater, N., & Richardson, L. B. (2003). Similarity as transformation. *Cognition*, 87:1–32.

Hall, R. W. & Klingsberg, P. (2006). Asymmetric rhythms and tiling canons. *The American Mathematical Monthly*, 113/10:887–896.

Hall, R. W. & Tymoczko, D. (2012). Submajorization and the geometry of unordered collections. *The American Mathematical Monthly*, 119/4: 263–283.

Hall, R. W. (2008). Geometrical music theory. *Science*, 320:328–329.

Ham, J. J. (2017). An architectural approach to 3D spatial drum notation. *Proceedings of the International Conference on Technologies for Music Notation and Representation (TENOR)*. www.tenor-conference.org/proceedings.html.

Handel, S. (1988a). Space is to time as vision is to audition: Seductive but misleading. *Journal of Experimental Psychology: Human Perception and Performance*, 14/2:315–317.

Handel, S. (1998b). The interplay between metric and figural rhythmic organization. *Journal of Experimental Psychology: Human Perception and Performance*, 24/5:1546–1561.

Handel, S. (1992). The differentiation of rhythmic structure. *Perception and Psychophysics*, 52:492–507.

Handel, S. & Todd, P. (1981). Segmentation of sequential patterns. *Journal of Experimental Psychology: Human Perception and Performance*, 7/1:41–55.

Hansen, N. C. (2011). The legacy of Lerdahl & Jackendoff's a generative theory of tonal music: Bridging a significant event in the history of music theory and recent developments in cognitive music research. *Danish Yearbook of Musicology*, 38:33–55.

Hansen, N. C., Sadakata, M., & Pearce, M. (2016). Nonlinear changes in the rhythm of European art music: Quantitative support for historical musicology. *Music Perception*, 33/4:414–431.

Harris, M. A., & Reingold, E. M. (2004). Line drawing, leap years, and Euclid. *ACM Computing Surveys*, 36/1:68–80.

Hartley, R. I. & Schaffalitzky, F. (2004). L-infinity minimization in geometric reconstruction problems. Proceedings of the IEEE Conference on Computer Vision and Pattern Recognition Oxford University.

Hasson, U., Hendler, T., Ben Bashat, D., & Malach, R. (2001). Vase or face? A neural correlate of shape-selective grouping processes in the human brain. *Journal of Cognitive Neuroscience*, 13:744–753.

Hauke, J. & Kossowski, T. (2011). Comparison of values of Pearson's and Spearman's correlations coefficients on the same sets of data. *Quaestiones Geographicae*, 30/2:87–93.

Hennig, H., Fleischmann, R., & Geisel, T. (2012). Musical rhythms: The science of being slightly off. *Physics Today*, 65/7:64–65.

Herremans, D. & Chew, E. (2016). Music generation with structural constraints: An operations research approach. *Proceedings of the 30th Annual Conference of the Belgian Operations Research Society*, Louvain-La-Neive, Belgium, pp. 37–39.

Hitt, R. & Zhang, X.-M. (2001). Dynamic geometry of polygons. *Elemente der Mathematik*, 56:21–37.

Hoequist, C. J. (1983). The perceptual center and rhythm categories. *Language and Speech*, 26/4:367–376.

Hoesl, F. & Senn, O. (2018). Modelling perceived syncopation in popular music drum patterns: A preliminary study. *Music and Science*, 1:1–15.

Hofmann-Engl, L. (2002). Rhythmic similarity: A theoretical and empirical approach. In: *Proceedings of the Seventh International Conference on Music Perception and Cognition*, C. Stevens, D. Burnham, G. McPherson, E. Schubert, & Renwick, J. (eds). Sidney, Australia. pp. 564–567.

Honing, H. (2002). Structure and interpretation of rhythm and timing. *Tijdschrift voor Muziektheorie*, 7/3:227–232.

Honing, H. (2012). Without it no music: Beat induction as a fundamental musical trait. *Annals of the New York Academy of Sciences*, 1252:85–91.

Honingh, A. K. (2006). Convexity and compactness as models for the preferred intonation of chords. *Proceedings of the 9th International Conference on Music Perception and Cognition*, University of Bologna, Bologna, Italy, August 22–26, pp. 225–230.

Honingh, A. K. & Bod, R. (2005). Convexity and the well-formedness of musical objects. *Journal of New Music Research*, 34/3:293–303.

Honingh, A. K. & Bod, R. (2011). In search of universal properties of musical scales. *Journal of New Music Research*, 40/1:81–89.

Hook, J. L. (1998). Rhythm in the music of Messiaen: An algebraic study and an application in the "Turangalîla Symphony." *Music Theory Spectrum*, 20/1:97–120.

Hook, J. (2006). Exploring musical space. *Science*, 313:49–50.

von Hornbostel, E. M. (1928). African Negro music. *Africa* 1:30–61.

Horowitz, D. (1994). Generating rhythms with genetic algorithms. *Proceedings of the 12th National Conference of the American Association of Artificial Intelligence*, Seattle, WA, p. 1459.

Hosemann, R. & Bagchi, S. N. (1954). On homometric structures. *Acta Crystallographica*, 7:237–241.

Hughes, D. W. (2000). No nonsense: The logic and power of acoustic-iconic mnemonic systems. *British journal of Ethnomusicology*, 9/2:93–120.

Hunter, D. J. & von Hippel, P. T. (2003). How rare is symmetry in musical 12-tone rows? *The American Mathematical Monthly*, 110/2:124–132.

Huron, D. & Ollen, J. (2003). Agogic contrast in French and English themes: Further support for Patel and Daniele (2003). *Music Perception*, 21:267–272.

Huron, D. & Ommen, A. (2006). An empirical study of syncopation in American popular music. *Music Theory Spectrum*, 28/2:211–231.

Huson, D. H. (1998). SplitsTree: Analyzing and visualizing evolutionary data. *Bioinformatics*, 14:68–73.

Huson, D. H. & Bryant, D. (2006). Application of phylogenetic networks in evolutionary studies. *Molecular Biology and Evolution*, 23/2:254–267.

Hutchinson, W. & Knopoff, L. (1987). The clustering of temporal elements in melody. *Music Perception*, 4/3:281–303.

Iglesias, J. E. (1981). On Patterson's cyclotomic sets and how to count them. *Zeitschrift f¨ur Kristallographie*, 156:187–196.

Imhausen, A. (2006). Ancient Egyptian mathematics: New perspectives on old sources. *The Mathematical Intelligencer*, 28/1:19–27.

Impagliazzo, J. (1989). Music, scales, and dozens. Part I: Mathematical considerations and Pythagorean scales. *The Duodecimal Bulletin*, 33/1:16–20.

Isaacson, E. J. (1990). Similarity of interval-class content between pitch-class sets: The IcVSIM relation. *Journal of Music Theory*, 34:1–28.

Iversen, J. R., Patel, A. D., & Ohgushi, K. (2008). Perception of rhythmic grouping depends on auditory experience. *Journal of the Acoustical Society of America*, 124/4:2263–2271.

Iyer, V. (2002). Embodied mind, situated cognition, and expressive microtiming in African-American music. *Music Perception: An Interdisciplinary Journal*, 19/3:387–414.

Jacoby, N. & McDermott, J. H. (2017). Integer ratio priors on musical rhythm revealed cross-culturally by iterated reproduction. *Current Biology*, 27:359–370.

Jacquemard, F., Ycart, A., & Sakai, M. (2017). Generating equivalent rhythmic notations based on rhythm tree languages. *Proceedings of the International Conference on Technologies for Music Notation and Representation (TENOR)*. www.tenor-conference.org/proceedings.html.

Jankowsky, R. C. (2013). Rhythmic elasticity and metric transformation in Tunisian Stambēlī. *Analytical Approaches to World Music*, 3/1:34–61.

Jaromczyk, J. W. & Toussaint, G. T. (1992). Relative neighborhood graphs and their relatives. *Proceedings of the IEEE*, 80/9:1502–1517.

Jeffreys, M. D. W. (1966/1967). Review article on the book: Africa and Indonesia: The evidence of the xylophone and other musical and cultural factors. *African Music*, 4/1:66–73.

Jekel, J. F. (1977). Should we stop using the p-value in descriptive studies? *Pediatrics*, 60/1:124–126.

Ji, K. Q. & Wilf, H. S. (2008). Extreme palindromes. *The American Mathematical Monthly*, 115/5:447–451.

Jiang, M. (2008). On a sum of distances along a circle. *Discrete Mathematics*, 308:2038–2045.

Jiang, M. (2009). A linear-time algorithm for Hamming distance with shifts. *Theory of Computing Systems*, 44/3:349–355.

Johansson, M. (2017). Non-isochronous musical meters: Towards a multidimensional model. *Ethnomusicology*, 61/1:31–51.

Johnson, H. S. F. & Chernoff, J. M. (1991). Basic conga drum rhythms in African-American musical styles. *Black Music Research Journal*, 11/1:55–73.

Jones, A. M. (1949). African music. *African Affairs*, 48/193:290–297.

Jones, A. M. (1954a). African rhythm. *Africa: Journal of the International African Institute*, 24/1:26–47.

Jones, A. M. (1954b). East and west, north and south. *African Music*, 1/1:57–62.

Jones, A. M. (1973/1974). Luo music and its rhythm. *African Music*, 5/3:43–54.

Jones, M. R. & Boltz, M. (1989). Dynamic attending and responses to time. *Psychological Review*, 96, 459–491.

de Jong, N. (2010). The tambú of Curaçao: Historical projections and the ritual map of experience. *Black Music Research Journal*, 30/2:197–214.

Jylhä, A. & Erkut, C. (2008). Inferring the hand configuration from hand clapping sounds. *Proceedings of the 11th International Conference on Digital Audio Effects (DAFx-08)*, Espoo, Finland.

Jylhä, A., Erkut, C., Şimşekli, U., & Cemgil, A. T. (2012). Sonic handprints: Person identification with hand clapping sounds by a model-based method. *Proceedings of the Audio Engineering Society's 45th International Conference*, Helsinki, Finland.

Kanaya, S., Kariya, K., & Fijisaki, W. (2016). Cross-modal correspondence among vision, audition, and touch in natural objects: An investigation of the perceptual properties of wood. *Perception*, 45/10:1099–1114.

Kappraff, J. (2010). Ancient harmonic law. *Symmetry in the History of Science, Art and Technology. Part II*, 21/1-3:207–228.

Katz, B. F. (2004). A measure of musical preference. *Journal of Consciousness Studies*, 11/3-4:28–57.

Kauffman, R. (1980). African rhythm: A reassessment. *Ethnomusicology*, 24/3:393–415.

Keller, P. E. & Repp, B. H. (2005). Staying offbeat: Sensorimotor syncopation with structured and unstructured auditory sequences. *Psychological Research*, 69:292–309.

Kendall, M. G. & Smith, B. B. (1940). On the method of paired comparisons. *Biometrika*, 31:324–45.

Kenyon, M. (1947). Modern meters. *Music and Letters*, 28\2:168–174.

Kim-Boyle, D. (2017). The 3-D score. *Proceedings of the International Conference on Technologies for Music Notation and Representation (TENOR)*. www.tenor-conference.org/proceedings.html.

King, A. (1960). Employments of the "Standard Pattern" in Yoruba music. *African Music*, 2/3:51–54.

Kirk, J. & Nicholson, N. (2016). Visualizing Euclidean rhythms using tangle theory. *Polymath: An Interdisciplinary Arts & Sciences Journal*, 6/1:1–10.

Kleppinger, S. V. (2003). On the influence of jazz rhythm in the music of Aaron Copland. *American Music*, 21/1:74–111.

Klette, R. & Rosenfeld, A. (2004). Digital straightness - a review. *Discrete Applied Mathematics*, 139:197–230.

Knight, R. (1974). Mandinka drumming. *African Arts*, 7/4:24–35.

Knopoff, L. & Hutchinson, W. (1983). Entropy as a measure of style: The influence of sample length. *Journal of Music Theory*, 27:75–97.

Koetting, J. (1970). Analysis and notation of West African drum ensemble music. *Publications of the Institute of Ethnomusicology*, 1/3:115–147.

Koetting, J. & Knight, R. (1986). What do we know about African Rhythm? *Ethnomusicology*, 30/1:58–63.

Kolata, G. B. (1978). Singing styles and human cultures: How are they related? *Science*, New Series, 200/4339:287–288.

Kolinski, M. (1973). A cross-cultural approach to metro-rhythmic patterns. *Ethnomusicology*, 17/3:494–506.

Krenek, E. (1937). Musik und mathematik. *Uber neue Musik*, Verlag der Ringbuchhandlung, Vienna, 71–89.

Ku, L. H. (1981). Quintuple meter in Korean instrumental music, *Asian Music*, 13/1:119–129.

Kubik, G. (1962). The phenomenon of inherent rhythms in East and Central African instrumental music. *African Music*, 3\1:33–42.

Kubik, G. (1975–1976). Musical bows in South-Western Angola, 1965. *African Music*, 5/4:98–104.

Kubik, G. (1998). Analogies and differences in African-American musical cultures across the hemisphere: Interpretive models and research strategies. *Black Music Research Journal*, 18/1–2:203–227.

Kvifte, T. (2007). Categories and timing: On the perception of meter. *Ethnomusicology*, 51/1:64–84.

Ladzepko, S. K. & Pantalleoni, H. (1970). Takada drumming. *African Music*, 4/4:6–31.

Lamb, E. (2012). Uncommon time: What makes Dave Brubeck's unorthodox jazz stylings so appealing? *Scientific American*, December 11. Retrieved July 22, 2017.

Lawergren, B. (1988). The origin of musical instruments and sounds. *Anthropos*, 83/1–3:31–45.

Leake, J. (2007). The modes of the standard African 12/8 bell. *Percussive Notes*, 38:38–40.

Leake, J. (2009). 3+3+2: The world's most famous rhythm structure. *Percussive Notes*, June:12–14.

Lee, H.-K. (1981). Quintuple meter in Korean instrumental music. *Asian Music*, 13/1:119–129.

Lempel, A. & Ziv, J. (1976). On the complexity of finite sequences. *IEEE Transactions on Information Theory*, 22/1:75–81.

Lemström, K. & Pienimäki, A. (2007). On comparing edit distance and geometric frameworks in content-based retrieval of symbolically encoded polyphonic music. *Musicae Scientiae*, 4A:135–152.

Levenshtein, V. I. (1966). Binary codes capable of correcting deletions, insertions, and reversals. *Soviet Physics - Doklady*, 6:707–710.

Levitin, D. J., Chordia, P., & Menon, V. (2012). Musical rhythm spectra from Bach to Joplin obey a 1/f power law. *Proceedings of the National Academy of Sciences*, 109/10:3716–3720.

Lewin, D. (1959). Intervallic relations between two collections of notes. *Journal of Music Theory*, 3/2:298–301.

Lewin, D. (1960). The intervallic content of a collection of notes, intervallic relations between a collection of notes and its complement: An application to Schoenberg's hexachordal pieces. *Journal of Music Theory*, 4/1:98–101.

Lewin, D. (1976). On the interval content of invertible hexachords. *Journal of Music Theory*, 20/2:185–188.

Lewin, D. (1977). Forte's interval vector, my interval function, and Regener's common-note function. *Journal of Music Theory*. 21/2:194–237.

Lewin, D. (1982). On extended Z-triples. *Theory and Practice*, 7:38–39.

Liebermann, P. & Liebermann, R. (1990). Symmetry in question and answer sequences in music. *Computers and Mathematics with Applications*, 19/7:59–66.

Liebman, E., Ornoy, E., & Chor, B. (2012). A phylogenetic approach to performance analysis. *Journal of New Music Research*, 41/2:215–242.

Lin, X., Li, C., Wang, H., & Zhang, Q. (2009). Analysis of music rhythm based on Bayesian theory. *Proceedings of the International Forum on Computer Science - Technology and Applications*, Chongqing, China, pp. 296–299.

List, G. (1978). The distribution of a melodic formula: Diffusion or polygenesis? *Yearbook of the International Folk Music Council*, 10:33–52.

Liu, Y. & Toussaint, G. T. (2010a). Mathematical notation, representation, and visualization of musical rhythm: A comparative perspective. *Proceedings of the International Conference on Computer and Computational Intelligence*, Nanning, China, pp. 28–32.

Liu, Y. & Toussaint, G. T. (2010b). Categories of repetition in the geometric meander art of Greek and Roman mosaics. *Hyperseeing*, Spring, pp. 25–38.

Liu, Y. & Toussaint, G. T. (2010c). Unraveling Roman mosaic meander patterns: A simple algorithm for their generation. *Journal of Mathematics and the Arts*, 4/1:1–11.

Liu, Y. & Toussaint, G. T. (2011). The marble frieze patterns of the Cathedral of Siena: Geometric structure, multistable perception, and types of repetition. *Journal of Mathematics and the Arts*, 5/3:115–127.

Lo-Bamijoko, J. N. (1987). Classification of Igbo musical instruments in Nigeria. *African Music*, 6/4:19–41.

Locke, D. (1982). Principles of offbeat timing and cross-rhythm in Southern Ewe dance drumming. *Ethnomusicology*, 26:217–246.

Logan, O. (1879). The ancestry of brudder bones. *Harper's New Monthly Magazine*, pp. 687–698.

Logan, W. (1984). The ostinato idea in Black improvised music: A preliminary investigation. *The Black Perspective in Music*, 12/2:193–215.

Lomax, A. (1959). Folk song style. *American Anthropologist*, 61:927–954.

Lomax, A. (1972). Brief progress report: Cantometrics-choreometrics projects. *Yearbook of the International Folk Music Council*, 4:142–145.

Lomax, A. & Grauer, V. (1964). Cantometrics. *Journal of American Folklore Supplement*, April:37–38.

London, J. (1995). Some examples of complex meters and their implications for models of metric perception. *Music Perception: An Interdisciplinary Journal*, 13/1:59–77.

London, J. (2000). Hierarchical representations of complex meters. *Proceedings of the 6th International Conference on Music, Perception and Cognition*, Keele University, Keele, UK.

London, J. (2002). Some non-isomorphisms between pitch and time. *Journal of Music Theory*, Spring/Fall, 46\1–2:127–151.

London, J. (2012). Three things linguists need to know about rhythm and time in music. *Empirical Musicology Review*, 7/1–2:5–11.

London, J. & Cogsdill, E. (2012). Tapping rate affects tempo judgments for some listeners. *Abstracts of the Conference on Perspectives on Rhythm and Timing*, Glasgow, July, 19–21 2012.

Long, G. M. & Olszweski, A. D. (1999). To reverse or not to reverse: When is an ambiguous figure not ambiguous? *American Journal of Psychology*, 112/1:41–71.

Longuet-Higgins, H. C. & Lee, C. S. (1982). The perception of musical rhythms. *Perception*, 11:115–128.

Longuet-Higgins, H. & Lee, C. (1984). The rhythmic interpretation of monophonic music. *Music Perception*, 1/4:424–441.

Longuet-Higgins, H. C., Webber, B., Cameron, W., Bundy, A., Hudson, R., Hudson, L., Ziman, J., Sloman, A., Sharples, M., & Dennett, D. (1994). Artificial intelligence and musical cognition. *Philosophical Transactions: Physical Sciences and Engineering*, 349/1689:103–113.

Lowe, C. E. (1942). What is rhythm? *The Musical Times*, 83/1193:202–203.

Lowrance, R. & Wagner, R. A. (1975). An extension of the string-to-string correction problem. *Journal of the Association for Computing Machinery*, 22:177–183.

Lubiw, A. & Tanur, L. (2004). Pattern matching in polyphonic music as a weighted geometric translation problem. *Proceedings of the International Symposium on Music Information Retrieval (ISMIR)*, Barcelona.

MacGregor, J. N. (1985). A measure of temporal patterns. *Perception and Psychophysics*, 38:97–100.

Madison, G. & Sioros, G. (2014). What musicians do to induce the sensation of groove in simple and complex melodies, and how listeners perceive it. *Frontiers in Psychology*. doi: 10.3389/fpsyg.2014.00894.

Magill, J. M. & Pressing, J. L. (1997). Asymmetric cognitive clock structures in West African rhythms. *Music Perception*, 15/2:189–222.

Mandereau, J., Ghisi, D., Amiot, E., Andreatta, M., & Agon, C. (2011). Z-relation and homometry in musical distributions. *Journal of Mathematics and Music*, 5/2:83–98.

Mantel, N. (1967). The detection of disease clustering and a generalized regression approach. *Cancer Research*, 27:209–220.

Manuel, P. (1985). The anticipated bass in Cuban popular music. *Latin American Music Review/Revista de Música Latinoamericana*, 6/2:249–261.

Manuel, P. (2004). The 'Guajira' between Cuba and Spain: A study in continuity and change. *Latin American Music Review/Revista de Música Latinoamericana*, 25/2:137–162.

Manuel, P. (2009). From contradanza to son: New perspectives on the prehistory of Cuban popular music. *Latin American Music Review/Revista de Música Latinoamericana*. 30/2:184–212.

Mardix, S. (1990). Homometrism in close-packed structures. *Acta Crystallographica*, A46:133–138.

Marsden, A. (2012). Counselling a better relationship between mathematics and musicology. *Journal of Mathematics and Music*, 6/2:145–153.

Marvin, E. W. (1991). The perception of rhythm in non-tonal music: Rhythmic contours in the music of Edgard Varèse. *Music Theory Spectrum*, 13/1:61–78.

Mathiesen, T. J. (1985). Rhythm and meter in ancient Greek music. *Music Theory Spectrum*, 7/Spring:159–180.

Matthews, L. J., Edmonds, J., Wildman, W. J., & Nunn, C. L. (2012). Cultural inheritance or cultural diffusion of religious violence? A quantitative case study of the Radical Reformation, *Religion, Brain and Behavior*, 2/3:1–13.

Maturi, T. A. & Elsayigh, A. (2010). A comparison of correlation coefficients via a three-step bootstrap approach. *Journal of Mathematics Research*, 2/2:3–10.

McAdams, S. & Bregman, A. S. (1979). Hearing musical streams. *Computer Music Journal*, 3/4:26–43.

McCartin, B. J. (1998). Prelude to musical geometry. *The College Mathematics Journal*, 29/5:354–370.

McCartin, B. J. (2007). A geometric property of the octatonic scale. *International Mathematical Forum*, 2/49:2417–2436.

McCartin, B. J. (2012). Geometric demonstration of the unique intervallic multiplicity property of the diatonic musical scale. *International Mathematical Forum*, 9/18:891–906.

McCartin, B. J. (2014). Geometric demonstration of the generalized unique intervallic multiplicity theorem. *International Mathematical Forum*, 9/18:891–906.

McCartin, B. J. (2015). Geometric proofs of the common tones theorem. *International Mathematical Forum*, 10/6:289–299.

McCartin, B. J. (2016). Geometric proofs of the complementary chords theorem. *International Mathematical Forum*, 11/1:27–39.

McGowan, R. W. & Levitt, A. G. (2011). A comparison of rhythm in English dialects and music. *Music Perception: An Interdisciplinary Journal*, 28/3:307–314.

McLachlan, N. (2000). A spatial theory of rhythmic resolution. *Leonardo*, 10:61–67.

Meisters, G. H. (1975). Polygons have ears. *The American Mathematical Monthly*, 82/6:648–651.

Meyer, L. B. (1998). A universe of universals. *The Journal of Musicology*, 6/1:3–25.

Middleton, R. (1983). 'Play It Again Sam': Some notes on the productivity of repetition in popular music. *Popular Music*, 3:235–270.

Migeod, F. W. H. & Johnston, H. (1915). Notes on West Africa according to Ptolemy. *Journal of the Royal African Society*, 14/56:414–426.

Miller, G. A. & Heise, G. (1950). The thrill threshold. *Journal of the Acoustical Society of America*, 22:637–638.

Milne, A. J. & Dean, R. T. (2016). Computational creation and morphing of multilevel rhythms by control of evenness. *Computer Music Journal*, 40/1:35–53.

Milne, A. J., Herff, S. A., Bulger, D., Sethares, W. A., & Dean, R. (2016). XronoMorph: Algorithmic generation of perfectly balanced and well-formed rhythms. *Proceedings of the International Conference on New Interfaces for Musical Expression*, Griffith University, Brisbane, Australia, July 11–15, 2016.

Milne, J., Bulger, D. & Herff, S. A. (2017). Exploring the space of perfectly balanced rhythms and scales. *Journal of Mathematics and Music*, 11:2–3, 101–133.

Minsky, M. (1981). Music, mind, and meaning. *Computer Music Journal*, 5/3:28–44.

Miranda, E. R. (2004). At the crossroads of evolutionary computation and music: Self-programming synthesizers, swarm orchestras and the origins of melody. *Evolutionary Computation*, 12/2:137–158.

Miranda-Medina, J. F. & Tro, J. (2014). The geometry of the Peruvian Landó and its diverse rhythmic patterns. *Proceedings of the 9th Conference on Interdisciplinary Musicology*, Berlin, Germany, pp. 218–221.

Mohamad, M., Rappaport, D., & Toussaint, G. T. (2011). Minimum many-to-many matchings for computing the distance between two sequences. *Proceedings of the 23rd Canadian Conference on Computational Geometry*, Toronto, Canada, pp. 49–54.

Monahan, C. B. & Carterette, E. C. (1985). Pitch and duration as determinants of musical space. *Music Perception: An Interdisciplinary Journal*, 3/1:1–32.

Mongeau, M. & Sankoff, D. (1990). Comparison of musical sequences. *Computers and the Humanities*, 24/3:161–175.

Mongoven, C. & Carbon, C.-C. (2016). Acoustic Gestalt: On the perceptibility of melodic symmetry. *Musicæ Scientiæ*, 21/1:41–59.

Monin, B. & Oppenheimer, D. M. (2005). Correlated averages vs. averaged correlations: Demonstrating the warm glow heuristic beyond aggregation. *Social Cognition*, 23/3:257–278.

Monro, G. & Pressing, J. (1998). Sound visualization using embedding: The art and science of auditory autocorrelation. *Computer Music Journal*, 22/2:20–34.

Montagu, J. (2004). How old is music? *The Galpin Society Journal*, 57:171–182.

Mont-Reynaud, B. & Goldstein, M. (1985). On finding rhythmic patterns in musical lines. *Proceedings of the International Computer Music Conference*, San Francisco, CA, pp. 391–397.

Morris, R. (1979–1980). A similarity index for pitch-class sets. *Perspectives of New Music*, 18/1–2:445–460.

Morris, R. D. (1990). Pitch-class complementation and its generalizations. *Journal of Music Theory*, 34/2:175–245.

Morris, R. D. (1993). New directions in the theory and analysis of musical contour. *Music Theory Spectrum*, 15/2:205–228.

Morris, R. (1998). Sets, scales, and rhythmic cycles: A classification of talas in Indian music. *21st Annual Meeting of the Society for Music Theory*, Chapel Hill, NC.

Müllensiefen, D. & Frieler, C. (2004). Cognitive adequacy in the measurement of melodic similarity: Algorithmic vs. human judgments. *Computing in Musicology*, 13:147–176.

Munyaradzi, G. & Zimidzi, W. (2012). Comparison of Western music and African music. *Creative Education*, 3/2:193–195.

Murdock, D. J., Tsai, Y.-L., & Adcock, J. (2008). P-values are random variables. *The American Statistician*, 62/3:242–245.

Murphy, D. (2011). Quantization revisited: A mathematical and computational model. *Journal of Mathematics and Music*, 15/1:21–34.

Murphy, S. (2016). Cohn's Platonic model and the regular irregularities of recent popular multimedia. *Music Theory Online*, 22/3.

Nauert, P. (1994). A theory of complexity to constrain the approximation of arbitrary sequences of timepoints. *Perspectives of New Music*, 32/2:226–263.

Navarro, G. (2001). A guided tour to approximate string matching. *ACM Computing Surveys*, 33/1:31–88.

Ngumu, P.-C. & A. Tracey (1980). Rows of squares: A standard model for transcribing traditional African music in Cameroon. *African Music*, 6/1:59–61.

Noll, T. (2007). Musical intervals and special linear transformations. *Journal of Mathematics and Music: Mathematical and Computational Approaches to Music Theory, Analysis, Composition and Performance*, 1/2:121–137.

Nzewi, O. (2000). The technology and music of the Nigerian Igbo Ogene Anuka bell orchestra. *Leonardo Music Journal*, 10:25–31.

Olsen, P. R. (1967). La musique Africaine dans le Golfe Persique. *Journal of the International Folk Music Council*, 19:28–36.

ÓMaidín, D. (1998). A geometrical algorithm for melodic difference. *Computing in Musicology*, 11:65–72.

Oomen, B. J. (1995). String alignment with substitution, insertion, deletion, squashing, and expansion operations. *Information Sciences*, 83:89–107.

Orpen, K. S. & Huron, D. (1992). Measurement of similarity in music: A quantitative approach for non-parametric representations. *Computers in Music Research*, 4:1–44.

Ozaki, K., Shimbo, M., Komachi, M., & Matsumoto, Y. (2011). Using the mutual k-nearest neighbor graphs for semi-supervised classification of natural language data. *Proceedings of the Fifteenth Conference on Computational Natural Language Learning*, Association for Computational Linguistics, Portland, OR, June 23–24, 2011, pp. 154–162.

Paiement, J.-F., Bengio, S., & Eck, D. (2008). A distance model for rhythms. *Proceedings of the 25th International Conference on Machine Learning*, Helsinki, Finland.

Paiement, J.-F., Bengio, S., Grandvalet, Y., & Eck, D. (2008). A generative model for rhythms. Neural Information Processing Systems, Workshop on Brain, Music and Cognition.

Palmer, C. & Krumhansl, C. L. (1990). Mental representations for musical meter. *Journal of Experimental Psychology: Human Perception and Performance*, 16:728–741.

Papari, G. & Petkov, N. (2005). Algorithm that mimics human perceptual grouping of dot patterns. *Proceedings First International Symposium on Brain, Vision and Artificial Intelligence*, Naples, Italy, October 19–21, Lecture Notes in Computer Science, Vol. 3704, Springer-Verlag, Berlin, Heidelberg, pp. 497–506.

Parizet, E. (2002). Paired comparison listening tests and circular error rates. *Acta Acustica*, 88:594–598.

Parncutt, R. (1994). A perceptual model of the pulse salience and metrical accent in musical rhythms. *Music Perception: An Interdisciplinary Journal*, 11/4:409–464.

Parncutt, R. (2007). Systematic musicology and the history and future of Western musical scholarship. *Journal of Interdisciplinary Music Studies*, 1/1:1–32.

Patel, A. D. & Daniele, J. R. (2003a). An empirical comparison of rhythm in language and music. *Cognition*, 87:B35–B45.

Patel, A. D. & Daniele, J. R. (2003b). Stress-timed vs. syllable-timed music? A comment on Huron and Ollen (2003). *Music Perception: An Interdisciplinary Journal*, 21/2:273–276.

Patel, A. D. (2003). Rhythm in language and music: Parallels and differences. *Annals of the New York Academy of Sciences*, 999:140–143.

Patel, A. D. (2006). Musical rhythm, linguistic rhythm, and human evolution. *Music Perception*, 24/1:99–104.

Patsopoulos, D & Patronis, T. (2006). The theorem of Thales: A study of the naming of theorems in school geometry textbooks. *The International Journal for the History of Mathematics Education*, 1/1:57–68.

Patterson, A. L. (1944). Ambiguities in the X-ray analysis of crystal structures. *Physical Review*, 65/5–6:195–201.

Pearce, M. & Müllensiefen, D. (2017). Compression-based modelling of musical similarity perception. *Journal of New Music Research*, 46/2:135–155.

Pearsall, E. (1997). Interpreting music durationally: A set-theory approach to rhythm. *Perspectives of New Music*, 35/1:205–230.

Pearson, K. (1920). Notes on the history of correlation. *Biometrika*, 13:25–45.

Peltola, L., Erkut, C., Cook, P. R., & Välimäki, V. (2007). Synthesis of hand clapping sounds. *IEEE Transactions on Audio, Speech and Language Processing*, 15\3: 1021–1029.

Penrose, R. (1992). On the cohomology of impossible figures. *Leonardo*, 25/3–4:245–247.

Pérez Fernández, R. A. (2007). El mito del carácter invariable de las lineas temporales. *Transcultural Music Review*, 11, (article No. 11) [Accessed October 27, 2009].

Perkins, D. N. & Howard, V. A. (1976). Toward a notation for rhythm perception. *Interface*, 5/1–2:69–86.

Petrov, B. (2012). Bulgarian rhythms: Past, present and future. *Dutch Journal of Music Theory*, 17/3:157–167.

Phillips, I. (2008). Perceiving temporal properties. *European Journal of Philosophy*, 18/2:176–202.

Pickover, C. A. (1990). On the aesthetics of Sierpinski gaskets formed from large Pascal's triangles. *Leonardo*, 23/4:411–417.

Polak, R. (2015). Pattern and variation in the timing of Aksak meter: Commentary on Goldberg. *Empirical Musicology Review*, 10/4:329–340.

Polak, R. & London, J. (2014). Timing and meter in Mande drumming from Mali. *Music Theory Online*, 20/1:1–22.

Polansky, L. (1996). Morphological metrics. *Journal of New Music Research*, 25:289–368.

Poole, A. (2018). Comparing timeline rhythms in pygmy and bushmen music. *Empirical Musicology Review*, 12/3–4:172–193.

Pöppel, E. (1989). The measurement of music and the cerebral clock: A new theory. *Leonardo*, 22/1:83–89.

Post, O. & Toussaint, G. T. (2011). The edit distance as a measure of perceived rhythmic similarity. *Empirical Musicology Review*, 6/3:164–179.

Povel, D. J. (1984). A theoretical framework for rhythm perception. *Psychological Research*, 45:315–337.

Prasad, A. (1999). John McLaughlin: Spheres of influence. *Interviews - Music Without Borders*, www.innerviews. org/inner/mclaughlin.html.

Pressing, J. (1983). Cognitive isomorphisms between pitch and rhythm in world musics: West Africa, the Balkans and Western tonality. *Studies in Music*, 17:38–61.

Pressing, J. (1997). Cognitive complexity and the structure of musical patterns. *Proceedings of the Fourth Conference of the Australasian Cognitive Science Society*, Newcastle, Australia. [CD-ROM].

Pressing, J. (2002). Black Atlantic rhythm: Its computational and transcultural foundations. *Music Perception*, 19/3:285–310.

Proca-Ciortea, V. (1969). On rhythm in Rumanian folk dance. *Yearbook of the International Folk Music Council*, 1:176–199.

Quinn, I. (1997). Fuzzy extensions to the theory of contour. *Music Theory Spectrum*, 19/2:232–263.

Quinn, I. (1999). The combinatorial model of pitch contour. *Music Perception: An Interdisciplinary Journal*, 16/4:439–456.

Quintero-Rivera, A. G. & Márquez, R. (2003). Migration and worldview in salsa music. *Latin American Music Review/Revista de Música Latinoamericana*, 24/2:210–232.

Rahn, J. (1987). Asymmetrical ostinatos in Sub-Saharan music: Time, pitch, and cycles reconsidered. *In Theory Only: Journal of the Michigan Music Theory Society*, 9/7:23–37.

Rahn, J. (1995). A non-numerical predicate of wide applicability for perceived intervallic relations. *Muzica*, 4/2:23–33.

Rahn, J. (1996). Turning the analysis around: African-derived rhythms and Europe-derived music theory. *Black Music Research Journal*, 16/1:71–89.

Rappaport, D. (2005). Geometry and harmony. *Proceedings of the 8th Annual International Conference of BRIDGES: Mathematical Connections in Art, Music, and Science*, Banff, Canada, pp. 67–72.

Ravi, S. S., Rosenkrants, D. J., & Tayi, G. K., (1994). Heuristic and special case algorithms for dispersion problems. *Operations Research*, 42/2:299–310.

Ravignani, A. & Madison, G. (2017). The paradox of isochrony in the evolution of human rhythm. *Frontiers of Psychology*. doi: 10.3389/fpsyg.2017.01820.

Ravignani, A. & Norton, P. (2017). Measuring rhythmic complexity: A primer to quantify and compare temporal structure in speech, movement, and animal vocalizations. *Journal of Language Evolution*, 2/1:4–19.

Ravignani, A. (2014). The evolutionary origins of rhythm: A top-down/bottom-up approach to patterning in music and language. *Procedia—Social and Behavioral Sciences*, 126:113–114.

Ravignani, A. (2017). Visualizing and interpreting rhythmic patterns using phase space plots. *Music Perception*, 34/5:557–568.

Read, R. C. (1997). Combinatorial problems in the theory of music. *Discrete Mathematics*, 167–168:543–551.

Regener, R. (1974). On Allen Forte's theory of chords. *Perspectives of New Music*, 13/1:191–212.

Reinach, T. (1893). La musique des hymnes de delphes. *Bulletin de Correspondance Hellénique*, 17:584–610.

Reiner, D. L. (1985). Enumeration in music theory. *American Mathematical Monthly*, 92:51–54.

Reinhardt, C. (2005). Taxi cab geometry: History and applications. *The Montana Mathematics Enthusiast*, 1/1:38–55.

Repp, B. H. (1987). The sound of two hands clapping: An exploratory study. *Journal of the Acoustical Society of America*, 81/4:1100–1109.

Repp, B. H. & Marcus, R. J. (2010). No sustained sound illusion in rhythmic sequences. *Music Perception: An Interdisciplinary Journal*, 28/2:121–134.

Repp, B. H., Windsor, W. L., & Desain, P. (2002). Effects of tempo on the timing of simple musical rhythms. *Music Perception: An Interdisciplinary Journal*, 19/4:565–593.

Rey, M. (2006). The rhythmic component of Afrocubanismo in the art music of Cuba. *Black Music Research Journal*, 26/2:181–212.

Rivière, H. (1993). On rhythmical marking in music. *Ethnomusicology*, 37/2:243–250.

Robbins, J. (1990). The Cuban 'son' as form, genre, and symbol. *Latin American Music Review/Revista de Música Latinoamericana*, 11/2:182–200.

Robson, E. (2001). Neither Sherlock Holmes nor Babylon: A reassessment of Plimpton 322. *Historia Mathematica*, 28/3:167–206.

Robson, E. (2002). Words and pictures: New light on Plimpton 322. *American Mathematical Monthly, Mathematical Association of America*, 109/2:105–120.

Roche, D. (2000). The dhāk, Devi Amba's hourglass drum in tribal Southern Rajasthan, India. *Asian Music*, 32/1:59–99.

Roeder, J. (1994). Interacting pulse streams in Schoenberg's atonal polyphony. *Music Theory Spectrum*, 16/2:231–249.

Roeder, J. (1998). Review of Christopher F. Hasty, Meter as Rhythm. *Music Theory Online*, 4/4:6.

Rogers, D. W. (1999). A geometric approach to PC-set similarity. *Perspectives of New Music*, 37/1:77–90.

Ruskey, F. & Sawada, J. (1999). An efficient algorithm for generating necklaces with fixed density. *SIAM Journal on Computing*, 29/2:671–684.

Sachs, C. (1952). Rhythm and tempo: An introduction. *The Musical Quarterly*, 38/3:384–398.

Safavian, S. R. & Landgrebe, D. (1991). A survey of decision tree classifier methodology. *IEEE Transactions on Systems, Man and Cybernetics*, 21/3:660–674.

Saltini, R. A. (1983). Structural levels and choice of beat-class sets in Steve Reich's phase shifting music. *Intégral*, 7:149–178.

Sampaio, P. A., Ramalho, G., & Tedesco, P. (2008). CinBalada: A Multiagent rhythm factory. *Journal of the Brazilian Computer Society*, Campinas, 14/3:31–49.

Sandroni, C. (2000). Le tresillo: Rythme et "Métissage" dans la musique populaire latino-américaine imprimée au xixe siècle. *Cahiers de Musiques Traditionnelles*, 13:55–64.

Savage, P. E., Brown, S., Sakai, E., & Currie, T. E. (2015). Statistical universals reveal the structures and functions of human music. *PNAS*, 112/29:8987–8992.

Scavone, G. P., Lakatos, S., & Harbke, C. R. (2002). The Sonic Mapper: An interactive program for obtaining similarity ratings with auditory stimuli. *Proceedings of the International Conference on Auditory Display*, Kyoto, Japan, pp. 1–4.

Scheirer, E. D., Watson, R. B., & Vercoe, B. L. (2000). On the perceived complexity of short musical segments. *Proceedings of the International Conference on Music Perception and Cognition*, Keele, UK.

Scherzinger, M. (2010). Temporal geometries of an African music: A preliminary sketch. *Music Theory Online*, 16/4.

Scherzinger, M. (2013). Fractal harmonies of Southern Africa. *Analytical Approaches to World Music*, 3/1:62–90.

Schönemann, P. H. (1983). Some theory and results for metrics for bounded response scales. *Journal of Mathematical Psychology*, 27:311–324.

Schultz, R. (2008). Melodic contour and nonretrogradable structure in the birdsong of Olivier Messiaen. *Music Theory Spectrum*, 30/1:89–137.

Schulze, H.-H. (1989). Categorical perception of rhythmic patterns. *Psychological Research*, 51:10–15.

Schutz, M. (2008). Seeing music? What musicians need to know about vision. *Empirical Musicology Review*, 3/3:83–108.

Scott, D. & Isaacson, E. J. (1998). The interval angle: A similarity measure for pitch-class Sets. *Perspectives of New Music*, 36/2:107–142.

Senechal, M. (2008). A point set puzzle revisited. *European Journal of Combinatorics*, 21/8:1933–1944.

Series, C. (1985). The geometry of Markoff numbers. *The Mathematical Intelligencer*, 7/3:20–29.

Serwadda, M. & Pantaleoni, H. (1968). A possible notation for African dance drumming. *African Music*, 4/2:47–52.

Sethares, W. A. & Toussaint, G. T. (2014). Expressive timbre and timing in rhythmic performance: Analysis of Steve Reich's *Clapping Music*. *Journal of New Music Research*, 44/1:11–24.

Shmulevich, I. & Povel, D.-J. (1998). Rhythm complexity measures for music pattern recognition. *Proceedings of the IEEE 2nd Workshop on Multimedia Signal Processing*, Redondo Beach, CA.

Shmulevich, I. & Povel, D.-J. (2000a). Measures of temporal pattern complexity. *Journal of New Music Research*, 29/1:61–69.

Šimundža, M. (1987). Messiaen's rhythmical organization and classical Indian theory of rhythm (I). *International Review of the Aesthetics and Sociology of Music*, 18/1:117–144.

Šimundža, M. (1988). Messiaen's rhythmical organization and classical Indian theory of rhythm (II). *International Review of the Aesthetics and Sociology of Music*, 19/1:53–73.

Singer, S. (1974). The metrical structure of Macedonian dance. *Ethnomusicology*, 18/3:379–404.

Sinha, P. & Russell, R. (2011). A perceptually-based comparison of image-similarity metrics. *Perception*, 40/11:1269–1281.

Sioros, G. & Guedes, C. (2011). Complexity driven recombination of midi loops. *Proceedings of the 12th International Symposium on Music Information Retrieval*, Miami, FL, October 24–28, pp. 381–386.

Skiena, S. S., Smith, W. D., & Lemke, P. (1990). Reconstructing sets from inter-point distances. *Proceedings of the Sixth Annual ACM Symposium on Computational Geometry*, Berkley, CA, pp. 332–339.

Smith, L. & Honing, H. (2006). Evaluating and extending computational models of rhythmic syncopation in music. *Proceedings of the International Computer Music Conference*, San Francisco, CA, pp. 688–691.

Smith, L. A., McNab, R. J., & Witten, I. H. (1998). Sequence-based melodic comparison: A dynamic programming approach. *Computing in Musicology*, 11:101–117.

Snyder, J. L. (1990). Entropy as a measure of musical style: The influence of a priori assumptions. *Music Theory Spectrum*, 12:121–160.

Söderberg, B. (1953). Musical instruments used by the Babembe. *The African Music Society Newsletter*, 1/6:46–56.

Soderberg, S. (1995). Z-related sets as dual inversions. *Journal of Music Theory*, 39:77–100.

Song, C., Pearce, M., & Harte, C. (2015). SynPy: A python toolkit for syncopation modelling. *Proceedings of the 12th Sound and Music Computing Conference*, Maynooth, Ireland, July 26th–August 1st.

Spearman, C. (1904). The proof and measurement of association between two things. *American Journal of Psychology*, 15:72–101.

Srinivasamurthy, A., Repetto, R. C., Sundar, H., & Serra, X., (2014). Transcription and recognition of syllable based percussion patterns: The case of Beijing Opera. *Proceedings of the 15th International Society for Music Information Retrieval (ISMIR) Conference*, Taipei, Taiwan, pp. 431–436. http://compmusic.upf.edu/bo-perc-patterns, Accessed December 30, 2017.

Standifer, J. A. (1988). The Tuareg: Their music and dances. *The Black Perspective in Music*. 16/1:45–62.

Stanford, E. T. (1972). The Mexican son. *Yearbook of the International Folk Music Council*, 4:66–86.

Stewart, G. (1989). Soukous - Birth of the beat. *The Beat*, 8/6:18–21.

Stewart, J. (2010). Articulating the African diaspora through rhythm: Diatonic patterns, nested looping structures, and the music of Steve Coleman. *Intermediality*, 16:167–184.

Stewart, J. (2018). Articulating the African diaspora through rhythm: Diatonic patterns, nested looping structures, and the music of Steve Coleman. *Intermédialités*, 16:167–184.

Stone, R. M. (1985). In search of time in African music. *Music Theory Spectrum*, 7:139–148.

Stone, R. M. (2007). Shaping time and rhythm in African music: Continuing concerns and emergent issues in motion and motor action. *Transcultural Music Review*, 11, (article No. 4) [Accessed October 27, 2009].

Swindle, P. F. (1913). The inheritance of rhythm. *The American Journal of Psychology*, 24/2:180–203.

Tamir, A. (1998). Comments on the paper: 'Heuristic and special case algorithms for dispersion problems' by S. S. Ravi, D. J. Rosenkrants, & G. K. Tayi. *Operations Research*, 46/1:157–158.

Tangian, A. (2003). Constructing rhythmic canons. *Perspectives of New Music*, 41/2:66–94.

Tanguiane, A. S. (1994). A principle of correlativity of perception and its application to music recognition. *Music Perception: An Interdisciplinary Journal*, 11/4:465–502.

Tehrani, J. & Collard, M. (2002). Investigating cultural evolution through biological phylogenetic analyses of Turkmen textiles. *Journal of Anthropological Archaeology*, 21/4:443–463.

Temperley, D. (2000). Meter and grouping in African music: A view from music theory. *Ethnomusicology*, 44/1:65–96.

Tenney, J. & Polansky, L. (1980). Temporal Gestalt perception in music. *Journal of Music Theory*, 24/2:205–241.

Thaut, M. H., Trimarchi, P. D., & Parsons, L. M. (2014). Human brain besis of musical rhythm perception: Common and distinct neural substrates for meter, tempo, and pattern. *Brain Sciences*, 4:428–452.

Thul, E. & Toussaint, G. T. (2008a). Rhythm complexity measures: A comparison of mathematical models of human perception and performance. *Proceedings of the 9th International Conference on Music Information Retrieval*, Philadelphia, PA, September 14–18, pp. 663–668.

Thul, E. & Toussaint, G. T. (2008b). A comparative phylogenetic analysis of African timelines and North Indian talas. *Proceedings of BRIDGES: Mathematics, Music, Art, Architecture, and Culture*, Leeuwarden, The Netherlands.

Thul, E. & Toussaint, G. T. (2008c). Analysis of musical rhythm complexity measures in a cultural context. In: *Proceedings of the Canadian Conference on Computer Science and Software Engineering*, Desai, B. C. (ed.). Concordia University, Montreal. pp. 1–9.

Thurlow, W. (1957). An auditory figure-ground effect. *American Journal of Psychology*, 70:653–654.

Todd, N. P. (1994). The auditory "primal sketch": A multiscale model of rhythmic grouping. *Journal of New Music Research*, 23:25–70.

Toiviainen, P. & Eerola, T. (2003). Where is the beat? Comparison of finnish and South-African listeners. In: *Proceedings of the 5th Triennial ESCOM Conference*. R. Kopiez, A. C. Lehmann, I. Wolther & C. Wolf (eds.). Hanover University of Music and Drama, Germany. pp. 501–504.

Tóth, L. F. (1956). On the sum of distances determined by a pointset. *Acta Mathematica Academiae Scientiarum Hungaricae*, 7:397–401.

Tóth, L. F. (1959). Über eine Punktverteilung auf der Kugel. *Acta Mathematica Academiae Scientiarum Hungaricae*, 10:13–19.

Tousman, S. A., Pastore, R. E., & Schwartz, S. (1989). Source characteristics: A study of hand clapping. *Journal of the Acoustical Society of America*, 85:S53.

Toussaint, G. T. & Beltran, J. F. (2013). Subsymmetries predict auditory and visual pattern complexity. *Perception*, 42/10:1095–1100.

Toussaint, G. T. & Berzan, C. (2012). Proximity-graph instance-based learning, support vector machines, and high dimensionality: An empirical comparison. *Proceedings of the Eighth International Conference on Machine Learning and Data Mining*, Berlin, Germany, July 16–19.

Toussaint, G. T. (1970). On a simple Minkowski metric classifier. *IEEE Transactions on Systems Science and Cybernetics*, 6:360–362.

Toussaint, G. T. (1974). Generalizations of π: Some applications. *The Mathematical Gazette*, 58/406:289–291.

Toussaint, G. T. (1975). Sharper lower bounds for discrimnation information in terms of variation. *IEEE Transactions on Information Theory*, 21/1:99–100.

Toussaint, G. T. (1978). The use of context in pattern recognition. *Pattern Recognition*, 10:189–204.

Toussaint, G. T. (1980). The relative neighborhood graph of a finite planar set. *Pattern Recognition*, 12:261–268.

Toussaint, G. T. (1985). A historical note on convex hull finding algorithms. *Pattern Recognition Letters*, 3:21–28.

Toussaint, G. T. (1991). Anthropomorphic polygons. *The American Mathematical Monthly*, 98/1:31–35.

Toussaint, G. T. (1993). A new look at Euclid's second proposition. *The Mathematical Intelligencer*, 15/3:12–23.

Toussaint, G. T. (2002). A mathematical analysis of African, Brazilian, and Cuban clave rhythms. *Proceedings of BRIDGES: Mathematical Connections in Art, Music and Science*, Townson University, Towson, MD, July 27–29, pp. 157–168.

Toussaint, G. T. (2003a). Algorithmic, geometric, and combinatorial problems in computational music theory. *Proceedings of X Encuentros de Geometria Computacional*, University of Sevilla, Sevilla, Spain, June 16–17, pp. 101–107.

Toussaint, G. T. (2003b). Classification and phylogenetic analysis of African ternary rhythm timelines. *Proceedings of BRIDGES: Mathematical Connections in Art, Music, and Science, University of Granada*, Granada, Spain July 23–27, pp. 25–36.

Toussaint, G. T. (2004a). Computational geometric aspects of musical rhythm. *Abstracts 14th Fall Workshop on Computational Geometry*, Massachusetts Institute of Technology, Cambridge, MA, pp. 47–48.

Toussaint, G. T. (2004b). A comparison of rhythmic similarity measures. In *Proceedings of the 5th International Conference on Music Information Retrieval*, Barcelona, Spain, pp. 242–245.

Toussaint, G. T. (2005a). The geometry of musical rhythm. In: *Proceeding of Japan Conference on Discrete and Computational Geometry*, J. Akiyama, M. Kano, & X. Tan, (eds), vol. 3742. Lecture Notes in Computer Science, Springer-Verlag, Berlin/Heidelberg, pp. 198–212.

Toussaint, G. T. (2005b). Mathematical features for recognizing preference in Sub-Saharan African traditional rhythm timelines. *Proceedings of the 3rd International Conference on Advances in Pattern Recognition*, University of Bath, Bath, UK, August 22–25, pp. 18–27.

Toussaint, G. T. (2005c). The Euclidean algorithm generates traditional musical rhythms. *Proceedings of BRIDGES: Mathematical Connections in Art, Music, and Science,* Banff, Alberta, Canada, pp. 47–56.

Toussaint, G. T. (2005d). Geometric proximity graphs for improving nearest neighbor methods in instance-based learning and data mining. *International Journal of Computational Geometry and Applications,* 15\2:101–150.

Toussaint, G. T. (2005e). The Erdős-Nagy theorem and its ramifications. *Computational Geometry: Theory and Applications,* 31/3:219–236.

Toussaint, G. T. (2006a). Interlocking rhythms, duration interval content, cyclotomic sets, and the haxachordal theorem. *Fourth International Workshop on Computational Music Theory,* Universidad Politecnica de Madrid, Escuela Universitaria de Informatica, July 24–28.

Toussaint, G. T. (2006b). A comparison of rhythmic dissimilarity measures. *FORMA,* 21/2:129–149.

Toussaint, G. T. (2007). Elementary proofs of the hexachordal theorem. Special Session on Mathematical Techniques in Music Analysis - I, *113th Annual Meeting of the American Mathematical Society,* New Orleans, LA, January 6.

Toussaint, G. T. (2010). Generating "good" musical rhythms algorithmically. *Proceedings of the 8th International Conference on Arts and Humanities,* Honolulu, Hawaii, January 13–16, pp. 774–791.

Toussaint, G. T. (2011). The rhythm that conquered the world: What makes a "good" rhythm good? *Percussive Notes,* November Issue:52–59.

Toussaint, G. T. (2012b). The edit distance as a measure of rhythm complexity. *Proceedings of the 2nd Stochastic Modeling Techniques and Data Analysis International Conference,* Chania, Greece, June 5–8.

Toussaint, G. T. (2012c). The pairwise variability index as a tool in musical rhythm analysis. *Proceedings of the 12th International Conference on Music Perception and Cognition (ICMPC), and 8th Triennial Conference of the European Society for the Cognitive Sciences of Music (ESCOM),* Thessaloniki, Greece, July 23–28.

Toussaint, G. T. (2013b). Recognition of hand clapping sounds - The significance of timbre in Steve Reich's Clapping Music. *International Journal of Computer Science and Electronics Engineering (IJCSEE),* 1/1:141–143.

Toussaint, G. T. (2013c). The pairwise variability index as a measure of rhythm complexity. *Analytical Approaches to World Music,* 2/2:1–42.

Toussaint, G. T. (2015a). Objective stimulus features for predicting human judgments of visual pattern goodness: An empirical comparison. Recent Advances in Computer Science, *Proceedings of the 19th International Conference on Computers (Part of CSCC'15),* Zakynthos Island, Greece, July 16–20, 2015, pp. 86–91.

Toussaint, G. T. (2015b). Objective stimulus predictors of perceptual and performance complexities of temporal patterns. *Rhythm Production and Perception Workshop 2015,* Amsterdam, July 6–8, 2015.

Toussaint, G. T. (2015c). Quantifying musical meter: How similar are African and Western rhythm? *Analytical Approaches to World Music Journal,* 4/2:1–30.

Toussaint, G. T. (2016). Measuring the perceptual similarity of Middle Eastern rhythms: A cross-cultural empirical study. *Proceedings of the Fourth International Conference on Analytical Approaches to World Music,* The New School, New York, June 8–11, 2016.

Toussaint, G. T. (2017). Rhythmic grouping with the mutual nearest-neighbor graph and its application to predicting perception of rhythm similarity. *Conference of the Society for Music Perception and Cognition,* San Diego, CA, July 30–August 3.

Toussaint, G. T., Campbell, M., & Brown, N. (2011). Computational models of symbolic rhythm similarity: Correlation with human judgments. *Analytical Approaches to World Music Journal,* 1/2:380–430.

Toussaint, G. T., Matthews, L., Campbell, M., & Brown, N. (2012). Measuring musical rhythm similarity: Transformation versus feature-based methods. *Journal of Interdisciplinary Music Studies,* 6/1:23–53.

Toussaint, G. T., Onea, N., & Vuong, Q. (2015). Measuring the complexity of two-dimensional patterns: Sub-symmetries versus Papentin complexity. *Proceedings of the 14th IAPR Conference on Machine Vision Applications (MVA 2015),* Tokyo, Japan, pp. 80–83.

Toussaint, G. T., & Oh, S. M. (2016). Measuring musical rhythm similarity: Further experiments with the many-to-many minimum-weight matching distance. Journal of Computer and Communications, 4/15:117–125.

Tracey, A. (1961). Time Out. The Dave Brubeck Quartet. CBS/ALD 6504.12" L.P. *African Music,* 2/4:112–113.

Tversky, A. (1977). Features of similarity. *Psychological Review,* 84:327–352.

Tymoczko, D. (2006). The geometry of musical chords. *Science,* 313/5783:72–74.

Tymoczko, D. (2009). Three conceptions of musical distance. In: *Mathematics and Computation in Music,* E. Chew, A. Childs, & C.-H. Chuan (eds), CCIS 38. Springer-Verlag, Berlin, Heidelberg. pp. 258–272. *Proceedings of the Second International Conference, John Clough Memorial Conference,* New Haven, CT.

Typke, R., Giannopoulos, P., Veltkamp, R. C., Wiering, F., & van Oostrum, R. (2003). Using transportation distances for measuring melodic similarity. *Proceedings of the 4th International Conference on Music Information Retrieval,* Baltimore, MD, pp. 107–114.

Tzanetakis, G., Kapur, A., Schloss, W. A., & Wright, M. (2007). Computational ethnomusicology. *Journal of Interdisciplinary Music Studies,* 1/2, 1–24.

Van der Aa, J., Honing, H., & Ten Cate, C. (2015). The perception of regularity in an isochronous stimulus in zebra finches (Taeniopygia guttata) and humans. *Behavioural Processes,* 115:37–45.

Van der Lee, P. (1995). Zarabanda: Esquemas rítmicos de acompañamiento en 6/8. *Latin American Review,* 16/2:199–220.

Van der Lee, P. (1998). Sitars and bossas: World music influences. *Popular Music*, 17/1:45–70.

Van der Sluis, F., Van den Broek, E., Glassey, R. J., Van Dijk, De Jong, E. M. A. G. (2014). When complexity becomes interesting. *Journal of the Association for Information Science and Technology*, 65/7:1478–1500.

Varney, J. (2001). An introduction to the Colombian. *Bambuco. Latin American Music Review*, 22/2:123–156.

Vecera, S. P. & O'Reilly, R. C. (1998). Figure-ground organization and object recognition processes: An interactive account. *Journal of Experimental Psychology: Human Perception and Performance*, 24/2:441–462.

Velasco, M. J. & Large, E. W. (2011). Pulse detection in syncopated rhythms using neural oscillators. *Proceedings of the 12th International Society for Music Information Retrieval Conference*, Miami, FL, pp. 185–190.

Vetterl, K. (1965). The method of classification and grouping of folk melodies. *Studia Musicologica Academiae Scientiarum Hungaricae*, 7–1/4:349–355.

Vitz, P. C. (1968). Information, run structure and binary pattern complexity. *Perception and Psychophysics*, 3/4A:275–280.

Voloshinov, A. V. (1996). Symmetry as a superprinciple of science and art. *Leonardo*, 29/2:109–113.

Vurkaç, M. (2011). Clave-direction analysis: A new arena for educational and creative applications of music technology. *Journal of Music, Education and Technology*, 4/1:27–46.

Vurkaç, M. (2012). A cross-cultural grammar for temporal harmony in Afro-Latin musics: Clave, partido-alto and other timelines. *Current Musicology*, 94:37–65.

Vuust, P. & Witek, M. A. G. (2014). Rhythmic complexity and predictive coding: A novel approach to modeling rhythm and meter perception in music. *Frontiers of Psychology*, 1111. Published online October 1, 2014. doi: 10.3389/fpsyg.2014.01111.

Wagner, R. A. & Fischer, M. J. (1974). The string-to-string correction problem. *Journal of the Association for Computing Machinery*, 21:68–173.

Wang, H.-M. & Huang, S. C. (2014). Musical rhythms affect heart rate variability: Algorithms and models. *Advances in Electrical Engineering*, 2014:14. doi: 10.1155/2014/851796.

Wang, H.-M., Lin, S.-H., Huang, Y.-C., Chen, I.-C., Chou, L.-C., Lai, Y.-L., Chen, Y.-F., Huang, S.-C., & Jan, M.-Y. (2009). A computational model of the relationship between musical rhythm and heart rhythm. *Proceedings of the IEEE International Symposium on Circuits and Systems*, Taipei, Taiwan, pp. 3102–3105.

Ward, W. E. (1927). Music in the Gold Coast. *Gold Coast Review*, 3:199–223.

Washburne, C. (1997). The clave of jazz: A Caribbean contribution to the rhythmic foundation of an African-American music. *Black Music Research Journal*, 17/1:59–80.

Washburne, C. (1998). Clave: The African roots of salsa. *CLAVE*, 1/1:24–37.

Waterman, R. A. (1948). Hot rhythm in negro music. *Journal of the American Musicological Society*, 1/1:24–37.

Wegner, U. (1993). Cognitive aspects of *amadinda* xylophone music from Buganda: Inherent patterns reconsidered. *Ethnomusicology*, 37/2:201–241.

Weil, H. (1893). Nouveaux fragments d'hymnes accompagné de notes de musique. *Bulletin de Correspondance Hellénique*, 17:569–583.

Wells, M. St. J. (1991). Rhythm and phrasing in Chinese tune-title lyrics: Old eight-beat and its 3-2-3-meter. *Asian Music*, 23/1:119–183.

Wen, O. X. & Krumhansl, C. L. (1916). Isomorphism of pitch and time. In: *Proceedings of the 9th International Conference of Students of Systematic Musicology* (SysMus16), Jyväskylä, Finland, 8th - 10th June 2016. B. Burger, J. Bamford, & E. Carlson (eds).

Werman, M., Peleg, S., Melter, R., & Kong, T. Y. (1986). Bi-partite graph matching for points on a line or a circle. *Journal of Algorithms*, 7:277–284.

Wiggins, G. A. (2012). Music, mind and mathematics: Theory, reality and formality. *Journal of Mathematics and Music*, 6/2:111–123.

Wild, J. (2007). Flat interval distribution and other generalizations of the all-interval property. *Lecture at the Mathematics and Computation in Music Conference*, Berlin, Germany.

Wild, J. (2009). Pairwise well-formed scales and a bestiary of animals on the hexagonal lattice. In: *Mathematics and Computation in Music*, E. Chew, A. Childs, & C.-H. Chuan (eds), CCIS 38, Springer-Verlag, Berlin, Heidelberg. pp. 273–285. *Proceedings of the Second International Conference, John Clough Memorial Conference*, New Haven, CT.

Wilson, D. (1986). Symmetry and its "love-hate" role in music. *Computers and Mathematics with Applications*, 12B/1–2:101–112.

Wilson, K. M. (1927). What is rhythm? *Music and Letters*, 8/1:2–12.

Winckelgren, I. (1992). How the brain 'sees' borders where there are none. *Science*, 256:1520–1521.

Windram, H., Charlston, T., & Howe, C. J. (2014). A phylogenetic analysis of Orlando Gibbons' Prelude in G. *Early Music*, 42:515–528.

Winlock, H. E. (1940). The origin of the ancient Egyptian calendar. *Proceedings of the American Philosophical Society*, 83/3:447–464.

Witek, M. A. G., Clarke, E. F., Wallentin, M., Kringelbach, M. L., & Vuust, P. (2014). Syncopation, body-movement and pleasure in groove music. *PLoS ONE*, 9/4:e94446.

Wolpert, D. H. & Macready, W. (2007). Using self-dissimilarity to quantify complexity. *Complexity*, 12/3:77–85.

Wright, O. (1995). A preliminary version of the kitāb al-Adwār. *Bulletin of the School of Oriental and African Studies*, 58/3:455–478. University of London, Published by Cambridge University Press on behalf of School of Oriental and African Studies.

Wright, M., Schloss, W. A., & Tzanetakis, G. (2008). Analyzing Afro-Cuban rhythm using rotation-aware clave template matching with dynamic programming. *Proceedings of the Ninth International Symposium on Music Information Retrieval, Drexel University*, Philadelphia, PA, September 14–18, pp. 647–652.

Wu, H.-H. & Bello, J. P. (2010). Audio-based music visualization for music structure analysis. *Proceedings of Sound and Music Computing Conference (SMC)*, Barcelona, Spain, pp. 1–6.

Yust, J. (2009). The geometry of melodic, harmonic, and metrical hierarchy. In: *Mathematics and Computation in Music*, E. Chew, A. Childs, & C.-H. Chuan (eds), CCIS 38, Springer-Verlag, Berlin, Heidelberg. pp. 180–192. *Proceeding of Second International Conference, John Clough Memorial Conference*, New Haven, CT.

Zabrodsky, H., Peleg, S., & Avnir, D. (1992). A measure of symmetry based on shape similarity. *Proceedings of the IEEE Symposium on Computer Vision and Pattern Recognition, Champaign, IL*, pp. 703–706.

Zahn, C. T. (1971). Graph-theoretical methods for detecting and describing gestalt clusters. *IEEE Transactions on Computers*, 20:68–86.

Zigel, L. J. (1994). Constricting the clave: The United States, Cuban music, and the new world order. *The University of Miami Inter-American Law Review*, 26/1:129–180.

Index

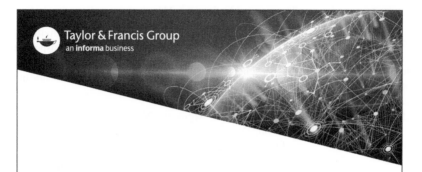

Printed and bound by CPI Group (UK) Ltd, Croydon, CR0 4YY

17/10/2024

01775697-0017